Wire *...rity*

The Newnes Know It All Series

PIC Microcontrollers: Know It All
Lucio Di Jasio, Tim Wilmshurst, Dogan Ibrahim, John Morton, Martin Bates, Jack Smith, D.W. Smith, and Chuck Hellebuyck
ISBN: 978-0-7506-8615-0

Embedded Software: Know It All
Jean Labrosse, Jack Ganssle, Tammy Noergaard, Robert Oshana, Colin Walls, Keith Curtis, Jason Andrews, David J. Katz, Rick Gentile, Kamal Hyder, and Bob Perrin
ISBN: 978-0-7506-8583-2

Embedded Hardware: Know It All
Jack Ganssle, Tammy Noergaard, Fred Eady, Lewin Edwards, David J. Katz, Rick Gentile, Ken Arnold, Kamal Hyder, and Bob Perrin
ISBN: 978-0-7506-8584-9

Wireless Networking: Know It All
Praphul Chandra, Daniel M. Dobkin, Alan Bensky, Ron Olexa, David A. Lide, and Farid Dowla
ISBN: 978-0-7506-8582-5

RF & Wireless Technologies: Know It All
Bruce Fette, Roberto Aiello, Praphul Chandra, Daniel Dobkin, Alan Bensky, Douglas Miron, David Lide, Farid Dowla, and Ron Olexa
ISBN: 978-0-7506-8581-8

Electrical Engineering: Know It All
Clive Maxfield, Alan Bensky, John Bird, W. Bolton, Izzat Darwazeh, Walt Kester, M.A. Laughton, Andrew Leven, Luis Moura, Ron Schmitt, Keith Sueker, Mike Tooley, D.F. Warne, and Tim Williams
ISBN: 978-1-85617-528-9

Audio Engineering: Know It All
Ian Sinclair, Richard Brice, Don Davis, Ben Duncan, John Linsley Hood, Eugene Patronis, Andrew Singmin, and John Watkinson
ISBN: 978-1-85617-526-5

Circuit Design: Know It All
Darren Ashby, Bonnie Baker, Stuart Ball, John Crowe, Barrie Hayes-Gill, Ian Grout, Ian Hickman, Walt Kester, Ron Mancini, Robert A. Pease, Mike Tooley, Tim Williams, Peter Wilson, and Bob Zeidman
ISBN: 978-1-85617-527-2

Test and Measurement: Know It All
Jon Wilson, Stuart Ball, G.M.S de Silva, Tony Fischer-Cripps, Dogan Ibrahim, Kevin James, Walt Kester, Michael Laughton, Chris Nadovich, Alex Porter, Ed Ramsden, Steve Scheiber, Douglas Warne, and Tim Williams
ISBN: 978-1-85617-530-2

Wireless Security: Know It All
Praphul Chandra, Alan Bensky, Tony Bradley, Chris Hurley, Steve Rackley, James Ransome, John Rittinghouse, Timothy Stapko, George Stefanek, Frank Thornton, and Jon Wilson
ISBN: 978-1-85617-529-6

For more information on these and other Newnes titles visit: www.newnespress.com

Wireless Security

Praphul Chandra
Alan Bensky
Tony Bradley
Chris Hurley
Steve Rackley
John Rittinghouse
James F. Ransome
Timothy Stapko
George L. Stefanek
Frank Thornton
Jon Wilson

ELSEVIER

AMSTERDAM • BOSTON • HEIDELBERG • LONDON
NEW YORK • OXFORD • PARIS • SAN DIEGO
SAN FRANCISCO • SINGAPORE • SYDNEY • TOKYO

Newnes is an imprint of Elsevier

Newnes

Newnes is an imprint of Elsevier
30 Corporate Drive, Suite 400, Burlington, MA 01803, USA
Linacre House, Jordan Hill, Oxford OX2 8DP, UK

Library of Congress Cataloging-in-Publication Data
Application submitted

British Library Cataloguing-in-Publication Data
A catalogue record for this book is available from the British Library.

ISBN: 978-1-85617-529-6

For information on all Newnes publications
visit our Web site at www.elsevierdirect.com

Typeset by Charon Tec Ltd., A Macmillan Company. (www.macmillansolutions.com)

Transferred to Digital Printing, 2010
Printed and bound in the United Kingdom

Contents

About the Authors

Steven Arms (Chapter 7) is a contributor to *Sensor Technology Handbook*. Mr. Arms received his Master's Degree in Mechanical Engineering at the University of Vermont in 1983. He has been awarded 25 US patents, and has over 10 pending. He has contributed to 18 journal publications as well as 44 abstracts/presentations in areas of advanced instrumentation, wireless sensing, and energy harvesting. Mr. Arms is founder and President of MicroStrain, Inc., a Vermont manufacturer of micro-displacement sensors, inertial sensing systems, and wireless data logging nodes for recording and transmitting strain, vibration, temperature, and orientation data. MicroStrain has been recognized as an innovator in the sensors industry – as of 2008 the firm has received nine (9) Best of Sensors Expo Gold awards for its new products. MicroStrain has received funding from the US Navy and US Army to develop wireless sensor networks which use strain energy harvesting to eliminate battery maintenance.

Alan Bensky, MScEE, (Chapter 4) is an electronics engineering consultant with over 30 years of experience in analog and digital design, management, and marketing. Specializing in wireless circuits and systems, Bensky has carried out projects for varied military and consumer applications and led the development of three patents on wireless distance measurement. Bensky has taught electrical engineering courses and gives lectures on radio engineering topics. He is the author of *Short-range Wireless Communication, Second Edition*, published by Elsevier in 2004, and has written several articles in international and local publications. He also wrote a recently published book on wireless location technologies and applications. Bensky is a senior member of IEEE.

Tony Bradley, CISSP-ISSAP (Chapter 22 and Appendix B) author of *Essential Computer Security*. He is the Guide for the Internet/Network Security site on About.com, a part of The New York Times Company. He has written for a variety of other Web sites and publications, including *PC World*, SearchSecurity.com, WindowsNetworking.com, *Smart Computing* magazine, and *Information Security* magazine. Currently a security architect and consultant for a Fortune 100 company, Tony has driven security policies and technologies for antivirus and incident response for Fortune 500 companies, and he has been network administrator and technical support for smaller companies. Tony is a CISSP

(Certified Information Systems Security Professional) and ISSAP (Information Systems Security Architecture Professional). He is Microsoft Certified as an MCSE (Microsoft Certified Systems Engineer) and MCSA (Microsoft Certified Systems Administrator) in Windows 2000 and an MCP (Microsoft Certified Professional) in Windows NT. Tony is recognized by Microsoft as an MVP (Most Valuable Professional) in Windows security.

Praphul Chandra (Chapters 1, 17, 18, 19) author of *Bulletproof Wireless Security* works as a Research Scientist at HP Labs, India in the Access Devices Group. He joined HP Labs in April 2006. Prior to joining HP he was a senior design engineer at Texas Instruments (USA) where he worked on Voice over IP with specific focus on Wireless Local Area Networks. He is also the author of *Wi-Fi Telephony: Challenges and Solutions for Voice over WLANs*. He is an Electrical Engineer by training, though his interest in social science and politics has prompted him to concurrently explore the field of Public Policy. He maintains his personal website at www.thecofi.net.

Chris Hurley (Chapters 20 and 21) is the author of *WarDriving: Drive, Detect, and Defend: A Guide to Wireless Security.* Chris is a Principal Information Security Engineer working in the Washington, DC area. He is the founder of the WorldWide WarDrive, an effort by information security professionals and hobbyists to generate awareness of the insecurities associated with wireless networks. Primarily focusing his efforts on vulnerability assessments, he also performs penetration testing, forensics, and incident response operations on both wired and wireless networks. He has spoken at several security conferences, been published in numerous online and print publications, and been the subject of several interviews and stories regarding the WorldWide WarDrive. He is the primary organizer of the WarDriving Contest held at the annual DefCon hacker conference. Chris holds a bachelors degree in computer science from Angelo State University. He lives in Maryland with his wife of 14 years, Jennifer, and their daughter Ashley.

Steve Rackley (Chapters 2, 3 and 6) is the author of *Wireless Networking Technology*. He is an Independent Consultant.

James F. Ransome, CISM, CISSP (Chapters 1, 8, 13, 14, 15 and Appendix A) is the co-author of *Wireless Operational Security*. He is the Director of Information Technology Security at Qwest Communications.

John Rittinghouse, PhD, CISM (Chapters 1, 8, 13, 14, 15 and Appendix A) is the co-author of *Wireless Operational Security*. He is a Managing Partner at Rittinghouse Consulting Group, LLC.

Timothy Stapko (Chapter 9, 10, 11, 12, and 23) Engineering Manager, Digi International, CA, USA. Tim holds a MS degree in Computer Science from the University of California, Davis, and has over 8 years of experience in embedded systems software development. He was the sole developer on a ground-up implementation of Transport Layer Security (TLS/SSL) for the 8-bit Rabbit 3000 microprocessor and is also the author of *Practical Embedded Security*, a book focusing on the implementation of security for inexpensive embedded systems.

George L. Stefanek (Chapter 16) is the author of *Information Security Best Practices*. He is currently a consultant/developer of medical and scientific database systems.

Frank Thornton (Chapter 24) is the lead author of *RFID Security* and co-authored *WarDriving: Drive, Detect, and Defend: A Guide to Wireless Security*. Frank runs his own technology consulting firm, Blackthorn Systems, which specializes in wireless networks. His specialties include wireless network architecture, design, and implementation, as well as network troubleshooting and optimization. An interest in amateur radio helped him bridge the gap between computers and wireless networks. Frank was also a law enforcement officer for many years. As a detective and forensics expert he has investigated approximately one hundred homicides and thousands of other crime scenes.

Combining both professional interests, he was a member of the workgroup that established ANSI Standard "ANSI/NIST-CSL 1-1993 Data Format for the Interchange of Fingerprint Information."

Chris Townsend (Chapter 7) is a contributor to *Sensor Technology Handbook*. He is Vice President of Engineering for MicroStrain, Inc., a manufacturer of precision sensors and wireless sensing instrumentation. Chris's current main focus is on research and development of a new class of ultra low power wireless sensors for industry. Chris has been involved in the design of a number of products, including the world's smallest LVDT, inertial orientation sensors, and wireless sensors. He holds over 25 patents in the area of advanced sensing. Chris has a degree in Electrical Engineering from the University of Vermont.

Jon Wilson (Chapter 7) is the author of *Sensor Technology Handbook*. Principal Consultant, Jon S. Wilson Consulting, LLC, Chandler, AZ. BSME, MAutoE, MSE(IE). Test Engineer, Chrysler Corporation; Test Engineer, ITT Cannon Electric Co.; Environmental Lab Manager, Motorola Semiconductor Products; Applications

Engineering Manager and Marketing Manager, Endevco Corporation; Principal Consultant, Jon S. Wilson Consulting, LLC. Fellow Member, Institute of Environmental Sciences and Technology; Sr. Member ISA; Sr. Member, SAE. He has authored several text books and short course handbooks on testing and instrumentation, and many magazine articles and technical papers. Regular presenter of measurement and testing short courses for Endevco, Technology Training, Inc. and International Telemetry Conference as well as commercial and government organizations.

Part 1

Wireless Technology

CHAPTER 1

Wireless Fundamentals

Praphul Chandra
James F. Ransome
John Rittinghouse

What is it that makes a wireless medium so unique? What are the problems of operating in the wireless medium and how are they overcome? What are the different types of wireless networks in use today? How does each one of them work and how do they differ from each other? The aim of this chapter is to answer these questions so as to establish a context in which wireless security can be studied in the following chapters.

The first successful commercial use of wireless telecommunication was the deployment of cellular phones (mobile phones). In this book, we refer to these networks as *traditional wireless networks* (TWNs). These networks were designed with the aim of extending the existing wired *public switched telephone network* (PSTN) to include a large number of mobile nodes. The deployment of TWNs allowed users to be mobile and still make voice calls to any (fixed or mobile) phone in the world. In other words, TWNs were designed as a *wide area network* (WAN) technology enabling voice communication. These networks have evolved over time to support both voice and data communication but the underlying feature of the TWNs being a WAN technology is still true.

For a long time, TWNs were the predominant example of wireless telecommunication. In the late 1990s, another wireless technology emerged: wireless local area networks (WLANs). Unlike TWNs, WLANs were designed primarily with the aim of enabling data communication in a limited geographical area (local area network). Though this aim may seem counterintuitive at first (why limit the geographical coverage of a network?), this principle becomes easier to understand when we think of WLANs as a wireless Ethernet technology. Just as Ethernet (IEEE 802.3) provides the backbone of wired local area networks (LANs) today, IEEE 802.11 provides the backbone of wireless LANs.

Just as TWNs were initially designed for voice and over time evolved to support data, WLANs were initially designed for data and are now evolving to support voice.

Probably the most prominent difference between the two standards is that the former is a WAN technology and the latter is a LAN technology. TWNs and WLANs are today the two most dominant wireless telecommunication technologies in the world. Analysts are predicting the convergence (and co-existence) of the two networks in the near future.

Finally, we are today seeing the emergence of wireless mobile ad-hoc networks (MANETs). Even though this technology is still in an early phase, it promises to have significant commercial applications. As the name suggests, MANETs are designed with the aim of providing ad hoc communication. Such networks are characterized by the absence of any infrastructure and are formed on an as-needed (ad hoc) basis when wireless nodes come together within each others' radio transmission range.

We begin by looking at some of the challenges of the wireless medium.

1.1 The Wireless Medium

1.1.1 Radio Propagation Effects

The wireless medium is a harsh medium for signal propagation. Signals undergo a variety of alterations as they traverse the wireless medium. Some of these changes are due to the distance between the transmitter and the receiver, others are due to the physical environment of the propagation path and yet others are due to the relative movement between the transmitter and the receiver. We look at some of the most important effects in this section.

Attenuation refers to the drop in signal strength as the signal propagates in any medium. All electromagnetic waves suffer from attenuation. For radio waves, if r is the distance of the receiver from the transmitter, the signal attenuation is typically modeled as $1/r^2$ at short distances and $1/r^4$ at longer distances; in other words, the strength of the signal decreases as the square of the distance from the transmitter when the receiver is "near" the transmitter and as the fourth power when the receiver is "far away" from the transmitter. The threshold value of r where distances go from being "near" to being "far away" is referred to as the reference distance. It is important to emphasize that this is radio modeling we are talking about. Such models are used for simulation and analysis. Real-life radio propagation is much harsher and the signal strength and quality at any given point depends on a lot of other factors too.

Attenuation of signal strength predicts the average signal strength at a given distance from the transmitter. However, the instantaneous signal strength at a given distance has to take into account many other effects. One of the most important considerations that determine the instantaneous signal strength is, not surprisingly, the operating environment. For example, rural areas with smooth and uniform terrain are much more conductive to radio waves than the much more uneven (think tall buildings) and varying (moving automobiles, people and so on) urban environment. The effect of the operating environment on radio propagation is referred to as shadow fading (slow fading). The term refers to changes in the signal strength occurring due to changes in the operating environment. As an example, consider a receiver operating in an urban environment. The path from the transmitter to the sender may change drastically as the receiver moves over a range of tens of meters. This can happen if, for example, the receiver's movement resulted in the removal (or introduction) of an obstruction (a tall building perhaps) in the path between the transmitter and the receiver. Shadow fading causes the instantaneous received signal strength to be lesser than (or greater than) the average received signal strength.

Another propagation effect that strongly affects radio propagation is Raleigh fading (fast fading). Unlike slow fading which effects radio propagation when the distance between the transmitter and the receiver changes of the order of tens of meters, fast fading describes the changes in signal strength due to the relative motion of the order of a few centimeters. To understand how such a small change in the relative distance may affect the quality of the signal, realize that radio waves (like other waves) undergo wave phenomena like diffraction and interference. In an urban environment like the one shown in Figure 1.1a, these phenomena lead to multipath effects; in other words, a signal from the transmitter may reach the receiver from multiple paths. These multiple signals then

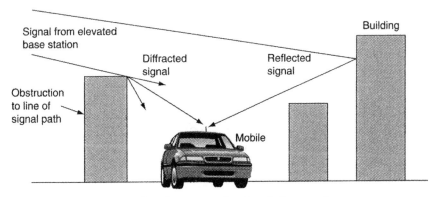

Figure 1.1a: Signal propagation in wireless environment

interfere with each other at the receiver. Since this interference can be either constructive or destructive, these signals may either reinforce each other or cancel each other out. Whether the interference is constructive or destructive depends on the path length (length the signal has traveled) and a small change in the path length can change the interference from a constructive to a destructive one (or vice versa). Thus, if either of the transmitter or the receiver move even a few centimeters, relative to each other, this changes the interference pattern of the various waves arriving at the receiver from different paths. This means that a constructive interference pattern may be replaced by a destructive one (or vice versa) if the receiver moves by as much as a few centimeters. This fading is a severe challenge in the wireless medium since it implies that even when the average signal strength at the receiver is high there are instances when the signal strength may drop dramatically.

Another effect of multipath is inter-symbol interference. Since the multiple paths that the signal takes between the transmitter and the receiver have different path lengths, this means that the arrival times between the multiple signals traveling on the multiple paths can be of the order of tens of microseconds. If the path difference exceeds 1-bit (symbol) period, symbols may interfere with each other and this can result in severe distortion of the received signal.

1.1.2 Hidden Terminal Problem

Wireless is a medium that must be shared by all terminals that wish to use it in a given geographical region. Also, wireless is inherently a broadcast medium since radio transmission cannot be "contained." These two factors create what is famously known as the *hidden terminal problem* in the wireless medium. Figure 1.1b demonstrates this problem.

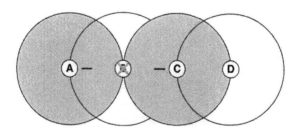

Figure 1.1b: Hidden node problem

Figure 1.1b shows three wireless terminals: A, B and C. The radio transmission range of each terminal is shown by a circle around the terminal. As is clear, terminal B lies within the radio transmission range of both terminals A and C. Consider now what happens if both A and C want to communicate with B. Most media access rules for a shared medium require that before starting transmission, a terminal "senses" the medium to ensure that the medium is idle and therefore available for transmission. In our case, assume that A is already transmitting data to B. Now, C also wishes to send data to B. Before beginning transmission, it senses the medium and finds it idle since it is beyond the transmission range of A. It therefore begins transmission to B, thus leading to collision with A's transmission when the signals reach B. This problem is known as the hidden terminal problem since, in effect, A and C are hidden from each other in terms of radio detection range.

1.1.3 Exposed Terminal Problem

The exposed terminal problem is at the opposite end of the spectrum from the hidden terminal problem. To understand this problem, consider the four nodes in Figure 1.1c.

In this example, consider what happens when B wants to send data to A and C wants to send data to D. As is obvious, both communications can go on simultaneously since they do not interfere with each other. However, the carrier sensing mechanism raises a false alarm in this case. Suppose B is already sending data to A. If C wishes to start sending data to D, before beginning it senses the medium and finds it busy (due to B's ongoing transmission). Therefore C delays its transmission unnecessarily. This is the exposed terminal problem.

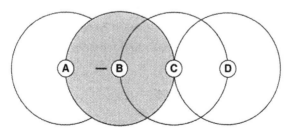

Figure 1.1c: Exposed node problem

1.1.4 Bandwidth

"The Queen is dead.
Long Live the Queen."

Bandwidth is one of the most important and one of the most confusing topics in telecommunications today. If you keep up to date with the telecommunication news, you would have come across conflicting reports regarding bandwidth. There are a lot of people claiming "bandwidth is cheap" and probably as many people claiming "it is extremely important to conserve bandwidth." So, what's the deal? Do networks today have enough bandwidth or not?

The problem is there is no single correct answer to that. The answer depends on where you are in the network. Consider the core of the IP and the PSTN networks: the two most widely deployed networks today. The bandwidth available at the core of these networks is much more than required: bandwidth therefore is cheap at the core of the network. Similarly the dawn of 100 Mbps and Gigabit Ethernet has made bandwidth cheap even in the access network (the part of the network that connects the end-user to the core). The wireless medium, however, is a little different and follows a simple rule: bandwidth is always expensive. This stems from the fact that in almost all countries the wireless spectrum is controlled by the government. Only certain bands of this spectrum are allowed for commercial use, thus making bandwidth costly in the wireless world. All protocols designed for the wireless medium therefore revolve around this central constraint.

1.1.5 Other Constraints

Bandwidth is not the only constraint for wireless networks. One of the prominent "features" of wireless networks is that they allow the end user (and hence the end node or terminal) to be mobile. This usually means that the wireless nodes which are being used need to be small enough to be mobile. This in turn means additional constraints: small devices mean limited processing power and limited battery life. Also using small nodes means that they are easy to lose or steal, thus having security implications.

1.2 Wireless Networking Basics

Organizations are adopting wireless networking technologies to address networking requirements in ever-increasing numbers. The phenomenal rate of change in wireless

standards and technology complicates the decision of how to implement or update
the technology. In this section, we delve into the inner workings of the wireless LAN
(WLAN) environment of today. We start by concentrating on the specific topic areas
related to wireless networking technology. We explain how LANs and WLANs work,
coexist, and interoperate. We discuss the varying wireless standards that exist today and
look at some future standards. Mobile workforces drive the need for strong understanding
of the issues surrounding mobile security, and we cover those issues in fairly extensive
detail. Because mobile security requires encryption to work, we cover the various
encryption standards in use today and explain the strengths and weaknesses of each. At
the end of this chapter, you should have a fundamental grasp of the essentials of wireless
networking. Now, let's begin by taking a look at WLANs.

1.2.1 Wireless Local Area Networks

In 1997, the IEEE ratified the 802.11 WLAN standards, establishing a global standard
for implementing and deploying WLANs. The throughput for 802.11 is 2 Mbps, which
was well below the IEEE 802.3 Ethernet counterpart. Late in 1999, the IEEE ratified
the 802.11b standard extension, which raised the throughput to 11 Mbps, making this
extension more comparable to the wired equivalent. The 802.11b standard also supports
the 2-Mbps data rate and operates on the 2.4-GHz band in radio frequency for high-speed
data communications.

WLANs are typically found within a small client node-dense locale (e.g., a campus or
office building), or anywhere a traditional network cannot be deployed for logistical
reasons. Benefits include user mobility in the coverage area, speed and simplicity of
physical setup, and scalability. Being a military spinoff, WLANs also provide security
features such as encryption, frequency hopping, and firewalls. Some of these features
are intrinsic to the protocol, making some aspects of WLANs at least as secure as
wired networks and usually more so. The drawbacks are high initial cost (mostly
hardware), limited range, possibility of mutual interference, and the need to security-
enable clients.

1.2.1.1 LANs and WLANs Explained

A LAN physically links computers in an organization together via network cabling,
often specified as Ethernet CAT 5 cabling. With a wired LAN, computers are connected
by a fixed network of wires that can be difficult to move and expensive to change.

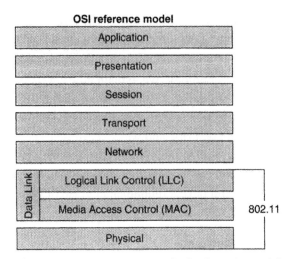

Figure 1.2: How 802.11 works in the OSI model

(see Chapter 2 for more detail on the OSI model)

These wires run throughout the walls, ceilings, and floors of the organization's buildings. LANs allow data and applications to be shared on multiple computers. A LAN provides a means for applications and files to be accessed on a central server via wired (or wireless) connections.

A WLAN enables a local network of computers to exchange data or other information without the use of cables. It can either replace or extend a wired LAN. The data can be transmitted as radio signals through the air, walls, ceilings, and so on without wired cabling. Using WLANs, mobile computing devices may be carried from place to place in an organization while remaining continuously connected. Any device that is within the range of an access point can potentially connect to the WLAN. This provides greatly increased freedom and flexibility compared to a wired network. WLANs are typically built using two key components:

1. An access point, or base station, that is usually physically connected to a LAN.

2. A wireless card that is either built into or added to a handheld, laptop, or desktop computer.

Using WLANs, it is easy to provide access to additional users. Access points can be added as necessary to manage the capacity and growth of an organization.

The most common wireless standard, 802.11b, has a data transfer rate of 11 Mbps—much slower than current wired LANs, which operate at 100 Mbps. Newly installed wired networks now operate at up to 1,000 Mbps (1Gb). 802.11b devices are often branded with a wireless fidelity (Wi-Fi) mark to indicate interoperability. A WLAN has sufficient bandwidth to handle a wide range of applications and services; however, it has a limited ability to deliver multimedia applications at sufficient quality, and a wired LAN is probably needed to access these. Ongoing advances in wireless standards continue to increase the data rate achievable with new equipment.

1.2.1.2 How WLANs Work

In a typical WLAN, a transmitter/receiver (transceiver) device, called an access point, is normally physically connected to the wired network using standard cabling. At a minimum, the access point receives, buffers, and transmits data between the WLAN and the wired network infrastructure, using radio frequencies to transmit data to each user. A single access point can support a small group of users, normally within a range of up to 100 meters depending on the local environment.

To access a WLAN via the access point, users need a computer with a wireless network interface card (NIC) either built in or installed. A wireless NIC can be inserted into the PCMCIA slot of a computer, laptop, or handheld device. Wireless NIC cards can also be installed into the PCI expansion slot of a desktop computer. In either case, the necessary software drivers should also be installed. It is also possible to create a peer-to-peer network—joining two or more computers together simply by installing a wireless NIC in each machine. Equally, it is possible for one of the computers in the peer-to-peer network to be joined to a LAN and to provide Internet access for the other computers in the network.

1.2.1.3 Advantages and Disadvantages of WLANs

WLANs have advantages and disadvantages when compared with wired LANs. A WLAN will make it simple to add or move workstations and to install access points to provide connectivity in areas where it is difficult to lay cable. Temporary or semipermanent buildings that are in range of an access point can be wirelessly connected to a LAN to give these buildings connectivity. Where computer labs are used in schools, the computers (laptops) could be put on a mobile cart and wheeled from classroom to classroom, provided they are in range of access points. Wired network points would be needed for each of the access points. A WLAN has some specific advantages:

- It is easier to add or move workstations.

- It is easier to provide connectivity in areas where it is difficult to lay cable.

- Installation is fast and easy, and it can eliminate the need to pull cable through walls and ceilings.

- Access to the network can be from anywhere within range of an access point.

- Portable or semipermanent buildings can be connected using a WLAN.

- Although the initial investment required for WLAN hardware can be similar to the cost of wired LAN hardware, installation expenses can be significantly lower.

- When a facility is located on more than one site (such as on two sides of a road), a directional antenna can be used to avoid digging trenches under roads to connect the sites.

- In historic buildings where traditional cabling would compromise the façade, a WLAN can avoid the need to drill holes in walls.

- Long-term cost benefits can be found in dynamic environments requiring frequent moves and changes.

WLANs also have some disadvantages:

- As the number of computers using the network increases, the data transfer rate to each computer will decrease accordingly.

- As standards change, it may be necessary to replace wireless cards and/or access points.

- Lower wireless bandwidth means some applications such as video streaming will be more effective on a wired LAN.

- Security is more difficult to guarantee and requires configuration.

- Devices will only operate at a limited distance from an access point, with the distance determined by the standard used and buildings and other obstacles between the access point and the user.

- A wired LAN is most likely to be required to provide a backbone to the WLAN; a WLAN should be a supplement to a wired LAN and not a complete solution.

- Long-term cost benefits are harder to achieve in static environments that require few moves and changes.

1.2.1.4 Current WLAN Standards

The Institute of Electrical and Electronics Engineers (IEEE) 802.11b specification is the most common standard for WLANs. 802.11a has recently been cleared to operate without a license in the 5-GHz frequency range. Several other standards are still under development, including 802.11g, 802.11h, and Ultra Wideband (UWB).

IEEE 802.11a

802.11a operates in the 5-GHz waveband. Because they operate in different frequencies, 802.11b and 802.11a are incompatible. This means organizations that have already deployed 802.11b networks, but want the faster speeds available through 802.11a, will either have to build a new WLAN alongside the existing one or purchase hardware that allows for both types of wireless cards to coexist. Several manufacturers currently provide, or are planning to release, dual-mode access points, which can hold both 802.11a and 802.11b cards. The power output of 802.11a is restricted, and the range tapers off more quickly than in 802.11b.

At present, 802.11a experiences few of the signal interference and congestion problems occasionally experienced by 802.11b users, although this may change now that license restrictions on the 5-GHz frequency band have been lifted and use increases. We can expect to see an increase in dual-platform access points to ensure compatibility between equipment.

IEEE 802.11b

This type of equipment is most common for establishing a WLAN in business. Not all 802.11b equipment is compatible, so the Wi-Fi compatibility mark has been developed. A Wi-Fi mark indicates compatibility between 802.11b products from different manufacturers. 802.11b equipment that is not Wi-Fi branded should be treated with caution. The 802.11b standard operates in the 2.4-GHz spectrum and has a nominal data transfer rate of 11 Mbps; however, in practice, the data transmission rate is approximately 4 to 7 Mbps, around one-tenth the speed of a current wired LAN. This is still adequate for accessing most data or applications, including Internet access, but would be insufficient for multimedia applications or for instances when a large number of simultaneous users want to access data in a single WLAN. Some manufacturers are producing 802.11b solutions with additional functionality that offers throughput at 22 Mbps speed. It is unclear if this type of equipment will be compatible across vendors.

1.2.1.5 Future WLAN Standards

Several standards are in development, all of which have a nominal data transfer rate of 54 Mbps, with actual data throughput probably in the vicinity of 27 to 30 Mbps. These standards, briefly discussed as follows, are known as IEEE 802.11g and IEEE 802.11h.

IEEE 802.11g

The IEEE is in the process of developing the 802.11g standard. Products have been launched by manufacturers such as Apple, D-link, and Buffalo, which are based on the draft 802.11g specification. The standard was expected to be finalized by IEEE around mid to late 2003. 802.11g operates in the 2.4-GHz waveband, the same as the 802.11b standard. If 802.11g becomes available, its advantage is that it offers some degree of backward compatibility with the existing 802.11b standard.

Potential users must be aware that any equipment installed before a final standard is ratified may not conform to the final standard. Organizations are advised to gain the necessary assurances from their supplier or a technical authority, and it is recommended that they seek up-to-date advice on the current licensing situation before making any financial commitment.

IEEE 802.11h

This is a modification of 802.11a being developed to comply with European regulations governing the 5-GHz frequency. 802.11h includes transmit power control (TPC) to limit transmission power and dynamic frequency selection (DFS) to protect sensitive frequencies. These changes protect the security of military and satellite radar networks sharing some of this waveband. It is likely that if 802.11h equipment meets with approval, current regulations will be reviewed, and 802.11h may supersede 802.11a. This may mean that no new 802.11a equipment could be installed and used, and existing equipment could be used only if no changes are made to the network.

1.2.1.6 Bluetooth and Wireless Personal Area Networks

Bluetooth-enabled devices allow creation of point-to-point or multipoint wireless personal area networks (WPANs) or "piconets" on an ad hoc or as-needed basis.

Bluetooth is intended to provide a flexible network topology, low energy consumption, robust data capacity, and high-quality voice transmission. Bluetooth is a low-cost radio solution that can provide links between devices at distances of up to 10 meters. Bluetooth has access speeds of 1 Mbps, considerably slower than the various 802.11 standards; however, the Bluetooth chips required to connect to the WLAN are considerably cheaper and have lower power requirements. Bluetooth technology is embedded in a wide range of devices. The Bluetooth wireless transport specification is meant to be used for bridging mobile computing devices, fixed telecom, tethered LAN computing, and consumer equipment together using low-cost, miniaturized RF components. Transport of both data and voice is supported in Bluetooth. Originally, Bluetooth was envisioned as a methodology to connect cellular or PCS telephones to other devices without wires. Examples of these other applications include USB "dongles," peripheral device connections, and PDA extensions.

Bluetooth, as a WPAN, should not be confused with a WLAN because it is not intended to do the same job. Bluetooth is primarily used as a wireless replacement for a cable to connect a handheld, mobile phone, MP3 player, printer, or digital camera to a PC or to each other, assuming the various devices are configured to share data. The Bluetooth standard is relatively complex; therefore, it is not always easy to determine if any two devices will communicate. Any device should be successfully tested before a purchase is made. Bluetooth is an open specification made available to Bluetooth SIG members on a royalty-free basis. More than 2,000 companies worldwide have declared interest in Bluetooth-enabled devices (for more information about the Bluetooth SIG, see http://www.bluetooth.org).

According to a *New Scientist* article about Bluetooth gadgetry written by Will Knight (August 11, 2003), Ollie Whitehouse, a researcher based in the United Kingdom and employed by the computer security company @Stake, created an eavesdropping tool to demonstrate just how vulnerable Bluetooth-enabled devices such as laptops, mobile phones, and handheld computers are to someone attempting to pilfer data from them. Whitehouse claimed his "Red Fang" program can be used in any setting where it is common to see users with mobile devices, such as while riding airplanes or trains or sitting in coffee shops. The Red Fang system can be used to scan these devices and determine if they are unprotected. Whitehouse sees his program as being similar to those used for "war-driving" activities (where people seek out poorly secured 802.11 wireless networks.) Ollie claims many people do not even know Bluetooth wireless technology is installed on their mobile devices, let alone the fact that the corresponding Bluetooth security settings are often turned off by default.

A report issued by research firm Gartner in September 2002 indicated that many people do not activate Bluetooth security features, and this inaction potentially exposes their devices to exploitation by hackers. Red Fang was unveiled in August 2003 at the Defcon computer security conference held in Las Vegas, Nevada. Bruce Potter, a security expert with U.S. think tank the Shmoo Group, improved the tool by making it more user friendly, according to Knight in his aforementioned article, which went on to state that "Potter expects Bluetooth security to become a growing concern, considering the high penetration of Bluetooth-enabled devices and the lack of knowledge about Bluetooth security in corporate security departments."

1.2.1.7 Ultra Wideband

UWB is a high-bandwidth (100 Mbps and up), short-range (less than 10 meters) specification for wireless connection designed to connect computers, or computers and consumer-electronics devices such as handheld and digital cameras, with the potential to become a high-bandwidth connection. There are some doubts that UWB equipment will become commercially available.

1.2.1.8 Security for WLANs

The IEEE has issued a new security standard for incorporation into 802.11a, 802.11b, and any new 802.11 standards that are approved for use, such as 802.11g. These enhancements, known as 802.11i, replace wired equivalent privacy (WEP) encryption with a new, more secure security encryption protocol called temporal key integrity protocol (TKIP), Advanced Encryption Standard (AES), and 802.1 x authentication. 802.11i was incorporated into the IEEE 802.11-2007 standard. While waiting for this standard to be completed, the Wi-Fi Alliance created an interim standard called Wi-Fi protected access (WPA), to allow the introduction of secure wireless products prior to finalization of 802.11i.

Anyone with a compatible wireless device can detect the presence of a WLAN. If appropriate security mechanisms are put in place, this does not mean that the data can be accessed, however. The WLAN should be configured so that anyone trying to access the WLAN has at least the same access restrictions as they would if they sat down at a wired network workstation. All of the following suggestions are practical steps that can be put in place to improve WLAN security. A combination of methods can be selected to meet company needs.

Wired Equivalent Privacy (WEP)

WLANs that adhere to the WiFi standard have a built-in security mechanism called WEP, an encryption protocol. When WLAN equipment is bought, WEP encryption is often turned off by default. In order to be effective, this security mechanism should be turned on. Originally, WEP encryption used a 40-bit or 64-bit encryption key and later a 128-bit key to increase security. All access devices must, however, be set to use the same encryption key in order to communicate. Using encryption will slow down the data transfer rate by up to 30 percent because it takes up some of the available bandwidth.

SSID

SSID is a method wireless networks use to identify or name an individual WLAN. Short for Service Set Identifier, the SSID is a 32-character unique identifier attached to the header of packets sent over a WLAN. This identifier acts as a device-level password when a mobile device tries to connect to the basic service set (BSS). The SSID enables a differentiation from one WLAN to another. All access points and all devices attempting to connect to a specific WLAN must use the same SSID. A device is not permitted to join the BSS unless it can provide the unique SSID. Because an SSID can be sniffed in plain text from a packet, it does not supply any security to the network. Sometimes, the SSID is referred to as a *network name*. It is a good practice to change the default SSID (network name) and passwords when setting up an *access point*. Access points may be factory-set to broadcast SSID data to allow users to configure access. This feature should be turned off. Here are some other tips to secure your WLAN:

- Install personal firewall software on all connected PCs and laptops.

- Utilize the Access Control List (ACL) that is standard in WLAN equipment. All devices have a unique MAC address; the ACL can deny access to any device not authorized to access the network.

- Remove rogue unauthorized access points from the network.

- Use password protection on all sensitive folders and files.

- Avoid wireless accessibility outside buildings where it is not required; directional aerials can be obtained to restrict the signal to 180 or 90 degrees from the access point.

- Switching off the power to the access point during off hours makes the WLAN unavailable at those times.

All WLAN systems should run on their own LAN segment, and a firewall should isolate them from the corporate LAN. Users wishing to access company databases through the wireless network should use a Virtual Private Network (VPN), encrypting all WLAN traffic and, thereby, protecting the system from hacking.

1.2.1.9 WLAN Performance

It is important to note that transmission speeds for all WLANs vary with file size, number of users, and distance from the access point. The performance of each standard varies considerably. 802.11b has a nominal data transfer of 11 Mbps, but in practice, the data rate is approximately 4 to 7 Mbps, depending on the equipment used. Performance figures vary depending on the number of users, the local environment, and any obstructions that are in the way. Buildings with many girders, thick walls, and concrete will often shorten the effective range, and there may be areas that are "dead zones." For that reason, it is always wise to try out a few different products to see how well they perform in your environment.

1.2.1.10 WLAN Implementation Concerns

Planning

To determine the location of access points, it is essential that a site survey be undertaken by a specialist in this area. A site survey will also determine the number of access points required to give the desired coverage, the range of each access point, and its channel designation. The access point, or the antenna that is attached to the access point, will usually be mounted high in a room; however, an access point may be mounted anywhere that is practical as long as the desired radio coverage is obtained. Larger spaces generally require more access points. Before a site survey is undertaken, it is advisable to prepare a floor plan to show where coverage is required. It is also essential that technical staff assist with the site survey. There may be a charge for the site survey, but the supplier will often refund the charge for the site survey if the equipment is subsequently purchased from them.

Capacity

Planning a wireless LAN should take into account the maximum number of people who will be using wireless devices in the range of any access point. The recommended number of devices accessing each access point is 15 to 20. Where light usage of the network occurs, such as with Web browsing or accessing applications that require little exchange of data, this could rise to around 30 users. It is relatively easy to scale wireless LANs by adding more access points, but for an 802.11b network, no more than three access points may operate in any one coverage area because of frequency congestion. An 802.11a network can operate up to eight access points in one area, each using a different channel.

With single-story facilities, the access points will usually be spread far apart, making channel assignment relatively straightforward. You can easily use a drawing to identify the position of your access points in relation to each other and assign each access point to a channel. The key with a larger 802.11b network is to make sure channels are assigned in a way that minimizes the overlap of signals. If access points will be located on multiple floors, then the task is slightly more difficult because planning will need to be done three-dimensionally.

Performance

Each 802.11b access point can potentially be accessed from up to 350 meters away outdoors. This distance is very much reduced indoors, the maximum distance attainable in ideal conditions being 100 meters. The range depends on the extent to which the radio signal is dissipated by transmission through walls and other obstructions and interference from reflection. The amount of data carried by a WLAN also falls as the distance increases; as signal strength becomes reduced, data rate reductions from 11 Mbps to 5.5, 2, or 1 Mbps will automatically occur in order to maintain the connection. It is possible that there may be issues of interference when using a 802.11b WLAN and where there are devices with Bluetooth connectivity, such as handheld devices and mobile phones, in the same area.

Post-installation coverage and performance may also vary as a result of building work, metal storage cabinets introduced into the coverage area, or other obstructions or reflectors that are moved or introduced. Microwave ovens and fish tanks have also been known to affect the performance of 802.11b networks. In the case of microwave ovens, they use the same frequency band as WLANs. If a different standard is implemented after installation of an existing 802.11b wireless network, a further site survey may be required because the coverage is likely to be different when compared to 802.11b. More access points may be required to maintain or improve data rates.

Power Consumption

The power output of 802.11a access points is limited, and the range from each access point may vary depending on the specific frequency used within the 5-GHz spectrum. Wireless cards typically have several different settings related to their power usage. For example, standby mode, where the transmission of data occurs only when required, uses relatively little power and leads to improved battery performance. This works similarly to a mobile phone that only connects to a network when making or receiving a call. Wireless cards that have recently come to market tend to consume less power. This trend is expected to continue as newer generations of cards become available.

Power over Ethernet

Access points require a power supply in addition to a wired network connection. Access points are often sited high on walls and away from existing power outlets. Some products on the market will supply power to the access point via the network cable, removing the need to install additional power cabling.

CHAPTER 2

Wireless Network Logical Architecture

Steve Rackley

The logical architecture of a network refers to the structure of standards and protocols that enable connections to be established between physical devices, or nodes, and which control the routing and flow of data between these nodes.

Since logical connections operate over physical links, the logical and physical architectures rely on each other, but the two also have a high degree of independence, as the physical configuration of a network can be changed without changing its logical architecture, and the same physical network can in many cases support different sets of standards and protocols.

The logical architecture of wireless networks will be described in this chapter with reference to the OSI model.

2.1 The OSI Network Model

The Open Systems Interconnect (OSI) model was developed by the International Standards Organization (ISO) to provide a guideline for the development of standards for interconnecting computing devices. The OSI model is a framework for developing these standards rather than a standard itself—the task of networking is too complex to be handled by a single standard.

The OSI model breaks down device to device connection, or more correctly application to application connection, into seven so-called layers of logically related tasks (see Table 2.1). An example will show how these layers combine to achieve a task such as sending and receiving an e-mail between two computers on separate local area networks (LANs) that are connected via the Internet.

Table 2.1: The seven layers of the OSI model

Layer	Description	Standards and Protocols
7 — Application layer	Standards to define the provision of services to applications—such as checking resource availability, authenticating users, etc.	HTTP, FTP, SNMP, POP3, SMTP
6 — Presentation layer	Standards to control the translation of incoming and outgoing data from one presentation format to another.	SSL
5 — Session layer	Standards to manage the communication between the presentation layers of the sending and receiving computers. This communication is achieved by establishing, managing and terminating "sessions".	ASAP, SMB
4 — Transport layer	Standards to ensure reliable completion of data transfers, covering error recovery, data flow control, etc. Makes sure all data packets have arrived.	TCP, UDP
3 — Network layer	Standards to define the management of network connections — routing, relaying and terminating connections between nodes in the network.	IPv4, IPv6, ARP
2 — Data link layer	Standards to specify the way in which devices access and share the transmission medium (known as *Media Access Control* or *MAC*) and to ensure reliability of the physical connection (known as *Logical Link Control* or *LLC*).	ARP Ethernet (IEEE 802.3), Wi-Fi (IEEE 802.11), Bluetooth (802.15.1)
1 — Physical layer	Standards to control transmission of the data stream over a particular medium, at the level of coding and modulation methods, voltages, signal durations and frequencies.	Ethernet, Wi-Fi, Bluetooth, WiMAX

The process starts with the sender typing a message into a PC e-mail application (Figure 2.1). When the user selects "Send," the operating system combines the message with a set of Application layer (Layer 7) instructions that will eventually be read and actioned by the corresponding operating system and application on the receiving computer.

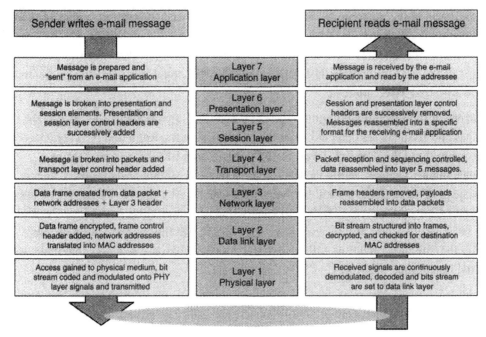

Figure 2.1: The OSI model in practice—an e-mail example

The message plus Layer 7 instructions are then passed to the part of sender's operating system that deals with Layer 6 presentation tasks. These include the translation of data between application layer formats as well as some types of security such as secure sockets layer (SSL) encryption. This process continues down through the successive software layers, with the message gathering additional instructions or control elements at each level.

By Layer 3—the Network layer—the message will be broken down into a sequence of data packets, each carrying a source and destination IP address. At the data-link layer, the IP address is "resolved" to determine the physical address of the first device that the sending computer needs to transmit frames to—the so-called MAC or media access control address. In this example, this device may be a network switch that the sending computer is connected to or the default gateway to the Internet from the sending computer's LAN. At the physical layer, also called the PHY layer, the data packets are encoded and modulated onto the carrier medium—a twisted wire pair in the case of a wired network, or electromagnetic radiation in the case of a wireless network—and transmitted to the device with the MAC address resolved at Layer 2.

Transmission of the message across the Internet is achieved through a number of device-to-device hops involving the PHY and data link layers of each routing or relaying device in the chain. At each step, the data link layer of the receiving device determines the MAC address of the next immediate destination, and the PHY layer transmits the packet to the device with that MAC address.

On arrival at the receiving computer, the PHY layer will demodulate and decode the voltages and frequencies detected from the transmission medium, and pass the received data stream up to the data link layer. Here the MAC and LLC elements, such as a message integrity check, will be extracted from the data stream and executed, and the message plus instructions passed up the protocol stack. At Layer 4, a protocol such as Transport Control Protocol (TCP), will ensure that all data frames making up the message have been received and will provide error recovery if any frames have gone missing. Finally the e-mail application will receive the decoded ASCII characters that make up the original transmitted message.

Standards for many layers of the OSI model have been produced by various organizations such as the Institute of Electrical and Electronics Engineers (IEEE). Each standard details the services that are provided within the relevant layer and the protocols or rules that must be followed to enable devices or other layers to call on those services. In fact, multiple standards are often developed for each layer, and they either compete until one emerges as the industry "standard" or else they peacefully coexist in niche areas.

The logical architecture of a wireless network is determined principally by standards that cover the data link (LLC plus MAC) and PHY layers of the OSI model. The following sections will give a preliminary introduction to these standards and protocols, while more detailed descriptions will be found in Parts III to V where local area (LAN), personal area (PAN) and metropolitan area (MAN) wireless networking technologies are described, respectively.

The next section starts this introductory sketch one layer higher—at the Network layer—not because this layer is specific to wireless networking, but because of the fundamental importance of its addressing and routing functions and of the underlying Internet Protocol (IP).

2.2 Network Layer Technologies

The Internet protocol (IP) is responsible for addressing and routing each data packet within a session or connection set up under the control of transport layer protocols such as TCP or UDP. The heart of the Internet Protocol is the IP address, a 32-bit number that is attached to each data packet and is used by routing software in the network or Internet to establish the source and destination of each packet.

While IP addresses, which are defined at the Network layer, link the billions of devices connected to the Internet into a single virtual network, the actual transmission of data frames between devices relies on the MAC addresses of the network interface cards (NICs), rather than the logical IP addresses of each NIC's host device. Translation between the Layer 3 IP address and the Layer 2 MAC address is achieved using Address Resolution Protocol (ARP), which is described in Section 2.2.4.

2.2.1 IP Addressing

The 32-bit IP address is usually presented in "dot decimal" format as a series of four decimal numbers between 0 and 255; for example, 200.100.50.10. This could be expanded in full binary format as 11001000.01100100.00110010.00001010.

As well as identifying a computer or other networked device, the IP address also uniquely identifies the network that the device is connected to. These two parts of the IP address are known as the host ID and the network ID. The network ID is important because it allows a device transmitting a data packet to know what the first port of call needs to be in the route to the packet's destination.

If a device determines that the network ID of the packet's destination is the same as its own network ID, then the packet does not need to be externally routed, for example through the network's gateway and out onto the Internet. The destination device is on its own network and is said to be "local" (Table 2.2). On the other hand, if the destination network ID is different from its own, the destination is a remote IP address and the packet will need to be routed onto the Internet or via some other network bridge to reach its destination. The first stage in this will be to address the packet to the network's gateway.

This process uses two more 32-bit numbers, the "subnet mask" and the "default gateway." A device determines the network ID for a data packet destination by doing a "logical AND" operation on the packet's destination IP address and its own subnet mask. The device determines its own network ID by doing the same operation using its own IP address and subnet mask.

Table 2.2: Local and remote IP addresses

Sending Device		
IP Address:	200.100.50.10	11001000.01100100.00110010.00001010
Subnet Mask:	255.255.255.240	11111111.11111111.11111111.11110000
Network ID:	200.100.50.000	11001000.01100100.00110010.00000000
Local IP address		
IP Address:	200.100.50.14	11001000.01100100.00110010.00001110
Subnet Mask:	255.255.255.240	11111111.11111111.11111111.11110000
Network ID:	200.100.50.000	11001000.01100100.00110010.00000000
Remote IP address		
IP Address:	200.100.50.18	11001000.01100100.00110010.00010010
Subnet Mask:	255.255.255.240	11111111.11111111.11111111.11110000
Network ID:	200.100.50.016	11001000.01100100.00110010.00010000

Table 2.3: Private IP address ranges

Class	Private address range start	Private address range end
A	10.0.0.0	10.255.255.255
B	172.16.0.0	172.31.255.255
C	192.168.0.0	192.168.255.255

2.2.2 Private IP Addresses

In February 1996, the Network Working Group requested industry comments on RFC 1918, which proposed three sets of so-called private IP addresses (Table 2.3) for use within networks that did not require Internet connectivity. These private addresses were intended to conserve IP address space by enabling many organizations to reuse the same sets of addresses within their private networks. In this situation it did not matter that a computer had an IP address that was not globally unique, provided that that computer did not need to communicate via the Internet.

Subsequently, the Internet Assigned Numbers Authority (IANA) reserved addresses 169.254.0.0 to 169.254.255.255 for use in Automatic Private IP Addressing (APIPA). If a computer has its TCP/IP configured to obtain an IP address automatically from a DHCP server, but is unable to locate such a server, then the operating system will automatically assign a private IP address from within this range, enabling the computer to communicate within the private network.

2.2.3 Internet Protocol Version 6 (IPv6)

With 32 bits, a total of 2^{32} or 4.29 billion IP addresses are possible—more than enough, one would think, for all the computers that the human population could possibly want to interconnect.

However, the famous statements that the world demand for computers would not exceed five machines, probably incorrectly attributed to Tom Watson Sr., chairman of IBM in 1943, or the statement of Ken Olsen, founder of Digital Equipment Corporation (DEC), to the 1977 World Future Society convention that "there is no reason for any individual to have a computer in his home," remind us how difficult it is to predict the growth and diversity of computer applications and usage.

The industry is now working on IP version 6, which will give 128-bit IP addresses based on the thinking that a world population of 10 billion by 2020 will eventually be served by many more than one computer each. IPv6 will give a comfortable margin for future growth, with 3.4×10^{38} possible addresses—that is, 3.4×10^{27} for each of the 10 billion population, or 6.6×10^{23} per square meter of the earth's surface.

It seems doubtful that there will ever be a need for IPv7, although, to avoid the risk of joining the short list of famously mistaken predictions of trends in computer usage, it may be as well to add the caveat "on this planet."

2.2.4 Address Resolution Protocol

As noted above, each PHY layer data transmission is addressed to the (Layer 2) MAC address of the network interface card of the receiving device, rather than to its (Layer 3) IP address. In order to address a data packet, the sender first needs to find the MAC address that corresponds to the immediate destination IP address and label the data packet with this MAC address. This is done using address resolution protocol (ARP).

Conceptually, the sending device broadcasts a message on the network that requests the device with a certain IP address to respond with its MAC address. The TCP/IP software operating in the destination device replies with the requested address and the packet can be addressed and passed on to the sender's data link layer.

In practice, the sending device keeps a record of the MAC addresses of devices it has recently communicated with, so it does not need to broadcast a request each time. This ARP table or "cache" is looked at first and a broadcast request is only made if the destination IP address is not in the table. In many cases, a computer will be sending the packet to its default gateway and will find the gateway's MAC address from its ARP table.

2.2.5 Routing

Routing is the mechanism that enables a data packet to find its way to a destination, whether that is a device in the next room or on the other side of the world.

A router compares the destination address of each data packet it receives with a table of addresses held in memory—the router table. If it finds a match in the table, it forwards the packet to the address associated with that table entry, which may be the address of another network or of a "next-hop" router that will pass the packet along towards its final destination.

If the router can't find a match, it goes through the table again looking at just the network ID part of the address (extracted using the subnet mask as described above). If a match is found, the packet is sent to the associated address or, if not, the router looks for a default next-hop address and sends the packet there. As a final resort, if no default address is set, the router returns a "Host Unreachable" or "Network Unreachable" message to the sending IP address. When this message is received it usually means that somewhere along the line a router has failed.

What happens if, or when, this elegantly simple structure breaks down? Are there packets out there hopping forever around the Internet, passing from router to router and never finding their destination? The IP header includes a control field that prevents this from happening. The time-to-live (TTL) field is initialized by the sender to a certain value, usually 64, and reduced by one each time the packet passes through a router. When TTL get down to zero, the packet is discarded and the sender is notified using an Internet control message protocol (ICMP) "time-out" message.

2.2.5.1 Building Router Tables

The clever part of a router's job is building its routing table. For simple networks a static table loaded from a start-up file is adequate but, more generally, *dynamic routing* enables tables to be built up by routers sending and receiving broadcast messages.

These can be either ICMP Router Solicitation and Router Advertisement messages which allow neighboring routers to ask, "Who's there?" and respond, "I'm here," or more useful RIP (Router Information Protocol) messages, in which a router periodically broadcasts its complete router table onto the network.

Other RIP and ICMP messages allow routers to discover the shortest path to an address, to update their tables if another router spots an inefficient routing and to periodically update routes in response to network availability and traffic conditions.

A major routing challenge occurs in mesh or mobile ad-hoc networks (MANETs), where the network topology may be continuously changing.

2.2.6 Network Address Translation

As described in Section 2.2.2 RFC 1918 defined three sets of private IP addresses for use within networks that do not require Internet connectivity.

However, with the proliferation of the Internet and the growing need for computers in these previously private networks to go online, the limitation of this solution to conserving IP addresses soon became apparent. How could a computer with a private IP address ever get a response from the Internet, when its IP address would not be recognized by any router out in the Internet as a valid destination? *Network address translation* (NAT) provides the solution to this problem.

When a computer sends a data packet to an IP address outside a private network, the gateway that connects the private network to the Internet will replace the private IP source address (192.168.0.1 in Table 2.4), by a public IP address (e.g., 205.55.55.1). The receiving server and Internet routers will recognize this as a valid destination address and route the data packet correctly. When the originating gateway receives a returning data packet it will replace the destination address in the data packet with the original private IP address of the initiating computer. This process of private to public IP address translation at the Internet gateway of a private network is known as *network address translation*.

Table 2.4: Example of a simple static NAT table

Private IP address	Public IP address
192.168.0.1	205.55.55.1
192.168.0.2	205.55.55.2
192.168.0.3	205.55.55.3
192.168.0.4	205.55.55.4

2.2.6.1 Static and Dynamic NAT

In practice, similar to routing, NAT can be either static or dynamic. In static NAT, every computer in a private network that requires Internet access has a public IP address assigned to it in a prescribed NAT table. In dynamic NAT, a pool of public IP addresses are available and are mapped to private addresses as required.

Needless to say, dynamic NAT is by far the most common, as it is automatic and requires no intervention or maintenance.

2.2.7 Port Address Translation

One complication arises if the private network's gateway has only a single public IP address available to assign, or if more computers in a private network try to connect than there are IP addresses available to the gateway. This will often be the case for a small organization with a single Internet connection to an ISP. In this case, it would seem that only one computer within the private network would be able to connect to the Internet at a time. Port address translation (PAT) overcomes this limitation by mapping private IP addresses to different port numbers attached to the single public IP address.

When a computer within the private network sends a data packet to be routed to the Internet, the gateway replaces the source address with the single public IP address together with a random port number between 1024 and 65536 (Figure 2.2). When a data packet is returned with this destination address and port number, the PAT table (Table 2.5) enables the gateway to route the data packet to the originating computer in the private network.

2.3 Data Link Layer Technologies

The data link layer is divided into two sub-layers—*logical link control* (LLC) *and media access control* (MAC). From the data link layer down, data packets are addressed

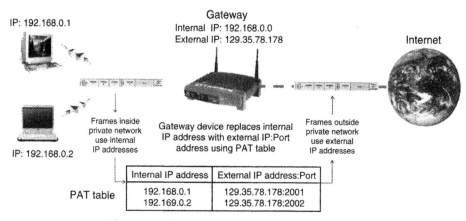

Figure 2.2: Port address translation in practice

Table 2.5: Example of a simple PAT table

Private IP address	Public IP address: Port
192.168.0.1	129.35.78.178:2001
192.168.0.2	129.35.78.178:2002
192.168.0.3	129.35.78.178:2003
192.168.0.4	129.35.78.178:2004

using MAC addresses to identify the specific physical devices that are the source and destination of packets, rather than the IP addresses, URLs or domain names used by the higher OSI layers.

2.3.1 Logical Link Control

Logical link control (LLC) is the upper sub-layer of the data link layer (Figure 2.3), and is most commonly defined by the IEEE 802.2 standard. It provides an interface that enables the Network layer to work with any type of media access control layer.

A frame produced by the LLC and passed down to the MAC layer is called an *LLC protocol data unit* (LPDU), and the LLC layer manages the transmission of LPDUs between the link layer service access points of the source and destination devices. A link layer *service access point* (SAP) is a port or logical connection point to a Network layer protocol (Figure 2.4). In a network supporting multiple Network layer protocols, each

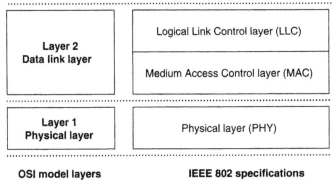

OSI model layers   **IEEE 802 specifications**

Figure 2.3: OSI layers and IEEE 802.2 specifications

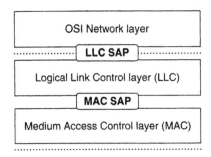

Figure 2.4: Logical location of LLC and MAC service access points

will have specific source SAP (SSAP) and destination SAP (DSAP) ports. The LPDU includes the 8-bit DSAP and SSAP addresses to ensure that each LPDU is passed on receipt to the correct Network layer protocol.

The LLC layer defines connectionless (Type 1) and connection oriented (Type 2) communication services and, in the latter case, the receiving LLC layer keeps track of the sequence of received LPDUs. If an LPDU is lost in transit or incorrectly received, the destination LLC requests the source to restart the transmission at the last received LPDU.

The LLC passes LPDUs down to the MAC layer at a logical connection point known as the *MAC service access point* (MAC SAP). The LPDU is then called a *MAC service data unit* (MSDU) and becomes the data payload for the MAC layer.

2.3.2 Media Access Control

The second sub-layer of the data link layer controls how and when a device is allowed to access the PHY layer to transmit data; this is the *media access control* or *MAC* layer.

In the following sections, the addressing of data packets at the MAC level is first described. This is followed by a brief look at MAC methods applied in wired networks, which provides an introduction to the more complex solutions required for media access control in wireless networks.

2.3.2.1 MAC Addressing

A receiving device needs to be able to identify those data packets transmitted on the network medium that are intended for it—this is achieved using MAC addresses. Every network adapter, whether it is an adapter for Ethernet, wireless or some other network technology, is assigned a unique serial number called its MAC address when it is manufactured.

The Ethernet address is the most common form of MAC address and consists of six bytes, usually displayed in hexadecimal, such as 00-D0-59-FE-CD-38. The first three bytes are the manufacturer's code (00-D0-59 in this case is Intel) and the remaining three are the unique serial number of the adapter. The MAC address of a network adapter on a Windows PC can be found in Windows 95, 98 or Me by clicking Start→Run, and then typing "winipcfg", and selecting the adapter, or in Windows NT, 2000, and XP by opening a DOS Window (click Start→Programs→Accessories→Command Prompt) and typing "ipconfig/all".

When an application such as a web browser sends a request for data onto the network, the Application layer request comes down to the MAC SAP as an MSDU. The MSDU is extended with a MAC header that includes the MAC address of the source device's network adapter. When the requested data is transmitted back onto the network, the original source address becomes the new destination address and the network adapter of the original requesting device will detect packets with its MAC address in the header, completing the round trip.

As an example, the overall structure of the IEEE 802.11 MAC frame, or MAC protocol data unit (MPDU) is shown in Figure 2.5.

The elements of the MPDU are shown in Table 2.6.

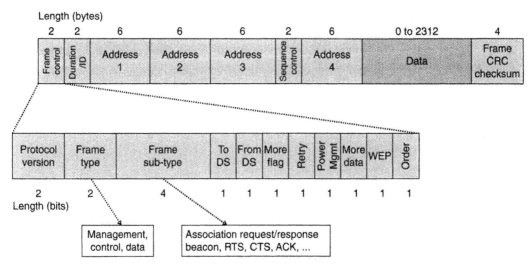

Figure 2.5: MAC frame structure

Table 2.6: Elements of the 802.11 MPDU frame structure

MPDU element	Description
Frame control	A sequence of flags to indicate the protocol version (802.11 a/b/g), frame type (management, control, data), sub-frame type (e.g., probe request, authentication, association request, etc.), fragmentation, retries, encryption, etc.
Duration	Expected duration of this transmission. Used by waiting stations to estimate when the medium will again be idle.
Address 1 to Address 4	Destination and source, plus optional to and from addresses within the distribution system.
Sequence	Sequence number to identify frame fragments or duplicates.
Data	The data payload passed down as the MSDU.
Frame check sequence	A CRC-32 checksum to enable transmission errors to be detected.

2.3.3 Media Access Control in Wired Networks

If two devices transmit at the same time on a network's shared medium, whether wired or wireless, the two signals will interfere and the result will be unusable to both devices. Access to the shared medium therefore needs to be actively managed to ensure that the available bandwidth is not wasted through repeated collisions of this type. This is the main task of the MAC layer.

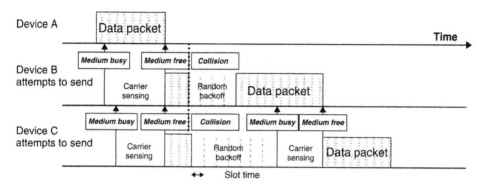

Figure 2.6: Ethernet CSMA/CD timing

2.3.3.1 Carrier Sense Multiple Access/Collision Detection (CSMA/CD)

The most commonly used MAC method to control device transmission, and the one specified for Ethernet based networks, is *carrier sense multiple access/collision detection* (CSMA/CD) (Figure 2.6). When a device has a data frame to transmit onto a network that uses this method, it first checks the physical medium (carrier sensing) to see if any other device is already transmitting. If the device senses another transmitting device it waits until the transmission has finished. As soon as the carrier is free, it begins to transmit data, while at the same time continuing to listen for other transmissions.

If it detects another device transmitting at the same time (collision detection), it stops transmitting and sends a short jam signal to tell other devices that a collision has occurred. Each of the devices that were trying to transmit then computes a random back-off period within a range 0 to t_{max}, and tries to transmit again when this period has expired. The device that by chance waits the shortest time will be the next to gain access to the medium, and the other devices will sense this transmission and go back into carrier sensing mode.

A very busy medium may result in a device experiencing repeated collisions. When this happens t_{max} is doubled for each new attempt, up to a maximum of 10 doublings, and if the transmission is unsuccessful after 16 attempts the frame is dropped and the device reports an "excessive collision error."

2.3.3.2 Other Wired Network MAC Methods

Another common form of media access control for wired networks, defined by the IEEE 802.5 standard, involves passing an electronic "token" between devices on the network in a pre-defined sequence. The token is similar to a baton in a relay race in that a device can only transmit when it has captured the token.

If a device does not need control of the media to transmit data it passes the token on immediately to the next device in the sequence, while if it does have data to transmit it can do so once it receives the token. A device can only keep the token and continue to use the media for a specific period of time, after which it has to pass the token on to the next device in the sequence.

2.3.4 Media Access Control in Wireless Networks

The collision detection part of CSMA/CD is only possible if the PHY layer transceiver enables the device to listen to the medium while transmitting. This is possible on a wired network, where invalid voltages resulting from collisions can be detected, but is not possible for a radio transceiver since the transmitted signal would overload any attempt to receive at the same time. In wireless networks such as 802.11, where collision detection is not possible, a variant of CSMA/CD known as *CSMA/CA* is used, where the CA stands for *collision avoidance*.

Apart from the fact that collisions are not detected by the transmitting device, CSMA/CA has some similarities with CSMA/CD. Devices sense the medium before transmitting and wait if the medium is busy. The duration field in each transmitted frame (see preceding Table 2.6) enables a waiting device to predict how long the medium will be busy.

Once the medium is sensed as being idle, waiting devices compute a random time period, called the contention period, and attempt to transmit after the contention period has expired. This is a similar mechanism to the back off in CSMA/CD, except that here it is designed to avoid collisions between stations that are waiting for the end of another station's transmitted frame rather than being a mechanism to recover after a detected collision.

2.4 Physical Layer Technologies

When the MPDU is passed down to the PHY layer, it is processed by the *PHY Layer Convergence Procedure* (PLCP) and receives a preamble and header, which depend on the specific type of PHY layer in use. The PLCP preamble contains a string of bits that enables a receiver to synchronize its demodulator to the incoming signal timing.

The preamble is terminated by a specific bit sequence that identifies the start of the header, which in turn informs the receiver of the type of modulation and coding scheme to be used to decode the upcoming data unit.

The assembled PLCP protocol data unit (PPDU) is passed to the physical medium dependent (PMD) sub-layer, which transmits the PPDU over the physical medium, whether that is twisted-pair, fiber-optic cable, infrared or radio.

PHY layer technologies determine the maximum data rate that a network can achieve, since this layer defines the way the data stream is coded onto the physical transmission medium. However, the MAC and PLCP headers, preambles and error checks, together with the idle periods associated with collision avoidance or back off, mean that the PMD layer actually transmits many more bits than are passed down to the MAC SAP by the data link layer.

The next sections look at some of the PHY layer technologies applied in wired networks and briefly introduce the key features of wireless PHY technologies.

2.4.1 Physical Layer Technologies—Wired Networks

Most networks that use wireless technology will also have some associated wired networking elements, perhaps an Ethernet link to a wireless access point, a device-to-device FireWire or USB connection, or an ISDN based Internet connection. Some of the most common wired PHY layer technologies are described in this section, as a precursor to the more detailed discussion of local, personal and metropolitan area wireless network PHY layer technologies provided later.

2.4.1.1 Ethernet (IEEE 802.3)

The first of these, Ethernet, is a data link layer LAN technology first developed by Xerox and defined by the IEEE 802.3 standard. Ethernet uses carrier sense multiple access with collision detection (CSMA/CD), described above, as the media access control method.

Ethernet variants are known as "*A*" *Base*-"*B*" networks, where "A" stands for the speed in Mbps and "B" identifies the type of physical medium used. 10 Base-T is the standard Ethernet, running at 10 Mbps and using an *unshielded twisted-pair copper wire* (UTP), with a maximum distance of 500 meters between a device and the nearest hub or repeater.

The constant demand for increasing network speed has meant that faster varieties of Ethernet have been progressively developed. 100 Base-T, or Fast Ethernet operates at

100 Mbps and is compatible with 10 Base-T standard Ethernet as it uses the same twisted-pair cabling and CSMA/CD method. The trade-off is with distance between repeaters, a maximum of 205 meters being achievable for 100 Base-T. Fast Ethernet can also use other types of wiring—100 Base-TX, which is a higher-grade twisted-pair, or 100 Base-FX, which is a two strand fiber-optic cable. Faster speeds to 1 Gbps or 10 Gbps are also available.

The PMD sub-layer is specified separately from the Ethernet standard, and for UTP cabling this is based on the twisted pair-physical medium dependent (TP-PMD) specification developed by the ANSI X3T9.5 committee.

The same frame formats and CSMA/CD technology are used in 100 Base-T as in standard 10 Base-T Ethernet, and the 10-fold increase in speed is achieved by increasing the clock speed from 10 MHz to 125 MHz, and reducing the interval between transmitted frames, known as the Inter-Packet Gap (IPG), from 9.6 ms to 0.96 ms. A 125-MHz clock speed is required to deliver a 100 Mbps effective data rate because of the 4B/5B encoding described below.

To overcome the inherent low-pass nature of the UTP physical medium, and to ensure that the level of RF emissions above 30 MHz comply with FCC regulations, the 100 Base-T data encoding scheme was designed to bring the peak power in the transmitted data signal down to 31.25 MHz (close to the FCC limit) and to reduce the power in high frequency harmonics at 62.5 MHz, 125 MHz and above.

4B/5B encoding is the first step in the encoding scheme (Figure 2.7). Each 4-bit nibble of input data has a 5th bit added to ensure there are sufficient transitions in the transmitted bit stream to allow the receiver to synchronize for reliable decoding. In the second

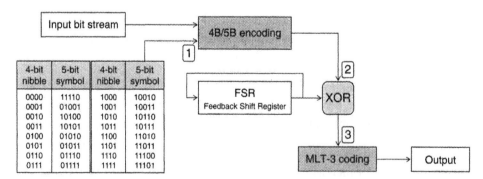

Figure 2.7: 100 Base-T Ethernet data encoding scheme

step an 11-bit feedback shift register (FSR) produces a repeating pseudo-random bit pattern which is XOR'd with the 4B/5B output data stream. The effect of this pseudo-randomization is to minimize high frequency harmonics in the final transmitted data signal. The same pseudo-random bit stream is used to recover the input data in a second XOR operation at the receiver.

The final step uses an encoding method called Multi-Level Transition 3 (MLT-3) to shape the transmitted waveform in such a way that the center frequency of the signal is reduced from 125 MHz to 31.25 MHz.

MLT-3 is based on the repeating pattern 1, 0, −1, 0. As shown in Figure 2.8, an input 1-bit causes the output to transition to the next bit in the pattern while an input 0-bit causes no transition, i.e., the output level remaining unchanged. Compared to the Manchester Phase Encoding (MPE) scheme used in 10 Base-T Ethernet, the cycle length of the output signal is reduced by a factor of 4, giving a signal peak at 31.25 MHz instead of 125 MHz. On the physical UTP medium, the 1, 0 and −1 signal levels are represented by line voltages of +0.85, 0.0 and −0.85 volts.

2.4.1.2 ISDN

ISDN, which stands for Integrated Services Digital Network, allows voice and data to be transmitted simultaneously over a single pair of telephone wires. Early analog phone networks were inefficient and error prone as a long distance data communication medium and, since the 1960s, have gradually been replaced by packet-based digital switching systems.

The International Telephone and Telegraph Consultative Committee (CCITT), the predecessor of the International Telecommunications Union (ITU), issued initial

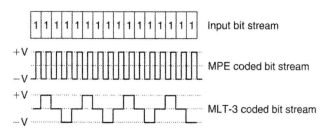

Figure 2.8: Ethernet MPE and fast Ethernet MLT-3 encoding

guidelines for implementing ISDN in 1984, in CCITT Recommendation I.120. However, industry-wide efforts to establish a specific implementation for ISDN only started in the early 1990s when US industry members agreed to create the National ISDN 1 standard (NI-1). This standard, later superseded by National ISDN 2 (NI-2), ensured the interoperability of end user and exchange equipment.

Two basic types of ISDN service are defined—basic rate interface (BRI) and primary rate interface (PRI). ISDN carries voice and user data streams on "bearer" (B) channels, typically occupying a bandwidth of 64 kbps, and control data streams on "demand" (D) channels, with a 16 kbps or 64 kbps bandwidth depending on the service type.

BRI provides two 64 kbps B channels, which can be used to make two simultaneous voice or data connections or can be combined into one 128 kbps connection. While the B channels carry voice and user data transmission, the D channel is used to carry Data Link and Network layer control information.

The higher capacity PRI service provides 23 B channels plus one 64 kbps D channel in the US and Japan, or 30 B channels plus one D channel in Europe. As for BRI, the B channels can be combined to give data bandwidths of 1472 kbps (US) or 1920 kbps (Europe).

As noted above, telephone wires are not ideal as a digital communication medium. The ISDN PHY layer limits the effect of line attenuation, near-end and far-end crosstalk and noise by using pulse amplitude modulation (PAM) technology to achieve a high data rate at a reduced transmission rate on the line.

This is achieved by converting multiple (often two or four) binary bits into a single multilevel transmitted symbol. In the US, the 2B1Q method is used, which converts two binary bits (2B) into a single output symbol, known as a "quat" (1Q), which can have one of four values (Figure 2.9 and Table 2.7). This effectively halves the transmission rate on the line, so that a 64 kbps data rate can be transmitted at a symbol rate of 32 ksps, achieving higher data rates within the limited bandwidth of the telephone system.

As well as defining a specific PHY layer, ISDN also specifies Data Link and Network layer operation. LAP-D (link access protocol D-channel) is a Data Link protocol, defined in ITU-T Q.920/921, which ensures error free transmission on the PHY layer. Two Network layer protocols are defined in ITU-T Q.930 and ITU-T Q.931 to establish, maintain and terminate user-to-user, circuit-switched, and packet-switched network connections.

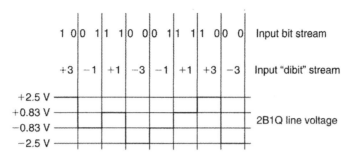

Figure 2.9: 2B1Q line code using in ISDN

Table 2.7: 2B1Q line code used in ISDN

Input "DIBIT"	Output "QUAT"	Line voltage
10	+3	+2.5
11	+1	+.833
00	−1	−.833
01	−3	−2.5

2.4.1.3 FireWire

FireWire, also known as IEEE 1394 or i.Link, was developed by Apple Computer Inc. in the mid-1990s as a local area networking technology. At that time it provided a 100 Mbps data rate, well above the Universal Serial Bus (USB) speed of 12 Mbps, and it was soon taken up by a number of companies for applications such as connecting storage and optical drives.

FireWire is now supported by many electronics and computer companies, often under the IEEE 1394 banner, because of its ability to reliably and inexpensively transmit digital video data at high speeds, over single cable lengths of up to 4.5 meters. The standard data rate is 400 Mbps, although a faster version is also available delivering 800 Mbps and with plans for 3.2 Gbps. Range can be extended up to 72 meters using signal repeaters in a 16-link daisy chain, and FireWire to fiber transceivers are also available that replace the copper cable by optical fiber and can extend range to 40 km. A generic FireWire topology is shown in Figure 2.10.

The FireWire standard defines a serial input/output port and bus, a 4 or 6 wire dual-shielded copper cable that can carry both data and power, and the related Data Link, Network

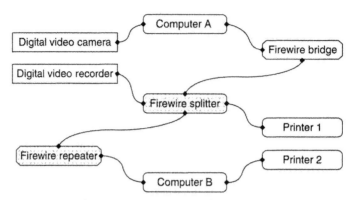

Figure 2.10: FireWire network topology: daisy-chain and tree structures

and Transport layer protocols. FireWire is based on the control and status register management (CSR) architecture, which means that all interconnected devices appear as a single memory of up to 256 Terabytes (256×10^{12} bytes). Each transmitted packet of data contains three elements: a 10-bit bus ID that is used to determine which FireWire bus the data packet originated from, a 6-bit ID that identifies which device or node on that bus sent the data packet, and a 48-bit offset that is used to address registers and memory in a node.

While primarily used for inter-device communication, The Internet Society has combined IP with the FireWire standard to produce a standard called IP over IEEE 1394, or IP 1394. This makes it possible for networking services such as FTP, HTTP and TCP/IP to run on the high speed FireWire PHY layer as an alternative to Ethernet.

An important feature of FireWire is that the connections are "hot-swappable", which means that a new device can be connected, or an existing device disconnected, while the connection is live. Devices are automatically assigned node IDs, and these IDs can change as the network topology changes. Combining the node ID variability of FireWire with the IP requirement for stable IP addresses of connected devices, presents one of the interesting problems in enabling IP connections over FireWire. This is solved using a special Address Resolution Protocol (ARP) called 1394 ARP.

In order to uniquely identify a device in the network, 1394 ARP uses the 64-bit Extended Unique Identifier (EUI-64), a unique 64-bit number that is assigned to every FireWire device on manufacture. This is an extended version of the MAC address that is used to

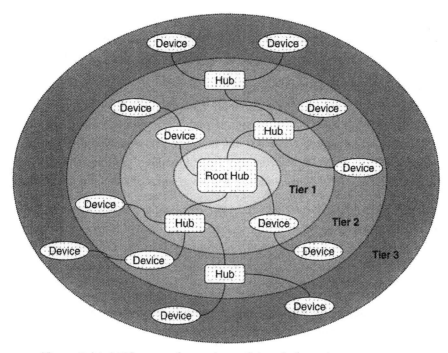

Figure 2.11: USB network topology: daisy-chain and tree structures

address devices other than network interfaces. A 48-bit MAC address can be converted into a 64-bit EUI-64 by prefixing the two hexadecimal octets "FF-FF".

2.4.1.4 Universal Serial Bus

The *Universal Serial Bus* (USB) was introduced in the mid-1990s to provide a hot-swappable "plug-and-play" interface that would replace different types of peripheral interfaces (parallel ports, serial ports, PS/2, MIDI, etc.) for devices such as joysticks, scanners, keyboards and printers. The maximum bandwidth of USB 1.0 was 12 Mbps, but this has since increased to a FireWire matching 480 Mbps with USB 2.0.

USB uses a host-centric architecture, with a host controller dealing with the identification and configuration of devices connected either directly to the host or to intermediate hubs (Figure 2.11). The USB specification supports both isochronous and asynchronous transfer types over the same connection. Isochronous transfers require guaranteed bandwidth and low latency for applications such as telephony and media streaming, while asynchronous transfers are delay-tolerant and are able to wait for available bandwidth. USB control protocols are designed specifically to give a low protocol overhead, resulting in highly effective utilization of the available bandwidth.

This available bandwidth is shared among all connected devices and is allocated using "pipes," with each pipe representing a connection between the host and a single device. The bandwidth for a pipe is allocated when the pipe is established, and a wide range of different device bit rates and device types can be supported concurrently. For example, digital telephony devices can be concurrently accommodated ranging from 1 "bearer" plus 1 "demand" channel (64 kbps—see ISDN above) up to T1 capacity (1.544 Mbps).

USB employs NRZI (non return to zero inverted) as a data encoding scheme. In NRZI encoding, a 1-bit is represented by no change in output voltage level and a 0-bit is represented by a change in voltage level (Figure 2.12). A string of 0-bits therefore causes the NRZI output to toggle between states on each bit cycle, while a string of 1-bits causes a period with no transitions in the output.

NRZI has the advantage of a somewhat improved noise immunity compared with the straight encoding of the input data stream as output voltages.

2.4.2 Physical Layer Technologies—Wireless Networks

The PHY layer technologies that provide the Layer 1 foundation for wireless networks will be described further in Parts III, IV and V, where LAN, PAN and MAN technologies and their implementations will be covered in detail.

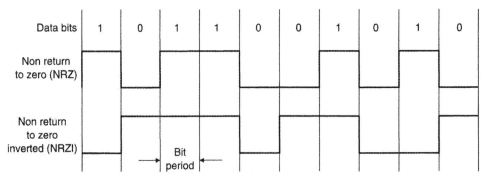

Figure 2.12: USB NRZI data encoding scheme

Table 2.8: Aspects of PHY and data link layer wireless technologies

Technology aspect	Issues and considerations
Spectrum	What part of the electromagnetic spectrum is used, what is the overall bandwidth available, how is this segmented into channels? What mechanisms are available to control utilized bandwidth to ensure coexistence with other users of the same spectrum?
Propagation	What power levels are permitted by regulatory authorities in the spectrum in question? What mechanisms are available to control the transmitted power or propagation pattern to minimize co-channel interference for other users, maximize effective range or utilize spatial diversity to increase throughput?
Modulation	How is encoded data carried on the physical medium, for example by modulating one or more carriers in phase and/or amplitude, or by modulating pulses in amplitude and/or position?
Data encoding	How are the raw bits of a data frame coded into symbols for transmission? What functions do these coding mechanisms serve, for example increasing robustness to noise or increasing the efficient use of available bandwidth?
Media access	How is access to the transmission medium controlled to ensure that the bandwidth available for data transmission is maximized and that contention between users is efficiently resolved? What mechanisms are available to differentiate media access for users with differing service requirements?

Each wireless PHY technology, from Bluetooth to ZigBee, will be described in terms of a number of key aspects, as summarized in Table 2.8.

The range and significance of the issues vary depending on the type of technology (Ir, RF, Near-field) and its application (PAN, LAN or WAN).

2.5 Operating System Considerations

In order to support networking, an operating system needs as a minimum to implement networking protocols, such as TCP/IP, and the device drivers required for network hardware. Early PC operating systems, including Windows versions prior to Windows 95, were not designed to support networking. However, with the rise of the Internet and other networking technologies, virtually every operating system today qualifies as a network operating system (NOS).

Individual network operating systems offer additional networking features such as firewalls, simplified set-up and diagnostic tools, remote access, and inter-connection with networks running other operating systems, as well as support for network administration tasks such as enforcing common settings for groups of users.

The choice of a network operating system will not be covered in detail here, but should be based on a similar process to that for selecting WLAN and WPAN technologies. Start by determining networking service requirements such as security, file sharing, printing and messaging. The two main network operating systems are the Microsoft Windows and Novell NetWare suites of products. A key differentiator between these two products may be a requirement for interoperability support in networks that include other operating systems such as UNIX or Linux. NetWare is often the preferred NOS in mixed operating-system networks, while simplicity of installation and administration makes Windows the preferred product suite in small networks where technical support may be limited.

2.6 Summary

The OSI network model provides the conceptual framework to describe the logical operation of all types of networks, from a wireless PAN link between a mobile phone and headset to the global operation of the Internet.

The key features that distinguish different networking technologies, particularly wired and wireless, are defined at the Data Link (LLC and MAC) and physical (PHY) layers.

CHAPTER 3

Wireless Network Physical Architecture

Steve Rackley

3.1 Wired Network Topologies—A Refresher

The topology of a wired network refers to the physical configuration of links between networked devices or nodes, where each node may be a computer, an end-user device such as a printer or scanner, or some other piece of network hardware such as a hub, switch or router.

The building block from which different topologies are constructed is the simple point-to-point wired link between two nodes, shown in Figure 3.1. Repeating this element results in the two simplest topologies for wired networks—bus and ring.

For the ring topology, there are two possible variants depending on whether the inter-node links are simplex (one-way) or duplex (two-way). In the simplex case, each inter-node link has a transmitter at one end and a receiver at the other, and messages circulate in

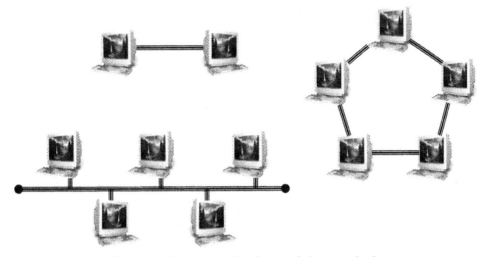

Figure 3.1: Point-to-point, bus and ring topologies

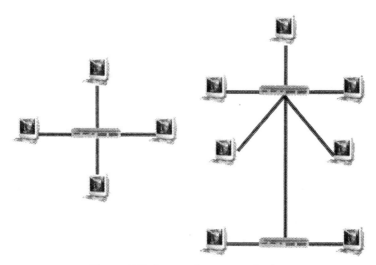

Figure 3.2: Star and tree topologies

one direction around the ring, while in the duplex case each link has both transmitter and receiver (a so-called transceiver) at each end, and messages can circulate in either direction.

Bus and ring topologies are susceptible to single-point failures, where a single broken link can isolate sections of a bus network or halt all traffic in the case of a ring.

The step that opens up new possibilities is the introduction of specialized network hardware nodes designed to control the flow of data between other networked devices. The simplest of these is the passive hub, which is the central connection point for LAN cabling in star and tree topologies, as shown in Figure 3.2. An active hub, also known as a repeater, is a variety of passive hub that also amplifies the data signal to improve signal strength over long network connections.

For some PAN technologies, such as USB, star and tree topologies can be built without the need for specialized hardware, because of the daisy-chaining capability of individual devices.

An active or passive hub in a star topology LAN transmits every received data packet to every connected device. Each device checks every packet and decodes those identified by the device's MAC address. The disadvantage of this arrangement is that the bandwidth of the network is shared among all devices, as shown in Figure 3.3. For example, if two PCs are connected through a 10 Mbps passive hub, each will have on average 5 Mbps of bandwidth available to it.

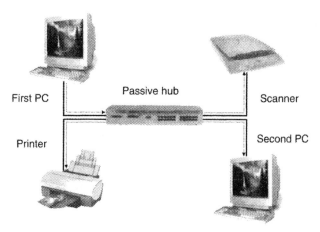

Figure 3.3: A passive hub in a physical star network

If the first PC is transmitting data, the hub relays the data packets on to all other devices in the network. Any other device on the network will have to wait its turn to transmit data.

A switching hub (or simply a switch) overcomes this bandwidth sharing limitation by only transmitting a data packet to the device to which it is addressed. Compared to a non-switching hub, this requires increased memory and processing capability, but results in a significant improvement in network capacity.

The first PC (Figure 3.4) is transmitting data stream A to the printer and the switch directs these data packets only to the addressed device. At the same time, the scanner is sending data stream B to the second PC. The switch is able to process both data streams concurrently, so that the full network bandwidth is available to every device.

3.2 Wireless Network Topologies

3.2.1 Point-to-Point Connections

The simple point-to-point connection shown in Figure 3.1 is probably more common in wireless than in wired networks, since it can be found in a wide variety of different wireless situations, such as:

- peer-to-peer or ad-hoc Wi-Fi connections
- wireless MAN back-haul provision

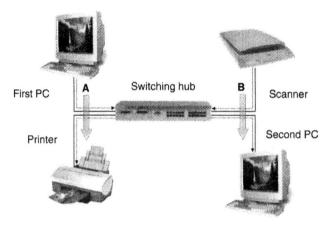

Figure 3.4: Switching hub in physical star network

- LAN wireless bridging
- Bluetooth
- IrDA

3.2.2 Star Topologies in Wireless Networks

In wireless networks the node at the center of a star topology (Figure 3.5), whether it is a WiMAX base station, Wi-Fi access point, Bluetooth Master device or a ZigBee PAN coordinator, plays a similar role to the hub in a wired network. The different wireless networking technologies require and enable a wide range of different functions to be performed by these central control nodes.

The fundamentally different nature of the wireless medium means that the distinction between switching and nonswitching hubs is generally not relevant for control nodes in wireless networks, since there is no direct wireless equivalent of a separate wire to each device. The wireless LAN switch or controller (Figure 3.6), described in Section 3.3.3, is a wired network device that switches data to the access point that is serving the addressed destination station of each packet.

The exception to this general rule arises when base stations or access point devices are able to spatially separate individual stations or groups of stations using sector or array

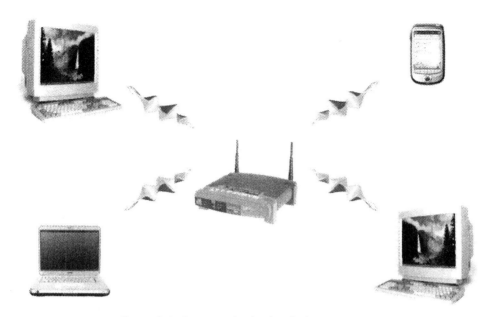

Figure 3.5: Star topologies in wireless networks

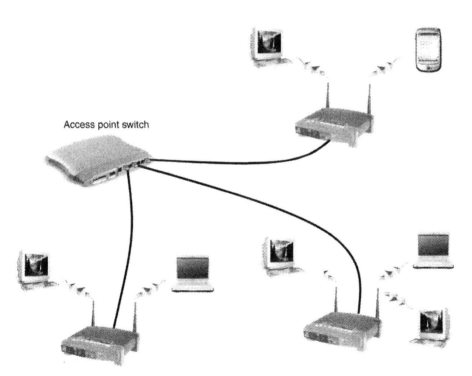

Access point switch

Figure 3.6: A tree topology using a wireless access point switch

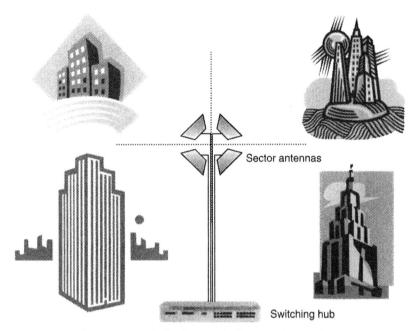

Sector antennas

Switching hub

Figure 3.7: Switched star wireless MAN topology

antennas. Figure 3.7 shows a wireless MAN example, with a switch serving four base station transmitters each using a 90° sector antenna.

With this configuration, the overall wireless MAN throughput is multiplied by the number of transmitters, similar to the case of the wired switching hub shown in Figure 3.4.

In the wireless LAN case, a similar spatial separation can be achieved using a new class of device called an access point array, described in Section 3.3.5, which combines a wireless LAN controller with an array of sector antennas to multiply network capacity. The general technique of multiplying network throughput by addressing separate spatial zones or propagation paths is known as space division multiplexing and finds its most remarkable application in MIMO radio.

3.2.3 Mesh Networks

Mesh networks, also known as mobile ad hoc networks (MANETs), are local or metropolitan area networks in which nodes are mobile and communicate directly with

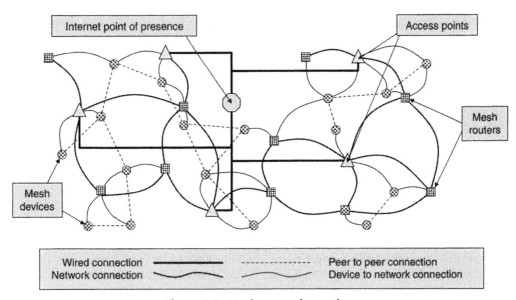

Figure 3.8: Mesh network topology

adjacent nodes without the need for central controlling devices. The topology of a mesh, shown generically in Figure 3.8, can be constantly changing, as nodes enter and leave the network, and data packets are forwarded from node-to-node towards their destination in a process called hopping.

The data routing function is distributed throughout the entire mesh rather than being under the control of one or more dedicated devices. This is similar to the way that data travels around the Internet, with a packet hopping from one device to another until it reaches its destination, although in mesh networks, the routing capabilities are included in every node rather than just in dedicated routers.

This dynamic routing capability requires each device to communicate its routing information to every device it connects with, and to update this as nodes move within, join and leave the mesh.

This distributed control and continuous reconfiguration allows for rapid rerouting around overloaded, unreliable or broken paths, allowing mesh networks to be self-healing and very reliable, provided that the density of nodes is sufficiently high to allow alternative paths. A key challenge in the design of the routing protocol is to achieve this continuous reconfiguration capability with a manageable overhead in terms of data bandwidth taken up by routing information messages.

The multiplicity of paths in a mesh network has a similar impact on total network throughput as the multiple paths shown in Figures 3.4 and 3.7 for the case of wired network switches and sectorized wireless networks. Mesh network capacity will grow as the number of nodes, and therefore the number of usable alternative paths, grow, so that capacity can be increased simply by adding more nodes to the mesh.

As well as the problem of efficiently gathering and updating routing information, mesh networks face several additional technical challenges such as:

- Wireless link reliability—a packet error rate that may be tolerable over a single hop in an hub and spokes configuration will quickly compound over multiple hops, limiting the size to which a mesh can grow and remain effective.

- Seamless roaming—seamless connection and reconnection of moving nodes has not been a requirement in most wireless network standards, although 802.11 Task Groups TGr and TGs are addressing this.

- Security—how to authenticate users in a network with no stable infrastructure?

From a practical standpoint, the self-configuring, self-optimizing and self-healing characteristics of mesh networks eliminate many of the management and maintenance tasks associated with large-scale wireless network deployments.

ZigBee is one standard which explicitly supports mesh networks and the IEEE 802.11 Task Group TGs is in the process of developing a standard which addresses WLAN mesh networks. Two industry bodies have already been established to promote 802.11s mesh proposals, the Wi-Mesh Alliance and SEEMesh (Simple, Efficient and Extensible Mesh).

3.3 Wireless LAN Devices

3.3.1 Wireless Network Interface Cards

The wireless network interface card (NIC) turns a device such as a PDA, laptop or desktop computer into a wireless station and enables the device to communicate with other stations in a peer-to-peer network or with an access point.

Figure 3.9: A variety of wireless NIC forms

(courtesy of Belkin Corporation, D-Link (Europe) Ltd. and Linksys (a division of Cisco Systems Inc.))

Wireless NICs are available in a variety of form factors (Figure 3.9), including PC (Type II PCMCIA) and PCI cards, as well as external USB devices and USB dongles, or compact flash for PDAs. Most wireless NICs have integrated antennas, but a few manufacturers provide NICs with an external antenna connection or detachable integrated antenna which can be useful to attach a high-gain antenna when operating close to the limit of wireless range.

There are few features to distinguish one wireless NIC from another. Maximum transmitter power is limited by local regulatory requirements and, for standards based equipment, certification by the relevant body (such as Wi-Fi certification for 802.11) will ensure interoperability of equipment from different manufacturers. The exception will be proprietary extensions or equipment released prior to standard ratification, such as "pre-n" hardware announced by some manufacturers in advance of 802.11n ratification.

High-end mobile products, particularly laptop computers, are increasingly being shipped with integrated wireless NICs, and with Intel's Centrino® technology the wireless LAN interface became part of the core chipset family.

3.3.2 Access Points

The access point (AP) is the central device in a wireless local area network (WLAN) that provides the hub for wireless communication with the other stations in the network. The access point is usually connected to a wired network and provides a bridge between wired and wireless devices.

The first generation of access points, now termed "fat" access points, began to appear after the ratification of the IEEE 802.11b standard in 1999, and provided a full range of processing and control functions within each unit, including:

- security features, such as authentication and encryption support

- access control based on lists or filters

- SNMP configuration capabilities

Transmit power level setting, RF channel selection, security encryption and other configurable parameters required user configuration of the access point, typically using a web-based interface.

As well as providing this basic functionality, access points designed for home or small office wireless networking typically include a number of additional networking features, as shown in Table 3.1.

Table 3.1: Optional access point functionality

Feature	Description
Internet gateway	Supporting a range of functions such as: routing, Network Address Translation, DHCP server providing dynamic IP addresses to client stations, and Virtual Private Network (VPN) passthrough.
Switching hub	Several wired Ethernet ports may be included that provide switching hub capabilities for a number of Ethernet devices.
Wireless bridge or repeater	Access point that can function as a relay station, to extend the operating range of another access point, or as a point-to-point wireless bridge between two networks.
Network storage server	Internal hard drives or ports to connect external storage, providing centralized file storage and back-up for wireless stations.

Figure 3.10: First generation wireless access points

(courtesy of Belkin Corporation, D-Link (Europe) Ltd. and Linksys (a division of Cisco Systems Inc.))

Figure 3.10 illustrates a range of access point types, including weatherproofed equipment for outdoor coverage.

In contrast to the first generation "fat" access point described above, slimmed-down "thin" access points are also available that limit access point capabilities to the essential RF communication functions and rely on the centralisation of control functions in a wireless LAN switch.

3.3.3 Wireless LAN Switches or Controllers

In a large wireless network, typically in a corporate environment with tens and perhaps hundreds of access points, the need to individually configure access points can make WLAN management a complicated task. Wireless LAN switches simplify the deployment and management of large-scale WLANs. A wireless LAN switch (also known as a wireless LAN controller or access router), is a networking infrastructure device designed to handle a variety of functions on behalf of a number of dependent, or "thin", access points (Figure 3.11).

As shown in Table 3.2, this offers several advantages for large-scale WLAN implementations, particularly those supporting voice services.

The driver behind the development of the wireless switch is to enable the task of network configuration and management, which becomes increasingly complex and time

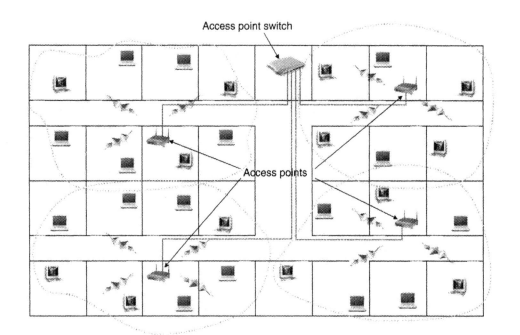

Figure 3.11: Wireless LAN topology using a wireless switch

Table 3.2: "Thin" access point advantages

Advantage	Description
Lower cost	A "thin" access point is optimized to cost effectively implement wireless communication functions only, reducing initial hardware cost as well as future maintenance and upgrade costs.
Simplified access point management	Access point configuration, including security functions, is centralized in order to simplify the network management task.
Improved roaming performance	Roaming handoffs are much faster than with conventional access points, which improves the performance of voice services.
Simplified network upgrades	The centralized command and control capability makes it easier to upgrade the network in response to evolving WLAN standards, since upgrades only have to be applied at the switch level, and not to individual access points.

consuming as wireless networks grow. A wireless switch provides centralized control of configuration, security, performance monitoring and troubleshooting, which is essential in an enterprise scale wireless LAN.

Taking security as an example, with WEP, WPA, and 802.11i, all potentially in use at the same time in a large WLAN deployment, if security configuration has to be managed for individual access points, the routine management of encryption keys and periodic upgrade of security standards for each installed access point quickly becomes unmanageable. With a centralized security architecture provided by a wireless switch, these management tasks only need to be completed once.

WLAN switches also provide a range of additional features, not found in first generation access points, as described in Table 3.3.

3.3.4 Lightweight Access Point Protocol

Centralizing command and control into a wireless LAN switch device introduces the need for a communication protocol between the switch and its dependent access points, and the need for interoperability requires that this protocol is based on an industry standard.

Table 3.3: Wireless LAN switch features

Feature	Description
Layout planning	Automated site survey tools that allow import of building blueprints and construction specifications and determine optimal access point locations.
RF management	Analysis of management frames received from all access points enables RF signal related problems to be diagnosed and automatically corrected, by adjusting transmit power level or channel setting of one or more access points.
Automatic configuration	Wireless switches can provide automatic configuration by determining the best RF channel and transmit power settings for individual access points.
Load balancing	Maximizing network capacity by automatic load balancing of users across multiple access points.
Policy-based access control	Access policies can be based on access point groupings and client lists that specify which access points or groups specific client stations are permitted to connect to.
Intrusion detection	Rogue access points and unauthorized users or ad hoc networks can be detected and located, either by continuous scanning or by scheduled site surveys.

The Lightweight Access Point Protocol (LWAPP) standardizes communications between switches or other hub devices and access points, and was initially developed by the Internet Engineering Task Force (IETF).

The IETF specification describes the goals of the LWAPP protocol as follows:

- To reduce the amount of protocol code being executed at the access point so that efficient use can be made of the access point's computing power, by applying this to wireless communication rather than to bridging, forwarding and other functions.

- To use centralized network computing power to execute the bridging, forwarding, authentication, encryption and policy enforcement functions for a wireless LAN.

- To provide a generic encapsulation and transport mechanism for transporting frames between hub devices and access points, which will enable multi-vendor interoperability and ensure that LWAPP can be applied to other access protocols in the future.

The main communication and control functions that are achieved using LWAPP are summarized in Table 3.4.

Although the initial draft specification for LWAPP expired in March 2004, a new IETF working group called Control and Provisioning of Wireless Access Points (CAPWAP) was formed, with most working group members continuing to recommend LWAPP over alternatives such as Secure Light Access Point Protocol (SLAPP), Wireless LAN Control Protocol (WICOP) and CAPWAP Tunneling Protocol (CTP). It seems likely that LWAPP will be the basis of an eventual CAPWAP protocol.

3.3.5 Wireless LAN Arrays

The so-called "3rd generation" architecture for WLAN deployment uses a device called an access point array, which is the LAN equivalent of the sectorized WMAN base station illustrated in Figure 3.7.

A single access point array incorporates a wireless LAN controller together with 4, 8 or 16 access points, which may combine both 802.11a and 802.11b/g radio interfaces. A typical example uses four access points for 802.11a/g coverage, employing 180° sector antennas offset by 90°, and 12 access points for 802.11a covering, with 60° sector antennas offset by 30°, as illustrated in Figure 3.12.

LWAPP function	Description
Access point device discovery and information exchange	An access point sends a Discovery Request frame and any receiving access router responds with a Discovery Reply frame. The access point selects a responding access router and associates by exchanging Join Request and Join Reply frames.
Access point certification, configuration, provisioning and software control	After association, the access router will provision the access point, providing a Service Set Identifier (SSID), security parameters, operating channel and data rates to be advertised. The access router can also configure MAC operating parameters (e.g., number of transmission attempts for a frame), transmit power, DSSS or OFDM parameters and antenna configuration in the access point. After provisioning and configuration, the access point is enabled for operation.
Data and management frame encapsulation, fragmentation and formatting	LWAPP encapsulates data and management frames for transport between the access point and access router. Fragmentation of frames and reassembly of fragment will be handled if the encapsulated data or management frames exceed the Maximum Transmission Unit (MTU) supported between the access point and access router.
Communication control and management between access points and associated devices	LWAPP enables the access router to request statistical reports from its access points, including data about the communication between the access point and its associated devices (e.g., retry counts, RTS/ACK failure counts).

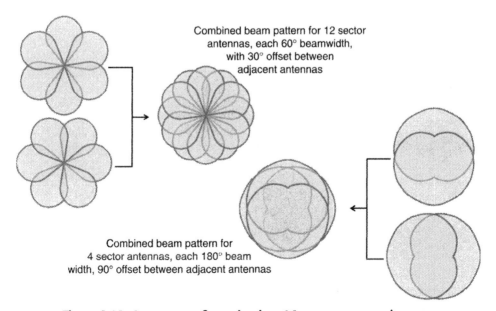

Combined beam pattern for 12 sector antennas, each 60° beamwidth, with 30° offset between adjacent antennas

Combined beam pattern for 4 sector antennas, each 180° beam width, 90° offset between adjacent antennas

Figure 3.12: Antenna configuration in a 16-sector access point array

This type of device, with 16 access points operating 802.11a and g networks at an individual headline data rate of 54 Mbps, offers a total wireless LAN capacity of 864 Mbps. The increased gain of the sector antennas also means that the operating range of an access point array can be double or more the range of a single access point with an omnidirectional antenna.

For high capacity coverage over a larger operating area, multiple access point arrays, controlled by a second tier of WLAN controllers would create a tree topology, as shown in Figure 3.13, with multi-gigabit total WLAN capacity.

3.3.6 Miscellaneous Wireless LAN Hardware

3.3.6.1 Wireless Network Bridging

Wireless bridge components that provide point-to-point WLAN or WMAN links are available from a number of manufacturers, packaged in weather proof enclosures for outdoor use (Figure 3.14). The D-Link DWL 1800 is one example which bundles a 16 dBi flat panel antenna with a 2.4 GHz radio providing a transmit power of 24 dBm (under FCC) or 14 dBm (under ETSI regulations), to deliver a range of 25 km under FCC or 10 km under ETSI.

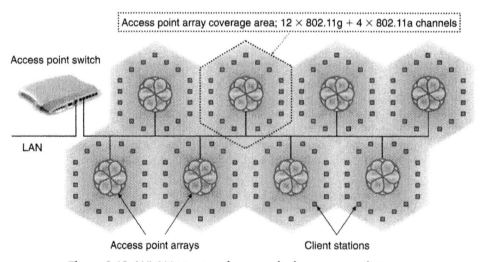

Figure 3.13: WLAN tree topology employing access point arrays

Figure 3.14: Outdoor wireless bridges
(courtesy of D-Link (Europe) Ltd. and Linksys (a division of Cisco Systems Inc.))

Many simple wireless LAN access points also support network bridging, or can be upgraded with a firmware upgrade to provide this capability. Configuring these devices simply involves entering the MAC address of the other endpoint into each station's access control list, so that each station only decodes packets transmitted by the other endpoint of the bridge.

3.3.6.2 Wireless Printer Servers

A wireless printer server allows a printer to be flexibly shared among a group of users in the home or office without the need for the printer to be hosted by one computer or to be connected to a wired network.

Typically, as well as wired Ethernet and wireless LAN interfaces, this device may include one or more different types of printer connections, such as USB or parallel printer ports, as well as multiple ports to enable multiple printers to be connected—such as a high-speed black and white laser and a separate color printer.

A printer server for home or small office wireless networking may also be bundled with a 4-port switch to enable other wired network devices to share the printer and use the

Figure 3.15: Wireless printer servers

(courtesy of Belkin Corporation, D-Link (Europe) Ltd. and Links (a division
of Cisco Systems Inc.))

wireless station as a bridge to other devices on the wireless network. Figure 3.15 shows a range of wireless printer servers.

3.3.7 Wireless LAN Antennas

3.3.7.1 Traditional Fixed Gain Antennas

Antennas for 802.11b and 11g networks, operating in the 2.4 GHz ISM band, are available to achieve a variety of coverage patterns. The key features that dictate the choice of antenna for a particular application are gain, measured in dBi, and angular beamwidth, measured in degrees.

The most common WLAN antenna, standard in all NICs and in most access points, is the omnidirectional antenna, which has a gain in the range from 0 to 7 dBi and a beamwidth, perpendicular to the antenna axis, of a full 360°. A range of WLAN antennas is shown in Figure 3.16 and Table 3.5 summarizes typical parameters. For sector antennas with a given horizontal beamwidth, the trade-off for higher gain is a narrower vertical beamwidth, which will result in a smaller coverage area at a given distance and will require more precise alignment.

A further important feature of an antenna is its polarization, which refers to the orientation of the electric field in the electromagnetic wave emitted by the antenna. Most common antennas, including all those listed in Table 3.5, produce linearly polarized

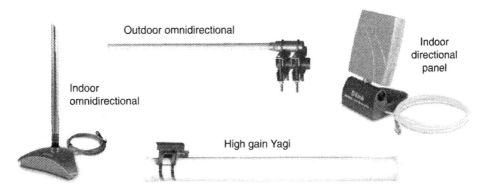

Figure 3.16: Wireless LAN antenna types

(courtesy of D-Link (Europe) Ltd.)

Table 3.5: Typical wireless LAN antenna parameters for 2.4 GHz operation

Antenna type	Sub-type	Beamwidth (Degrees)	Gain (dBi)
Omnidirectional		360	0–15
Patch/Panel		15–75	8–20
Sector		180	8–15
		120	9–20
		90	9–20
		60	10–17
Directional	Yagi	10–30	8–20
	Parabolic reflector	5–25	14–30

waves, with the electrical field oriented either vertically or horizontally—hence vertical or horizontal polarization. WLAN antennas that produce circular polarization are also available (helical antennas) but are less common.

It is important that the polarizations of transmitting and receiving antennas are matched, since a vertical polarized receiving antenna will be unable to receive a signal transmitted by a horizontally polarized transmitting antenna, and vice versa. It is equally important for antennas to be correctly mounted, as rotating an antenna 90° about the direction of propagation will change its polarization by the same angle (e.g., from horizontal to vertical).

Although WLAN operation in the 5-GHz band has developed more recently than in the 2.4-GHz band, a similar selection of antennas is available for the higher-frequency band. A variety of dual band omnidirectional and patch antennas are also available to operate in both WLAN bands.

3.3.7.2 Smart Antennas

The data throughput of a wireless network that uses a traditional antenna of the type described above is limited because only one network node at a time can use the medium to transmit a data packet. Smart antennas aim to overcome this limitation by allowing multiple nodes to transmit simultaneously, significantly increasing network throughput. There are two varieties of smart antenna—switched beam and adaptive array.

A switched beam antenna consists of an array of antenna elements each having a predefined beam pattern with a narrow main lobe and small side-lobes (Figure 3.17). Switching between beams allows one array element to be selected that provides the best gain in the direction of a target node, or the lowest gain towards an interfering source.

The simplest form of switched beam antenna is the pair of diversity receiver antennas often implemented in wireless LAN access points to reduce multipath effects in indoor environments. The receiver senses which of the two antennas is able to provide the highest signal strength and switches to that antenna.

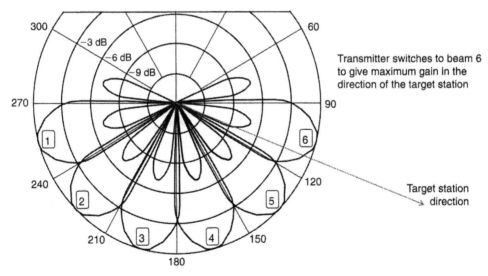

Figure 3.17: Beam pattern of a six element switched beam array

Adaptive beams or beam-forming antennas consist of two or more antenna elements in an array and a so-called beam-forming algorithm, which assigns a specific gain and phase shift to the signal sent to or received from each antenna element. The result is an adjustable radiation pattern that can be used to steer the main lobe of the beam in the direction of the desired maximum gain (Figure 3.18).

As well as focusing its beam pattern towards a particular node, the adaptive beam antenna can also place a "null" or zero gain point in the direction of a source of interference. Because the gain and phase shift applied to individual array elements is under real-time software control (Figure 3.19), the antenna can dynamically adjust its beam pattern to compensate for multipath and other sources of interference and noise. Like adaptive beam arrays, multi-input multi-output or MIMO, radio also uses multiple antennas to increase network capacity. The key difference between these two techniques is that MIMO radio exploits multipath propagation between a single transmitter and receiver, while adaptive beam arrays use multiple antennas to focus a single spatial channel. Some other differences are described in Table 3.6.

Also under development is a new type of switched beam wireless LAN antenna called a plasma antenna, which uses a solid-state plasma (an ionized region in a silicon layer)

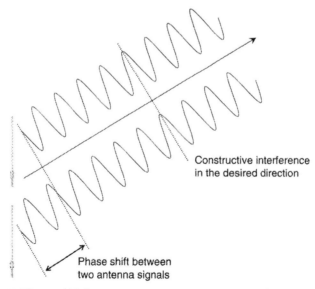

Figure 3.18: Phase shift between two antennas resulting in a directed beam

as a reflector to focus and direct the emitted RF beam. A plasma antenna will be able to switch a medium gain (10–15 dBi) beam with approximately 10° beamwidth to 1 of 36 directions within a full 360° coverage, with a switching time that is less than the gap between transmitted frames.

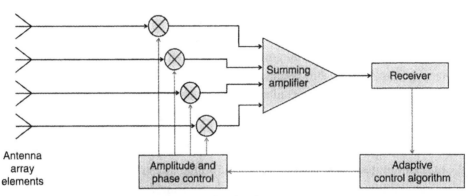

Figure 3.19: Adaptive beam antenna

Table 3.6: Adaptive beam arrays and MIMO radio compared

	Adaptive beam array	MIMO radio
Objective	Focus the propagation pattern along a single desired spatial direction, to allow multiple access, reduce interference or increase range.	Exploit multipath propagation to increase data capacity by multiplexing data streams over several spatial channels.
Antenna configuration	Two or more antenna elements at Tx and/or Rx. Tx and Rx configurations are independent.	Typically 2 × 2 or 4 × 4 (Tx × Rx). Tx and Rx are linked by the digital signal processing algorithm.
Spatial diversity	Single spatial channel focused between transmitter and receiver.	Multiple spatial channels, exploiting multipath propagation.
Data multiplexing	Single bit stream encoded to all transmit antennas.	Data stream multiplexed over spatial channels.
Signal processing	Simple phase and gain modification for each antenna.	Complex processing algorithm to decode signals over multiple spatial channels.
Application example	3rd generation WLAN access points—Section 3.3	PHY layer for the 802.11n standard

3.4 Wireless PAN Devices

3.4.1 Wireless PAN Hardware Devices

In this section the wide range of PAN technologies will be described, from Bluetooth, which is most commonly identified with personal area networking, to ZigBee, a technology primarily aimed at networking home and industrial control devices.

3.4.1.1 Bluetooth Devices

With high powered (Class 1) Bluetooth radios, which can equal the range of Wi-Fi devices, the boundary between personal and local area networking is blurred and, within the limitations of achievable data rates, most of the WLAN devices described in Section 3.3 could equally well be built using Bluetooth technology.

The most common types of Bluetooth devices and some of their key features are summarized in Table 3.7 and shown in Figure 3.20.

Table 3.7: Bluetooth devices and features

Bluetooth device	Key features
Mobile phone	Interface with a Bluetooth hands-free headset. Connect to PDA or PC to transfer or back-up files. Exchange contact details (business cards), calendar entries, photos, etc. with other Bluetooth devices.
PDA	Connect to PC to transfer or back-up files. Connect to the Internet via a Bluetooth access point. Exchange contact details (business cards), calendar entries, photos, etc. with other Bluetooth devices.
Headset or headphones	Hands-free mobile telephony. Audio streaming from PC, TV, MP3 player or hi-fi system.
Audio transceiver	Audio streaming from a PC or hi-fi system to Bluetooth headphones.
Access point	Extend a LAN to include Bluetooth enabled devices. Internet connectivity for Bluetooth devices.
Bluetooth adapters	Bluetooth enable a range of devices, such as laptops or PDAs. As for WLAN NICS, they are available in a range of form factors, with USB dongles being the most popular. Serial adapter for plug-and-play connectivity to any serial RS-232 device.

Continued

Table 3.7: (*Continued*)

Bluetooth device	Key features
Print adapter	Print files or photos from Bluetooth enabled device.
PC input devices	Wireless connectivity to a PC mouse or keyboard.
GPS receiver	Provide satellite navigation capabilities to Bluetooth-enabled devices loaded with required navigation software.
Dial-up modem	Provide wireless connectivity from a PC to a dial-up modem.

Figure 3.20: A variety of Bluetooth devices

(courtesy of Belkin Corporation, D-Link (Europe) Ltd., Linksys

(a division of Cisco Systems Inc.) and Zoom Technologies, Inc.)

As developing wireless PAN technologies such as wireless USB mature, a comparable range of devices will be developed to support these networks. Novel capabilities inherent in these new technologies will also result in new device types offering new services, an example being the capability of the multi-band OFDM radio to spatially locate a wireless USB station, offering the potential for devices that rely on location based services.

3.4.1.2 ZigBee Devices

ZigBee is a low data rate, very low power, wireless networking technology that initially has focused on home automation but is likely to find a wide range of applications, including a low cost replacement for Bluetooth in applications that do not require higher data rates.

The key features of a range of currently available and expected ZigBee devices is summarized in Table 3.8, and some of these devices are shown in Figure 3.21.

Table 3.8: ZigBee devices and features

ZigBee device	Key features
PC input devices	Wireless connectivity to a PC mouse or keyboard.
Automation devices	Wireless control devices for home and industrial automation functions such as heating, lighting and security.
Wireless remote control	Replacing Ir remote for TV etc., and eliminating the line-of-sight and alignment restriction.
Sensor modem	Provides a wireless networking interface for a number of existing current loop sensors for home or industrial automation.
Ethernet gateway	A ZigBee network coordinator that enables command of ZigBee end devices or routers from an Ethernet network.

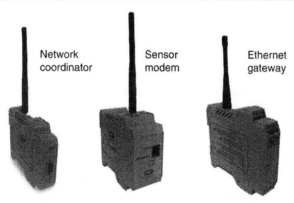

Figure 3.21: A variety of ZigBee devices

(courtesy of Cirronet Inc.)

3.4.2 Wireless PAN Antennas

In practice, since PAN operating range is generally under ten meters, Bluetooth and other PAN devices will typically use simple integrated omnidirectional antennas. However, for PANs such as Bluetooth that share the 2.4-GHz ISM radio band with 802.11b/g WLANs, the wide range of external WLAN antennas described above are also available to enable PAN devices to be operated with extended range.

3.5 Wireless MAN Devices

While wireless LANs and PANs present a wide diversity of topologies and device types, to date wireless MAN devices have serviced only fixed point-to-point and point to multi-point topologies, requiring in essence only two device types, the base station and the client station.

However, following the ratification of the 802.16e standard (also designated 802. 16-2005), broadband Internet access will soon be widely available to mobile devices and a range of new mobile wireless MAN devices is emerging, driving the convergence of mobile phones and PDAs.

3.5.1 Fixed Wireless MAN Devices

Wireless networking devices for fixed wireless MAN applications, essentially for last mile broadband Internet access, fall into two categories—base station equipment and customer premises equipment (CPE).

Some examples of base station equipment to support wireless MANs of differing scales are shown in Figure 3.22. The macro scale base station shown can potentially support thousands of subscribers in dense metropolitan area deployments, while the micro scale equipment is designed to support lower user numbers in sparse rural areas. Some of the types and key features of base station and CPE hardware are summarized in Table 3.9.

Figure 3.23 shows a variety of different wireless MAN CPE equipment.

3.5.2 Fixed Wireless MAN Antennas

Antennas for fixed wireless MAN applications similarly divide into base station and CPE. The general types of antennas summarized earlier in Table 3.5 for LANs are equally

Figure 3.22: Micro and macro WMAN base station equipment

(courtesy of Aperto Networks Inc.)

Table 3.9: Wireless MAN devices and features

WMAN device	Typical features
Basic self installed indoor CPE	Basic WMAN connectivity to a customer PC or network. Multiple diversity or adaptive array antennas to improve non line-of-sight reception.
Outdoor CPE	External antenna and radio. Provides higher antenna gain and longer range.
Base station equipment	Modular and scalable construction. Macro and micro configurations for dense metropolitan or sparse rural installations. Flexible RF channel usage, from one channel over multiple antenna sectors to multiple channels per antenna sector.
Integrated network gateway	MAN interface with network gateway functions (Routing, NAT and firewall capabilities). Optionally with integrated wireless LAN access point. Integrated CPU to support additional WISP services such as VoIP telephony.

Figure 3.23: Fixed wireless MAN CPE equipment

(courtesy of Aperto Networks Inc.)

Table 3.10: Factors determining wireless MAN antenna choice

Location	Antenna type	Application
CPE	Omnidirectional	Low gain requirement—for subscribers located close to the base station.
	Patch	Intermediate gain—mid range equipment should be applicable to most subscribers.
	Directional (Yagi or parabolic reflector)	High gain, high cost equipment to maximize data rate at the edge of the operating area.
Base station	Sector, intermediate gain	Wide area coverage close to the base station. Wider vertical beamwidth required to provide coverage close to the base station.
	Sector, high gain	Wide area coverage at a distance from the base station. High gain adds range with narrower vertical beamwidth.
	Directional	High gain antenna for point-to-point applications, such as backhaul, bridging between base stations, etc.

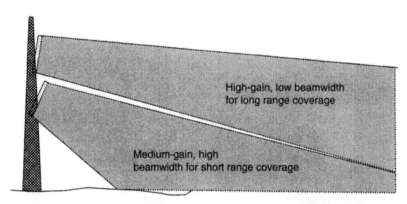

Figure 3.24: WMAN base station sector antenna configuration

applicable to MAN installations, with appropriate housings for outdoor service and mountings designed for wind and ice loading. The factors that determine the choice of antenna are summarized in Table 3.10.

Depending on its elevation relative to the target area, a base station may comprise two sets of sector antennas as illustrated in Figure 3.24. A set of intermediate gain antennas,

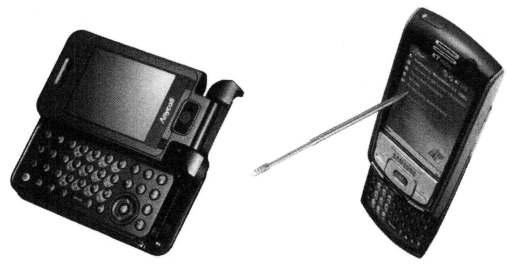

Figure 3.25: Wireless MAN enabled phones

(courtesy of Samsung Electronics)

with higher vertical beamwidth, provide coverage over short-to-medium distances, with a second set of high-gain, low vertical beamwidth antennas providing coverage at longer range.

3.5.3 Mobile Wireless MAN Devices

The first implementation of mobile wireless MAN services and devices, delivering broadband Internet access to the user on the move, has been in the South Korean market, driven by the rapid development of the WiBro standard (a sub-set of 802.16e). Commercial uptake has also been speeded by the use of licensed spectrum, and the granting in 2005 of operating licenses to three telecom companies.

The form factor for devices to deliver mobile Internet services reflects the need to combine telephony and PDA capabilities—a larger screen and QWERTY input to overcome the limitations experienced with WAP phones. Figure 3.25 shows two early WiBro phones developed by Samsung.

CHAPTER 4

Radio Communication Basics

Alan Bensky

In this chapter on radio communication basics, we first will take a look at the phenomena that lets us transfer information from one point to another without any physical medium—the propagation of radio waves. If you want to design an efficient radio communication system, even for operation over relatively short distances, you should understand the behavior of the wireless channel in the various surroundings where this communication is to take place. While the use of "brute force"—increasing transmission power—could overcome inordinate path losses, limitations imposed on design by required battery life, or by regulatory authorities, make it imperative to develop and deploy short-range radio systems using solutions that a knowledge of radio propagation can give.

The overall behavior of radio waves is described by Maxwell's equations. In 1873, the British physicist James Clerk Maxwell published his *Treatise on Electricity and Magnetism* in which he presented a set of equations that describe the nature of electromagnetic fields in terms of space and time. Heinrich Rudolph Hertz performed experiments to confirm Maxwell's theory, which led to the development of wireless telegraph and radio. Maxwell's equations form the basis for describing the propagation of radio waves in space, as well as the nature of varying electric and magnetic fields in conducting and insulating materials, and the flow of waves in waveguides. From them, you can derive the skin effect equation and the electric and magnetic field relationships very close to antennas of all kinds. A number of computer programs on the market, based on the solution of Maxwell's equations, help in the design of antennas, anticipate electromagnetic radiation problems from circuit board layouts, calculate the effectiveness of shielding, and perform accurate simulation of ultra-high-frequency and microwave circuits. While you don't have to be an expert in Maxwell's equations to use these programs (you do in order to write them!), having some familiarity with the equations may take the mystery out of the operation of the software and give an appreciation for its range of application and limitations.

4.1 Mechanisms of Radio Wave Propagation

Radio waves can propagate from transmitter to receiver in four ways: through ground waves, sky waves, free space waves, and open field waves.

Ground waves exist only for vertical polarization, produced by vertical antennas, when the transmitting and receiving antennas are close to the surface of the earth. The transmitted radiation induces currents in the earth, and the waves travel over the earth's surface, being attenuated according to the energy absorbed by the conducting earth. The reason that horizontal antennas are not effective for ground wave propagation is that the horizontal electric field that they create is short circuited by the earth. Ground wave propagation is dominant only at relatively low frequencies, up to a few megahertz, so it needn't concern us here.

Sky wave propagation is dependent on reflection from the ionosphere, a region of rarified air high above the earth's surface that is ionized by sunlight (primarily ultraviolet radiation). The ionosphere is responsible for long-distance communication in the high-frequency bands between 3 and 30 MHz. It is very dependent on time of day, season, longitude on the earth, and the multi-year cyclic production of sunspots on the sun. It makes possible long-range communication using very low-power transmitters. Most short-range communication applications use VHF, UHF, and microwave bands, generally above 40 MHz. There are times when ionospheric reflection occurs at the low end of this range, and then sky wave propagation can be responsible for interference from signals originating hundreds of kilometers away. However, in general, sky wave propagation does not affect the short-range wireless applications that we are interested in.

The most important propagation mechanism for short-range communication on the VHF and UHF bands is that which occurs in an open field, where the received signal is a vector sum of a direct line-of-sight signal and a signal from the same source that is reflected off the earth. We discuss below the relationship between signal strength and range in line-of-sight and open-field topographies.

The range of line-of-sight signals, when there are no reflections from the earth or ionosphere, is a function of the dispersion of the waves from the transmitter antenna.

In this free-space case the signal strength decreases in inverse proportion to the distance away from the transmitter antenna. When the radiated power is known, the field strength is given by equation (4-1):

$$E = \frac{\sqrt{30 \cdot P_t \cdot G_t}}{d} \tag{4-1}$$

where P_t is the transmitted power, G_t is the antenna gain, and d is the distance. When P_t is in watts and d is in meters, E is volts/meter.

To find the power at the receiver (P_r) when the power into the transmitter antenna is known, use (4-2):

$$P_r = \frac{P_t G_t G_r \lambda^2}{(4\pi d)^2} \tag{4-2}$$

G_t and G_r are the transmitter and receiver antenna gains, and λ is the wavelength.

Range can be calculated on this basis at high UHF and microwave frequencies when high-gain antennas are used, located many wavelengths above the ground. Signal strength between the earth and a satellite, and between satellites, also follows the inverse distance law, but this case isn't in the category of short-range communication! At microwave frequencies, signal strength is also reduced by atmospheric absorption caused by water vapor and other gases that constitute the air.

4.2 Open Field Propagation

Although the formulas in the previous section are useful in some circumstances, the actual range of a VHF or UHF signal is affected by reflections from the ground and surrounding objects. The path lengths of the reflected signals differ from that of the line-of-sight signal, so the receiver sees a combined signal with components having different amplitudes and phases. The reflection causes a phase reversal. A reflected signal having a path length exceeding the line-of-sight distance by exactly the signal wavelength or a multiple of it will almost cancel completely the desired signal ("almost" because its amplitude will be slightly less than the direct signal amplitude). On the other hand, if the path length of the reflected signal differs exactly by an odd multiple of half the wavelength, the total signal will be strengthened by "almost" two times the free space direct signal.

In an open field with flat terrain there will be no reflections except the unavoidable one from the ground. It is instructive and useful to examine in depth the field strength versus distance in this case."

In Figure 4.1 we see transmitter and receiver antennas separated by distance d and situated at heights h_1 and h_2. Using trigonometry, we can find the line of sight and reflected signal path lengths d_1 and d_2. Just as in optics, the angle of incidence equals the angle of reflection θ. We get the relative strength of the direct signal and reflected signal using the inverse path length relationship. If the ground were a perfect mirror, the relative reflected signal strength would exactly equal the inverse of d_2. In this case, the reflected signal phase would shift 180 degrees at the point of reflection. However, the ground is not a perfect reflector. Its characteristics as a reflector depend on its conductivity, permittivity, the polarization of the signal and its angle of incidence. Accounting for polarization, angle of incidence, and permittivity, we can find the reflection coefficient, which approaches -1 as the distance from the transmitter increases. The signals reaching the receiver are represented as complex numbers since they have both phase and amplitude. The phase is found by subtracting the largest interval of whole wavelength multiples from the total path length and multiplying the remaining fraction of a wavelength by 2π radians, or 360 degrees.

Figure 4.2 gives a plot of relative open field signal strength versus distance using the following parameters:

Polarity—horizontal

Frequency—300 MHz

Antenna heights—both 3 meters

Relative ground permittivity—15

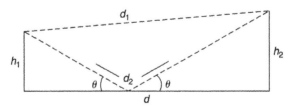

Figure 4.1: Open field signal paths

Also shown is a plot of free space field strength versus distance (dotted line). In both plots, signal strength is referenced to the free space field strength at a range of 3 meters.

Notice in Figure 4.2 that, up to a range of around 50 meters, there are several sharp depressions of field strength, but the signal strength is mostly higher than it would be in free space. Beyond 100 meters, signal strength decreases more rapidly than for the free space model. Whereas there is an inverse distance law for free space, in the open field beyond 100 meters (for these parameters) the signal strength follows an inverse square law. Increasing the antenna heights extends the distance at which the inverse square law starts to take effect. This distance, d_m, can be approximated by

$$d_m = (12 \times h_1 \times h_2)/\lambda \tag{4-3}$$

where h_1 and h_2 are the transmitting and receiving antenna heights above ground and λ is the wavelength, all in the same units as the distance d_m.

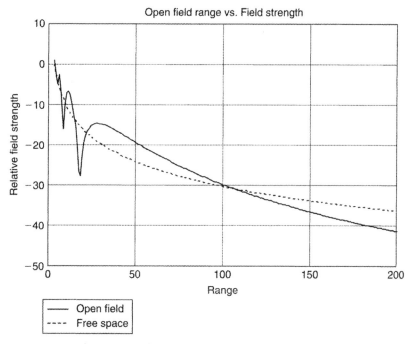

Figure 4.2: Field strength vs. range at 300 MHz

In plotting Figure 4.2, we assumed horizontal polarization. Both antenna heights, h_1 and h_2, are 3 meters. When vertical polarization is used, the extreme local variations of signal strengths up to around 50 meters are reduced, because the ground reflection coefficient is less at larger reflection angles. However, for both polarizations, the inverse square law comes into effect at approximately the same distance. This distance in Figure 4.2 where λ is 1 meter is, from Eq. (4-3): $d_m = (12 \times 3 \times 3)/\lambda = 108$ meters. In Figure 4.2 we see that this is approximately the distance where the open-field field strength falls below the free-space field strength.

4.3 Diffraction

Diffraction is a propagation mechanism that permits wireless communication between points where there is no line-of-sight path due to obstacles that are opaque to radio waves. For example, diffraction makes it possible to communicate around the earth's curvature, as well as beyond hills and obstructions. It also fills in the spaces around obstacles when short-range radio is used inside buildings. Figure 4.3 is an illustration of diffraction geometries, showing an obstacle whose extremity has the shape of a knife edge. The obstacle should be seen as a half plane whose dimension is infinite into and out of the

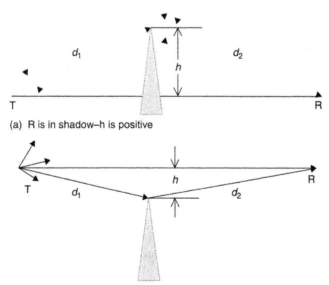

(a) R is in shadow–h is positive

(b) T and R are line of sight–h is negative

Figure 4.3: Knife-edge diffraction geometry

paper. The field strength at a receiving point relative to the free-space field strength without the obstacle is the diffraction gain. The phenomenon of diffraction is due to the fact that each point on the wave front emanating from the transmitter is a source of a secondary wave emission. Thus, at the knife edge of the obstacle, as shown in Figure 4.3a, there is radiation in all directions, including into the shadow.

The diffraction gain depends in a rather complicated way on a parameter that is a function of transmitter and receiver distances from the obstacle, d_1 and d_2, the obstacle dimension h, and the wavelength. Think of the effect of diffraction in an open space in a building where a wide metal barrier extending from floor to ceiling exists between the transmitter and the receiver. In our example, the space is 12 meters wide and the barrier is 6 meters wide, extending to the right side. When the transmitter and receiver locations are fixed on a line at a right angle to the barrier, the field strength at the receiver depends on the perpendicular distance from the line-of-sight path to the barrier's edge. Figure 4.4 is a plot of diffraction gain when transmitter and receiver are each 10 meters from the edge of the obstruction and on either side of it. The dimension "h" varies between −6 meters and 6 meters—that is, from the left side of the space where the dimension "h" is considered negative, to the right side where "h" is positive and fully in the shadow of the barrier. Transmission frequency for the plot is 300 MHz. Note that the barrier affects the received signal strength even when there is a clear line of sight between the transmitter and receiver ("h" is negative as shown in Figure 4.3b). When the barrier edge is on the line of

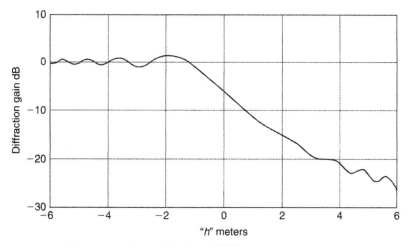

Figure 4.4: Example plot of diffraction gain vs. "h"

sight, diffraction gain is approximately $-6\,\text{dB}$, and as the line-of-sight path gets farther from the barrier (to the left in this example), the signal strength varies in a cyclic manner around $0\,\text{dB}$ gain. As the path from transmitter to receiver gets farther from the barrier edge into the shadow, the signal attenuation increases progressively.

Admittedly, the situation depicted in Figure 4.4 is idealistic, since it deals with only one barrier of very large extent. Normally there are several partitions and other obstacles near or between the line of sight path and a calculation of the diffraction gain would be very complicated, if not impossible. However, a knowledge of the idealistic behavior of the defraction gain and its dependence on distance and frequency can give qualitative insight.

4.4 Scattering

A third mechanism affecting path loss, after reflection and diffraction, is scattering. Rough surfaces in the vicinity of the transmitter do not reflect the signal cleanly in the direction determined by the incident angle, but diffuse it, or scatter it in all directions. As a result, the receiver has access to additional radiation and path loss may be less than it would be from considering reflection and diffraction alone.

The degree of roughness of a surface and the amount of scattering it produces depends on the height of the protuberances on the surface compared to a function of the wavelength and the angle of incidence. The critical surface height h_c is given by [Gibson, p. 360]

$$h_c = \frac{\lambda}{8\cos\theta_i} \tag{4-4}$$

where λ is the wavelength and θ_i is the angle of incidence. It is the dividing line between smooth and rough surfaces when applied to the difference between the maximum and the minimum protuberances.

4.5 Path Loss

The numerical path loss is the ratio of the total radiated power from a transmitter antenna times the numerical gain of the antenna in the direction of the receiver to the power available at the receiver antenna. This is the ratio of the transmitter power delivered to a lossless antenna with numerical gain of 1 ($0\,\text{dB}$) to that at the output of a $0\,\text{dB}$ gain receiver antenna. Sometimes, for clarity, the ratio is called the *isotropic* path loss. An isotropic radiator is an ideal antenna that radiates equally in all directions and therefore has a gain of

0 dB. The inverse path loss ratio is sometimes more convenient to use. It is called the path gain and when expressed in decibels is a negative quantity. In free space, the isotropic path gain *PG* is derived from (4-2), resulting in

$$PG = \frac{\lambda^2}{(4\pi d)^2} \qquad (4\text{-}5)$$

We have just examined several factors that affect the path loss of VHF-UHF signals—ground reflection, diffraction, and scattering. For a given site, it would be very difficult to calculate the path loss between transmitters and receivers, but empirical observations have allowed some general conclusions to be drawn for different physical environments. These conclusions involve determining the exponent, or range of exponents, for the distance *d* related to a short reference distance d_0. We then can write the path gain as dependent on the exponent *n*:

$$PG = k \left(\frac{d_0}{d} \right)^n \qquad (4\text{-}6)$$

where *k* is equal to the path gain when $d = d_0$. Table 4.1 shows path loss for different environments.

As an example of the use of the path gain exponent, let's assume the open field range of a security system transmitter and receiver is 300 meters. What range can we expect for their installation in a building?

Figure 4.5 shows curves of path gain versus distance for free-space propagation, open field propagation, and path gain with exponent 6, a worst case taken from Table 4.1 for

Table 4.1: Path loss exponents for different environments

[Gibson, p. 362]

Environment	Path gain exponent
Free space	2
Open field (long distance)	4
Cellular radio — urban area	2.7–4
Shadowed urban cellular radio	5–6
In building line of sight	1.6–1.8
In building — obstructed	4–6

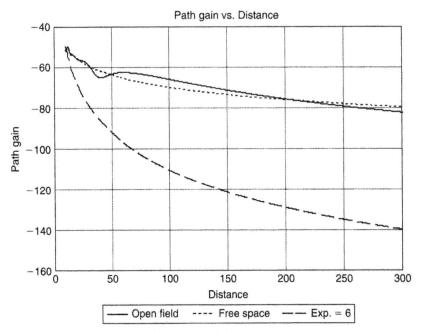

Figure 4.5: Path gain

"In building, obstructed." Transmitter and receiver heights are 2.5 meters, polarization is vertical, and the frequency is 915 MHz. The reference distance is 10 meters, and for all three curves the path gain at 10 meters is taken to be that of free space. For an open field distance of 300 meters, the path gain is -83 dB. The distance on the curve with exponent $n = 6$ that gives the same path gain is 34 meters. Thus, a wireless system that has an outdoor range of 300 meters may be effective only over a range of 34 meters, on the average, in an indoor installation.

The use of an empirically derived relative path loss exponent gives an estimate for average range, but fluctuations around this value should be expected. The next section shows the spread of values around the mean that occurs because of multipath radiation.

4.6 Multipath Phenomena

We have seen that reflection of a signal from the ground has a significant effect on the strength of the received signal. The nature of short-range radio links, which are very often installed indoors and use omnidirectional antennas, makes them accessible to a multitude of reflected rays, from floors, ceilings, walls, and the various furnishings and people that are invariably present near the transmitter and receiver. Thus, the total signal strength at the receiver is the vector sum of not just two signals, as we studied in Section 4.2, but of many signals traveling over multiple paths. In most cases indoors, there is no direct line-of-sight path, and all signals are the result of reflection, diffraction and scattering.

From the point of view of the receiver, there are several consequences of the multipath phenomena.

(a) *Variation of signal strength.* Phase cancellation and strengthening of the resultant received signal causes an uncertainty in signal strength as the range changes, and even at a fixed range when there are changes in furnishings or movement of people. The receiver must be able to handle the considerable variations in signal strength.

(b) *Frequency distortion.* If the bandwidth of the signal is wide enough so that its various frequency components have different phase shifts on the various signal paths, then the resultant signal amplitude and phase will be a function of sideband frequencies. This is called frequency selective fading.

(c) *Time delay spread.* The differences in the path lengths of the various reflected signals causes a time delay spread between the shortest path and the longest path. The resulting distortion can be significant if the delay spread time is of the order of magnitude of the minimum pulse width contained in the transmitted digital signal. There is a close connection between frequency selective fading and time-delay distortion, since the shorter the pulses, the wider the signal bandwidth. Measurements in factories and other buildings have shown multipath delays ranging from 40 to 800 ns (Gibson, p. 366).

(d) *Fading.* When the transmitter or receiver is in motion, or when the physical environment is changing (tree leaves fluttering in the wind, people moving around), there will be slow or rapid fading, which can contain amplitude and frequency distortion, and time delay fluctuations. The receiver AGC and demodulation circuits must deal properly with these effects.

4.7 Flat Fading

In many of the short-range radio applications covered in this book, the signal bandwidth is narrow and frequency distortion is negligible. The multipath effect in this case is classified as *flat fading*. In describing the variation of the resultant signal amplitude in a multipath environment, we distinguish two cases: (1) there is no line-of-sight path and the signal is the resultant of a large number of randomly distributed reflections; (2) the random reflections are superimposed on a signal over a dominant constant path, usually the line of sight.

Short-range radio systems that are installed indoors or outdoors in built-up areas are subject to multipath fading essentially of the first case. Our aim in this section is to determine the signal strength margin that is needed to ensure that reliable communication can take place at a given probability. While in many situations there will be a dominant signal path in addition to the multipath fading, restricting ourselves to an analysis of the case where all paths are the result of random reflections gives us an upper bound on the required margin.

4.7.1 Rayleigh Fading

The first case can be described by a received signal $R(t)$, expressed as

$$R(t) = r \cdot \cos(2\pi \cdot f_c \cdot t + \theta) \qquad (4\text{-}7)$$

where r and θ are random variables for the peak signal, or envelope, and phase. Their values may vary with time, when various reflecting objects are moving (people in a room, for example), or with changes in position of the transmitter or receiver which are small in respect to the distance between them. We are not dealing here with the large-scale path gain that is expressed in Eq. (4-5) and (4-6). For simplicity, Eq. (4-7) shows a CW (continuous wave) signal as the modulation terms are not needed to describe the fading statistics.

The envelope of the received signal, r, can be statistically described by the Rayleigh distribution whose probability density function is

$$p(r) = \frac{r}{\sigma^2} e^{-r^2/2\sigma^2} \qquad (4\text{-}8)$$

where σ^2 represents the variance of $R(t)$ in Eq. (4-7), which is the average received signal power. This function is plotted in Figure 4.6. We normalized the curve with σ equal to 1. In this plot, the average value of the signal envelope, shown by a dotted vertical line, is 1.253. Note that it is not the most probable value, which is 1 (σ). The area of the curve between any two values of signal strength r represents the probability that the signal strength will be in that range. The average for the Rayleigh distribution, which is not symmetric, does not divide the curve area in half. The parameter that does this is the *median*, which in this case equals 1.1774. There is a 50% probability that a signal will be below the median and 50% that it will be above.

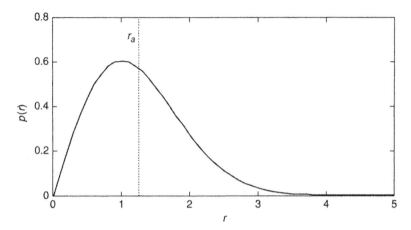

Figure 4.6: Rayleigh probability density function

As stated above, the Rayleigh distribution is used to determine the signal margin required to give a desired communication reliability over a fading channel with no line of sight. The curve labeled "1 Channel" in Figure 4.7 is a cumulative distribution function with logarithmic axes. For any point on the curve, the probability of fading below the margin indicated on the abscissa is given as the ordinate. The curve is scaled such that "0 dB" signal margin represents the point where the received signal equals the mean power of the fading signal, σ^2, making the assumption that the received signal power with no fading equals the average power with fading. Some similar curves in the literature use the median power, or the power corresponding to the average envelope signal level, r_a, as the reference, "0 dB" value.

An example of using the curve is as follows. Say you require a communication reliability of 99%. Then the minimum usable signal level is that for which there is a 1% probability of fading below that level. On the curve, the margin corresponding to 1% is 20 dB. Thus, you need a signal strength 20 dB larger than the required signal if there was no fading. Assume you calculated path loss and found that you need to transmit 4 mW to allow reception at the receiver's sensitivity level. Then, to ensure that the signal will be received 99% of the time during fading, you'll need 20 dB more power or 6 dBm (4 mW) plus 20 dB equals 26 dBm or 400 mW. If you don't increase the power, you can expect loss of communication 63% of the time, corresponding to the "0 dB" margin point on the "Channel 1" curve of Figure 4.7.

Table 4.2 shows signal margins for different reliabilities.

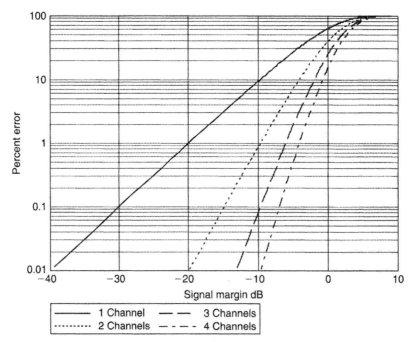

Figure 4.7: Fading margins

Table 4.2: Signal margins for different reliabilities

Reliability, Percent	Fading Margin, dB
90	10
99	20
99.9	30
99.99	40

4.8 Diversity Techniques

Communication reliability for a given signal power can be increased substantially in a multipath environment through diversity reception. If signals are received over multiple, independent channels, the largest signal can be selected for subsequent processing and use. The key to this solution is the independence of the channels. The multipath effect

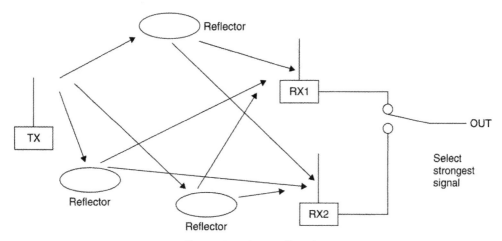

Figure 4.8: Space diversity

of nulling and of strengthening a signal is dependent on transmitter and receiver spatial positions, on wavelength (or frequency) and on polarity. Let's see how we can use these parameters to create independent diverse channels.

4.8.1 Space Diversity

A signal that is transmitted over slightly different distances to a receiver may be received at very different signal strengths. For example, in Figure 4.2 the signal at 17 meters is at a null and at 11 meters at a peak. If we had two receivers, each located at one of those distances, we could choose the strongest signal and use it. In a true multipath environment, the source, receiver, or the reflectors may be constantly in motion, so the nulls and the peaks would occur at different times on each channel. Sometimes Receiver 1 has the strongest signal, at other times Receiver 2. Figure 4.8 illustrates the paths to two receivers from several reflectors. Although there may be circumstances where the signals at both receiver locations are at around the same level, when it doesn't matter which receiver output is chosen, most of the time one signal will be stronger than the other. By selecting the strongest output, the average output after selection will be greater than the average output of one channel alone. To increase even more the probability of getting a higher average output, we could use three or more receivers. From Figure 4.7 you can find the required fading margin using diversity reception having 2, 3, or 4 channels. Note that the plots in Figure 4.7 are based on completely independent channels. When the channels are not completely independent, the results will not be as good as indicated by the plots.

It isn't necessary to use complete receivers at each location, but separate antennas and front ends must be used, at least up to the point where the signal level can be discerned and used to decide on the switch position.

4.8.2 Frequency Diversity

You can get a similar differential in signal strength over two or more signal channels by transmitting on separate frequencies. For the same location of transmitting and receiving antennas, the occurrences of peaks and nulls will differ on the different frequency channels. As in the case of space diversity, choosing the strongest channel will give a higher average signal-to-noise ratio than on either one of the channels. The required frequency difference to get near independent fading on the different channels depends on the diversity of path lengths or signal delays. The larger the difference in path lengths, the smaller the required frequency difference of the channels.

4.8.3 Polarization Diversity

Fading characteristics are dependent on polarization. A signal can be transmitted and received separately on horizontal and vertical antennas to create two diversity channels. Reflections can cause changes in the direction of polarization of a radio wave, so this characteristic of a signal can be used to create two separate signal channels. Thus, cross-polarized antennas can be used at the receiver only. Polarization diversity can be particularly advantageous in a portable handheld transmitter, since the orientation of its antenna will not be rigidly defined.

Polarization diversity doesn't allow the use of more than two channels, and the degree of independence of each channel will usually be less than in the two other cases. However, it may be simpler and less expensive to implement and may give enough improvement to justify its use, although performance will be less than can be achieved with space or frequency diversity.

4.8.4 Diversity Implementation

In the descriptions above, we talked about selecting or switching to the channel having the highest signal level. A more effective method of using diversity is called "maximum ratio combining." In this technique, the outputs of each independent channel are added together after the channel phases are made equal and channel gains are adjusted for equal signal levels. Maximum ratio combining is known to be optimum as it gives the best statistical reduction of fading of any linear diversity combiner. In applications where accurate amplitude estimation is difficult, the channel phases only may be equalized and the outputs added without weighting the gains. Performance in this case is almost as good as in maximum ratio combining. [Gibson, p. 170, 171]

Space diversity has the disadvantage of requiring significant antenna separation, at least in the VHF and lower UHF bands. In the case where multipath signals arrive from all directions, antenna spacing on the order of .5λ to .8λ is adequate in order to have reasonably independent, or decorrelated, channels. This is at least one-half meter at 300 MHz. When the multipath angle spread is small—for example, when directional antennas are used—much larger separations are required.

Frequency diversity eliminates the need for separate antennas, but the simultaneous use of multiple frequency channels entails increased total power and spectrum utilization. Sometimes data are repeated on different frequencies so that simultaneous transmission doesn't have to be used. Frequency separation must be adequate to create decorrelated channels. The bandwidths allocated for unlicensed short-range use are rarely adequate, particularly in the VHF and UHF ranges (transmitting simultaneously on two separate bands can and has been done). Frequency diversity to reduce the effects of time delay spread is achieved with frequency hopping or direct sequence spread spectrum modulation, but for the spreads encountered in indoor applications, the pulse rate must be relatively high—of the order of several megabits per second—in order to be effective. For long pulse widths, the delay spread will not be a problem anyway, but multipath fading will still occur and the amount of frequency spread normally used in these cases is not likely to solve it.

When polarity diversity is used, the orthogonally oriented antennas can be close together, giving an advantage over space diversity when housing dimensions relative to wavelength are small. Performance may not be quite as good, but may very well be adequate, particularly when used in a system having portable hand-held transmitters, which have essentially random polarization.

Although we have stressed that at least two independent (decorrelated) channels are needed for diversity reception, sometimes shortcuts are taken. In some low-cost security systems, for example, two receiver antennas—space diverse or polarization diverse—are commutated directly, usually by diode switches, before the front end or mixer circuits.

Thus, a minimum of circuit duplication is required. In such applications the message frame is repeated many times, so if there happens to be a multipath null when the front end is switched to one antenna and the message frames are lost, at least one or more complete frames will be correctly received when the switch is on the other antenna, which is less affected by the null. This technique works for slow fading, where the fade doesn't change much over the duration of a transmission of message frames. It doesn't appear to give any advantage during fast fading, when used with moving hand-held transmitters, for example. In that case, a receiver with one antenna will have a better chance of decoding at least one of many frames than when switched antennas are used and only half the total number of frame repetitions are available for each. In a worst-case situation with fast fading, each antenna in turn could experience a signal null.

4.8.5 Statistical Performance Measure

We can estimate the performance advantage due to diversity reception with the help of Figure 4.7. Curves labeled "2 Channels" through "4 Channels" are based on the selection combining technique.

Let's assume, as before, that we require communication reliability of 99 percent, or an error rate of 1 percent. From probability theory the probability that two independent channels would both have communication errors is the product of the error probabilities of each channel. Thus, if each of two channels has an error probability of 10 percent, the probability that both channels will have signals below the sensitivity threshold level when selection is made is .1 times .1, which equals .01, or 1 percent. This result is reflected in the curve "2 Channels". We see that the signal margin needed for 99 percent reliability (1 percent error) is 10 dB. Using diversity reception with selection from two channels allows a reliability margin of only 10 dB instead of 20 dB, which is required if there is no diversity. Continuing the previous example, we need to transmit only 40 mW for 99 percent reliability instead of 400 mW. Required margins by selection among three channels and four channels is even less—6 dB and 4 dB, respectively.

Remember that the reliability margins using selection combining diversity as shown in Figure 4.7 are ideal cases, based on the Rayleigh fading probability distribution and independently fading channels. However, even if these premises are not realized in practice, the curves still give us approximations of the improvement that diversity reception can bring.

4.9 Noise

The ultimate limitation in radio communication is not the path loss or the fading. Weak signals can be amplified to practically any extent, but it is the noise that bounds the range we can get or the communication reliability that we can expect from our radio system. There are two sources of receiver noise—interfering radiation that the antenna captures along with the desired signal, and the electrical noise that originates in the receiver circuits. In either case, the best signal-to-noise ratio will be obtained by limiting the bandwidth to what is necessary to pass the information contained in the signal. A further improvement can be had by reducing the receiver noise figure, which decreases the internal receiver noise, but this measure is effective only as far as the noise received through the antenna is no more than about the same level as the receiver noise. Finally, if the noise can be reduced no further, performance of digital receivers can be improved by using error correction coding up to a point, which is designated as channel capacity. The capacity is the maximum information rate that the specific channel can support, and above this rate communication is impossible. The channel capacity is limited by the noise density (noise power per hertz) and the bandwidth.

Figure 4.9 shows various sources of noise over a frequency range of 20 kHz up to 10 GHz. The strength of the noise is shown in microvolts/meter for a bandwidth of 10 kHz, as received by a half-wave dipole antenna at each frequency. The curve labeled "equivalent receiver noise" translates the noise generated in the receiver circuits into an equivalent field strength so that it can be compared to the external noise sources. Receiver noise figures on which the curve is based vary in a log-linear manner with 2 dB at 50 MHz and 9 dB at 1 GHz. The data in Figure 4.9 are only a representative example of radiation and receiver noise, taken at a particular time and place.

Note that all of the noise sources shown in Figure 4.9 are dependent on frequency. The relative importance of the various noise sources to receiver sensitivity depends on their strength relative to the receiver noise. Atmospheric noise is dominant on the low radio frequencies but is not significant on the bands used for short-range communication—above around 40 MHz. Cosmic noise comes principally from the sun and from the center of our galaxy. In the figure, it is masked out by man-made noise, but in locations where man-made noise is a less significant factor, cosmic noise affects sensitivity up to 1 GHz.

Man-made noise is dominant in the range of frequencies widely used for short-range radio systems—VHF and low to middle UHF bands. It is caused by a wide range of

Noise and its sources

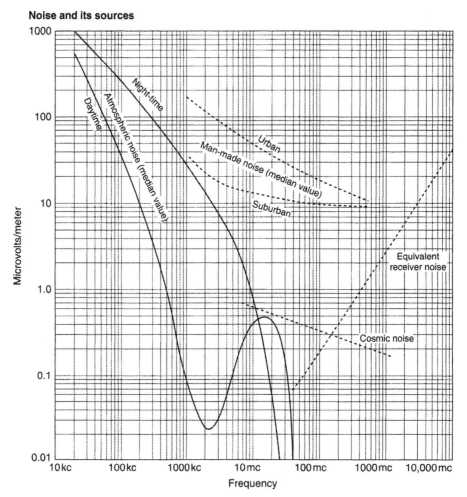

Figure 4.9: External noise sources

(*Reference Data for Radio Engineers*, Fourth Edition)

ubiquitous electrical and electronic equipment, including automobile ignition systems, electrical machinery, computing devices and monitors. While we tend to place much importance on the receiver sensitivity data presented in equipment specifications, high ambient noise levels can make the sensitivity irrelevant in comparing different devices. For example, a receiver may have a laboratory measured sensitivity of −105 dBm for a signal-to-noise ratio of 10 dB. However, when measured with its antenna in a known

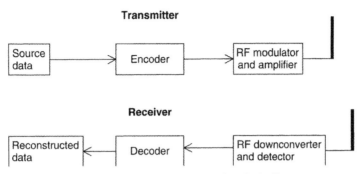

Figure 4.10: Radio communication link diagram

electric field and accounting for the antenna gain, −95 dBm may be required to give the same signal-to-noise ratio.

From around 800 MHz and above, receiver sensitivity is essentially determined by the noise figure. Improved low-noise amplifier discrete components and integrated circuit blocks produce much better noise figures than those shown in Figure 4.9 for high UHF and microwave frequencies. Improving the noise figure must not be at the expense of other characteristics—intermodulation distortion, for example, which can be degraded by using a very high-gain amplifier in front of a mixer to improve the noise figure. Intermodulation distortion causes the production of inband interfering signals from strong signals on frequencies outside of the receiving bandwidth.

External noise will be reduced when a directional antenna is used. Regulations on unlicensed transmitters limit the peak radiated power. When possible, it is better to use a high-gain antenna and lower transmitter power to achieve the same peak radiated power as with a lower gain antenna. The result is higher sensitivity through reduction of external noise. Manmade noise is usually less with a horizontal antenna than with a vertical antenna.

4.10 Communication Protocols and Modulation

A simple block diagram of a digital wireless link is shown in Figure 4.10. The link transfers information originating at one location, referred to as source data, to another location where it is referred to as reconstructed data. A more concrete implementation of a wireless system, a security system, is shown in Figure 4.11.

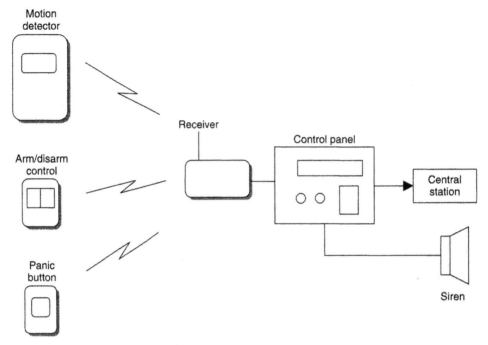

Figure 4.11: Security system

4.10.1 Baseband Data Format and Protocol

Let's first take a look at what information we may want to transfer to the other side. This is important in determining what bandwidth the system needs.

4.10.1.1 Change-of-State Source Data

Many short-range systems only have to relay information about the state of a contact. This is true of the security system of Figure 4.11 where an infrared motion detector notifies the control panel when motion is detected. Another example is the push-button transmitter, which may be used as a panic button or as a way to activate and deactivate the control system, or a wireless smoke detector, which gives advance warning of an impending fire. There are also what are often referred to as "technical" alarms—gas detectors, water level detectors, and low and high temperature detectors—whose function is to give notice of an abnormal situation.

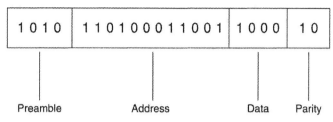

Figure 4.12: Message frame

All these examples are characterized as very low-bandwidth information sources. Change of state occurs relatively rarely, and when it does, we usually don't care if knowledge of the event is signaled tens or even hundreds of milliseconds after it occurs. Thus, required information bandwidth is very low—several hertz.

It would be possible to maintain this very low bandwidth by using the source data to turn on and off the transmitter at the same rate the information occurs, making a very simple communication link. This is not a practical approach, however, since the receiver could easily mistake random noise on the radio channel for a legitimate signal and thereby announce an intrusion, or a fire, when none occurred. Such false alarms are highly undesirable, so the simple on/off information of the transmitter must be coded to be sure it can't be misinterpreted at the receiver.

This is the purpose of the encoder shown in Figure 4.10. This block creates a group of bits, assembled into a frame, to make sure the receiver will not mistake a false occurrence for a real one. Figure 4.12 is an example of a message frame. The example has four fields. The first field is a preamble with start bit, which conditions the receiver for the transfer of information and tells it when the message begins. The next field is an identifying address. This address is unique to the transmitter and its purpose is to notify the receiver from where, or from what unit, the message is coming. The data field follows, which may indicate what type of event is being signaled, followed, in some protocols, by a parity bit or bits to allow the receiver to determine whether the message was received correctly.

Address Field
The number of bits in the address field depends on the number of different transmitters there may be in the system. Often the number of possibilities is far greater than this, to prevent confusion with neighboring, independent systems and to prevent the statistically possible chance that random noise will duplicate the address. The number of possible

addresses in the code is 2^{L1}, where $L1$ is the length of the message field. In many simple security systems the address field is determined by dip switches set by the user. Commonly, eight to ten dip switch positions are available, giving 256 to 1024 address possibilities. In other systems, the address field, or device identity number, is a code number set in the unit micro-controller during manufacture. This code number is longer than that produced by dip switches, and may be 16 to 24 bits long, having 65,536 to 16,777,216 different codes. The longer codes greatly reduce the chances that a neighboring system or random event will cause a false alarm. On the other hand, the probability of detection is lower with the longer code because of the higher probability of error. This means that a larger signal-to-noise ratio is required for a given probability of detection.

In all cases, the receiver must be set up to recognize transmitters in its own system. In the case of dip-switch addressing, a dip switch in the receiver is set to the same address as in the transmitter. When several transmitters are used with the same receiver, all transmitters must have the same identification address as that set in the receiver. In order for each individual transmitter to be recognized, a subfield of two to four extra dip switch positions can be used for this differentiation. When a built-in individual fixed identity is used instead of dip switches, the receiver must be taught to recognize the identification numbers of all the transmitters used in the system; this is done at the time of installation. Several common ways of accomplishing this are:

(a) *Wireless "learn" mode.* During a special installation procedure, the receiver stores the addresses of each of the transmitters which are caused to transmit during this mode;

(b) *Infrared transmission.* Infrared emitters and detectors on the transmitter and receiver, respectively, transfer the address information;

(c) *Direct key-in.* Each transmitter is labeled with its individual address, which is then keyed into the receiver or control panel by the system installer;

(d) *Wired learn mode.* A short cable temporarily connected between the receiver and transmitter is used when performing the initial address recognition procedure during installation.

Advantages and Disadvantages of the Two Addressing Systems

Some advantages and disadvantages of the two types of addressing system are shown in Tables 4.3 and 4.4. Table 4.3 shows dip switch addressing and Table 4.4 shows internal fixed code identity addressing.

Table 4.3: Dip switch

Advantages	Disadvantages
Unlimited number of transmitters can be used with a receiver.	Limited number of bits increases false alarms and interference from adjacent systems.
Can be used with commercially available data encoders and decoders.	Device must be opened for coding during installation.
Transmitter or receiver can be easily replaced without recoding the opposite terminal.	Multiple devices in a system are not distinguishable in most simple systems.
	Control systems are vulnerable to unauthorized operation since the address code can be duplicated by trial and error.

Table 4.4: Internal fixed code identity

Advantages	Disadvantages
Large number of code bits reduces possibility of false alarms.	Longer code reduces probability of detection.
System can be set up without opening transmitter.	Replacing transmitter or receiver involves redoing the code learning procedure.
Each transmitter is individually recognized by receiver.	Limited number of transmitters can be used with each receiver.
	Must be used with a dedicated microcontroller. Cannot be used with standard encoders and decoders.

Code-hopping Addressing

While using a large number of bits in the address field reduces the possibility of false identification of a signal, there is still a chance of purposeful duplication of a transmitter code to gain access to a controlled entry. Wireless push buttons are used widely for access control to vehicles and buildings. Radio receivers exist, popularly called "code grabbers," which receive the transmitted entry signals and allow retransmitting them for fraudulent access to a protected vehicle or other site. To counter this possibility, addressing techniques were developed that cause the code to change every time the push button is pressed, so that even if the transmission is intercepted and recorded, its repetition by a would-be intruder will not activate the receiver, which is now expecting a different code. This method is variously called *code rotation*, *code hopping*, or *rolling code addressing.* In order to make it virtually impossible for a would-be intruder to guess or try various combinations to arrive at the correct code, a relatively large number of address bits are used. In some devices, 36-bit addresses are employed, giving a total of over 68 billion possible codes.

In order for the system to work, the transmitter and receiver must be synchronized. That is, once the receiver has accepted a particular transmission, it must know what the next transmitted address will be. The addresses cannot be sequential, since that would make it too easy for the intruder to break the system. Also, it is possible that the user might press the push button to make a transmission but the receiver may not receive it, due to interference or the fact that the transmitter is too far away. This could even happen several times, further unsynchronizing the transmitter and the receiver. All of the code-hopping systems are designed to prevent such unsynchronization.

Following is a simplified description of how code hopping works, aided by Figure 4.13.

Both the receiver and the transmitter use a common algorithm to generate a pseudorandom sequence of addresses. This algorithm works by manipulating the address bits in a certain fashion. Thus, starting at a known address, both sides of the link will create the same next address. For demonstration purposes, Figure 4.13 shows the same sequence of two-digit decimal numbers at the transmitting side and the receiving side. The solid transmitter arrow points to the present transmitter address and the solid receiver arrow points to the expected receiver address. After transmission and reception, both transmitter and receiver calculate their next addresses, which will be the same. The arrows are synchronized to point to the same address during a system set-up procedure.

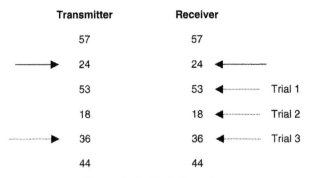

Figure 4.13: Code hopping

As long as the receiver doesn't miss a transmission, there is no problem, since each side will calculate an identical next address. However, if one or more transmissions are missed by the receiver, when it finally does receive a message, its expected address will not match the received address. In this case it will perform its algorithm again to create a new address and will try to match it. If the addresses still don't match, a new address is calculated until either the addresses match or a given number of trials have been made with no success. At this point, the transmitter and receiver are unsynchronized and the original setup procedure has to be repeated to realign the transmitter and receiver addresses.

The number of trials permitted by the receiver may typically be between 64 and 256. If this number is too high, the possibility of compromising the system is greater (although with a 36-bit address a very large number of trials would be needed for this) and with too few trials, the frequency of inconvenient resynchronization would be greater. Note that a large number of trials takes a lot of time for computations and may cause a significant delay in response.

Several companies make rolling code components, among them Microchip, Texas Instruments, and National Semiconductor.

Data Field

The next part of the message frame is the data field. Its number of bits depends on how many pieces of information the transmitter may send to the receiver. For example, the motion detector may transmit three types of information: motion detection, tamper detection, or low battery.

Parity Bit Field

The last field is for error detection bits, or parity bits. As discussed later, some protocols have inherent error detection features so the last field is not needed.

Baseband Data Rate

Once we have determined the data frame, we can decide on the appropriate baseband data rate. For the security system example, this rate will usually be several hundred hertz up to a maximum of a couple of kilohertz. Since a rapid response is not needed, a frame can be repeated several times to be more certain it will get through. Frame repetition is needed in systems where space diversity is used in the receiver. In these systems, two separate antennas are periodically switched to improve the probability of reception. If signal nulling occurs at one antenna because of the multipath phenomena, the other antenna will produce a stronger signal, which can be correctly decoded. Thus, a message frame must be sent more often to give it a chance to be received after unsuccessful reception by one of the antennas.

Supervision

Another characteristic of digital event systems is the need for link supervision. Security systems and other event systems, including medical emergency systems, are one-way links. They consist of several transmitters and one receiver. As mentioned above, these systems transmit relatively rarely, only when there is an alarm or possibly a low-battery condition. If a transmitter ceases to operate, due to a component failure, for example, or if there is an abnormal continuing interference on the radio channel, the fact that the link has been broken will go undetected. In the case of a security system, the installation will be unprotected, possibly, until a routine system inspection is carried out. In a wired system, such a possibility is usually covered by a normally energized relay connected through closed contacts to a control panel. If a fault occurs in the device, the relay becomes unenergized and the panel detects the opening of the contacts. Similarly, cutting the connecting wires will also be detected by the panel. Thus, the advantages of a wireless system are compromised by the lower confidence level accompanying its operation.

Many security systems minimize the risk of undetected transmitter failure by sending a supervisory signal to the receiver at a regular interval. The receiver expects to receive a signal during this interval and can emit a supervisory alarm if the signal is not received. The supervisory signal must be identified as such by the receiver so as not to be mistaken for an alarm.

The duration of the supervisory interval is determined by several factors:

- Devices certified under FCC Part 15 paragraph 15.231, which applies to most wireless security devices in North America, may not send regular transmissions more frequently than one per hour.

- The more frequently regular supervision transmissions are made, the shorter the battery life of the device.

- Frequent supervisory transmissions when there are many transmitters in the system raise the probability of a collision with an alarm signal, which may cause the alarm not to get through to the receiver.

- The more frequent the supervisory transmissions, the higher the confidence level of the system.

While it is advantageous to notify the system operator at the earliest sign of transmitter malfunction, frequent supervision raises the possibility that a fault might be reported when it doesn't exist. Thus, most security systems determine that a number of consecutive missing supervisory transmissions must be detected before an alarm is given. A system which specifies security emissions once every hour, for example, may wait for eight missing supervisory transmissions, or eight hours, before a supervisory alarm is announced. Clearly, the greater the consequences of lack of alarm detection due to a transmitter failure, the shorter the supervision interval must be.

4.10.1.2 Continuous Digital Data

In other systems flowing digital data must be transmitted in real time and the original source data rate will determine the baseband data rate. This is the case in wireless LANs and wireless peripheral-connecting devices. The data is arranged in message frames, which contain fields needed for correct transportation of the data from one side to the other, in addition to the data itself.

An example of a frame used in *synchronous data link control* (SDLC) is shown in Figure 4.14. It consists of beginning and ending bytes that delimit the frame in the message, address and control fields, a data field of undefined length, and check bits or parity bits for letting the receiver check whether the frame was correctly received. If it is, the receiver sends a short acknowledgment and the transmitter can continue with the next frame. If no acknowledgment is received, the transmitter repeats the message again and again until it is received. This is called an ARQ (automatic repeat query) protocol. In high-noise environments, such as encountered on radio channels, the repeated transmissions can significantly slow down the message throughput.

More common today is to use a *forward error control* (FEC) protocol. In this case, there is enough information in the parity bits to allow the receiver to correct a small number

Beginning Flag-8 Bits	Address-8 Bits	Control-8 Bits	Information- Any no. of bits	Error Detection-16 Bits	Ending flag- 8 Bits

Figure 4.14: Synchronous data link control frame

of errors in the message so that it will not have to request retransmission. Although more parity bits are needed for error correction than for error detection alone, the throughput is greatly increased when using FEC on noisy channels.

In all cases, we see that extra bits must be included in a message to insure proper transmission, and the consequently longer frames require a higher transmission rate than what would be needed for the source data alone. This message overhead must be considered in determining the required bit rate on the channel, the type of digital modulation, and consequently the bandwidth.

4.10.1.3 Analog Transmission

Analog transmission devices, such as wireless microphones, also have a baseband bandwidth determined by the data source. A high-quality wireless microphone may be required to pass 50 to 15,000 Hz, whereas an analog wireless telephone needs only 100 to 3000 Hz. In this case determining the channel bandwidth is more straightforward than in the digital case, although the bandwidth depends on whether AM or FM modulation is used. In most short-range radio applications, FM is preferred—narrow-band FM for voice communications and wide-band FM for quality voice and music transmission.

4.10.2 Baseband Coding

The form of the information signal that is modulated onto the RF carrier we call here baseband coding. We refer below to both digital and analog systems, although strictly speaking the analog signal is not coded but is modified to obtain desired system characteristics.

4.10.2.1 Digital Systems

Once we have a message frame, composed as we have shown by address and data fields, we must form the information into signal levels that can be effectively transmitted, received, and decoded. Since we're concerned here with binary data transmission, the baseband coding selected has to give the signal the best chance to be decoded after it has been modified by noise and the response of the circuit and channel elements. This coding consists essentially of the way that zeros and ones are represented in the signal sent to the modulator of the transmitter.

There are many different recognized systems of baseband coding. We will examine only a few common examples.

These are the dominant criteria for choosing or judging a baseband code:

(a) *Timing*. The receiver must be able to take a data stream polluted by noise and recognize transitions between each bit. The bit transitions must be independent of the message content—that is, they must be identifiable even for long strings of zeros or ones.

(b) *DC content*. It is desirable that the average level of the message—that is, its DC level—remains constant throughout the message frame, regardless of the content of the message. If this is not the case, the receiver detection circuits must have a frequency response down to DC so that the levels of the message bits won't tend to wander throughout the frame. In circuits where coupling capacitors are used, such a response is impossible.

(c) *Power spectrum*. Baseband coding systems have different frequency responses. A system with a narrow frequency response can be filtered more effectively to reduce noise before detection.

(d) *Inherent error detection*. Codes that allow the receiver to recognize an error on a bit-by-bit basis have a lower possibility of reporting a false alarm when error-detecting bits are not used.

(e) *Probability of error*. Codes differ in their ability to properly decode a signal, given a fixed transmitter power. This quality can also be stated as having a lower probability of error for a given signal-to-noise ratio.

(f) *Polarity independence*. There is sometimes an advantage in using a code that retains its characteristics and decoding capabilities when inverted. Certain types of modulation and demodulation do not retain polarity information. Phase modulation is an example.

Now let's look at some common codes (Figure 4.15) and rate them according to the criteria above. It should be noted that coding considerations for event-type reporting are far different from those for flowing real-time data, since bit or symbol times are not so critical, and message frames can be repeated for redundancy to improve the probability of detection and reduce false alarms. In data flow messages, the data rate is important and sophisticated error detection and correction techniques are used to improve system sensitivity and reliability.

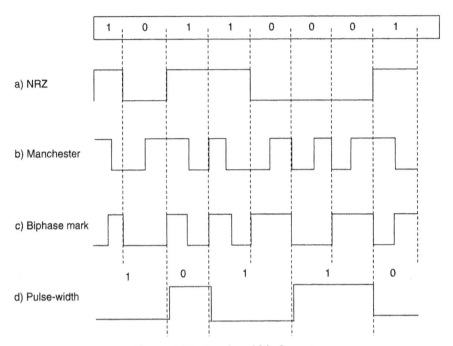

Figure 4.15: Baseband bit formats

(a) *Non-return to Zero (NRZ)—Figure 4.15a.*

This is the most familiar code, since it is used in digital circuitry and serial wired short-distance communication links, like RS-232. However, it is rarely used directly for wireless communication. Strings of ones or zeros leave it without defined bit boundaries, and its DC level is very dependent on the message content. There is no inherent error detection. If NRZ coding is used, an error detection or correction field is imperative. If amplitude shift keying (ASK) modulation is used, a string of zeros means an extended period of no transmission at all. In any case, if NRZ signaling is used, it should only be for very short frames of no more than eight bits.

(b) *Manchester code—Figure 4.15b.*

A primary advantage of this code is its relatively low probability of error compared to other codes. It is the code used in Ethernet local area networks. It gives good timing information since there is always a transition in the middle of a bit, which is decoded as zero if this is a positive transition and a one otherwise. The Manchester code has a constant DC component and its waveform doesn't change if it passes through a capacitor or transformer. However, a "training" code sequence should be inserted before the message information as a preamble to allow capacitors in the receiver detection circuit to reach charge equilibrium before the actual message bits appear. Inverting the Manchester code turns zeros to ones and ones to zeros. The frequency response of Manchester code has components twice as high as NRZ code, so a low-pass filter in the receiver must have a cut-off frequency twice as high as for NRZ code with the same bit rate.

(c) *Biphase Mark—Figure 4.15c.*

This code is somewhat similar to the Manchester code, but bit identity is determined by whether or not there is a transition in the middle of a bit. For biphase mark, a level transition in the middle of a bit (going in either direction) signifies a one, and a lack of transition indicates zero. Biphase space is also used, where the space character has a level transition. There is always a transition at the bit boundaries, so timing content is good. A lack of this transition gives immediate notice of a bit error and the frame should then be aborted. The biphase code has constant DC level, no matter what the message content, and a preamble should be sent to allow capacitor charge equalization before the message bits arrive. As with the Manchester code, frequency content is twice as much as for the NRZ code. The biphase mark or space code has the added advantage of being polarity independent.

(d) *Pulse width modulation—Figure 4.15d.*

As shown in the figure, a one has two timing durations and a zero has a pulse width of one duration. The signal level inverts with each bit so timing information for synchronization is good. There is a constant average DC level. Since the average pulse width varies with the message content, in contrast with the other examples, the bit rate is not constant. This code has inherent error detection capability.

(e) *Motorola MC145026-145028 coding—Figure 4.16.*

Knowledge of the various baseband codes is particularly important if the designer creates his own protocol and implements it on a microcontroller. However, there are several off-the-shelf integrated circuits that are popular for simple event transmission transmitters where a microcontroller is not needed for other circuit functions, such as in panic buttons or door-opening controllers.

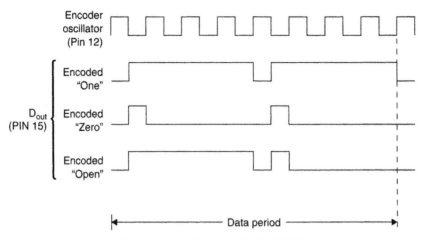

Figure 4.16: Motorola MC145026

The Motorola chips are an example of nonstandard coding developed especially for event transmission where three-state addressing is determined as high, low, or open connections to device pins. Thus, a bit symbol can be one of three different types. The receiver, MC145028, must recognize two consecutive identical frames to signal a valid message. The Motorola protocol gives very high reliability and freedom from false alarms. Its signal does have a broad frequency spectrum relative to the data rate and the receiver filter passband must be designed accordingly. The DC level is dependent on the message content.

4.10.2.2 Analog Baseband Conditioning

Wireless microphones and headsets are examples of short-range systems that must maintain high audio quality over the vagaries of changing path lengths and indoor environments, while having small size and low cost. To help them achieve this, they have a signal conditioning element in their baseband path before modulation. Two features used to achieve high signal-to-noise ratio over a wide dynamic range are pre-emphasis/de-emphasis and compression/expansion. Their positions in the transmitter/receiver chain are shown in Figure 4.17.

The transmitter audio signal is applied to a high-pass filter (pre-emphasis) which increases the high frequency content of the signal. In the receiver, the detected audio goes through a complementary low-pass filter (de-emphasis), restoring the signal to its original spectrum composition. However, in so doing, high frequency noise that entered the signal

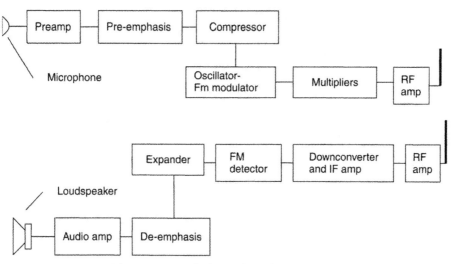

Figure 4.17: Wireless microphone system

path after the modulation process is filtered out by the receiver while the desired signal is returned to its original quality.

Compression of the transmitted signal raises the weak sounds and suppresses strong sounds to make the modulation more efficient. Reversing the process in the receiver weakens annoying background noises while restoring the signal to its original dynamic range.

4.10.3 RF Frequency and Bandwidth

There are several important factors to consider when determining the radio frequency of a short-range system:

- Telecommunication regulations
- Antenna size
- Cost
- Interference
- Propagation characteristics

When you want to market a device in several countries and regions of the world, you may want to choose a frequency that can be used in the different regions, or at least frequencies that don't differ very much so that the basic design won't be changed by changing frequencies.

The UHF frequency bands are usually the choice for wireless alarm, medical, and control systems. The bands that allow unlicensed operation don't require rigid frequency accuracy. Various components—SAWs and ICs—have been specially designed for these bands and are available at low prices, so choosing these frequencies means simple designs and low cost. Many companies produce complete RF transmitter and receiver modules covering the most common UHF frequencies, among them 315 MHz and 902 to 928 MHz (U.S. and Canada), and 433.92 MHz band and 868 to 870 MHz (European Community).

Antenna size may be important in certain applications, and for a given type of antenna, its size is proportional to wavelength, or inversely proportional to frequency. When spatial diversity is used to counter multipath interference, a short wavelength of the order of the size of the device allows using two antennas with enough spacing to counter the nulling that results from multipath reflections. In general, efficient built-in antennas are easier to achieve in small devices at short wavelengths.

From VHF frequencies and up, cost is directly proportional to increased frequency.

Natural and manmade background noise is higher on the lower frequencies. On the other hand, certain frequency bands available for short-range use may be very congested with other users, such as the ISM bands. Where possible, it is advisable to chose a band set aside for a particular use, such as the 868–870 MHz band available in Europe.

Propagation characteristics also must be considered in choosing the operating frequency. High frequencies reflect easily from surfaces but penetrate insulators less readily than lower frequencies.

The radio frequency bandwidth is a function of the baseband bandwidth and the type of modulation employed. For security event transmitters, the required bandwidth is small, of the order of several kilohertz. If the complete communication system were designed to take advantage of this narrow bandwidth, there would be significant performance advantages over the most commonly used systems having a bandwidth of hundreds of kilohertz. For given radiated transmitter power, the range is inversely dependent on the receiver bandwidth. Also, narrow-band unlicensed frequency allotments can be used where available in the different regions, reducing interference from other users. However, cost and complexity considerations tend to outweigh communication reliability for these systems, and manufacturers decide to make do with the necessary performance compromises. The bandwidth of the mass production security devices is thus determined by the frequency stability of the transmitter and receiver frequency determining elements, and not by the required signaling bandwidth. The commonly used SAW devices dictate a bandwidth of at least 200 kHz, whereas the signaling bandwidth may be only 2 kHz. Designing the receiver with a passband of 20 kHz instead of 200 kHz would increase sensitivity by 10 dB, roughly doubling the range. This entails using stable crystal oscillators in the transmitter and in the receiver local oscillator.

4.10.4 Modulation

Amplitude modulation (AM) and frequency modulation (FM), known from commercial broadcasting, have their counterparts in modulation of digital signals, but you must be careful before drawing similar conclusions about the merits of each. The third class of modulation is phase modulation, not used in broadcasting, but its digital counterpart is commonly used in high-end, high-data-rate digital wireless communication.

Digital AM is referred to as ASK—amplitude shift keying—and sometimes as OOK—on/off keying. FSK is frequency shift keying, the parallel to FM. Commercial AM has a bandwidth of 10 kHz whereas an FM broadcasting signal occupies 180 kHz. The high post-detection signal-to-noise ratio of FM is due to this wide bandwidth. However, on the negative side, FM has what is called a threshold effect, also due to wide bandwidth. Weak FM signals are unintelligible at a level that would still be usable for AM signals. When FM is used for two-way analog communication, narrow-band FM, which occupies a similar bandwidth to AM, also has comparable sensitivity for a given S/N.

4.10.4.1 Modulation for Digital Event Communication

For short-range digital communication we're not interested in high fidelity, but rather high sensitivity. Other factors for consideration are simplicity and cost of modulation and demodulation. Let's now look into the reasons for choosing one form of modulation or the other.

An analysis of error rates versus bit energy to noise density shows that there is no inherent advantage of one system, ASK or FSK, over the other. This conclusion is based on certain theoretical assumptions concerning bandwidth and method of detection. While practical implementation methods may favor one system over the other, we shouldn't jump to conclusions that FSK is necessarily the best, based on a false analogy to FM and AM broadcasting.

In low-cost security systems, ASK is the simplest and cheapest method to use. For this type of modulation we must just turn on and turn off the radio frequency output in accordance with the digital modulating signal. The output of a microcontroller or dedicated coding device biases on and off a single SAW-controlled transistor RF oscillator. Detection in the receiver is also simple. It may be accomplished by a diode detector in several receiver architectures, to be discussed later, or by the RSSI (received signal strength indicator) output of many superheterodyne receiver ICs employed today. Also, ASK must be used in the still widespread superregenerative receivers.

For FSK, on the other hand, it's necessary to shift the transmitting frequency between two different values in response to the digital code. More elaborate means is needed for this than in the simple ASK transmitter, particularly when crystal or SAW devices are used to keep the frequency stable. In the receiver, also, additional components are required for FSK demodulation as compared to ASK. We have to decide whether the additional cost and complexity is worthwhile for FSK.

In judging two systems of modulation, we must base our results on a common parameter that is a basis for comparison. This may be peak or average power. For FSK, the peak and average powers are the same. For ASK, average power for a given peak power depends on the duty cycle of the modulating signal. Let's assume first that both methods, ASK and FSK, give the same performance—that is, the same sensitivity—if the average power in both cases are equal. It turns out that in this case, our preference depends on whether we are primarily marketing our system in North America or in Europe. This is because of the difference in the definition of the power output limits between the telecommunication regulations in force in the US and Canada as compared to the common European regulations.

The US FCC Part 15 and similar Canadian regulations specify an *average* field strength limit. Thus, if the transmitter is capable of using a peak power proportional to the inverse of its modulation duty cycle, while maintaining the allowed average power, then under our presumption of equal performance for equal average power, there would be no reason to prefer FSK, with its additional complexity and cost, over ASK.

In Western Europe, on the other hand, the low-power radio specification, ETSI 300 220 limits the *peak* power of the transmitter. This means that if we take advantage of the maximum allowed peak power, FSK is the proper choice, since for a given peak power, the average power of the ASK transmitter will always be less, in proportion to the modulating signal duty cycle, than that of the FSK transmitter.

However, is our presumption of equal performance for equal average power correct? Under conditions of added white Gaussian noise (AWGN) it seems that it is. This type of noise is usually used in performance calculations since it represents the noise present in all electrical circuits as well as cosmic background noise on the radio channel. But in real life, other forms of interference are present in the receiver passband that have very different, and usually unknown, statistical characteristics from AWGN. On the UHF frequencies normally used for short-range radio, this interference is primarily from other transmitters using the same or nearby frequencies. To compare performance, we must examine how the ASK and FSK receivers handle this type of interference. This examination is pertinent in the US and Canada where we must choose between ASK and FSK when considering that the average power, or signal-to-noise ratio, remains constant. Some designers believe that a small duty cycle resulting in high peak power per bit is advantageous since the presence of the bit, or high peak signal, will get through a background of interfering signals better than another signal with the same average power but a lower peak. To check this out, we must assume a fair and equal basis of comparison. For a given data rate, the low-duty-cycle ASK signal will have shorter pulses than for the FSK case. Shorter pulses means higher baseband bandwidth and a higher cutoff frequency for the post detection bandpass filter, resulting in more broadband noise for the same data rate. Thus, the decision depends on the assumptions of the type of interference to be encountered and even then the answer is not clear cut.

An analysis of the effect of different types of interference is given by Anthes of RF Monolithics (see references). He concludes that ASK, which does not completely shut off the carrier on a "0" bit, is marginally better than FSK.

4.10.4.2 Continuous Digital Communication

For efficient transmission of continuous digital data, the modulation choices are much more varied than in the case of event transmission. We can see this in the three leading cellular digital radio systems, all of which have the same use and basic requirements. The system referred to as DAMPS or TDMA (time division multiple access) uses a type of modulation called Pi/4 DPSK (differential phase shift keying). The GSM network is based on GMSK (Gaussian minimum shift keying). The third major system is CDMA (code division multiple access). Each system claims that its choice is best, but it is clear that there is no simple cut-and-dried answer. We aren't going into the details of the cellular systems here, so we'll look at the relevant trade-offs for modulation methods in what we have defined as short-range radio applications. At the end of this section, we review the basic principles of digital modulation and spread-spectrum modulation.

For the most part, license-free applications specify ISM bands where signals are not confined to narrow bandwidth channels. However, noise power is directly proportional to bandwidth, so the receiver bandwidth should be no more than is required for the data rate used. Given a data rate and an average or peak power limitation, there are several reasons for preferring one type of modulation over another. They involve error rate, implementation complexity, and cost. A common way to compare performance of the different systems is by curves of bit error rate (BER) versus the signal-to-noise ratio, expressed as energy per bit divided by the noise density (defined below). The three most common types of modulation system are compared in Figure 4.18. Two of the modulation types were mentioned above. The third, phase shift keying, is described below.

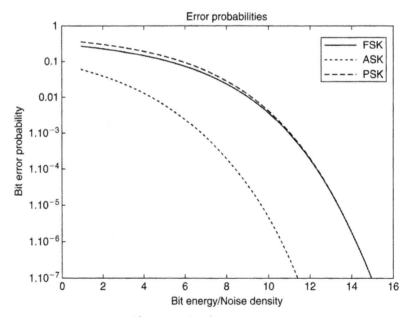

Figure 4.18: Bit error rates

Phase Shift Keying (PSK)

Whereas in amplitude shift keying and frequency shift keying the amplitude and frequency are varied according to the digital source data, in PSK it is the phase of the RF carrier that is varied. In its simplest form, the waveform looks like Figure 4.19. Note that the phase of the carrier wave shifts 180 deg. according to the data signal bits. Similar to FSK, the carrier remains constant, thus giving the same advantage that we mentioned for FSK—maximum signal-to-noise ratio when there is a peak power limitation.

Comparing Digital Modulation Methods

In comparing the different digital modulation methods, we need a common reference parameter that reflects the signal power and noise at the input of the receiver, and the bit rate of the data. This common parameter of signal to noise ratio for digital systems is expressed as the signal energy per bit divided by the noise density, E/N_o. We can relate this parameter to the more familiar signal to noise ratio, S/N, and the data rate R, as follows:

1. S/N = signal power/noise power. The noise power is the noise density N_o in watts/Hz times the transmitted signal bandwidth B_T in Hz:

$$S/N = S/(N_o B_T)$$

2. The signal energy in joules is the signal power in watts, S, times the bit time in seconds, which is $1/(\text{data rate}) = 1/R$, thus $E = S(1/R)$

3. The minimum transmitted bandwidth (Nyquist bandwidth) to pass a bit stream of R bits per second is $B_T = R$ Hz.

4. In sum:

$$E/N_o = (S/R)/N_o = S/N_o R = S/N$$

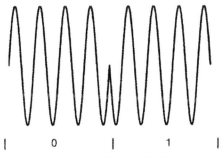

Figure 4.19: Phase shift keying

The signal power for this expression is the power at the input of the receiver, which is derived from the radiated transmitted power, the receiver antenna gain, and the path loss. The noise density N_o, or more precisely the one sided noise power spectral density, in units of watts/Hz, can be calculated from the expression:

$$N_o = kT = 1.38 \times 10^{-23} \times T(\text{Kelvin}) \text{ watt/hertz}$$

The factor k is Boltzmann's constant and T is the equivalent noise temperature that relates the receiver input noise to the thermal noise that is present in a resistance at the same temperature T. Thus, at standard room temperature of 290 degrees Kelvin, the noise power density is 4×10^{-21} watts/Hz, or -174 dBm/Hz.

The modulation types making up the curves in Figure 4.18 are:

Phase shift keying (PSK)

Noncoherent frequency shift keying (FSK)

Noncoherent amplitude shift keying (ASK)

We see from the curves that the best type of modulation to use from the point of view of lowest bit error for a given signal-to-noise ratio (E/N_o) is PSK. There is essentially no difference, according to the curves, between ASK and FSK. (This is true only when noncoherent demodulation is used, as in most simple short-range systems.) What then must we consider in making our choice?

PSK is not difficult to generate. It can be done by a balanced modulator. The difficulty is in the receiver. A balanced modulator can be used here too but one of its inputs, which switches the polarity of the incoming signal, must be perfectly correlated with the received signal carrier and without its modulation. The balanced modulator acts as a multiplier and when a perfectly synchronized RF carrier is multiplied by the received signal, the output, after low-pass filtering, is the original bit stream. PSK demodulation is shown in Figure 4.20.

There are several ways of generating the required reference carrier signal from the received signal. Two examples are the Costas loop and the squaring loop, which include three multiplier blocks and a variable frequency oscillator (VFO). [See Dixon.] Because of the complexity and cost, PSK is not commonly used in inexpensive short-range equipment, but it is the most efficient type of modulation for high-performance data communication systems.

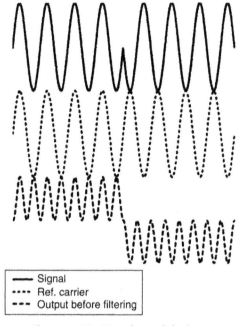

Figure 4.20: PSK demodulation

Amplitude shift keying is easy to generate and detect, and as we see from Figure 4.18, its bit error rate performance is essentially the same as for FSK. However, FSK is usually the modulation of choice for many systems. The primary reason is that peak power is usually the limitation, which gives FSK a 3-dB advantage, since it has constant power for both bit states, whereas ASK has only half the average power, assuming a 50% duty cycle and equal probability of marks and spaces. FSK has slightly more complexity than ASK, and that's probably why it isn't used in all short-range digital systems.

A modulation system which incorporates the methods discussed above but provides a high degree of interference immunity is spread spectrum, which we'll discuss later on in this chapter.

4.10.4.3 Analog Communication

For short-range analog communication—wireless microphones, wireless earphones, auditive assistance devices—FM is almost exclusively used. When transmitting high-quality audio, FM gives an enhanced post-detection signal-to-noise ratio, at the expense of greater bandwidth. Even for narrow band FM, which doesn't have post-detection signal-to-noise enhancement, its noise performance is better than that of AM, because a limiting IF amplifier can be used to reduce the noise. AM, being a linear modulation process, requires linear amplifiers after modulation in the transmitter, which are less efficient than the class C amplifiers used for FM. Higher power conversion efficiency gives FM an advantage in battery-operated equipment.

4.10.4.4 Advanced Digital Modulation

Two leading characteristics of wireless communication in the last few years are the need for increasing data rates and better utilization of the radio spectrum. This translates to higher speeds on narrower bandwidths. At the same time, much of the radio equipment is portable and operated by batteries. So what is needed is:

- high data transmission rates
- narrow bandwidth
- low error rates at low signal-to-noise ratios
- low power consumption

Breakthroughs have occurred with the advancement of digital modulation and coding systems. We deal here with digital modulation principles.

The types of modulation that we discussed previously, ASK, FSK, and PSK, involve modifying a radio frequency carrier one bit at a time. We mentioned the Nyquist bandwidth, which is the narrowest bandwidth of an ideal filter which permits passing a bit stream without intersymbol interference. As the bandwidth of the digital bit stream is further reduced, the bits are lengthened and interfere with the detection of subsequent bits. This minimum, or Nyquist, bandwidth equals one-half of the bit rate at baseband, but twice as much for the modulated signal. We can see this result in Figure 4.21. An alternating series of marks and spaces can be represented by a sine wave whose

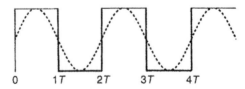

Figure 4.21: Nyquist bandwidth

frequency is one-half the bit rate: $f_{sin} = 1/2T$. An ideal filter with a lower cutoff frequency will not pass this fundamental frequency component and the data will not get through.

Any other combination of bits will create other frequencies, all of which are lower than the Nyquist frequency. It turns out then that the maximum number of bits per hertz of filter cutoff frequency that can be passed at baseband is two. Therefore, if the bandwidth of a telephone line is 3.4 kHz, the maximum binary bit rate that it can pass without intersymbol interference is 6.8k bits per second. We know that telephone line modems pass several times this rate. They do it by incorporating several bits in each *symbol* transmitted, for it is actually the symbol rate that is limited by the Nyquist bandwidth, and the problem that remains is to put several bits on each symbol in such a manner that they can be effectively taken off the symbol at the receiving end with as small as possible chance of error, given a particular *S/N*.

In Figure 4.22 we see three ways of combining bits with individual symbols, each of them based on one of the basic types of modulation—ASK, FSK, PSK. Each symbol duration *T* can carry one of four different values, or two bits. Using any one of the modulation types shown, the telephone line, or wireless link, can pass a bit rate twice as high as before over the same bandwidth. Combinations of these types are also employed,

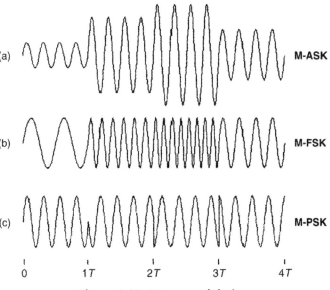

Figure 4.22: M-ary modulation

particularly of ASK and PSK, to put even more bits on a symbol. Quadrature amplitude modulation, QAM, sends several signal levels on four phases of the carrier frequency to give a high bandwidth efficiency—a high bit-rate relative to the signal bandwidth. It seems then that there is essentially no limit to the number of bits that could be compressed into a given bandwidth. If that were true, the 3.4-kHz telephone line could carry millions of bits per second, and the internet bottleneck to our homes would no longer exist. However, there is a very definite limit to the rate of information transfer over a transmission medium where noise is present, expressed in the Hartley-Shannon law:

$$C = W \log(1 + S/N)$$

This expression tells us that the maximum rate of information (the capacity C) that can be sent without errors on a communication link is a function of the bandwidth, W, and the signal-to-noise ratio, S/N.

In investigating the ways of modulating and demodulating multiple bits per symbol, we'll first briefly discuss a method not commonly used in short-range applications (although it could be). This is called M-ary FSK (Figure 4.22b) and in contrast to the aim we mentioned above of increasing the number of bits per hertz, it increases the required bandwidth as the number of bits per symbol is increased. "M" in "M-FSK" is the number of different frequencies that may be transmitted in each symbol period. The benefit of this method is that the required S/N per bit for a given bit error rate decreases as the number of bits per symbol increases. This is analogous to analog FM modulation, which uses a wideband radio channel, well in excess of the bandwidth of the source audio signal, to increase the resultant S/N. M-ary FSK is commonly used in point-to-point microwave transmission links for high-speed data communication where bandwidth limitation is no problem but the power limitation is.

Most of the high-data-rate bandwidth-limited channels use multiphase PSK or QAM. While there are various modulation schemes in use, the essentials of most of them can be described by the block diagram in Figure 4.23. This diagram is the basis of what may be called vector modulation, IQ modulation, or quadrature modulation. "I" stands for "in phase" and "Q" stands for "quadrature."

The basis for quadrature modulation is the fact that two completely independent data streams can be simultaneously modulated on the same frequency and carrier wave. This is possible if each data stream modulates coherent carriers whose phases are 90 degrees

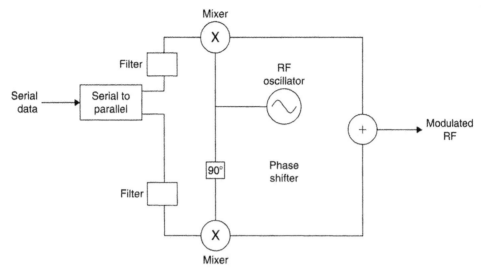

Figure 4.23: Quadrature modulation

apart. We see in the diagram that each of these carriers is created from the same source by passing one of them through a 90-degree phase shifter. Now it doesn't matter which method of modulation is used for each of the phase shifted carriers. Although the two carriers are added together and amplified before transmission, the receiver, by reversing the process used in the transmitter (see Figure 4.24), can completely separate the incoming signal into its two components.

The diagrams in Figures 4.23 and 4.24 demonstrate quadrature phase shift keying (QPSK) where a data stream is split into two, modulated independently bit by bit, then combined to be transmitted on a single carrier. The receiver separates the two data streams, then demodulates and combines them to get the original serial digital data. The phase changes of the carrier in response to the modulation is commonly shown on a vector or constellation diagram. The constellation diagram for QPSK is shown in Figure 4.25. The Xs on the plot are tips of vectors that represent the magnitude (distance from the origin) and phase of the signal that is the sum of the "I" and the "Q" carriers shown on Figure 4.23, where each is multiplied by −1 or +1, corresponding to bit values of 0 and 1. The signal magnitudes of all four possible combinations of the two bits are the same—$\sqrt{2}$ when the I and Q carriers have a magnitude of 1. I and Q bit value combinations corresponding to each vector are shown on the plot.

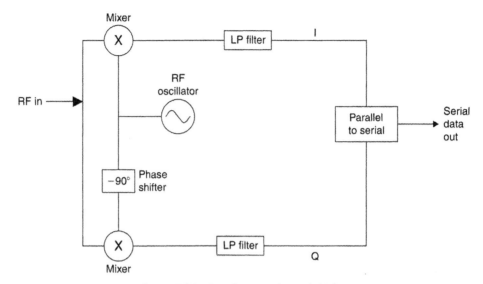

Figure 4.24: Quadrature demodulation

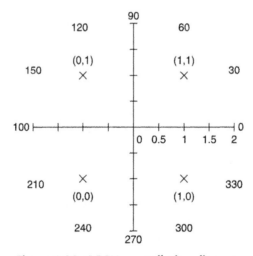

Figure 4.25: QPSK constellation diagram

We have shown that two data streams, derived from a data stream of rate R2, can be sent in parallel on the same RF channel and at the same bandwidth as a single data stream having half the bit rate, R1. At the receiver end, the demodulation process of each of the split bit streams is exactly the same as it would be for the binary phase shift modulation shown in Figure 4.20 above, and its error rate performance is the same as is shown for the BPSK curve in Figure 4.18. Thus, we've doubled the data rate on the same bandwidth channel while maintaining the same error rate as before. In short, we've doubled the efficiency of the communication.

However, there are some complications in adopting quadrature modulation. Similar to the basic modulation methods previously discussed, the use of square waves to modulate RF carriers causes unwanted sidebands that may exceed the allowed channel bandwidth. Thus, special low-pass filters, shown in Figure 4.23, are inserted in the signal paths before modulation. Even with these filters, the abrupt change in the phase of the RF signal at the change of data state will also cause increased sidebands. We can realize from Figure 4.22c that changes of data states may cause the RF carrier to pass through zero when changing phase. Variations of carrier amplitude make the signal resemble amplitude modulation, which requires inefficient linear amplifiers, as compared to nonlinear amplifiers that can be used for frequency modulation, for example.

Two variations of quadrature phase shift keying have been devised to reduce phase changes between successive symbols and to prevent the carrier from going through zero amplitude during phase transitions. One of these is offset phase shift keying. In this method, the I and Q data streams in the transmitter are offset in time by one-half of the bit duration, causing the carrier phase to change more often but more gradually. In other words, the new "I" bit will modulate the cosine carrier at half a bit time earlier (or later) than the time that the "Q" bit modulates the sine carrier.

The other variant of QPSK is called pi/4 DPSK. It is used in the US TDMA (time division multiple access) digital cellular system. In it, the constellation diagram is rotated pi/4 radians (45 degrees) at every bit time such that the carrier phase angle can change by either 45 or 135 degrees. This system reduces variations of carrier amplitude so that more efficient nonlinear power amplifiers may be used in the transmitter.

Another problem with quadrature modulation as described above is the need for a coherent local oscillator in the receiver in order to separate the in-phase and quadrature data streams. As for bipolar phase shift keying, this problem may be ameliorated by using differential modulation and by multiplying a delayed replica of the received signal by itself to extract the phase differences from symbol to symbol.

The principle of transmitting separate data streams on in phase and quadrature RF carriers may be extended so that each symbol on each carrier contains 2, 3, 4 or more bits. For example, 4 bits per each carrier vector symbol allows up to 16 amplitude levels per carrier and a total of 256 different states of amplitude and phase altogether. This type of modulation is called quadrature amplitude modulation (QAM), and in this example, 256-QAM, 8 data bits can be transmitted in essentially the same time and bandwidth as one bit sent by binary phase shift keying. The constellation of 16-QAM (4 bits per symbol) is shown in Figure 4.26.

Remember that the process of concentrating more and more bits to a carrier symbol cannot go on without limit. While the bit rate per hertz goes up, the required S/N of the channel for a given bit error rate (BER) also increases—that is, more power must be transmitted, or as a common alternative in modern digital communication systems, sophisticated coding algorithms are incorporated in the data protocol. However, the ultimate limit of data rate for a particular communication channel with noise is the Hartley-Shannon limit stated above.

We now turn to examine another form of digital modulation that is becoming very important in a growing number of short-range wireless applications—spread-spectrum modulation.

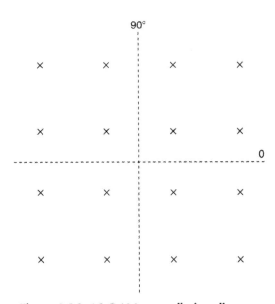

Figure 4.26: 16-QAM constellation diagram

4.10.4.5 Spread Spectrum

The regulations for unlicensed communication using unspecified modulation schemes, both in the US and in Europe, determine maximum power outputs ranging from tens of microwatts up to 10 milliwatts in most countries. This limitation greatly reduces the possibilities for wireless devices to replace wires and to obtain equivalent communication reliability. However, the availability of frequency bands where up to one watt may be transmitted in the US and 100 mW in Europe, under the condition of using a specified modulation system, greatly enlarges the possible scope of use and reliability of unlicensed short-range communication.

Spread spectrum has allowed the telecommunication authorities to permit higher transmitter powers because spread-spectrum signals can coexist on the same frequency bands as other types of authorized transmissions without causing undue interference or being unreasonably interfered with. The reason for this is evident from Figure 4.27. Figure 4.27a shows the spread-spectrum signal spread out over a bandwidth much larger

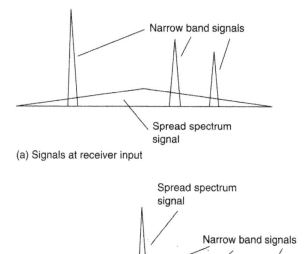

(a) Signals at receiver input

(b) Signals after despreading

Figure 4.27: Spread spectrum and narrow-band signals

than the narrow-band signals. Although its total power if transmitted as a narrow band signal could completely overwhelm another narrow-band signal on the same or adjacent frequency, the part of it occupying the narrow-band signal bandwidth is small related to the total, so it doesn't interfere with it. In other words, spreading the transmitted power over a wide frequency band greatly reduces the signal power in a narrow bandwidth and thus the potential for interference.

Figure 4.27b shows how spread-spectrum processing reduces interference from adjacent signals. The de-spreading process concentrates the total power of the spread-spectrum signal into a narrow-band high peak power signal, whereas the potentially interfering narrow-band signals are spread out so that their power in the bandwidth of the desired signal is relatively low.

These are some advantages of spread spectrum modulation:

- FCC rules allow higher power for nonlicensed devices
- Reduces co-channel interference—good for congested ISM bands
- Reduces multipath interference
- Resists intentional and unintentional jamming
- Reduces the potential for eavesdropping
- Permits code division multiplexing of multiple users on a common channel.

There is sometimes a tendency to compare spread spectrum with wide-band frequency modulation, such as used in broadcasting, since both processes achieve performance advantages by occupying a channel bandwidth much larger than the bandwidth of the information being transmitted. However, it's important to understand that there are principal differences in the two systems.

In wide band FM (WBFM), the bandwidth is spread out directly by the amplitude of the modulating signal and at a rate determined by its frequency content. The result achieved is a signal-to-noise ratio that is higher than that obtainable by sending the same signal over baseband (without modulation) and with the same noise density as on the RF channel. In WBFM, the post detection signal-to-noise ratio (S/N) is a multiple of the S/N at the input to the receiver, but that input S/N must be higher than a threshold value, which depends on the deviation factor of the modulation.

In contrast, spread spectrum has no advantage over baseband transmission from the point of view of signal-to-noise ratio. Usual comparisons of modulation methods are based on a channel having only additive wideband Gaussian noise. With such a basis for comparison, there would be no advantage at all in using spread spectrum compared to sending the same data over a narrow-band link. The advantages of spread spectrum are related to its relative immunity to interfering signals, to the difficulty of message interception by a chance eavesdropper, and to its ability to use code selective signal differentiation. Another often-stated advantage to spread spectrum—reduction of multipath interference—is not particularly relevant to short-range communication because the pulse widths involved are much longer than the delay times encountered indoors. (A spread-spectrum specialist company, Digital Wireless, claims a method of countering multipath interference over short distances.)

The basic difference between WBFM and spread spectrum is that the spreading process of the latter is completely independent of the baseband signal itself. The transmitted signal is spread by one (or more) of several different spreading methods, and then un-spread in a receiver that knows the spreading code of the transmitter.

The methods for spreading the bandwidth of the spread-spectrum transmission are frequency-hopping spread spectrum (FHSS), direct-sequence spread spectrum (DSSS), pulsed-frequency modulation or chirp modulation, and time-hopping spread spectrum. The last two types are not allowed in the FCC rules for unlicensed operation and after giving a brief definition of them we will not consider them further.

1. In frequency-hopping spread spectrum, the RF carrier frequency is changed relatively rapidly at a rate of the same order of magnitude as the bandwidth of the source information (analog or digital), but not dependent on it in any way. At least several tens of different frequencies are used, and they are changed according to a pseudo-random pattern known also at the receiver. The spectrum bandwidth is roughly the number of the different carrier frequencies times the bandwidth occupied by the modulation information on one hopping frequency.

2. The direct-sequence spread-spectrum signal is modulated by a pseudo-random digital code sequence known to the receiver. The bit rate of this code is much higher than the bit rate of the information data, so the bandwidth of the RF signal is consequently higher than the bandwidth of the data.

3. In chirp modulation, the transmitted frequency is swept for a given duration from one value to another. The receiver knows the starting frequency and duration so it can unspread the signal.

4. A time-hopping spread-spectrum transmitter sends low duty cycle pulses with pseudo-random intervals between them. The receiver unspreads the signal by gating its reception path according to the same random code as used in the transmitter.

Actually, all of the above methods, and their combinations that are sometimes employed, are similar in that a pseudo-random or arbitrary (in the case of chirp) modulation process used in the transmitter is duplicated in reverse in the receiver to unravel the wide-band transmission and bring it to a form where the desired signal can be demodulated like any narrowband transmission.

The performance of all types of spread-spectrum signals is strongly related to a property called process gain. It is this process gain that quantifies the degree of selection of the desired signal over interfering narrow-band and other wide-band signals in the same passband. Process gain is the difference in dB between the output S/N after unspreading and the input S/N to the receiver:

$$PG_{dB} = (S/N)_{out} - (S/N)_{in}$$

The process gain factor may be approximated by the ratio

$$PG_f = (\text{RF bandwidth})/(\text{rate of information})$$

A possibly more useful indication of the effectiveness of a spread-spectrum system is the jamming margin:

$$\text{Jamming Margin} = PG - (L_{sys} + (S/N)_{out})$$

where L_{sys} is system implementation losses, which may be of the order of 2 dB.

The jamming margin is the amount by which a potentially interfering signal in the receiver's passband may be stronger than the desired signal without impairing the desired signal's ability to get through.

Let's now look at the details of frequency hopping and direct-sequence spread spectrum.

Frequency Hopping

FHSS can be divided into two classes—fast hopping and slow hopping. A fast-hopping transmission changes frequency one or more times per data bit. In slow hopping, several bits are sent per hopping frequency. Slow hopping is used for fast data rates since the frequency synthesizers in the transmitter and receiver are not able to switch and settle to new frequencies fast enough to keep up with the data rate if one or fewer bits per hop are transmitted. The spectrum of a frequency-hopping signal looks like Figure 4.28.

Figure 4.29 is a block diagram of FHSS transmitter and receiver. Both transmitter and receiver local oscillator frequencies are controlled by frequency synthesizers. The receiver must detect the beginning of a transmission and synchronize its synthesizer to that of the transmitter. When the receiver knows the pseudo-random pattern of the transmitter, it can lock onto the incoming signal and must then remain in synchronization by changing frequencies at the same time as the transmitter. Once exact synchronization has been obtained, the IF frequency will be constant and the signal can be demodulated just as in a normal narrow-band superheterodyne receiver. If one or more of the frequencies that the transmission occupies momentarily is also occupied by an interfering signal, the bit or bits that were transmitted at that time may be lost. Thus, the transmitted message must contain redundancy or error correction coding so that the lost bits can be reconstructed.

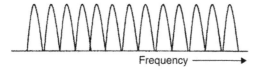

Figure 4.28: Spectrum of frequency-hopping spread spectrum signal

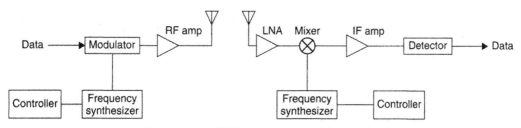

Figure 4.29: FHSS transmitter and receiver

Direct Sequence

Figure 4.30 is a diagram of a DSSS system. A pseudo-random spreading code modulates the transmitter carrier frequency, which is then modulated by the data. The elements of this code are called chips. The frequency spectrum created is shown in Figure 4.31. The required bandwidth is the width of the major lobe, shown on the drawing as $2 \times Rc$, or twice the chip rate. Due to the wide bandwidth, the signal-to-noise ratio at the receiver input is very low, often below 0 dB. The receiver multiplies a replica of the transmitter pseudo-random spreading code with the receiver local oscillator and the result is mixed with the incoming signal. When the transmitter and receiver spreading codes are the same and are in phase, a narrow-band IF signal results that can be demodulated in a conventional fashion.

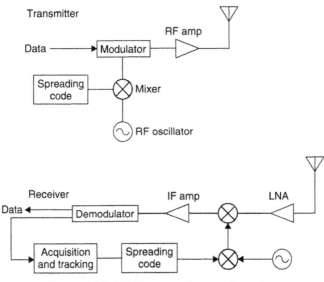

Figure 4.30: DSSS transmitter and receiver

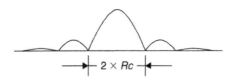

Figure 4.31: DSSS frequency spectrum

There are two stages to synchronizing transmitter and receiver pseudorandom codes—acquisition and tracking. In order to acquire a signal, the receiver multiplies the known expected code with the incoming RF signal. The output of this multiplication will be random noise when signals other than the desired signal exist on the channel. When the desired signal is present, it must be synchronized precisely in phase with the receiver's code sequence to achieve a strong output. A common way to obtain this synchronization is for the receiver to adjust its code to a rate slightly different from that of the transmitter. Then the phase difference between the two codes will change steadily until they are within one bit of each other, at which time the output of the multiplier increases to a peak when the codes are perfectly synchronized. This method of acquisition is called a sliding correlator.

The next stage in synchronization, tracking, keeps the transmitted and received code bits aligned for the complete duration of the message. In the method called "tau delta" the receiver code rate is varied slightly around the transmitted rate to produce an error signal that is used to create an average rate exactly equal to the transmitter's.

Other methods for code acquisition and tracking can be found in any book on spread spectrum—for example, Dixon's—listed in the references. Once synchronization has been achieved, the resulting narrow-band IF signal has a S/N equal to the received S/N plus the process gain, and it can be demodulated as any narrow-band signal. The difference in using spread spectrum is that interfering signals that would render the normal narrowband signal unintelligible are now reduced by the amount of the process gain.

Relative Advantages of DSSS and FHSS

The advantages of one method over the other are often debated, but there is no universally agreed-upon conclusion as to which is better, and both types are commonly used. They both are referenced in the IEEE specification 802.11 for wireless LANs. However some conclusions, possibly debatable, can be made:

- For a given data rate and output power, an FHSS system can be designed with less power consumption than DSSS.

- DSSS can provide higher data rates with less redundancy.

- Acquisition time is generally lower with FHSS (although short times are achieved with DSSS using matched filter and short code sequence).

- FHSS may have higher interference immunity, particularly when compared with DSSS with lowest allowed process gain.

- DSSS has more flexibility—frequency agility may be used to increase interference immunity.

4.10.5 RFID

A growing class of applications for short-range wireless is radio frequency identification (RFID). A basic difference between RFID and the applications discussed above is that RFID devices are not communication devices *per se* but involve interrogated transponders. Instead of having two separate transmitter and receiver terminals, an RFID system consists of a reader that sends a signal to a passive or active tag and then receives and interprets a modified signal reflected or retransmitted back to it.

The reader has a conventional transmitter and receiver having similar characteristics to the devices already discussed. The tag itself is special for this class of applications. It may be passive, receiving its power for retransmission from the signal sent from the reader, or it may have an active receiver and transmitter and tiny embedded long-life battery. Ranges of RFID may be several centimeters up to tens of meter. Operation frequencies range from 125 kHz up to 2.45 GHz. Frequencies are usually those specified for nonlicensed applications. Higher data rates demand higher frequencies.

In its most common form of operation, an RFID system works as follows. The reader transmits an interrogation signal, which is received by the tag. The tag may alter the incoming signal in some unique manner and reflect it back, or its transmitting circuit may be triggered to read its ID code residing in memory and to transmit this code back to the receiver. Active tags have greater range than passive tags. If the tag is moving in relation to the receiver, such as in the case of toll collection on a highway, the data transfer must take place fast enough to be complete before the tag is out of range.

These are some design issues for RFID:

- Tag orientation is likely to be random, so tag and reader antennas must be designed to give the required range for any orientation.

- Multiple tags within the transmission range can cause return message collisions. One way to avoid this is by giving the tags random delay times for response, which may allow several tags to be interrogated at once.

- Tags must be elaborately coded to prevent misreading.

4.11 Summary

This chapter has looked at various factors that affect the range of reliable wireless communication. Propagation of electromagnetic waves is influenced by physical objects in and near the path of line-of-sight between transmitter and receiver. We can get a first rough approximation of communication distance by considering only the reflection of the transmitted signal from the earth. We described several techniques of diversity reception that can reduce the required power for a given reliability when the communication link is subject to fading.

Noise was presented as the ultimate limitation on communication range. We saw that noise sources to be contended with depend on the operating frequency. The importance of low-noise receiver design depends on the relative intensity of noise received by the antenna to the noise generated in the receiver.

In this chapter we also examined various characteristics of short-range systems. We looked at the ways in which data is formatted into different information fields for transmission over a wireless link. We looked at several methods of encoding the one's and zero's of the baseband information before modulation in order to meet certain performance requirements in the receiver, such as constant DC level and minimum bit error probability. Analog systems were also mentioned, and we saw that pre-emphasis/de-emphasis and compression/expansion circuits in voice communication devices improve the signal-to-noise ratio and increase the dynamic range.

Reasons for preferring frequency or amplitude digital modulation were presented from the points of view of equipment complexity and of the different regulatory requirements in the US and in Europe. Similarly, there are several considerations in choosing a frequency band for a wireless system, among them background noise, antenna size, and cost.

The three basic modulation types involve impressing the baseband data on the amplitude, frequency, or phase of an RF carrier signal. Modern digital communication uses combinations and variations in the basic methods to achieve high bandwidth efficiency, or conversely, high signal-to-noise ratio with relatively low power. We gave an introduction to quadrature modulation and to the principles of spread-spectrum communication. The importance of advanced modulation methods is on the rise, and they can surely be expected to have an increasing influence on short-range radio design in the near future.

References

Anthes, John, "OOK, ASK and FSK Modulation in the Presence of an Interfering Signal," Application Note, RF Monolithics, Dallas, Texas.

Dixon, Robert C., *Spread Spectrum Systems*, John Wiley & Sons, New York, 1984 [4].

Gibson, Jerry D., Editor-in-chief, *The Mobile Communications Handbook*, CRC Press, Inc., 1996.

Rappaport, Theodore S., *Wireless Communications, Principles and Practice*, Prentice Hall, Upper Saddle River, NJ, 1996.

Spix, George J., "Maxwell's Electromagnetic Field Equations," unpublished tutorial, copyright 1995 (http://www.connectos.com/spix/rd/gj/nme/maxwell.htm).

Vear, Tim, "Selection and Operation of Wireless Microphone Systems," Shure Brothers Inc., 1998.

CHAPTER 5

Infrared Communication Basics

Steve Rackley

5.1 The Ir Spectrum

The infrared (Ir) part of the electromagnetic spectrum covers radiation having a wavelength in the range from roughly 0.78 μm to 1000 μm (1 mm). Infrared radiation takes over from extremely high frequency (EHF) at 300 GHz and extends to just below the red end of the visible light spectrum at around 0.76 μm wavelength. Unlike radio frequency radiation, which is transmitted from an antenna when excited by an oscillating electrical signal, infrared radiation is generated by the rotational and vibrational oscillations of molecules.

The infrared spectrum is usually divided into three regions, near, middle and far, where "near" means nearest to visible light (Table 5.1). Although all infrared radiation is invisible to the human eye, far infrared is experienced as thermal, or heat, radiation. Rather than using frequency as an alternative to wavelength, as is commonly done in the RF region, the wavenumber is used instead in the infrared region. This is the reciprocal of the wavelength and is usually expressed as the number of wavelengths per centimeter.

One aspect of wireless communication that becomes simpler outside the RF region is spectrum regulation, since the remit of the FCC and equivalent international agencies runs out at 300 GHz or 1 mm wavelength.

Table 5.1: Subdivision of the infrared spectrum

Infrared region	Wavelength (μm)	Wavenumber (/cm)
Near	0.78–2.5	12,800–4000
Middle	2.5–50	4000–200
Far	2.5–50	200–10

5.2 Infrared Propagation and Reception

The near infrared is the region used in data communications, largely as a result of the cheap availability of infrared emitting LEDs and optodetectors, solid-state devices that convert an electrical current directly into infrared radiation and vice versa. Infrared LEDs emit at discrete wavelengths in the range from 0.78 to 1.0 μm, the specific wavelength of the LED being determined by the particular molecular oscillation that is used to generate the radiation.

5.2.1 Transmitted Power Density – Radiant Intensity

Ir propagation is generally a simpler topic than RF propagation, although the same principles, such as the concept of a *link budget*, still apply. The link budget predicts how much transmitter power is required to enable the received data stream to be decoded at an acceptable BER. For Ir, the link budget calculation is far simpler than for RF, as terms like antenna gain, free space loss and multipath fading, discussed in the last chapter, no longer apply. As a result, propagation behavior can be more easily predicted for Ir than for RF.

The unit of infrared power intensity, or radiant intensity, is mW/sr, with sr being the abbreviation for steradian. The steradian is the unit of solid angle measure, and this is the key concept in understanding the link budget for Ir communication. As shown in Figure 5.1, the solid angle (S) subtended by an area A on the surface of a sphere of radius R is given as:

$$S = A/R^2 \text{ steradians} \qquad (5\text{-}1)$$

$$A = 2\pi R^2 (1 - \cos(a))$$

So

$$S = 2\pi(1 - \cos(a)) \qquad (5\text{-}2)$$

Note from Eq. 5-1 that at a distance of 1 meter a solid angle of 1 steradian subtends an area of 1 m². For small solid angles, the area A on the sphere can be approximated by the area of the flat circle of radius r, giving:

$$S = \pi r^2 / R^2 \text{ steradians}$$

As an example, the IrDA physical layer standard specifies a half angle (a) of between 15° and 30°. For 15°, $S = 2\pi(1 - \cos(15°)) = 0.214$ steradians.

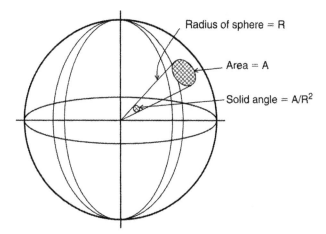

Figure 5.1: Solid angle subtended by an area

For an LED with a given emitter power density or radiant intensity, I_e, in mW/sr, the equivalent power density in mW/m^2 will be approximately given as:

$$P = I_e/R^2 \text{ mW/m}^2 \tag{5-3}$$

5.2.2 Emitter Beam Pattern

Similar to an RF antenna, an LED has a beam pattern in which radiated power drops off with increasing angle off-axis. In the example shown in Figure 5.2, the power density drops to about 85% of the on-axis value at an off-axis angle of 15°.

5.2.3 Inverse Square Loss

Equation 5-3 shows that the on-axis power density is inversely proportional to the square of the distance from the source. If R is doubled, the power density P drops by a factor of 4, as shown in Figure 5.3. This is the equivalent of the free space loss term in the RF link budget.

5.2.4 Ir Detector Sensitivity

The standard detection device for high-speed Ir communications is the photodiode, which has a detection sensitivity, or minimum threshold irradiance E_e, expressed in μW/cm^2. In standard power mode, the IrDA standard specifies a minimum emitter power of 40 mW/sr.

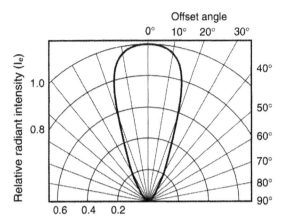

Figure 5.2: Typical LED emitted power polar diagram

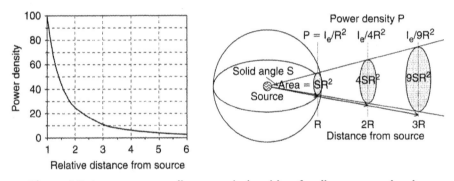

Figure 5.3: Inverse square distance relationship of radiant power density

From Eq. 5-3, the minimum power density at a receiving photodiode at a range of 1 meter will be $40\,\text{mW/m}^2$, or $4\,\mu\text{W/cm}^2$.

The sensitivity of a photo diode detector depends on the incident angle of the infrared source relative to the detector axis in a similar manner to that shown in Figure 5.2 for the beam pattern of an LED. Photodiode sensitivity also depends on the incident infrared wavelength as shown for example in Figure 5.4, and in any application a detector will be chosen with a peak spectral sensitivity close to the wavelength of the emitting device.

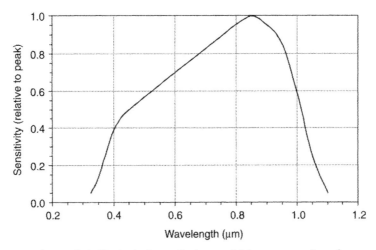

Figure 5.4: Typical photodiode sensitivity vs. wavelength

5.2.5 Ir Link Distance

The maximum link distance for an Ir link can be calculated as the distance R at which the equivalent power density (P) drops to the level of the detector's minimum threshold irradiance (E_e). Eq. 5-3 gives:

$$E_e = I_e / R^2 \text{ mW/m}^2$$

or

$$R = \left(I_e / E_e\right)^{1/2} \text{ m} \qquad\qquad (5\text{-}4)$$

The effective range of an Ir link can be increased substantially, up to several tens of meters, using lenses to collimate the transmitted beam and focus the beam onto the receiving photodiode. Alignment of the lenses and of the transmitting and receiving diodes will be critical to the effectiveness of such a system. As shown in Figure 5.5, a misalignment of approximately 1/3° would be sufficient to break a focused link over a range of 10 m.

Ir areal coverage may be increased in a home or small office environment by reflecting the Ir beam from a wall or ceiling in order to access a number of devices.

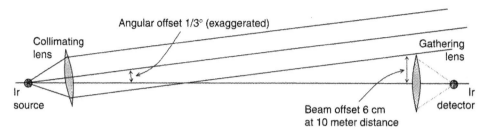

Figure 5.5: Focused Ir link alignment over a 10 m range

In order to preserve power in the reflection it will be important to use a high reflectance, low absorption material to reflect the beam. A white painted ceiling is a good reflector of sunlight, with a reflection coefficient for visible light of around 0.9, but is a poor reflector at Ir wavelengths, absorbing about 90% of the incident Ir radiation. To achieve a comparable 0.9 reflection coefficient for Ir, an aluminum or aluminum foil covered panel would be a suitable reflector.

5.3 Summary

Like the radio frequency technology discussed in the last chapter, the infrared communication technologies described in this chapter are at the heart of the physical layer of wireless networks. An understanding of the basics of these two technologies will provide a firm foundation from which to discuss the implementation of wireless networks, whether local, personal or metropolitan area (LAN, PAN or MAN).

In particular, the link budget calculation will be important in establishing power requirements and coverage in LAN and MAN applications. Spread spectrum and digital modulation techniques are key to understanding how a wide range of data rates can be accommodated within the limited available bandwidth, for example in the 2.4 and 5.8 GHz ISM bands.

Infrared communication links are perhaps the most "transparent" in terms of a very low requirement for user configuration, and to some extent this means that the user can be unconcerned about the underlying technology. However, even infrared links can be stretched to deliver performance over significant distances (tens of meters), given an understanding of the characteristics of Ir transmitters, detectors and infrared propagation.

CHAPTER 6

Wireless LAN Standards

Steve Rackley

6.1 The 802.11 WLAN Standards

6.1.1 Origins and Evolution

The development of wireless LAN standards by the IEEE began in the late 1980s, following the opening up of the three ISM radio bands for unlicensed use by the FCC in 1985, and reached a major milestone in 1997 with the approval and publication of the 802.11 standard. This standard, which initially specified modest data rates of 1 and 2 Mbps, has been enhanced over the years, the many revisions being denoted by the addition of a suffix letter to the original 802.11, as for example in 802.11a, b and g.

The 802.11a and 802.11b extensions were ratified in July 1999, and 802.11b, offering data rates up to 11 Mbps, became the first standard with products to market under the Wi-Fi banner. The 802.11g specification was ratified in June 2003 and raised the PHY layer data rate to 54 Mbps, while offering a degree of interoperability with 802.11b equipment with which it shares the 2.4 GHz ISM band.

Table 6.1 summarizes the 802.11 standard's relentless march through the alphabet, with various revisions addressing issues such as security, local regulatory compliance and mesh networking, as well as other enhancements that will lift the PHY layer data rate to 600 Mbps.

6.1.2 Overview of the Main Characteristics of 802.11 WLANs

The 802.11 standards cover the PHY and MAC layer definition for local area wireless networking. As shown in Figure 6.1, the upper part of the Data Link layer (OSI Layer 2) is provided by Logical Link Control (LLC) services specified in the 802.2 standard, which are also used by Ethernet (802.3) networks, and provide the link to the Network layer and higher layer protocols.

Table 6.1: The IEEE 802.11 standard suite

Standard	Key features
802.11a	High speed WLAN standard, supporting 54 Mbps data rate using OFDM modulation in the 5 GHz ISM band.
802.11b	The original Wi-Fi standard, providing 11 Mbps using DSSS and CCK on the 2.4 GHz ISM band.
802.11d	Enables MAC level configuration of allowed frequencies, power levels and signal bandwidth to comply with local RF regulations, thereby facilitating international roaming.
802.11e	Addresses quality of service (QoS) requirements for all 802.11 radio interfaces, providing TDMA to prioritize and error-correction to enhance performance of delay sensitive applications.
802.11f	Defines recommended practices and an Inter-Access Point Protocol to enable access points to exchange the information required to support distribution system services. Ensures inter-operability of access points from multiple vendors, for example to support roaming.
802.11g	Enhances data rate to 54 Mbps using OFDM modulation on the 2.4 GHz ISM band. Interoperable in the same network with 802.11b equipment.
802.11h	Spectrum management in the 5 GHz band, using dynamic frequency selection (DFS) and transmit power control (TPC) to meet European requirements to minimise interference with military radar and satellite communications.
802.11i	Addresses the security weaknesses in user authentication and encryption protocols. The standard employs advanced encryption standard (AES) and 802.1x authentication.
802.11j	Japanese regulatory extension to 802.11a adding RF channels between 4.9 and 5.0 GHz.
802.11k	Specifies network performance optimization through channel selection, roaming and TPC. Overall network throughput is maximized by efficiently loading all access points in a network, including those with weaker signal strength.
802.11n	Provides higher data rates of 150, 350 and up to 600 Mbps using MIMO radio technology, wider RF channels and protocol stack improvements, while maintaining backward compatibility with 802.11 a, b and g.
802.11p	Wireless access for the vehicular environment (WAVE), providing communication between vehicles or from a vehicle to a roadside access point using the licensed intelligent transportation systems (ITS) band at 5.9 GHz.
802.11r	Enables fast BSS to BSS (Basic Service Set) transitions for mobile devices, to support delay sensitive services such as VoIP on stations roaming between access points.
802.11s	Extending 802.11 MAC to support ESS (Extended Service Set) mesh networking. The 802.11s protocol will enable message delivery over self-configuring multi-hop mesh topologies.

Table 6.1: (Continued)

Standard	Key features
802.11T	Recommended practices on measurement methods, performance metrics and test procedures to assess the performance of 802.11 equipment and networks. The capital T denotes a recommended practice rather than a technical standard.
802.11u	Amendments to both PHY and MAC layers to provide a generic and standardized approach to inter-working with non-802.11 networks, such as Bluetooth, ZigBee and WiMAX.
802.11v	Enhancements to increase throughput, reduce interference and improve reliability through network management.
802.11w	Increased network security by extending 802.11 protection to management as well as data frames.

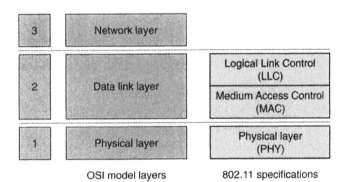

Figure 6.1: 802.11 logical architecture

802.11 networks are composed of three basic components: stations, access points and a distribution system, as described in Table 6.2.

In the 802.11 standard, WLANs are based on a cellular structure where each cell, under the control of an access point, is known as a basic service set (BSS). When a number of stations are working in a BSS it means that they all transmit and receive on the same RF channel, use a common BSSID, use the same set of data rates and are all synchronized to a common timer. These BSS parameters are included in "beacon frames" that are broadcast at regular intervals either by individual stations or by the access point.

Table 6.2: 802.11 network components

Component	Description
Station	Any device that implements the 802.11 MAC and PHY layer protocols.
Access point	A station that provides an addressable interface between a set of stations, known as a basic service set (BSS), and the distribution system.
Distribution system	A network component, commonly a wired Ethernet, that connects access points and their associated BSSs to form an extended service set (ESS).

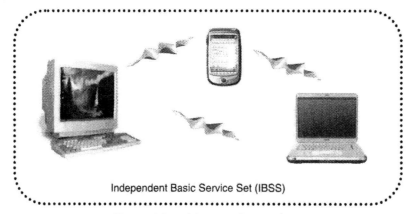

Independent Basic Service Set (IBSS)

Figure 6.2: Ad-hoc mode topology

The standard defines two modes of operation for a BSS: *ad-hoc mode* and *infrastructure mode*. An ad-hoc network is formed when a group of two or more 802.11 stations communicate directly with each other with no access point or connection to a wired network.

This operating mode (also known as *peer-to-peer* mode) allows wireless connections to be quickly established for data sharing among a group of wireless enabled computers (Figure 6.2). Under ad-hoc mode the service set is called an independent basic service set (IBSS), and in an IBSS all stations broadcast beacon packets, and use a randomly generated BSSID.

Infrastructure mode exists when stations are communicating with an access point rather than directly with each other. A home WLAN with an access point and several wired devices connected through an Ethernet hub or switch is a simple example of a

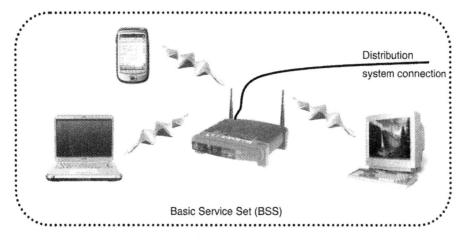

Figure 6.3: Infrastructure mode topology

BSS in infrastructure mode (Figure 6.3). All communication between stations in a BSS goes through the access point, even if two wireless stations in the same cell need to communicate with each other.

This doubling-up of communication within a cell (first from sending station to the access point, then from the access point to the destination station) might seem like an unnecessary overhead for a simple network, but among the benefits of using a BSS rather than an IBSS is that the access point can buffer data if the receiving station is in standby mode, temporarily out of range or switched off. In infrastructure mode, the access point takes on the role of broadcasting beacon frames.

The access point will also be connected to a distribution system which will usually be a wired network, but could also be a wireless bridge to other WLAN cells. In this case the cell supported by each access point is a BSS and if two or more such cells exist on a LAN the combined set is known as an extended service set (ESS).

In an ESS, access points (APs) will use the distribution system to transfer data from one BSS to another, and also to enable stations to move from one AP to another without any interruption in service. The transport and routing protocols that operate on the external network have no concept of mobility—of the route to a device changing rapidly—and within the 802.11 architecture the ESS provides this mobility to stations while keeping it invisible to the outside network.

Prior to 802.11k, support for mobility within 802.11 networks was limited to movement of a station between BSSs within a single ESS, so-called BSS transitions. With 802.11k, which will be described further in Section 6.4.3, the roaming of stations between ESSs is supported. When a station is sensed as moving out of range, an access point is able to deliver a site report that identifies alternative access points the station can connect to for uninterrupted service.

6.2 The 802.11 MAC Layer

The MAC layer is implemented in every 802.11 station, and enables the station to establish a network or join a pre-existing network and to transmit data passed down by Logical Link Control (LLC). These functions are delivered using two classes of services, station services and distribution system services, which are implemented by the transmission of a variety of management, control and data frames between MAC layers in communicating stations.

Before these MAC services can be invoked, the MAC first needs to gain access to the wireless medium within a BSS, with potentially many other stations also competing for access to the medium. The mechanisms to efficiently share access within a BSS are described in the next section.

6.2.1 Wireless Media Access

Sharing media access among many transmitting stations in a wireless network is more complex to achieve than in a wired network. This is because a wireless network station is not able to detect a collision between its transmission and the transmission from another station, since a radio transceiver is unable both to transmit and to listen for other stations transmitting at the same time.

In a wired network a network interface is able to detect collisions by sensing the carrier, for example the Ethernet cable, during transmission and ceasing transmission if a collision is detected. This results in a medium access mechanism known as carrier sense multiple access/ collision detection (CSMA/CD).

The 802.11 standard defines a number of MAC layer coordination functions to co-ordinate media access among multiple stations. Media access can either be contention-based, as in the mandatory 802.11 distributed coordination function (DCF), when all stations

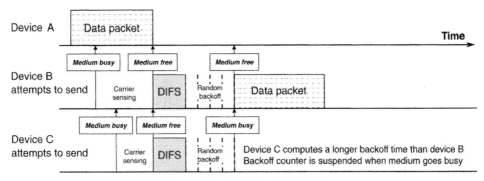

Figure 6.4: 802.11 CSMA/CA

essentially compete for access to the media, or contention free, as in the optional 802.11 point coordination function (PCF), when stations can be allocated specific periods during which they will have sole use of the media.

The media access method used by the distributed coordination function is carrier sense multiple access/collision avoidance (CSMA/CA), illustrated in Figure 6.4. In this mode a station that is waiting to transmit will sense the medium on the channel being used and wait until the medium is free of other transmissions. Once the medium is free, the station waits a predetermined period (the distributed inter-frame spacing or DIFS).

If the station senses no other transmission before the end of the DIFS period, it computes a random back-off time, between parameter values Cw_{min} and Cw_{max}, and commences its transmission if the medium remains free after this time has elapsed. The contention window parameter Cw is specified in terms of a multiple of a slot time that is 20μs for 802.11b or 9μs for 802.11a/g networks. The back-off time is randomized so that, if many stations are waiting, they will not all try again at the same time—one will have a shorter back-off and will succeed in starting its transmission. If a station has to make repeated attempts to transmit a packet, the computed back-off period is doubled with each new attempt, up to a maximum value Cw_{max} defined for each station. This ensures that, when many stations are competing for access, individual attempts are spaced out more widely to minimize repeated collisions.

If another station is sensed transmitting before the end of the DIFS period, this is because a short IFS (SIFS) can be used by a station that is waiting either to transmit certain control frames (CTS or ACK—see Figure 6.5) or to continue the transmission of parts of a data packet that has been fragmented to improve transmission reliability.

CSMA/CA is a simple media access protocol that works efficiently if there is no interference and if the data being transmitted across the network is not time critical. In the presence of interference, network throughput can be dramatically reduced as stations continually back-off to avoid collisions or wait for the medium to become idle.

CSMA/CA is a contention-based protocol, since all stations have to compete for access. With the exception of the SIFS mechanism noted above, no priorities are given and, as a result, no quality of service guarantees can be made.

The 802.11 standard also specifies an optional priority based media access mechanism, the point coordination function (PCF) which is able to provide contention free media access to stations with time critical requirements. This is achieved by allowing a station implementing PCF to use an interframe spacing (PIFS) intermediate between SIFS and DIFS, effectively giving these stations higher priority access to the medium. Once the point coordinator has control, it informs all stations of the length of the contention free period, to ensure that stations do not try to take control of the medium during this period. The coordinator then sequentially polls stations, giving any pollable station the opportunity to transmit a data frame.

Although it provides some limited capability for assuring quality of service, the PCF function has not been widely implemented in 802.11 hardware and it is only with the 802.11e enhancements, described below in the Section "Quality of Service (802.11e specification), p. 157", that quality of service (QoS) and prioritized access are more comprehensively incorporated into the 802.11 standard.

6.2.2 Discovering and Joining a Network

The first step for a newly activated station is to determine what other stations are within range and available for association. This can be achieved by either passive or active scanning.

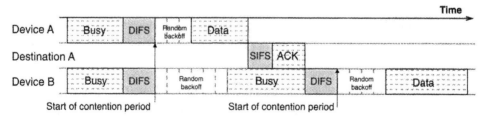

Figure 6.5: DCF transmission timing

In passive scanning the new station listens to each channel for a predetermined period and detects beacon frames transmitted by other stations. The beacon frame will provide a time synchronization mark and other PHY layer parameters, such as frequency hopping pattern, to allow the two stations to communicate.

If the new station has been set up with a preferred SSID name for association, it can use active scanning by transmitting a Probe frame containing this SSID and waiting for a Probe Response frame to be returned by the preferred access point. A broadcast Probe frame can also be sent, requesting all access points within range to respond with a Probe Response. This will provide the new station with a full list of access points available. The process of authentication and association can then start—either with the preferred access point or with another access point selected by the new station or by the user from the response list.

6.2.3 Station Services

MAC layer station services provide functions to send and receive data units passed down by the LLC and to implement authentication and security between stations, as described in Table 6.3.

Table 6.3: 802.11 MAC layer station services

Service	Description
Authentication	This service enables a receiving station to authenticate another station prior to association. An access point can be configured for either open system or shared key authentication. Open system authentication offers minimal security and does not validate the identity of other stations—any station that attempts to authenticate will receive authentication. Shared key authentication requires both stations to have received a secret key (e.g. a passphrase) via another secure channel such as direct user input.
Deauthentication	Prior to disassociation, a station will deauthenticate from the station that it intends to stop communication with. Both deauthentication and authentication are achieved by the exchange of management frames between the MAC layers of the two communicating stations.
Privacy	This service enables data frames and shared key authentication frames to be optionally encrypted before transmission, for example using wired equivalent privacy (WEP) or Wi-Fi protected access (WPA).

Continued

Table 6.3: (Continued)

Service	Description
MAC service data unit delivery	A MAC service data unit (MSDU) is a unit of data passed to the MAC layer by the logical link controller. The point at which the LLC accesses MAC services (at the "top" of the MAC layer) is termed the MAC service access point or SAP. This service ensures the delivery of MSDUs between these service access points. Control frames such as RTS, CTS and ACK may be used to control the flow of frames between stations, for example in 802.11b/g mixed-mode operation.

Table 6.4: 802.11 MAC layer distribution system services

Service	Description
Association	This service enables a logical connection to be made between a station and an access point. An access point cannot receive or deliver any data until a station has associated, since association provides the distribution system with the information necessary for delivery of data.
Disassociation	A station disassociates before leaving a network, for example when a wireless link is disabled, the network interface controller is manually disconnected or its host PC is powered down.
Reassociation	The reassociation service allows a station to change the attributes (such as supported data rates) of an existing association or to change its association from one BSS to another within an extended BSS. For example, a roaming station may change its association when it senses another access point transmitting a stronger beacon frame.
Distribution	The distribution service is used by a station to send frames to another station within the same BSS, or across the distribution system to a station in another BSS.
Integration	Integration is an extension of distribution when the access point is a portal to a non-802.11 network and the MSDU has to be transmitted across this network to its destination. The integration service provides the necessary address and media specific translation so that an 802.11 MSDU can be transmitted across the new medium and successfully received by the destination device's non-802.11 MAC.

6.2.4 Distribution System Services

The functionality provided by MAC distribution system services is distinct from station services in that these services extend across the distribution system rather than just between sending and receiving stations at either end of the air interface. The 802.11 distribution system services are described in Table 6.4.

6.3 802.11 PHY Layer

The initial 802.11 standard, as ratified in 1997, supported three alternative PHY layers; frequency hopping and direct sequence spread spectrum in the 2.4 GHz band as well as an infrared PHY. All three PHYs delivered data rates of 1 and 2 Mbps.

The infrared PHY specified a wavelength in the 800–900 nm range and used a diffuse mode of propagation rather than direct alignment of infrared transceivers, as is the case in IrDA for example (Section 10.5). A connection between stations would be made via passive ceiling reflection of the infrared beam, giving a range of 10–20 meters, depending on the height of the ceiling. Pulse position modulation was specified, 16-PPM and 4-PPM, respectively, for the 1 and 2 Mbps data rates.

Later extensions to the standard have focused on high rate DSSS (802.11b), OFDM (802.11a and g) and OFDM plus MIMO (802.11n). These PHY layers will be described in the following sections.

6.3.1 802.11a PHY Layer

The 802.11a amendment to the original 802.11 standard was ratified in 1999 and the first 802.11a compliant chipsets were introduced by Atheros in 2001. The 802.11a standard specifies a PHY layer based on orthogonal frequency division multiplexing (OFDM) in the 5 GHz frequency range. In the US, 802.11a OFDM uses the three unlicensed national information infrastructure bands (U-NII), with each band accommodating four non-overlapping channels, each of 20-MHz bandwidth. Maximum transmit power levels are specified by the FCC for each of these bands and, in view of the higher permitted power level, the four upper band channels are reserved for outdoor applications.

In Europe, in addition to the 8 channels between 5.150 and 5.350 GHz, 11 channels are available between 5.470 and 5.725 GHz (channels 100, 104, 108, 112, 116, 120, 124, 128, 132, 136, 140). European regulations on maximum power level and indoor versus outdoor use vary from country to country, but typically the 5.15–5.35 GHz band is reserved for indoor use with a maximum EIRP of 200 mW, while the 5.47–5.725 GHz band has an EIRP limit of 1 W and is reserved for outdoor use.

As part of the global spectrum harmonization drive following the 2003 ITU World Radio Communication Conference, the 5.470–5.725 GHz spectrum has also been available in the US since November 2003, subject to the implementation of the 802.11h spectrum management mechanisms described in Section 6.4.2.

Table 6.5: US FCC specified U-NII channels used in the 802.11a OFDM PHY

RF Band	Frequency Range (GHz)	Channel number	Centre frequency (GHz)	Maximum transmit power (mW)
U-NII lower band	5.150–5.250	36	5.180	50
		40	5.200	
		44	5.550	
		48	5.240	
U-NII middle band	5.250–5.350	52	5.260	250
		56	5.280	
		60	5.300	
		64	5.320	
U-NII upper band	5.725–5.825	149	5.745	1000
		153	5.765	
		157	5.785	
		162	5.805	

Each of the 20 MHz wide channels accommodates 52 OFDM subcarriers, with a separation of 312.5 kHz (=20 MHz/64) between centre frequencies. Four of the subcarriers are used as pilot tones, providing a reference to compensate for phase and frequency shifts, while the remaining 48 are used to carry data.

Four different modulation methods are specified, as shown in Table 6.6, which result in a range of PHY layer data rates from 6 Mbps up to 54 Mbps.

The coding rate indicates the error-correction overhead that is added to the input data stream and is equal to $m/(m + n)$ where n is the number of error correction bits applied to a data block of length m bits. For example, with a coding rate of 3/4 every 8 transmitted bits includes 6 bits of user data and 2 error correction bits.

The user data rate resulting from a given combination of modulation method and coding rate can be determined as follows, taking the 64-QAM, 3/4 coding rate line as an example. During one symbol period of 4 μs, which includes a guard interval of 800 ns between symbols, each carrier is encoded with a phase and amplitude represented by one point on the 64-QAM constellation. Since there are 64 such points, this encodes 6 code bits. The 48 subcarriers together therefore carry 6 × 48 = 288 code bits for each

Table 6.6: 802.11a OFDM modulation methods, coding and data rate

Modulation	Code bits per subcarrier	Code bits per OFDM symbol	Coding rate	Data bits per OFDM symbol	Data rate (Mbps)
BPSK	1	48	1/2	24	6
BPSK	1	48	3/4	36	9
QPSK	2	96	1/2	48	12
QPSK	2	96	3/4	72	18
16-QAM	4	192	1/2	96	24
16-QAM	4	192	3/4	144	36
64-QAM	6	288	2/3	192	48
64-QAM	6	288	3/4	216	54

symbol period. With a 3/4 coding rate, 216 of those code bits will be user data while the remaining 72 will be error correction bits. Transmitting 216 data bits every $4\,\mu s$ corresponds to a data rate of 216 data bits per OFDM symbol $\times$ 250 OFDM symbols per second $=$ 54 Mbps.

The 802.11a specifies 6, 12 and 24 Mbps data rates as mandatory, corresponding to 1/2 coding rate for BPSK, QPSK and 16-QAM modulation methods. The 802.11a MAC protocol allows stations to negotiate modulation parameters in order to achieve the maximum robust data rate.

Transmitting at 5 GHz gives 802.11a the advantage of less interference compared to 802.11b, operating in the more crowded 2.4 GHz ISM band, but the higher carrier frequency is not without disadvantages. It restricts 802.11a to near line-of-sight applications and, taken together with the lower penetration at 5 GHz, means that indoors more WLAN access points are likely to be required to cover a given operating area.

6.3.2 802.11b PHY Layer

The original 802.11 DSSS PHY used the 11-chip Barker spreading code together with DBPSK and DQPSK modulation methods to deliver PHY layer data rates of 1 and 2 Mbps respectively (Table 6.7).

The high rate DSSS PHY specified in 802.11b added complementary code keying (CCK), using 8-chip spreading codes.

Table 6.7: 802.11b DSSS modulation methods, coding and data rate

Modulation	Code length (Chips)	Code type	Symbol rate (Mbps)	Data bits per symbol	Data rate (Mbps)
BPSK	11	Barker	1	1	1
QPSK	11	Barker	1	2	2
DQPSK	8	CCK	1.375	4	5.5
DQPSK	8	CCK	1.375	8	11

The 802.11 standard supports dynamic rate shifting (DRS) or adaptive rate selection (ARS), allowing the data rate to be dynamically adjusted to compensate for interference or varying path losses. When interference is present, or if a station moves beyond the optimal range for reliable operation at the maximum data rate, access points will progressively fall back to lower rates until reliable communication is restored.

Conversely, if a station moves back within range for a higher rate, or if interference is reduced, the link will shift to a higher rate. Rate shifting is implemented in the PHY layer and is transparent to the upper layers of the protocol stack.

The 802.11 standard specifies the division of the 2.4 GHz ISM band into a number of overlapping 22 MHz channels. The FCC in the US and the ETSI in Europe have both authorized the use of spectrum from 2.400 to 2.4835 GHz, with 11 channels approved in the US and 13 in (most of) Europe. In Japan, channel 14 at 2.484 GHz is also authorized by the ARIB. Some countries in Europe have more restrictive channel allocations, notably France where only four channels (10 through 13) are approved. The available channels for 802.11b operation are summarized in Table 6.8.

The 802.11b standard also includes a second, optional modulation and coding method, packet binary convolutional coding (PBCC™–Texas Instruments), which offers improved performance at 5.5 and 11 Mbps by achieving an additional 3 dB processing gain. Rather than the 2 or 4 phase states or phase shifts used by BPSK/DQSK, PBCC uses 8-PSK (8 phase states) giving a higher chip per symbol rate. This can be translated into either a higher data rate for a given chipping code length, or a higher processing gain for a given data rate, by using a longer chipping code.

Table 6.8: International channel availability for 802.11b networks in the 2.4 GHz band

Channel number	Center frequency (GHz)	Geographical usage
1	2.412	US, Canada, Europe, Japan
2	2.417	US, Canada, Europe, Japan
3	2.422	US, Canada, Europe, Japan
4	2.427	US, Canada, Europe, Japan
5	2.432	US, Canada, Europe, Japan
6	2.437	US, Canada, Europe, Japan
7	2.442	US, Canada, Europe, Japan
8	2.447	US, Canada, Europe, Japan
9	2.452	US, Canada, Europe, Japan
10	2.457	US, Canada, Europe, Japan, France
11	2.462	US, Canada, Europe, Japan, France
12	2.467	Europe, Japan, France
13	2.472	Europe, Japan, France
14	2.484	Japan

6.3.3 802.11g PHY Layer

The 802.11g PHY layer was the third 802.11 standard to be approved by the IEEE standards board and was ratified in June 2003. Like 802.11b, 11g operates in the 2.4 GHz band, but increases the PHY layer data rate to 54 Mbps, as for 802.11a.

The 802.11g uses OFDM to add data rates from 12 Mbps to 54 Mbps, but is fully backward compatible with 802.11b, so that hardware supporting both standards can operate in the same 2.4 GHz WLAN. The OFDM modulation and coding scheme is identical to that applied in the 802.11a standard, with each 20 MHz channel in the 2.4 GHz band (as shown in Table 6.8) divided into 52 subcarriers, with 4 pilot tones and 48 data tones. Data rates from 6 to 54 Mbps are achieved using the same modulation methods and coding rates shown for 802.11a in Table 6.6.

Although 802.11b and 11g hardware can operate in the same WLAN, throughput is reduced when 802.11b stations are associated with an 11g network (so-called mixed-mode operation) because of a number of protection mechanisms to ensure interoperability, as described Table 6.9.

Table 6.9: 802.11b/g mixed-mode interoperability mechanisms

Mechanism	Description
RTS/CTS	Before transmitting, 11b stations request access to the medium by sending a request to send (RTS) message to the access point. Transmission can commence on receipt of the clear to send (CTS) response. This avoids collisions between 11b and 11g transmissions, but the additional RTS/CTS signaling adds a significant overhead that decreases network throughput.
CTS to self	The CTS to self option dispenses with the exchange of RTS/CTS messages and just relies on the 802.11b station to check that the channel is clear before transmitting. Although this does not provide the same degree of collision avoidance, it can increase throughput significantly when there are fewer stations competing for medium access.
Backoff time	802.11g back-off timing is based on the 802.11a specification (up to a maximum of $15 \times 9\,\mu s$ slots) but in mixed-mode an 802.11g network will adopt 802.11b backoff parameters (maximum $31 \times 20\,\mu s$ slots). The longer 802.11b backoff results in reduced network throughput.

Table 6.10: PHY and MAC SAP throughput comparison for 802.11a, b and g networks

Network standard and configuration	PHY data rate (Mbps)	Effective MAC SAP throughput (Mbps)	Effective Throughput versus 802.11b (%)
802.11b network	11	6	100
802.11g network with 802.11b stations (CTS/RTS)	54	8	133
802.11g network with 802.11b stations (CTS-to-self)	54	13	217
802.11g network with no 802.11b stations	54	22	367
802.11a	54	25	417

The impact of mixed mode operation on the throughput of an 802.11g network is shown in Table 6.10.

A number of hardware manufacturers have introduced proprietary extensions to the 802.11g specification to boost the data rate above 54 Mbps. An example is D-Link's proprietary "108G" which uses packet bursting and channel bonding to achieve a PHY layer data rate of 108 Mbps. Packet bursting, also known as frame bursting, bundles short data packets into fewer but larger packet to reduce the impact of gaps between transmitted packets.

Packet bursting as a data rate enhancement strategy runs counter to packet fragmentation as a strategy for improving transmission robustness, so packet bursting will only be effective when interference or high levels of contention between stations are absent.

Channel bonding is a method where multiple network interfaces in a single machine are used together to transmit a single data stream. In the 108 G example, two non-overlapping channels in the 2.4 GHz ISM band are used simultaneously to transmit data frames.

6.3.4 Data Rates at the PHY and MAC Layer

In considering the technical requirements for a WLAN implementation, it will be important to recognize the difference between the headline data rate of a wireless networking standard and the true effective data rate as seen by the higher OSI layers when passing data packets down to the MAC layer.

Each "raw" data packet passed to the MAC service access point (MAC SAP) will acquire a MAC header and a message integrity code and additional security related header information before being passed to the PHY layer for transmission. The headline data rate, for example 54 Mbps for 802.11a or 11g networks, measures the transmission rate of this extended data stream at the PHY layer.

The effective data rate is the rate at which the underlying user data is being transmitted if all the transmitted bits relating to headers, integrity checking and other overheads are ignored. For example, on average, every 6 bits of raw data passed to the MAC SAP of an 802.11b WLAN will gain an extra 5 bits of overhead before transmission, reducing a PHY layer peak data rate of 11 Mbps to an effective rate of 6 Mbps.

Table 6.10 shows the PHY and MAC SAP data rates for 802.11 WLANs. For 802.11g networks the MAC SAP data rate depends on the presence of 802.11b stations, as a result of the mixed-mode media access control mechanisms described in the previous section.

6.4 802.11 Enhancements

In the following sections some of the key enhancements to 802.11 network capabilities and performance will be described.

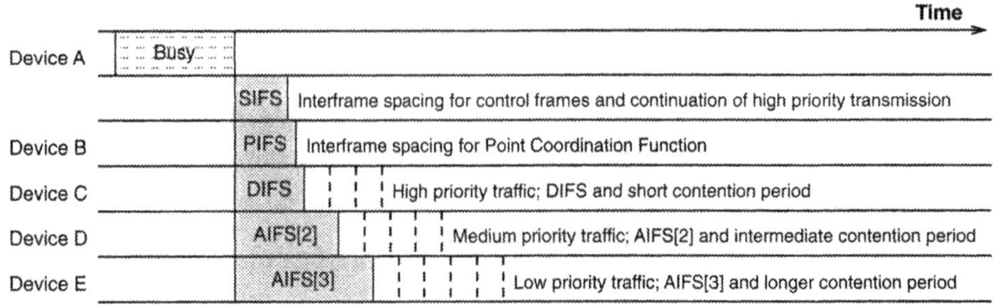

Figure 6.6: EDCF timing

6.4.1 Quality of Service (802.11e Specification)

The 802.11e specification provides a number of enhancements to the 802.11 MAC to improve the quality of service for time sensitive applications, such as streaming media and voice over wireless IP (VoWIP), and was approved for publication by the IEEE Standards Board in September 2005.

The 802.11e specification defines two new coordination functions for controlling and prioritizing media access, which enhance the original 802.11 DCF and PCF mechanisms described previously in Section 6.2.1. Up to eight traffic classes (TC) or access categories (AC) are defined, each of which can have specific QoS requirements and receive specific priority for media access.

The simplest of the 802.11e coordination functions is enhanced DCF (EDCF) which allows several MAC parameters determining ease of media access to be specified per traffic class. An arbitrary interframe space (AIFS) is defined which is equal to DIFS for the highest priority traffic class and longer for other classes. This provides a deterministic mechanism for traffic prioritization as shown in Figure 6.6.

The minimum back-off time Cw_{min} is also TC dependent, so that, when a collision occurs, higher priority traffic, with a lower Cw_{min}, will have a higher probability of accessing the medium.

Each station maintains a separate queue for each TC (Figure 6.7), and these behave as virtual stations, each with their individual MAC parameters. If two queues within a station reach the end of their back-off periods at the same time, data from the higher priority queue will be transmitted when the station gains access to the medium.

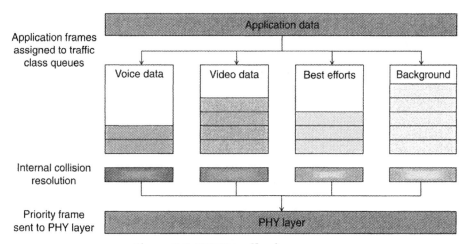

Figure 6.7: EDCF traffic class queues

Although the EDCF coordination mode does not provide a guaranteed service for any TC, it has the advantage of being simple to configure and implement as an extension of DCF.

The second enhancement defines a new hybrid coordination function (HCF) which complements the polling concept of PCF with an awareness of the QoS requirements of each station. Stations report the lengths of their queues for each traffic class and the hybrid coordinator uses this to determine which station will receive a transmit opportunity (TXOP) during a contention free transmission period. This HCF controlled channel access (HCCA) mechanism considers several factors in determining this allocation;

- The priority of the TC

- The QoS requirements of the TC (bandwidth, latency and jitter)

- Queue lengths per TC and per station

- The duration of the TXOP available to be allocated

- The past QoS given to the TC and station.

HCCA allows applications to schedule access according to their needs, and therefore enables QoS guarantees to be made. Scheduled access requires a client station to know its resource requirements in advance and scheduling concurrent traffic from multiple stations

Table 6.11: WMM access category descriptions

Access category	Description
WMM voice priority	Highest priority. Allows multiple concurrent VoWLAN calls, with low latency and quality equal to a toll voice call.
WMM video priority	Prioritizes video traffic above lower categories. One 802.11g or 802.11a channel can support 3 to 4 standard definition TV streams or 1 high definition TV stream.
WMM best effort priority	Traffic from legacy devices, from applications or devices that lack QoS capabilities, or traffic such as internet surfing that is less sensitive to latency but is affected by long delays.
WMM background priority	Low priority traffic, such as a file download or print job, that does not have strict latency or throughput requirements.

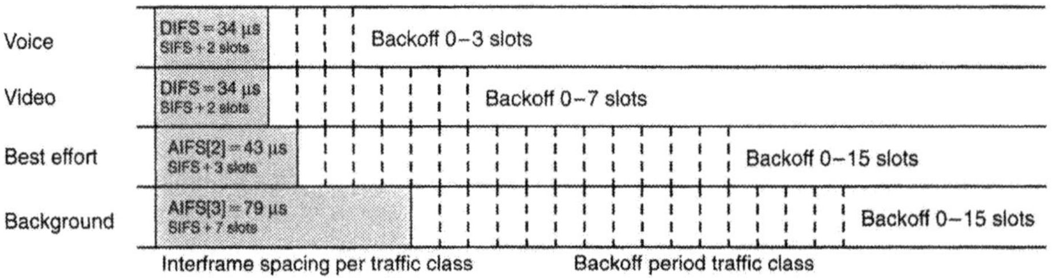

Figure 6.8: AIFS and back-off timing per WMM traffic class

also requires the access point to make certain assumptions regarding data packet sizes, data transmission rates and the need to reserve surplus bandwidth for transmission retries.

The Wi-Fi Alliance adopted a subset of the 802.11e standard in advance of the IEEE's September 2005 approval. This subset, called Wi-Fi multimedia (WMM), describes four access categories as shown in Table 6.11, with EDCF timings as shown in Figure 6.8.

The prioritization mechanism certified in WMM is equivalent to the EDCF coordination mode defined in 802.11e but did not initially include the scheduled access capability available through HCF and HCCA. This and other 802.11e capabilities are planned to be progressively included in the Wi-Fi Alliance's WMM certification program.

6.4.2 Spectrum Management at 5 GHz (802.11h)

The 802.11h standard supplements the 802.11 MAC with two additional spectrum management services, transmit power control (TPC), which limits the transmitted power to the lowest level needed to ensure adequate signal strength at the farthest station, and dynamic frequency selection (DFS), which enables a station to switch to a new channel if it detects other non-associated stations or systems transmitting on the same channel.

These mechanisms are required for 5 GHz WLANs operating under European regulations, in order to minimize interference with satellite communications (TPC) and military radar (DFS), and support for the 802.11h extensions was required from 2005 for all 802.11a compliant systems operating in Europe.

In the US, compliance with 802.11h is also required for 802.11a products operating in the 12 channels from 5.47 to 5.725 GHz. IEEE 802.11h compliant networks therefore have access to 24 non-overlapping OFDM channels, resulting in a potential doubling of overall network capacity.

6.4.2.1 Transmit Power Control

An 802.11h compliant station indicates its transmit power capability, including minimum and maximum transmit power levels in dBm, in the association or reassociation frame sent to an access point. An access point may refuse the association request if the station's transmit power capability is unacceptable, for example if it violates local constraints. The access point in return indicates local maximum transmit power constraints in its beacon and probe request frames.

An access point monitors signal strength within its BSS by requesting stations to report back the link margin for the frame containing the report request and the transmit power used to transmit the report frame back to the access point. This data is used by the access point to estimate the path loss to other stations and to dynamically adjust transmit power levels in its BSS in order to reduce interference with other devices while maintaining sufficient link margin for robust communication.

6.4.2.2 Dynamic Frequency Selection

When a station uses a Probe frame to identify access points in range, an access point will specify in the Probe Response frame that it uses DFS. When a station associates or re-associates with an access point that uses DFS, the station provides a list of supported channels that enables the access point to determine the best channel when a shift is required. As for TPC, an access point may reject an association request if a station's list of supported channels is considered unacceptable, for example if it is too limited.

To determine if other radio transmissions are present, either on the channel in use or on a potential new channel, an access point sends a measurement request to one or more stations identifying the channel where activity is to be measured, and the start time and duration of the measurement period. To enable these measurements, the access point can specify a quiet period in its beacon frames to ensure that all other associated stations stop transmission during the measurement period. After performing the requested measurement, stations send back a report on the measured channel activity to the access point.

When necessary, channel switching is initiated by the access point, which sends a channel switch announcement management frame to all associated stations. This announcement identifies the new channel, specifies the number of beacon periods until the channel switch takes effect, and also specifies whether or not further transmissions are allowed on the current channel. The access point can use the short interframe spacing (SIFS—see Section 6.2.1) to gain priority access to the wireless medium in order to broadcast a channel switch announcement.

Dynamic frequency selection is more complicated in an IBSS (ad-hoc mode) as there is no association process during which supported channel information can be exchanged, and no access point to coordinate channel measurement or switching. A separate DFS owner service is defined in 802.11h to address these complications, although channel switching remains inherently less robust in an IBSS than in an infrastructure mode BSS.

6.4.3 Network Performance and Roaming (802.11k and 802.11r)

A client station may need to make a transition between WLAN access points for one of three reasons, as described in Table 6.12.

The 802.11 Task Groups TGk and TGr are addressing issues relating to handoffs or transitions between access points that need to be fast and reliable for applications such as VoWLAN. TGk will standardize radio measurements and reports that will enable location-based services, such as a roaming station's choice of a new access point to connect to, while TGr aims to minimize the delay and maintain QoS during these transitions.

Table 6.12: Reasons for roaming in a WLAN

Roaming need	Description
Mobile client station	A mobile client station may move out of range of its current access point and need to transition to another access point with a higher signal strength.
Service availability	The QoS available at the current access point may either deteriorate or may be inadequate for a new service requirement, for example if a VoWLAN application is started.
Load balancing	An access point may redirect some associated clients to another available access point in order to maximize the use of available capacity within the network.

6.4.3.1 802.11k; Radio Resource Measurement Enhancements

The 802.11 Task Group TGk, subtitled Radio Resource Measurement Enhancements, began meeting in early 2003 with the objective of defining radio and network information gathering and reporting mechanisms to aid the management and maintenance of wireless LANs.

The 11k supplement will be compatible with the 802.11 MAC as well as implementing all mandatory parts of the 802.11 standards and specifications, and targets improved network management for all 802.11 networks. The key measurements and reports defined by the supplement are as follows;

- Beacon reports
- Channel reports
- Hidden station reports
- Client station statistics
- Site reports.

IEEE 802.11k will also extend the 802.11h TPC to cover other regulatory requirements and frequency bands.

Stations will be able to use these reports and statistics to make intelligent roaming decisions, for example eliminating a candidate access point if a high level of non-802.11 energy is detected in the channel being used. The 802.11k supplement only addresses the

Table 6.13: 802.11k measurements and reports

802.11k feature	Description
Beacon report	Access points will use a beacon request to ask a station to report all the access point beacons it detects on a specified channel. Details such as supported services, encryption types and received signal strength will be gathered.
Channel reports (noise histogram, medium sensing time histogram report and channel load report)	Access points can request stations to construct a noise histogram showing all non-802.11 energy detected on a specified channel, or to report data about channel loading (how long a channel was busy during a specified time interval as well as the histogram of channel busy and idle times).
Hidden station report	Under 802.11k, stations will maintain lists of hidden stations (stations that they can detect but are not detected by their access point). Access points can request a station to report this list and can use the information as input to roaming decisions.
Station statistic report and frame report	802.11k access points will be able to query stations to report statistics such as the link quality and network performance experienced by a station, the counts of packets transmitted and received, and transmission retries.
Site report	A station can request an access point to provide a site report—a ranked list of alternative access points based on an analysis of all the data and measurements available via the above reports.

measurement and reporting of this information and does not address the processes and decisions that will make use of the measurements.

The three roaming scenarios described above will be enabled by the TGk measurements and reports, summarized in Table 6.13.

For example, a mobile station experiencing a reduced RSSI will request a neighbor report from its current access point that will provide information on other access points in its vicinity. A smart roaming algorithm in the mobile station will then analyze channel conditions and the loading of candidate access points and select a new access point that is best able to provide the required QoS.

Once a new access point has been selected, the station will perform a BSS transition by disassociating from the current access point and associating with the new one, including authentication and establishing the required QoS.

6.4.3.2 802.11r; Fast BSS Transitions

The speed and security of transitions between access points will be further enhanced by the 802.11r specification which is also under development and is intended to improve WLAN support for mobile telephony via VoWLAN. IEEE 802.11r will give access points and stations the ability to make fast BSS to BSS transitions through a four-step process;

- Active or passive scanning for other access points in the vicinity,

- Authentication with one or more target access points,

- Reassociation to establish a connection with the target access point, and

- Pairwise temporal key (PTK) derivation and 802.1x based authentication via a 4-way handshake, leading to re-establishment of the connection with continuous QoS through the transition.

A key element of the process of associating with the new access point will be a pre-allocation of media reservations that will assure continuity of service—a station will not be in the position of having jumped to a new access point only to find it is unable to get the slot time required to maintain a time critical service.

The 802.11k and 802.11r enhancements address roaming within 802.11 networks, and are a step towards transparent roaming between different wireless networks such as 802.11, 3G and WiMAX. The IEEE 802.21 media independent handover (MIH) function will eventually enable mobile stations to roam across these diverse wireless networks.

6.4.4 MIMO and Data Rates to 600 Mbps (802.11n)

The IEEE 802.11 Task Group TGn started work during the second half of 2003 to respond to the demand for further increase in WLAN performance, and aims to deliver a minimum effective data rate of 100 Mbps through modifications to the 802.11 PHY and MAC layers.

This target data rate, at the MAC service entry point (MAC SAP), will require a PHY layer data rate in excess of 200 Mbps, representing a fourfold increase in throughput compared to 802.11a and 11g networks. Backward compatibility with 11a/b/g networks will ensure a smooth transition from legacy systems, without imposing excessive performance penalties on the high rate capable parts of a network.

Although there is still considerable debate among the supporters of alternative proposals, the main industry group working to accelerate the development of the 802.11n standard is the enhanced wireless consortium (EWC) which published Rev 1 of its MAC and PHY proposals in September 2005. The following description is based on the EWC proposals.

The two key technologies that are expected to be required to deliver the aspired 802.11n data rate are multi-input multi-output (MIMO) radio and OFDM with extended channel bandwidths.

MIMO radio, discussed in previous chapters, is able to resolve information transmitted over several signal paths using multiple spatially separated transmitter and receiver antennas. The use of multiple antennas provides an additional gain (the diversity gain) that increases the receiver's ability to decode the data stream.

The extension of channel bandwidths, most likely by the combination of two 20 MHz channels in either the 2.4 GHz or 5 GHz bands, will further increase capacity since the number of available OFDM data tones can be doubled.

To achieve a 100 Mbps effective data rate at the MAC SAP it is expected to require either a 2 transmitter $\times$ 2 receiver antenna system operating over a 40 MHz bandwidth or a 4 $\times$ 4 antenna system operating over 20 MHz, with respectively 2 or 4 spatially separated data streams being processed. In view of the significant increase in hardware and signal processing complexity in going from 2 to 4 data streams, the 40 MHz bandwidth solution is likely to be preferred where permitted by local spectrum regulations. To ensure backward compatibility, a PHY operating mode will be specified in which 802.11a/g OFDM is used in either the upper or lower 20 MHz of a 40 MHz channel.

Maximizing data throughput in 802.11n networks will require intelligent mechanisms to continuously adapt parameters such as channel bandwidth and selection, antenna configuration, modulation scheme and coding rate, to varying wireless channel conditions.

A total of 32 modulation and coding schemes are initially specified, in four groups of eight, depending on whether one to four spatial streams are used. Table 6.14 shows the modulation and coding schemes for the highest rate case—four spatial streams operating over 40 MHz bandwidth providing 108 OFDM data tones. For fewer spatial streams, the data rates are simply proportional to the number of streams.

Table 6.14: 802.11n OFDM modulation methods, coding and data rate

Modulation	Code bits per subcarrier (per stream)	Code bits per symbol (all streams)	Coding rate	Data bits per symbol (all streams)	Data rate (Mbps)
BPSK	1	432	1/2	216	54
QPSK	2	864	1/2	432	108
QPSK	2	864	3/4	648	162
16-QAM	4	1728	1/2	864	216
16-QAM	4	1728	3/4	1296	324
64-QAM	4	2592	2/3	1728	432
64-QAM	6	2592	3/4	1944	486
64-QAM	6	2592	5/6	2160	540

As for 802.11a/g, these data rates are achieved with a symbol period of 4.0 μs. A further data rate increase of 10/9 (e.g., from 540 to 600 Mbps) is achieved in an optional short guard interval mode, which reduces the symbol period to 3.6 μs by halving the inter-symbol guard interval from 800 ns to 400 ns.

MAC framing and acknowledgement overheads will also need to be reduced in order to increase MAC efficiency (defined as the effective data rate at the MAC SAP as a fraction of the PHY layer data rate). With the current MAC overhead, a PHY layer data rate approaching 500 Mbps would be required to deliver the target 100 Mbps data rate at the MAC SAP.

6.4.5 Mesh Networking (802.11s)

As described in Section 6.1, the 802.11 topology relies on a distribution system (DS) to link BSSs together to form an ESS. The DS is commonly a wired Ethernet linking access points (Figure 6.3), but the 802.11 standard also provides for a wireless distribution system between separated Ethernet segments by defining a four-address frame format that contains source and destination station addresses as well as the addresses of the two access points that these stations are connected to, as shown in Figure 6.9.

The objective of the 802.11s Task Group, which began working in 2004, is to extend the 802.11 MAC as the basis of a protocol to establish a wireless distribution system (WDS) that will operate over self-configuring multi-hop wireless topologies, in other words an ESS mesh.

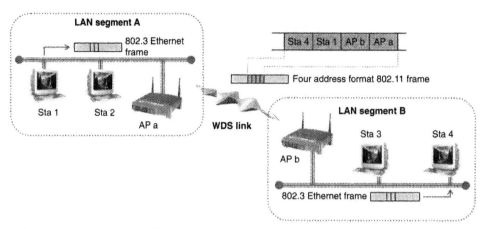

Figure 6.9: Wireless distribution system based on four-address format MAC frame

An ESS mesh is a collection of access points, connected by a WDS, that automatically learns about the changing topology and dynamically re-configures routing paths as stations and access points join, leave and move within the mesh. From the point of view of an individual station and its relationship with a BSS and ESS, an ESS mesh is functionally equivalent to a wired ESS.

Two industry alliances emerged during 2005 to promote alternative technical proposals for consideration by TGs; the Wi-Mesh Alliance and SEEMesh (for Simple, Efficient and Extensible Mesh).

The main elements of the Wi-Mesh proposal are a mesh coordination function (MCF) and a distributed reservation channel access protocol (DRCA) to operate alongside the HCCA and EDCA protocols (Figure 6.10). Some of the key features of the proposed Wi-Mesh MCF are summarized in Table 6.15.

The final ESS mesh specification is likely to include prioritized traffic handling based on 802.11e QoS mechanisms as well as security features and enhancements to the 802.11i standard.

The evolution of 802.11 security will be fully described in later chapters, but mesh networking introduces some security considerations in addition to those that have been progressively solved for non-mesh WLANs by WEP, WPA, WPA2 and 802.11i. In a mesh network additional security methods are needed to identify nodes that are authorized

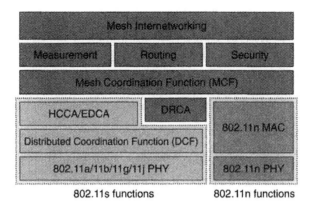

Figure 6.10: Wi-Mesh logical architecture

Table 6.15: Wi-Mesh mesh coordination function (MCF) features

Wi-Mesh MCF feature	Description
Media access coordination across multiple nodes	Media access coordination in a multi-hop network to avoid performance degradation and meet QoS guarantees.
Support for QoS	Traffic prioritization within the mesh; flow control over multi-hop paths; load control and contention resolution mechanisms.
Efficient RF frequency and spatial reuse	To mitigate performance loss resulting from hidden and exposed stations, and allow for concurrent transmissions to enhance capacity.
Scalability	Enabling different network sizes, topologies and usage models.
PHY independent	Independent of the number of radios, channel quality, propagation environment and antenna arrangement (including smart antennas).

to perform routing functions, in order to ensure a secure link for routing information messages. This will be more complicated to achieve in a mesh, where there will commonly be no centralized authentication server.

The work of the 802.11s TG is still in progress, and ratification of the final accepted proposal is not expected before 2008.

6.5 Other WLAN Standards

Although the wireless LAN landscape is now comprehensively dominated by the 802.11 family of standards, there was a brief period in the evolution of WLAN standards when that dominance was far from assured. From 1998 to 2000, equipment based on alternative standards briefly held sway. This short reign was brought to an end by the rapid market penetration of 802.11b products, with 10 million 802.11b based chipsets being shipped between 1999 and end-2001. The HomeRF and HiperLAN standards, which are now of mainly historical interest, are briefly described in the following sections.

6.5.1 HomeRF

The Home Radio Frequency (HomeRF) Working Group was formed in 1998 by a group of PC, consumer electronics and software companies, including Compaq, HP, IBM, Intel, Microsoft and Motorola, with the aim of developing a wireless network for the home networking market. The Working Group developed the specification for SWAP—Shared Wireless Access Protocol—which provided wireless voice and data networking.

SWAP was derived from the IEEE's 802.11 and ETSI's DECT (digitally enhanced cordless telephony) standards and includes MAC and PHY layer specifications with the main characteristics summarized in Table 6.16.

Although the HomeRF Working Group claimed some early market penetration of SWAP based products, by 2001, as SWAP 2.0 was being introduced with a 10 Mbps PHY layer data rate, the home networking market had been virtually monopolized by 802.11b products. The Working Group was finally disbanded in January 2003.

6.5.2 HiperLAN/2

HiperLAN stands for high performance radio local area network and is a wireless LAN standard that was developed by the European Telecommunications Standards Institute's Broadband Radio Access Networks (BRAN) project. The HiperLAN/2 Global Forum was

Table 6.16: Main characteristics of the HomeRF SWAP

SWAP specification	Main characteristics
MAC	TDMA for synchronous data traffic—up to 6 TDD voice conversations. CSMA/CA for asynchronous data traffic, with prioritization for streaming data. CSMA/CA and TDMA periods in a single SWAP frame.
PHY	FHSS radio in the 2.4 GHz ISM band. 50–100 hops per second. 2- and 4-FSK modulation deliver PHY layer data rates of 0.8 and 1.6 Mbps.

formed in September 1999 by Bosch, Ericsson, Nokia and others, as an open industry forum to promote HiperLAN/2 and ensure completion of the standard.

The HiperLAN/2 PHY layer is very similar to the 802.11a PHY, using OFDM in the 5 GHz band to deliver a PHY layer data rate of up to 54 Mbps. The key difference between 802.11a and HiperLAN/2 is at the MAC layer where, instead of using CSMA/CA to control media access, HiperLAN/2 uses time division multiple access (TDMA). Aspects of these two access methods are compared in Table 6.17.

The technical advantages of HiperLAN/2, namely QoS, European compatibility and higher MAC SAP data rate, have now to a large extent been superseded by 802.11 updates, such as the QoS enhancements introduced in 802.11e (see Section 6.4.1) and the 802.11h PHY layer enhancements specifically introduced to cater for European regulatory requirements (see Section 6.4.2). As a result, the previous support for HiperLAN/2 in the European industry has virtually disappeared.

Given the overwhelming industry focus on products based on the 802.11 suite of standards, it seems unlikely that HiperLAN/2 will ever establish a foothold in the wireless LAN market, the clearest indication of this being perhaps that Google News returns zero hits for HiperLAN/2!

6.6 Summary

Since the ratification of 802.11b in July 1999, the 802.11 standard has established a dominant and seemingly unassailable position as the basis of WLAN technology.

Table 6.17: CSMA/CA and TDMA media access compared

Media access method	Characteristics
CSMA/CA	Contention based access, collisions or interference result in indefinite back-off. QoS to support synchronous (voice and video) traffic only introduced with 802.11e. MAC efficiency reduced (54 Mbps at PHY = ca. 25 Mbps at MAC SAP).
TDMA	Dynamically assigned time slot based on a station's throughput. Support for synchronous traffic. Higher MAC efficiency (54 Mbps at PHY = ca. 40 Mbps at MAC SAP). Ability to interface with 3G as well as IP networks.

Table 6.18: 802.11a, b and g mandatory and optional modulation and coding schemes

Rate (Mbps)	802.11a		802.11b		802.11g	
	Mandatory	Optional	Mandatory	Optional	Mandatory	Optional
1 & 2			Barker		Barker	
5.5 & 11			CCK	PBCC	CCK	PBCC
6, 12 & 24	OFDM				OFDM	CCK-OFDM
9 & 18		OFDM				OFDM, CCK-OFDM
22 & 33						PBCC
36, 48 & 54		OFDM				OFDM, CCK-OFDM

The various 802.11 specifications draw on a wide range of applicable techniques, such as the modulation and coding schemes shown in Table 6.18, and continue to motivate the further development and deployment of new technologies, such as MIMO radio and the coordination and control functions required for mesh networking.

As future 802.11 Task Groups make a second pass through the alphabet, the further enhancement of WLAN capabilities will no doubt continue to present a rich and fascinating tapestry of technical developments.

This chapter has provided a grounding in the technical aspects and capabilities of current wireless LAN technologies. We will build on that foundation in future chapters.

CHAPTER 7

Wireless Sensor Networks

Chris Townsend
Steven Arms
Jon Wilson

7.1 Introduction to Wireless Sensor Networks

Sensors integrated into structures, machinery, and the environment, coupled with the efficient delivery of sensed information, could provide tremendous benefits to society. Potential benefits include: fewer catastrophic failures, conservation of natural resources, improved manufacturing productivity, improved emergency response, and enhanced homeland security [1]. However, barriers to the widespread use of sensors in structures and machines remain. Bundles of lead wires and fiber optic "tails" are subject to breakage and connector failures. Long wire bundles represent a significant installation and long term maintenance cost, limiting the number of sensors that may be deployed, and therefore reducing the overall quality of the data reported. Wireless sensing networks can eliminate these costs, easing installation and eliminating connectors.

The ideal wireless sensor is networked and scaleable, consumes very little power, is smart and software programmable, capable of fast data acquisition, reliable and accurate over the long term, costs little to purchase and install, and requires no real maintenance.

Selecting the optimum sensors and wireless communications link requires knowledge of the application and problem definition. Battery life, sensor update rates, and size are all major design considerations. Examples of low data rate sensors include temperature, humidity, and peak strain captured passively. Examples of high data rate sensors include strain, acceleration, and vibration.

Recent advances have resulted in the ability to integrate sensors, radio communications, and digital electronics into a single integrated circuit (IC) package. This capability is enabling networks of very low cost sensors that are able to communicate with each other using low power wireless data routing protocols. A wireless sensor network (WSN) generally consists of a basestation (or "gateway") that can communicate with a number of wireless sensors via a radio link. Data is collected at the wireless sensor node, compressed, and transmitted to the gateway directly or, if required, uses other wireless sensor nodes to forward data to the gateway. The transmitted data is then presented to the system by the gateway connection. The purpose of this chapter is to provide a brief technical introduction to wireless sensor networks and present a few applications in which wireless sensor networks are enabling.

7.2 Individual Wireless Sensor Node Architecture

A functional block diagram of a versatile wireless sensing node is provided in Figure 7.1. A modular design approach provides a flexible and versatile platform to address the needs of a wide variety of applications [2]. For example, depending on the sensors to be deployed, the signal conditioning block can be re-programmed or replaced. This allows for a wide variety of different sensors to be used with the wireless sensing node. Similarly, the radio link may be swapped out as required for a given applications'

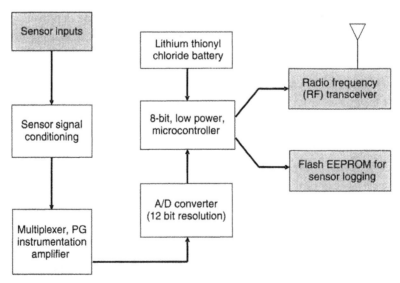

Figure 7.1: Wireless sensor node functional block diagram

wireless range requirement and the need for bidirectional communications. The use of flash memory allows the remote nodes to acquire data on command from a basestation, or by an event sensed by one or more inputs to the node. Furthermore, the embedded firmware can be upgraded through the wireless network in the field.

The microprocessor has a number of functions including:

1. managing data collection from the sensors

2. performing power management functions

3. interfacing the sensor data to the physical radio layer

4. managing the radio network protocol

A key feature of any wireless sensing node is to minimize the power consumed by the system. Generally, the radio subsystem requires the largest amount of power. Therefore, it is advantageous to send data over the radio network only when required. This sensor event-driven data collection model requires an algorithm to be loaded into the node to determine when to send data based on the sensed event. Additionally, it is important to minimize the power consumed by the sensor itself. Therefore, the hardware should be designed to allow the microprocessor to judiciously control power to the radio, sensor, and sensor signal conditioner.

7.3 Wireless Sensor Networks Architecture

There are a number of different topologies for radio communications networks. A brief discussion of the network topologies that apply to wireless sensor networks are outlined below.

7.3.1 Star Network (Single Point-to-Multipoint)

A star network (Figure 7.2) is a communications topology where a single base-station can send and/or receive a message to a number of remote nodes. The remote nodes can only send or receive a message from the single basestation, they are not permitted to send messages to each other. The advantage of this type of network for wireless sensor networks is in its simplicity and the ability to keep the remote node's power consumption to a minimum. It also allows for low latency communications between the remote node and the basestation. The disadvantage of such a network is that the basestation must be

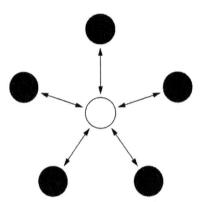

Figure 7.2: Star network topology

within radio transmission range of all the individual nodes and is not as robust as other networks due to its dependency on a single node to manage the network.

7.3.2 Mesh Network

A mesh network allows for any node in the network to transmit to any other node in the network that is within its radio transmission range. This allows for what is known as multihop communications; that is, if a node wants to send a message to another node that is out of radio communications range, it can use an intermediate node to forward the message to the desired node. This network topology has the advantage of redundancy and scalability. If an individual node fails, a remote node still can communicate to any other node in its range, which in turn, can forward the message to the desired location. In addition, the range of the network is not necessarily limited by the range in between single nodes, it can simply be extended by adding more nodes to the system. The disadvantage of this type of network is in power consumption for the nodes that implement the multihop communications are generally higher than for the nodes that don't have this capability, often limiting the battery life. Additionally, as the number of communication hops to a destination increases, the time to deliver the message also increases, especially if low power operation of the nodes is a requirement.

7.3.3 Hybrid Star-Mesh Network

A hybrid between the star and mesh network provides for a robust and versatile communications network, while maintaining the ability to keep the wireless sensor nodes

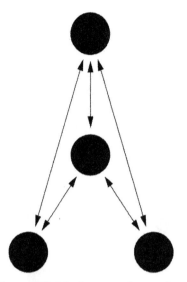

Figure 7.3: Mesh network topology

power consumption to a minimum. In this network topology, the lowest power sensor nodes are not enabled with the ability to forward messages. This allows for minimal power consumption to be maintained. However, other nodes on the network are enabled with multihop capability, allowing them to forward messages from the low power nodes to other nodes on the network. Generally, the nodes with the multi-hop capability are higher power, and if possible, are often plugged into the electrical mains line. This is the topology implemented by the up and coming mesh networking standard known as ZigBee.

7.4 Radio Options for the Physical Layer in Wireless Sensor Networks

The physical radio layer defines the operating frequency, modulation scheme, and hardware interface of the radio to the system. There are many low power proprietary radio integrated circuits that are appropriate choices for the radio layer in wireless sensor networks, including those from companies such as Atmel, MicroChip, Micrel, Melexis, and ChipCon. If possible, it is advantageous to use a radio interface that is standards based. This allows for interoperability among multiple companies networks. A discussion

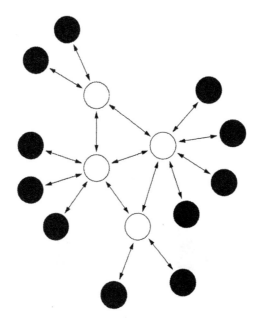

Figure 7.4: Hybrid star-mesh network topology

of existing radio standards and how they may or may not apply to wireless sensor networks is given next.

7.4.1 IEEE802.11x

IEEE802.11 is a standard that is meant for local area networking for relatively high bandwidth data transfer between computers or other devices. The data transfer rate ranges from as low as 1 Mbps to over 50 Mbps. Typical transmission range is 300 feet with a standard antenna; the range can be greatly improved with use of a directional high gain antenna. Both frequency hopping and direct sequence spread spectrum modulation schemes are available. While the data rates are certainly high enough for wireless sensor applications, the power requirements generally preclude its use in wireless sensor applications.

7.4.2 Bluetooth (IEEE802.15.1 and .2)

Bluetooth is a personal area network (PAN) standard that is lower power than 802.11. It was originally specified to serve applications such as data transfer from personal computers to peripheral devices such as cell phones or personal digital assistants. Bluetooth uses a star network topology that supports up to seven remote nodes communicating with a single basestation. While some companies have built wireless sensors based on Bluetooth, they have not been met with wide acceptance due to limitations of the Bluetooth protocol including:

1. Relatively high power for a short transmission range.

2. Nodes take a long time to synchronize to network when returning from sleep mode, which increases average system power.

3. Low number of nodes per network ($<=7$ nodes per piconet).

4. Medium access controller (MAC) layer is overly complex when compared to that required for wireless sensor applications.

7.4.3 IEEE 802.15.4

The 802.15.4 standard was specifically designed for the requirements of wireless sensing applications. The standard is very flexible, as it specifies multiple data rates and multiple transmission frequencies. The power requirements are moderately low; however, the hardware is designed to allow for the radio to be put to sleep, which reduces the power to a minimal amount. Additionally, when the node wakes up from sleep mode, rapid synchronization to the network can be achieved. This capability allows for very low average power supply current when the radio can be periodically turned off. The standard supports the following characteristics:

1. Transmission frequencies, 868 MHz/902–928 MHz/2.48–2.5 GHz.

2. Data rates of 20 Kbps (868 MHz Band) 40 Kbps (902 MHz band) and 250 Kbps (2.4 GHz band).

3. Supports star and peer-to-peer (mesh) network connections.

4. Standard specifies optional use of AES-128 security for encryption of transmitted data.

5. Link quality indication, which is useful for multi-hop mesh networking algorithms.

6. Uses direct sequence spread spectrum (DSSS) for robust data communications.

It is expected that of the three aforementioned standards, the IEEE 802.15.4 will become most widely accepted for wireless sensing applications. The 2.4-GHz band will be widely used, as it is essentially a worldwide license-free band. The high data rates accommodated by the 2.4-GHz specification will allow for lower system power due to the lower amount of radio transmission time to transfer data as compared to the lower frequency bands.

7.4.4 ZigBee

The ZigBee™ Alliance is an association of companies working together to enable reliable, cost-effective, low-power, wirelessly networked monitoring and control products based on an open global standard. The ZigBee alliance specifies the IEEE 802.15.4 as the physical and MAC layer and is seeking to standardize higher level applications such as lighting control and HVAC monitoring. It also serves as the compliance arm to IEEE802.15.4 much as the Wi-Fi alliance served the IEEE802.11 specification. The ZigBee network specification, ratified in 2004, will support both star network and hybrid star mesh networks. As can be seen in Figure 7.5, the ZigBee alliance encompasses the IEEE802.15.4 specification and expands on the network specification and the application interface.

7.4.5 IEEE1451.5

While the IEEE802.15.4 standard specifies a communication architecture that is appropriate for wireless sensor networks, it stops short of defining specifics about the sensor interface. The IEEE1451.5 wireless sensor working group aims to build on the efforts of previous IEEE1451 smart sensor working groups to standardize the interface of sensors to a wireless network. Currently, the IEEE802.15.4 physical layer has been chosen as the wireless networking communications interface, and at the time of this writing the group is in the process of defining the sensor interface.

7.5 Power Consideration in Wireless Sensor Networks

The single most important consideration for a wireless sensor network is power consumption. While the concept of wireless sensor networks looks practical and exciting on paper, if batteries are going to have to be changed constantly, widespread adoption will not occur. Therefore, when the sensor node is designed power consumption must be minimized. Figure 7.6 shows a chart outlining the major contributors to power

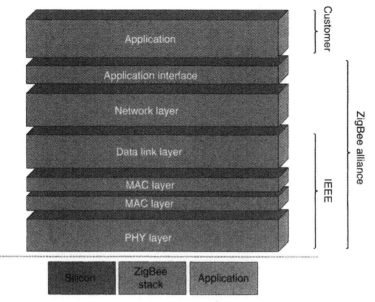

Figure 7.5: ZigBee stack

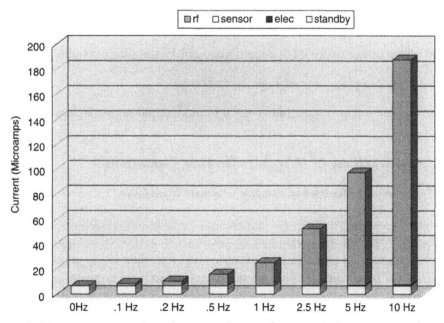

Figure 7.6: Power consumption of a 5000-ohm strain gage wireless sensor node

consumption in a typical 5000-ohm wireless strain gage sensor node versus transmitted data update rate. Note that by far the largest power consumption is attributable to the radio link itself.

There are a number of strategies that can be used to reduce the average supply current of the radio, including:

- Reduce the amount of data transmitted through data compression and reduction.
- Lower the transceiver duty cycle and frequency of data transmissions.
- Reduce the frame overhead.
- Implement strict power management mechanisms (power-down and sleep modes).
- Implement an event-driven transmission strategy; only transmit data when a sensor event occurs.

Power reduction strategies for the sensor itself include:

- Turn power on to sensor only when sampling.
- Turn power on to signal conditioning only when sampling sensor.
- Only sample sensor when an event occurs.
- Lower sensor sample rate to the minimum required by the application.

7.6 Applications of Wireless Sensor Networks

7.6.1 Structural Health Monitoring – Smart Structures

Sensors embedded into machines and structures enable condition-based maintenance of these assets [3]. Typically, structures or machines are inspected at regular time intervals, and components may be repaired or replaced based on their hours in service, rather than on their working conditions. This method is expensive if the components are in good working order, and in some cases, scheduled maintenance will not protect the asset if it was damaged in between the inspection intervals. Wireless sensing will allow assets to be inspected when the sensors indicate that there may be a problem, reducing the cost of maintenance and preventing catastrophic failure in the event that damage is detected.

Additionally, the use of wireless reduces the initial deployment costs, as the cost of installing long cable runs is often prohibitive.

In some cases, wireless sensing applications demand the elimination of not only lead wires, but the elimination of batteries as well, due to the inherent nature of the machine, structure, or materials under test. These applications include sensors mounted on continuously rotating parts [4], within concrete and composite materials [5], and within medical implants [6,7].

7.6.2 Industrial Automation

In addition to being expensive, leadwires can be constraining, especially when moving parts are involved. The use of wireless sensors allows for rapid installation of sensing equipment and allows access to locations that would not be practical if cables were attached. An example of such an application on a production line is shown in Figure 7.7. In this application, typically ten or more sensors are used to measure gaps where rubber seals are to be placed. Previously, the use of wired sensors was too cumbersome to be implemented in a production line environment. The use of wireless sensors in this application is enabling, allowing a measurement to be made that was not previously practical [8].

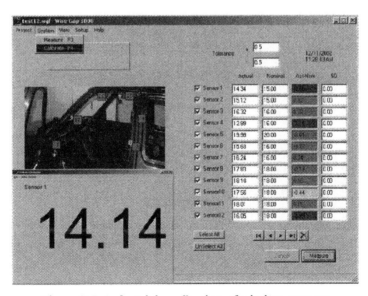

Figure 7.7: Industrial application of wireless sensors

Other applications include energy control systems, security, wind turbine health monitoring, environmental monitoring, location-based services for logistics, and health care.

7.6.3 Application Highlight – Civil Structure Monitoring

One of the most recent applications of today's smarter, energy-aware sensor networks is structural health monitoring of large civil structures, such as the Ben Franklin Bridge (Figure 7.8), which spans the Delaware River, linking Philadelphia and Camden, N.J [9,10]. The bridge carries automobile, train and pedestrian traffic. Bridge officials wanted to monitor the strains on the structure as high-speed commuter trains crossed over the bridge.

A star network of ten strain sensors was deployed on the tracks of the commuter rail train. The wireless sensing nodes were packaged in environmentally sealed NEMA rated enclosures. The strain gages were also suitably sealed from the environment and were spot welded to the surface of the bridge steel support structure. Transmission range of the sensors on this star network was approximately 100 meters.

The sensors operate in a low-power sampling mode where they check for presence of a train by sampling the strain sensors at a low sampling rate of approximately 6 Hz. When

Figure 7.8: Ben Franklin Bridge

a train is present the strain increases on the rail, which is detected by the sensors. Once detected, the system starts sampling at a much higher sample rate. The strain waveform is logged into local Flash memory on the wireless sensor nodes. Periodically, the waveforms are downloaded from the wireless sensors to the basestation. The basestation has a cell phone attached to it which allows for the collected data to be transferred via the cell network to the engineers' office for data analysis.

This low-power event-driven data collection method reduces the power required for continuous operation from 30 mA if the sensors were on all the time to less than 1 mA continuous. This enables a lithium battery to provide more than a year of continuous operation.

Resolution of the collected strain data was typically less than 1 microstrain. A typical waveform downloaded from the node is shown in Figure 7.9. Other performance specifications for these wireless strain sensing nodes have been provided in an earlier work [11].

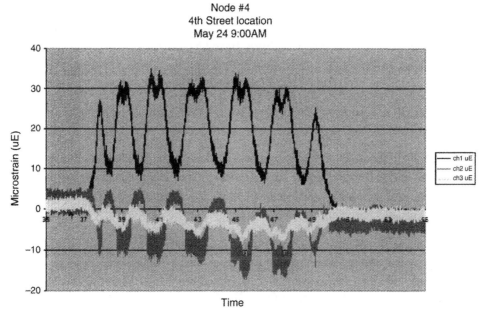

Figure 7.9: Bridge strain data

7.7 Future Developments

The most general and versatile deployments of wireless sensing networks demand that batteries be deployed. Future work is being performed on systems that exploit piezoelectric materials to harvest ambient strain energy for energy storage in capacitors and/or rechargeable batteries. By combining smart, energy saving electronics with advanced thin film battery chemistries that permit infinite recharge cycles, these systems could provide a long term, maintenance free, wireless monitoring solution [12].

References

1. Lewis, F.L., "Wireless Sensor Networks," *Smart Environments: Technologies, Protocols, and Applications*, ed. D.J. Cook and S.K. Das, John Wiley, New York, 2004.

2. Townsend, C.P, Hamel, M.J., Arms, S.W. (2001): "Telemetered Sensors for Dynamic Activity & Structural Performance Monitoring," SPIE's 8th Annual Int'l Conference on Smart Structures and Materials, Newport Beach, CA.

3. A. Tiwari, A., Lewis, F.L., Shuzhi S-G.; "Design & Implementation of Wireless Sensor Network for Machine Condition Based Maintenance," Int'l Conf. Control, Automation, Robotics, & Vision (ICARV), Kunming, China, 6–9 Dec. 2004.

4. Arms, S.A., Townsend, C.P.; "Wireless Strain Measurement Systems – Applications & Solutions," Proceedings of NSF-ESF Joint Conference on Structural Health Monitoring, Strasbourg, France, Oct 3–5, 2003.

5. Arms, S.W., Townsend, C.P., Hamel, M.J.; "Validation of Remotely Powered and Interrogated Sensing Networks for Composite Cure Monitoring," paper presented at the 8th International Conference on Composites Engineering (ICCE/8), Tenerife, Spain, August 7–11, 2001.

6. Townsend, C.P., and Arms, S.W., Hamel, M.J.; "Remotely Powered, Multichannel, Microprocessor-Based Telemetry systems for Smart Implantable Devices and Smart Structures," SPIE's 6th Annual Int'l Conference on Smart Structures and Materials, Newport Beach, CA, Mar 1–5, 1999.

7. Morris, B.A., D'Lima, D.D., Slamin, J., Kovacevic, N., Townsend, C.P., Arms, S.W., Colwell, C.W., e-Knee: The Evolution of the Electronic Knee Prosthesis: Telemetry Technology Development, Supplement to *Am. Journal of Bone & Joint Surgery*, January 2002.

8. Kohlstrand, K.M, Danowski, C, Schmadel, I, Arms, S.W; "Mind The Gap: Using Wireless Sensors to Measure Gaps Efficiently," *Sensors* Magazine, October 2003.

9. Galbreath, J.H, Townsend, C.P., Mundell, S.W., Hamel M.J., Esser B., Huston, D., Arms, S.W. (2003): Civil Structure Strain Monitoring with Power-Efficient High-Speed Wireless Sensor Networks, Proceedings International Workshop for Structural Health Monitoring, Stanford, CA.

10. Arms, S.W., Newhard, A.T., Galbreath, J.H., Townsend, C.P., "Remotely Reprogrammable Wireless Sensor Networks for Structural Health Monitoring Applications," ICCES International Conference on Computational and Experimental Engineering and Sciences, Medeira, Portugal, July 2004.

11. Arms, S.W., Townsend, C.P., Galbreath, J.H., Newhard, A.T.; "Wireless Strain Sensing Networks," Proceedings 2nd European Workshop on Structural Health Monitoring, Munich, Germany, July 7–9, 2004.

12. Churchill, D.L., Hamel, M.J., Townsend, C.P., Arms, S.W., "Strain Energy Harvesting for Wireless Sensor Networks," Proc. SPIE's 10th Int'l Symposium on Smart Structures & Materials, San Diego, CA. Paper presented March, 2003.

Part 2

Security Definitions and Concepts

CHAPTER 8

Attacks and Risks

John Rittinghouse
James F. Ransome

Regardless of whether computer and network data are transmitted on a wired or wireless medium, the basic security concepts remain much the same. Some of the content presented here has been excerpted from the *Cybersecurity Operations Handbook* [1] with the permission of Digital Press, an imprint of Elsevier Science.

For those among us who are tasked with managing business, and for that ever-shrinking number of Information Technology (IT) professionals who are not directly involved in the daily struggles coping with cybersecurity issues, one might be tempted to ask,

"What is the big deal about cybersecurity, really?"

"How does it affect our company infrastructure?"

"How does it affect users in our organization?"

"Is it something our management team should worry about?"

These are all legitimate questions. More and more today, IT professionals face an ever-growing and daunting task. Attacks occur *every* single day [2]. The only question to be asked in today's modern computing environment is, "Are we prepared to deal with an attack?" This book provides guidance on how to prepare for such assaults against organizational infrastructure. It will help network and systems administrators prepare to answer these types of questions and provide compelling information that can help even the most reluctant manager or administrator come to terms with the changed, threatening computing environment we face today.

8.1 Threats to Personal Privacy

Vast data stores in myriad organizations hold personal information about each of us. The accumulation of such large amounts of electronic information, combined with the increased ability of computers to monitor, process, and aggregate this information about people, creates a massive threat to our individual privacy. The reality of today is that all of this information and technology now available can be electronically linked together, allowing unknown entities to gain unabated access to even our most private information. This situation should give us reason to pause and ask ourselves if we have not created a modern information age with an unwanted byproduct some have often referred to as "Big Brother."

Although the magnitude and cost of the threat to our personal privacy is very difficult to determine, it is readily apparent that information technology is becoming powerful enough to warrant fears of the emergence of both government and corporate "Big Brothers." More awareness of the situation is needed at the organizational and personal level. With the increased accessibility of such information, we have created an ever-growing vulnerability that someone, such as a cyberterrorist, is likely to exploit. Another consideration of late is the recently legislated "Privacy Acts" that many different countries have enacted in order to try to protect the data assets of their citizenry. Such legislation has become an ever-growing part of this modern information age. All companies using computing resources today now need to be keenly aware of both these threats and the legal ramifications that ensue when they attempt to monitor, prevent, or provide access to their information resources.

8.2 Fraud and Theft

Computer systems can be exploited for conducting fraudulent activities and for outright theft. Such criminal acts are accomplished by "automating" traditional methods of fraud and by inventing and using new methods that are constantly being created by enterprising criminal minds. For example, individuals carrying out such criminal activity may use computers to transfer a company's proprietary customer data to computer systems that reside outside the company premises, or they may try to use or sell this valuable customer data to that company's competitors. Their motive may be profit or inflicting damage to the victimized company to compensate for some perceived injustice, or it may just be an act of malicious behavior for entertainment or bragging rights. Computer fraud and theft can be committed by both company insiders and outsiders, but studies have shown that most corporate fraud is committed by company insiders. [3]

In addition to the use of technology to commit fraud, computer hardware and software resources may be vulnerable to theft. Actual examples include the theft of unreleased software and storage of customer data in insecure places such as anonymous FTP accounts so that it can be accessed and stolen by outsiders. Data being exposed to these threats generates a secondary threat for a company: the loss of credibility and possible liability for damages as a result of premature release of information, exposure or loss of information, and so on. Preventive measures that should be taken here are quite simple, but are often overlooked. Implementation of efficient access control methodologies, periodic auditing, and firewall usage can, in most cases, prevent fraud from occurring or at least make it more easily detected.

8.3 Internet Fraud

The meteoric rise in fraud perpetrated over the Internet has brought about the classification of nine types of fraud, developed from the data reported to the Internet Fraud Complaint Center (IFCC) [4]. Analysts at the IFCC determine a fraud type for each Internet fraud complaint received. IFCC analysts sort complaints into one of the following nine fraud categories:

1. *Financial institution fraud*. Knowing misrepresentation of the truth or concealment of a material fact by a person to induce a business, organization, or other entity that manages money, credit, or capital to perform a fraudulent activity. [5] Credit/debit card fraud is an example of financial institution fraud that ranks among the most commonly reported offenses to the IFCC. Identity theft also falls into this category; cases classified under this heading tend to be those where the perpetrator possesses the complainant's true name identification (in the form of a social security card, driver's license, or birth certificate), but there has not been a credit or debit card fraud committed.

2. *Gaming fraud*. Risking something of value, especially money, for a chance to win a prize when there is a misrepresentation of the odds or events. [6] Sports tampering and claiming false bets are two examples of gaming fraud.

3. *Communications fraud*. A fraudulent act or process in which information is exchanged using different forms of media. Thefts of wireless, satellite, or landline services are examples of communications fraud.

4. *Utility fraud.* When an individual or company misrepresents or knowingly intends to harm by defrauding a government-regulated entity that performs an essential public service, such as the supply of water or electrical services. [7]

5. *Insurance fraud.* A misrepresentation by the provider or the insured in the indemnity against loss. Insurance fraud includes "padding" or inflating actual claims, misrepresenting facts on an insurance application, submitting claims for injuries or damage that never occurred, and staging accidents. [8]

6. *Government fraud.* A knowing misrepresentation of the truth, or concealment of a material fact, to induce the government to act to its own detriment. [9] Examples of government fraud include tax evasion, welfare fraud, and counterfeiting currency.

7. *Investment fraud.* Deceptive practices involving the use of capital to create more money, either through income-producing vehicles or through more risk-oriented ventures designed to result in capital gains. [10] Ponzi/Pyramid schemes and market manipulation are two types of investment fraud.

8. *Business fraud.* When a corporation or business knowingly misrepresents the truth or conceals a material fact. [11] Examples of business fraud include bankruptcy fraud and copyright infringement.

9. *Confidence fraud.* The reliance on another's discretion and/or a breach in a relationship of trust resulting in financial loss. A knowing misrepresentation of the truth or concealment of a material fact to induce another to act to his or her detriment. [12] Auction fraud and nondelivery of payment or merchandise are both types of confidence fraud and are the most reported offenses to the IFCC. The Nigerian Letter Scam is another offense classified under confidence fraud.

The Nigerian Letter Scam [13] has been around since the early 1980s. The scam is effected when a correspondence outlining an opportunity to receive nonexistent government funds from alleged dignitaries is sent to a "victim," but there is a catch. The scam letter is designed to collect advance fees from the victim. This most often requires payoff money to be sent from the victim to the "dignitary" in order to bribe government officials. Although other countries may be mentioned, the correspondence typically indicates "The Government of Nigeria" as the nation of origin. This scam is also referred to as "419 Fraud" after the relevant section of the Criminal Code of Nigeria, as well as "Advance Fee Fraud." Because of this scam, the country of Nigeria ranks second for total complaints reported at the IFCC on businesses by country. The IFCC has a policy of forwarding all Nigerian Letter Scam complaints to the U.S. Secret Service. The scam works as follows:

1. A letter, e-mail, or fax is sent from an alleged official representing a foreign government or agency.

2. The letter presents a business proposal to transfer millions of dollars in overinvoiced contract funds into your personal bank account. You are offered a certain percentage of the funds for your help.

3. The letter encourages you to travel overseas to complete the details.

4. The letter also asks you to provide blank company letterhead forms, banking account information, and telephone numbers.

5. Next, you receive various documents with official-looking stamps, seals, and logos testifying to the authenticity of the proposal.

6. Finally, they ask for upfront or advance fees for various taxes, processing fees, license fees, registration fees, attorney fees, and so on.

8.4 Employee Sabotage

Probably the easiest form of employee sabotage known to all system administrators would be "accidental" spillage. The act of intentionally spilling coffee or soda on a keyboard to make the computer unusable for some time is a criminal offense. Proving the spillage was deliberate, however, is next to impossible without the aid of hidden cameras or other surveillance techniques. Some administrators have even experienced severe cases where servers have been turned off over a weekend, resulting in unavailability, data loss, and the incurred, but needless cost, of hours of troubleshooting by someone. Employees are the people who are most familiar with their employer's computers and applications. They know what actions can cause damage, mischief, or sabotage. The number of incidents of employee sabotage is believed to be much smaller than the instances of theft, but the cost of such incidents can be quite high. [14]

As long as people feel unjustly treated, cheated, bored, harassed, endangered, or betrayed at work, sabotage will be used as a method to achieve revenge or a twisted sense of job satisfaction. Later in this book, we show how serious sabotage acts can be prevented by implementing methods of strict access control.

8.5 Infrastructure Attacks

Devastating results can occur from the loss of supporting infrastructure. This infrastructure loss can include power failures (outages, spikes, and brownouts), loss of communications, water outages and leaks, sewer problems, lack of transportation services, fire, flood, civil unrest, and strikes. A loss of infrastructure often results in system downtime, sometimes in the most unexpected ways. Countermeasures against loss of physical and infrastructure support include adding redundant systems and establishing recurring backup processes. Because of the damage these types of threats can cause, the Critical Infrastructure Protection Act was enacted.

8.6 Malicious Hackers

The term *malicious hacker* refers to those who break into computers without authorization. They can include both outsiders and insiders. The hacker threat should be considered in terms of past and potential future damage. Although current losses caused by hacker attacks are significantly smaller than losses caused by insider theft and sabotage, the hacker problem is widespread and serious. One example of malicious hacker activity is that directed against the public telephone system (which is, by the way, quite common, and the targets are usually employee voice mailboxes or special "internal-only" numbers allowing free calls to company insiders). Another common method is for hackers to attempt to gather information about internal systems by using port scanners and sniffers, password attacks, denial-of-service attacks, and various other attempts to break publicly exposed systems such as File Transfer Protocol (FTP) and World Wide Web (WWW) servers. By implementing efficient firewalls and auditing/alerting mechanisms, external hackers can be thwarted. Internal hackers are extremely difficult to contend with because they have already been granted access; however, conducting internal audits on a frequent and recurring basis will help organizations detect these activities.

8.7 Malicious Coders

Malicious code refers to viruses, worms, Trojan horses, logic bombs, and other "uninvited" software. Sometimes mistakenly associated just with personal computers, such types of malicious code can attack other platforms. The actual costs that have been attributed to the presence of malicious code most often include the cost of system outages and the cost of staff time for those who are involved in finding the *malware* and repairing the systems. Frequently, these costs are quite significant.

Today, we are subject to a vast number of virus incidents. This has generated much discussion about the issues of organizational liability and must be taken into account. Viruses are the most common case of malicious code. In today's modern computing platform, some form of antivirus software must be included in order to cope with this threat. To do otherwise can be extremely costly. In 1999, a virus named Melissa was released with devastating results. [15] The Melissa virus caused an estimated $80 million in damage and disrupted computer and network operations worldwide.

Melissa was especially damaging as viruses go because its author had deliberately created the virus to evade existing antivirus software and to exploit specific weaknesses in corporate and personal e-mail software, as well as server and desktop operating systems software. Melissa infected e-mail and propagated itself in that infected state to 50 other e-mail addresses it obtained from the existing e-mail address book it found on the victim's machine. It immediately began sending out these infectious e-mails from every machine it touched. The Melissa infection spread across the Internet at an exponential rate. Systems were literally brought down from overload as a result of exponential propagation.

8.8 Industrial Espionage

A company might be subject to industrial espionage simply because competitors share some level of sensitive customer information, which might be worth millions for interested parties ranging from governments to corporate and private entities. It is not only the press who would be willing to pay for information. This situation might be encouraging enough for many hackers to tempt fate and attempt to obtain such information. Internal staff might consider the risk minimal and give away such information. There could be active attempts to retrieve information without authorization by hacking, sniffing, and other measures. A case of espionage can have serious consequences for a company, in terms of incurring the cost of lawsuits and resulting damage awards. This situation can also devastate a company's reputation in the marketplace.

Formally defined, *industrial espionage* is the act of gathering proprietary data from private companies or governments to aid others. Industrial espionage can be perpetrated either by companies seeking to improve their competitive advantage or by governments seeking to aid their domestic industries. Foreign industrial espionage carried out by a government is often referred to as *economic espionage*. Because information is processed and stored on computer systems, computer security can help protect against such threats; it can do little, however, to reduce the threat of authorized employees selling that information.

Cases of industrial espionage are on the rise, especially after the end of the Cold War, when many intelligence agencies changed their orientation toward industrial targets. A 1992 study sponsored by the American Society for Industrial Security (ASIS) found that proprietary business information theft had increased 260 percent since 1985. The data indicated that 30 percent of the reported losses in 1991 and 1992 had foreign involvement. The study also found that 58% of thefts were perpetrated by current or former employees. The three most damaging types of stolen information were pricing information, manufacturing process information, and product development and specification information. Other types of information stolen included customer lists, basic research, sales data, personnel data, compensation data, cost data, proposals, and strategic plans.

Within the area of economic espionage, the Central Intelligence Agency (CIA) has stated that the main objective is obtaining information related to technology, but that information on U.S. government policy deliberations concerning foreign affairs and information on commodities, interest rates, and other economic factors is also a target. The Federal Bureau of Investigation (FBI) concurs that technology-related information is the main target, but also lists corporate proprietary information, such as negotiating positions and other contracting data, as targets.

Because of the increasing rise in economic and industrial espionage cases over the last decade, the *Economic and Espionage Act of 1996* was passed by the U.S. government. This law, coded as *18 U.S.C. §1832*, provides:

a. Whoever, with intent to convert a trade secret, that is related to or included in a product that is produced for or placed in interstate or foreign commerce, to the economic benefit of anyone other than the owner thereof, and intending or knowing that the offense will, injure any owner of that trade secret, knowingly

   1. steals, or without authorization appropriates, takes, carries away, or conceals, or by fraud, artifice, or deception obtains such information;
   2. without authorization copies, duplicates, sketches, draws, photographs, downloads, uploads, alters, destroys, photocopies, replicates, transmits, delivers, sends, mails, communicates, or conveys such information;

3. receives, buys, or possesses such information, knowing the same to have been stolen or appropriated, obtained, or converted without authorization;
4. attempts to commit any offense described in paragraphs (1) through (3); or
5. conspires with one or more other persons to commit any offense described in paragraphs (1) through (3), and one or more of such persons do any act to effect the object of the conspiracy, shall, except as provided in subsection (b), be fined under this title or imprisoned not more than 10 years, or both.

b. Any organization that commits any offense described in subsection (a) shall be fined not more than $5,000,000.

In a recent case, [16] against violators of *18 U.S.C. § 1832*, convictions were upheld in the appeal of Mr. Pin-Yen Yang and his daughter Hwei Chen Yang (Sally) for industrial espionage, among other crimes. Mr. Yang owned the Four Pillars Enterprise Company, Ltd., based in Taiwan. This company specialized in the manufacture of adhesives. Mr. Yang and his daughter conspired to illegally obtain trade secrets from their chief U.S. competitor, Avery Dennison Corporation, by hiring an ex-employee of Avery Dennison, a Dr. Lee. Lee was retained as a consultant by Yang, and the group conspired to pass confidential trade secrets from Avery to Four Pillars. When the FBI confronted Lee on the matter, he agreed to be videotaped in a meeting with Mr. Yang and his daughter. During the meeting, enough evidence was gathered to result in a conviction. [17]

Measures against industrial espionage consist of the same measures companies take to counter hackers, with the added security obtained by using data encryption technology. Where this is not possible because of government regulations (e.g., in France), proprietary compression or hashing algorithms can be used, which result in the same effect as encryption, but with a higher chance of being broken by a determined adversary. Legal protections exist, of course, but were once very difficult to dissect from the vast amount of legislation in Title 18 of the U.S. Code. Congress amended the many laws dotted throughout Title 18 into a comprehensive set of laws known as the 1996 National Information Infrastructure Protection Act.

8.9 Social Engineering

The weakest link in security will always be people, and the easiest way to break into a system is to engineer your way into it through the human interface. Almost every hacker group has engaged in some form of social engineering over the years, and in combination with other activities, they have been able to break into many corporations as a result. In this type of attack, the attacker chooses a mark he or she can scam to gain a password, user ID, or other usable information. Because most administrators and employees of companies are more concerned with providing efficiency and helping users, they may be unaware that the person they are speaking to is not a legitimate user. And because there are no formal procedures for establishing whether an end user is legitimate, the attacker often gains a tremendous amount of information in a very short time, and often with no way to trace the information leak back to the attacker.

Social engineering begins with a goal of obtaining information about a person or business and can range in activities from dumpster diving to cold calls or impersonations. As acknowledged in the movies, many hackers and criminals have realized that a wealth of valuable information often lays in the trash bins waiting to be emptied by a disposal company. Most corporations do not adequately dispose of information, and trash bins often contain information that may identify employees or customers. This information is not secured and is available to anyone who is willing to dive into the dumpster at night and look for it—hence, the term *dumpster diving*.

Other information is readily available via deception. Most corporations do not contain security measures that address deception adequately. What happens when the protocol is followed properly, but the person being admitted is not who he says he is? Many groups utilize members of their group in a fashion that would violate protocols to gather information about a corporate admittance policy. Often, the multiperson attack results in gaining admittance to the company and ultimately the information desired. Using the bathroom or going for a drink of water is always a great excuse for exiting from a meeting, and you often will not have an escort. Most corporations do not have terminal locking policies, and this is another way an attacker can gain access or load software that may pierce the company's firewall. So long as the people entering the corporation can act according to the role they have defined for their access and they look the part, it is unlikely that they will be detected.

Remotely, social engineering actually becomes less challenging. There are no visual expectations to meet, and people are very willing to participate with a little coaxing. As is often the case, giving away something free can always be a method for entry. Many social engineering situations involve sending along a free piece of software or something of value for free. Embedded within free software, Trojans, viruses, and worms can go undetected and can bypass system and network security. Because most security that protects the local machine has a hard time differentiating between real and fake software, it is often not risky for the attacker to deliver a keylogger or Trojan to the victim machine. Also equally effective, the customer support or employee support personnel can be duped into aiding a needy user with their passwords and access to information they do not necessarily know about.

8.9.1 Educate Staff and Security Personnel

According to NIST Publication SP800-12, [18] the purpose of computer security awareness, training, and education is to enhance security by

- Improving awareness of the need to protect system resources

- Developing skills and knowledge so computer users can perform their jobs more securely

- Building in-depth knowledge, as needed, to design, implement, or operate security programs for organizations and systems

By making computer system users aware of their security responsibilities and teaching them correct practices, it helps users change their behavior. It also supports individual accountability, which is one of the most important ways to improve computer security. Without knowing the necessary security measures (and how to use them), users cannot be truly accountable for their actions. The importance of this training is emphasized in the Computer Security Act, which requires training for those involved with the management, use, and operation of federal computer systems.

Awareness stimulates and motivates those being trained to care about security and reminds them of important security practices. By understanding what happens to an organization, its mission, customers, and employees when security fails, people are often motivated to take security more seriously. Awareness can take on different forms for particular audiences. Appropriate awareness for management officials might stress management's pivotal role in establishing organizational attitudes toward security. Appropriate awareness for other groups, such as system programmers or information analysts, should address the need for security as it relates to their jobs. In today's systems environment, almost everyone in an organization may have access to system resources and, therefore, may have the potential to cause harm.

Both dissemination and enforcement of policy are critical issues that are implemented and strengthened through training programs. Employees cannot be expected to follow policies and procedures of which they are unaware. In addition, enforcing penalties may be difficult if users can claim ignorance when they are caught doing something wrong. Training employees may also be necessary to show that a standard of due care has been taken in protecting information. Simply issuing policy, with no follow-up to implement that policy, may not suffice. Many organizations use acknowledgment statements that employees have read and understand computer security requirements.

Awareness is used to reinforce the fact that security supports the organization's mission by protecting valuable resources. If employees view security measures as just bothersome rules and procedures, they are more likely to ignore them. In addition, they may not make needed suggestions about improving security or recognize and report security threats and vulnerabilities. Awareness is also used to remind people of basic security practices, such as logging off a computer system or locking doors. A security awareness program can use many teaching methods, including videotapes, newsletters, posters, bulletin boards, flyers, demonstrations, briefings, short reminder notices at logon, talks, or lectures. Awareness is often incorporated into basic security training and can use any method that can change employees' attitudes. Effective security awareness programs need to be designed with the recognition that people tend to practice a tuning-out process (also known as *acclimation*). For example, after a while, a security poster, no matter how well designed, will be ignored; it will, in effect, simply blend into the environment. For this reason, awareness techniques should be creative and frequently changed.

Security education is more in-depth than security training and is targeted for security professionals and those whose jobs require expertise in security. Security education is normally outside the scope of most organizational awareness and training programs. It is more appropriately a part of employee career development. Security education is obtained through college or graduate classes or through specialized training programs. Because of this, most computer security programs focus primarily on awareness. An effective Computer Security Awareness and Training (CSAT) program requires proper planning, implementation, maintenance, and periodic evaluation. The following seven steps constitute one approach for developing a CSAT program:

Step 1: Identify program scope, goals, and objectives.

Step 2: Identify training staff.

Step 3: Identify target audiences.

Step 4: Motivate management and employees.

Step 5: Administer the program.

Step 6: Maintain the program.

Step 7: Evaluate the program.

8.9.2 Crafting Corporate Social Engineering Policy

When you begin the process of building a corporate policy for social engineering, several important considerations need to be included in the policy. Ensure that employees are aware of the data they are making available to others and what hackers might do with the knowledge they gain from that data. Train end users in the proper handling of social engineering tactics such as the following:

- Dumpster diving
- Phone calls
- E-mail
- Instant messaging
- On-site visits

8.9.2.1 Prevention

Teach employees how to prevent intrusion attempts by verifying identification, using secure communications methods, reporting suspicious activity, establishing procedures, and shredding corporate documents. It is important to define a simple, concise set of established procedures for employees to report or respond to when they encounter any of these types of attacks.

8.9.2.2 Audits

It is a good idea to periodically employ external consultants to perform audits and social engineering attempts to test employees and the network security readiness of your organization. Define the regularity of audits conducted by external consultants in a manner that cannot become predictable, such as a rotation of the month in each quarter an audit would occur. For example, if your external audits are conducted semiannually, the first audit of the year may occur in month one of quarter one. The next audit may occur in month three of quarter three. Then, when the next year comes around, you have rotated to another month or even changed to quarters two and four. The point is not which months and quarters audits are conducted, but that they are done in an unpredictable fashion that only you and your trusted few will know.

Endnotes

1. Rittinghouse, John W., and William M. Hancock. *Cybersecurity Operations Handbook*, First ed. New York: Digital Press, 2003.

2. See http://isc.sans.org/trends.html.

3. Computer Security Institute. "2002 CSI/FBI Computer Crime and Security Survey." Richard Power, 2003, at http://www.gocsi.com.

4. Internet Fraud Complaint Center. "IFCC Annual Internet Fraud Report, January 1, 2001 to December 31, 2001," at http://www1.ifccfbi.gov/strategy/statistics.asp.

5. Garner Bryan A., (ed.). *Black's Law Dictionary*, 7th ed. Cleveland, OH: West Group, 1999.

6. *Ibid.*

7. *Ibid.*

8. Wells Joseph T., and Gilbert Geis. *Fraud Examiners Manual*, 3rd ed., Vol 1. Association of Certified Fraud Examiners, Inc, 1998.

9. *See* note 5. See also, *The Merriam Webster Dictionary, Home and Office Edition.* Springfield, MA: Merriam-Webster, 1995.

10. Downes John, Jordan Elliot Goodman. *Barron's Dictionary of Finance and Investment Terms*, 5th ed. Hauppauge, NY: Barron's Educational Series, 1998.

11. *See* note 5.

12. *Ibid.*

13. *See* note 4.

14. U.S. Department of Justice. Press Release of February 26, 2002, "Former Computer Network Administrator at New Jersey High-Tech Firm Sentenced to 41 Months for Unleashing $10 Million Computer Time Bomb, at http://www.usdoj.gov/criminal/cybercrime/lloydSent.htm.

15. U.S. Department of Justice. Press Release of May 1, 2001, "Creator of Melissa Virus Sentenced to 20 Months in Federal Prison," at http://www.usdoj.gov/criminal/cybercrime/MelissaSent.htm.

16. U.S. Department of Justice. Electronic Citation: 2002 FED App. 0062P (6th Cir.), File Name: 02a0062p.06, decided and fled February 20, 2002, at http://www.usdoj.gov/criminal/cybercrime/4Pillars6thCir.htm.

17. The full text of this rendering can be reviewed at http://www.usdoj.gov/criminal/cybercrime/4Pillars6thCir.htm

18. NIST SP 800-12, *Computer Security Handbook*, author unknown. Washington, DC: U.S. Department of Commerce, 1996.

CHAPTER 9

Security Defined

Timothy Stapko

This chapter is by no means a complete treatment of the theory behind computer security—literally hundreds of books have been written on the subject—but it should at least provide a basic context for all readers. At the end of the chapter, we will provide a list of further reading for readers wanting a deeper treatment of the theory.

Computer security is a rapidly evolving field; every new technology is a target for hackers, crackers, spyware, trojans, worms, and malicious viruses. However, the threat of computer attacks dates back to the earliest days of mainframes used in the 1960s. As more and more companies turned to computer technology for important tasks, attacks on computer systems became more and more of a worry. In the early days of the personal computer, the worry was viruses. With the advent of the World Wide Web and the exponential expansion of the Internet in the late 1990s, the worry became hackers and denial of service attacks. Now, at the dawn of the new millennium, the worry has become spam, malware/spyware, email worms, and identity theft. All of this begs the question: How do we protect ourselves from this perpetual onslaught of ever-adapting attacks?

The answer, as you may have guessed, is to be vigilant, staying one step ahead of those who would maliciously compromise the security of your system. Utilizing cryptography, access control policies, security protocols, software engineering best practices, and good old common sense, we can improve the security of any system. As is stated by Matt Bishop,[1] computer security is both a science *and* an art. In this chapter, we will introduce this idea and review the basic foundations of computer security to provide a foundation for the information to come.

---

[1]*Author of* Computer Security: Art and Science.

9.1 What Is Security?

To begin, we need to define *security* in a fashion appropriate for our discussion. For our purposes, we will define computer security *as follows*:

> *Definition:* Computer Security. *Computer security is the protection of personal or confidential information and/or computer resources from individuals or organizations that would willfully destroy or use said information for malicious purposes.*

Another important point often overlooked in computer security is that the security does not need to be limited to simply the protection of resources from malicious sources—it could actually involve protection from the application itself. This is a topic usually covered in software engineering, but the concepts used there are very similar to the methods used to make an application secure. Building a secure computer system also involves designing a robust application that can deal with internal failures; no level of security is useful if the system crashes and is rendered unusable. A truly secure system is not only safe from external forces, but from internal problems as well. The most important point is to remember that *any* flaw in a system can be exploited for malicious purposes.

If you are not familiar with computer security, you are probably thinking, "What does 'protection' actually mean for a computer system?" It turns out that there are many factors that need to be considered, since any flaw in the system represents a potential vulnerability. In software, there can be buffer overflows, which potentially allow access to protected resources within the system. Unintended side effects and poorly understood features can also be gaping holes just asking for someone to break in. Use of cryptography does not guarantee a secure system either; using the strongest cryptography available does not help if someone can simply hack into your machine and steal that data directly from the source. Physical security also needs to be considered. Can a malicious individual gain access to an otherwise protected system by compromising the physical components of the system? Finally, there is the human factor. Social engineering, essentially the profession practiced by con artists, turns out to be a major factor in many computer system security breaches. This book will cover all of the above issues, except the human factor. There is little that can be done to secure human activities, and it is a subject best left to lawyers and politicians.

9.2 What Can We Do?

In the face of all these adversities, what can we do to make the system less vulnerable? We will look at the basics of computer security from a general level to familiarize the reader with the concepts that will be reiterated in the chapters to come. Even more experienced readers may find this useful as a review before delving into the specifics of network and Internet security.

9.3 Access Control and the Origins of Computer Security Theory

In their seminal computer security paper, "The Protection of Information and Computer Systems," (Saltzer 1976), Saltzer and Schroeder recorded the beginning concepts of access control, using the theory that it is better to deny access to all resources by default and instead explicitly allow access to those resources, rather than attempt to explicitly deny access rights.[2] The reason for this, which may be obvious to you, is that it is impossible to know *all* the possible entities that will attempt access to the protected resources, and the methods through which they gain this access. The problem is that it only takes one forgotten rule of denial to compromise the security of the entire system. Strict denial to all resources guarantees that only those individuals or organizations given explicit access to the resources will be able to have access. The system is then designed so that access to specific resources can be granted to specific entities. This control of resources is the fundamental idea behind computer security, and is commonly referred to as *access control.*

Over the years, computer scientists have formalized the idea of access control, building models and mathematically proving different policies. The most versatile and widely used model is called the *access control matrix*. Shown in Figure 9.1, the access control matrix is comprised of a grid, with resources on one axis and entities that can access those resources on the other. The entries in the grid represent the rights those entities have over the corresponding resources. Using this model, we can represent all security situations for any system. Unfortunately, the sheer number of possibilities makes it very difficult to use for any practical purposes in its complete form (representing all resources and possible

---

[2]This idea, by the authors' admission, had been around since at least 1965.

	Alice (manager)	Bob (IT admin)	Carl (normal user)	Donna (temporary user)
C:\Users\Alice	RWG	RWG		
C:\Users\Bob	R	RWG		
C:\Users\Carl	R	RWG	RWG	
C:\Users\Donna	R	RWG		RW

R = Can read from directory
W = Can write to directory
G = Can grant other users read, write, or grant permission

Figure 9.1: Access control matrix

users). We can, however, simplify the concept to represent larger ideas, simplifying the matrix for looking at systems in a consistent and logical manner. This is a concept that can be applied throughout the rest of the book to represent the resources and users that will be acting on the systems we are looking to secure. Having a logical and consistent representation allows us to compare and contrast different security mechanisms and policies as they apply to a given system.

In order to understand what an access control matrix can do for us, we will define a few rights that can be applied to users and resources. For our purposes, we will not give the access control matrix a complete formal treatment. We will instead focus on the rights and concepts that are directly applicable to the systems that we are looking at. For a more complete treatment of the theory behind access control matrices, see *Computer Security: Art and Science* by Matt Bishop.

The rights we are most interested in are *read, write,* and *grant*. These rights are defined as follows:

Read—The ability to access a particular resource to gain its current state, without any ability to change that state.

Write—The ability to change the state of a particular resource.

Grant—The ability of a user to give access rights (including grant privileges) to another user.

The rights defined here are a simplification of the full model, but will serve to help explain different security policies. The most important of these rights is *grant*. This right allows an expansion of rights to other users, and represents a possible security problem.

Given the rights defined as part of the access control matrix model, we can now analyze any given system and how secure it is, or is not. Using the matrix built from our system, we can mathematically guarantee certain states will or will not be entered. If we can prove that the only states the system enters are secure (that is, no unauthorized entities can get rights they are not entitled to, purposefully or inadvertently), then we can be sure that the system is secure. The problem with the access control matrix model, however, is that it has been proven this problem is undecidable in the general case when any user is given the *grant* right, since it opens the possibility of an inadvertent granting of a right to an unauthorized user. This does not mean the model does not have its uses. We will study different mechanisms and policies using this model because it simply and efficiently represents security concepts. In the next section, we are going to look at security policies and how they are designed and enforced in common applications.

9.4 Security Policies

The access control matrix provides a theoretical foundation for defining what security is, but what it does not do is provide a practical method for implementing security for a system.

For that, we need a security policy. The idea behind a security policy is simple: It is a set of rules that must be applied to and enforced by the system to guarantee some predefined level of security. Analogous to a legal system's penal code, a security policy defines what entities (people and other systems) should and should not do. When designing an application that you know will need security, part of the requirements should be a list of all the things in the system that need to be protected. This list should form the basis for the security policy, which should be an integral part of the design.

Each feature of the application must be accounted for in the policy, or there will be no security; as an example, think of the networked home thermostat example. If the security policy covers only the obvious features that may be threatened, such as a web interface, it might miss something more subtle, like someone opening the thermostat box and accessing the system directly. If the thermostat has the greatest Internet security in the world, but it sits wide open for someone to tinker with if he or she is physically there, it is probably in violation of the intended security policy, not the actual one. In this example, the security policy should include a rule, which can be as simple as a single statement that says the physical interface should have the same password scheme as the network interface. To take it a step further, the policy might also include a rule that the thermostat box should be physically locked and only certain personnel have access to the key.

Though it seems like there should be, there are no rules governing security policies in general. Certain organizations, such as corporations and the government, have certain guidelines and certifications that must be followed before an application is considered "secure," but even within a single organization, the security policies will likely differ. The problem is, as we will repeat throughout this book, that security is application-dependent. Although this means there is no official template to start from, a good starting place for a security policy would be the initial requirements document for the application. Each feature should be analyzed for its impact to the system should it ever be compromised. Other things to take into account are the hardware used, the development tools and language used, the physical properties of the application (where is it), and who is going to be using it.

When developing your security policy, think of your application as a castle. You need to protect the inhabitants from incoming attacks and make sure they are content (at least if they are not happy living in a castle); see Figure 9.2. Your policy is then a checklist of all the things that might allow the inhabitants to come to harm. There are the obvious things (get them out of the way first) like locking the castle gate and requiring proof of identity before allowing someone to enter (the password in an electronic application), or making sure the inhabitants can do their jobs (the application performs its tasks). However, to truly develop a useful policy, you have to think a little like the enemy. What are some other possibilities? Well, the enemy might not care about taking total control of the castle, and instead prevent it from functioning properly by stopping incoming traders from reaching the castle (denial of service attack). A more subtle attack might be something that is not noticed at first, but later becomes a serious problem, like hiring an inhabitant to sabotage the defenses (disgruntled employee making it easier for hackers), or poisoning the water supply (viruses). The enemy may also rely on cleverness, such as the mythical Trojan horse (Trojan horses, basically viruses that give hackers a doorway directly into a system). The enemy may not even be recognizable as an enemy, in the case of refugees from a neighboring war-ravaged country suddenly showing up and eating up all the available food and resources (worms like Blaster and Sasser come to mind). The possibilities go on and on, and it can be daunting to think of everything that might be a problem.

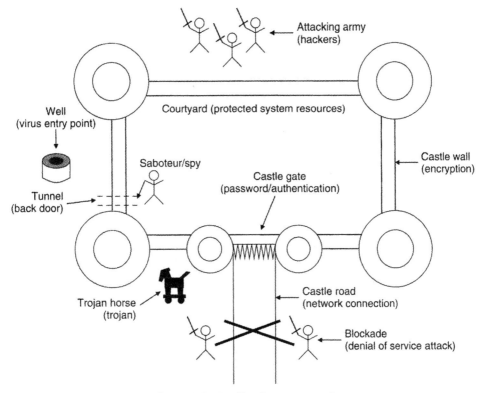

Figure 9.2: Application as a castle

In the castle example above, we paralleled some physical security problems with related computer security problems. This illustration may be useful to think of when developing a security policy, since many problems in security are very similar in vastly different contexts. It also helps to know what the most common attacks are, and weigh the likelihood of an attack versus the damage that the attack will wreak on your application should the attack be successful. One thing to try would be to think of the worst things your system could do (thermostat cranked up to 120°, opening the floodgates on the dam at the wrong time, etc.), and rank their severity on a scale (1 to 10, 1 to 100, etc.). Then think like a hacker and come up with attacks that each result in one or more of those scenarios, and assign a probability based upon the relative difficulty of the attack (this is very hard to do, see below). Take the probability of each attack that results in each of those scenarios and multiply it with the severity level. This will give you an idea of

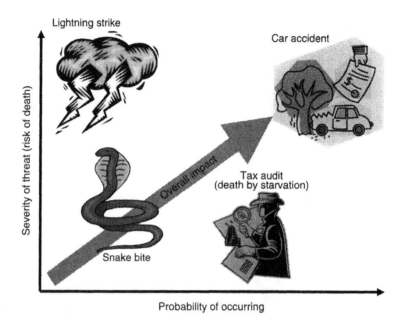

Figure 9.3: Severity of threats

the relative importance of protecting against a particular attack. The problem with this system is that the probability of an attack is very hard to determine (close to impossible for any sufficiently complex system). There is almost always someone smarter that can do something easily that you think is difficult (this is where feedback from colleagues is important). Furthermore, the probabilities you assign might change if another less "important" attack is successful. Given this, it still can be useful to assign relative weights to the things that can happen, and focus on those that would result in the most damage. In Figure 9.3, we see the relative impact of different threats in the real world, as compared to the probability of the threat occurring.

As can be inferred from the discussion above, producing a security policy is a creative process, much like painting a picture or writing music. Like other creative processes, it is therefore beneficial to recruit critics who will tell you what is wrong. In the case of security, the critic would likely be an expert security consultant who you hire to look over your application design (this is recommended for applications where security is truly important). If you think about it for a bit, it should be pretty easy to see that the more people that look at your policy, the better chance you have of catching problems.

It is also a time to be paranoid. In any case, more eyes will result in better security, even though this seems counterintuitive. Keeping things secret (security through obscurity) would seem to provide an additional level of security, and indeed it does, but it detracts from the overall security of the system. The reason for this is that it is unlikely that an individual (or even an organization) can think of all the possibilities for security breaches alone; it requires different viewpoints and experience to see what might lead to problems later.[3] Over and over, it has been proven that hiding the workings of security mechanisms does not work. Almost all of today's most widely used algorithms and protocols are wide open for review by all. A good security policy should be able to stand on its own without having to be kept secret—just don't share your passwords with anyone!

Throughout the rest of the book, keep the idea of implementing security policies in mind, since what we are really talking about is the enforcement of the rules put forth in a security policy. The enforcement of a rule may include cryptography, smart engineering, or basic common sense. This book is about the tools and mechanisms that can be employed for the enforcement of security policies. In the next section we will look at one of the most important tools in our security enforcement arsenal: cryptography.

9.4.1 Cryptography

In the last section, we looked at security policies and how they define the access that users have to resources. However, we did not look at the mechanisms that are used to enforce these policies. In this section, we introduce and describe the most important of these mechanisms: cryptography. *Cryptography* is the science of encoding data such that a person or machine cannot easily (or feasibly) derive the encoded information without the knowledge of some secret *key*, usually a large, difficult to calculate number. There are several forms of cryptography, some new, some dating back thousands of years. There is proof that ancient civilizations (namely, Egypt and Rome) used primitive forms of cryptography to communicate sensitive military information without the fear that the courier might be captured by the enemy. In modern history, the most famous form of encryption is also one of the most infamous—the Enigma machines used by Germany in World War II that were

---

[3]One possible exception to this is the US National Security Agency (NSA), since they employ a large number of security experts, giving them a definite advantage over other organizations. If you are the NSA, then feel free to keep everything secret; otherwise you may want to recruit some help.

Figure 9.4: Enigma Machine (from www.nsa.gov)

broken by the Allies. An example Enigma machine is shown in Figure 9.4. The breakdown in security allowed the US and Britain to gain important intelligence from the Germans, leading directly to defeat of Germany. The Enigma example also illustrates some important concepts about cryptography. Several things had to happen to break the system, such as the Allies obtaining an actual Enigma machine and German soldiers not following the security policy required by the system (essentially always using the same password). Those breaches are hard to guard against, so the system must be designed to minimize the impact of those attacks. The Enigma system was fundamentally flawed because these relatively minor breaches[4] led to complete failure of the system—as a result, the Allies won the war.

---

[4]Not to belittle the efforts of the allies to obtain an Enigma machine, but how do you protect a machine like that—one careless move and the enemy has one. In the Internet age, it is even easier to obtain a working "machine" for any number of security systems, since source code is easy to copy.

The newest form of encryption is *quantum cryptography*, a form of cryptography that utilizes the properties of subatomic particles and quantum mechanics to encode data in a theoretically unbreakable way. There are some actual practical applications of quantum cryptography in existence today, but they are severely limited and far too expensive for all but the most important applications.

9.4.2 Symmetric Cryptography

To start our discussion of cryptography, we will start with the oldest and most prevalent form of encryption: *symmetric-key cryptography*. This is the type of cryptography practiced by ancient civilizations and was the only true type of cryptography until the last century. Symmetric-key cryptography is characterized by the use of a single secret *key* to encrypt and decrypt secret information. This use of a single key is where the name *symmetric* came from, the same algorithm and key are used in both directions—hence, the entire operation is symmetric (we will see the opposite of symmetric cryptography, called *asymmetric cryptography,* in the next section).

To illustrate what symmetric cryptography is, we will use the classic computer security characters Alice, Bob, and Eve. Alice wishes to send a message to Bob, but Eve could benefit from having the information contained in the message, to Alice's detriment. To protect the message from Eve, Alice wants to employ symmetric-key cryptography. However, since the same key needs to be used for both encryption and decryption, Bob needs to have a copy of the key so he can read the message. This works fine *if* Alice and Bob met earlier to exchange copies of the keys they want to use. It would also work if they had a reliable and trustworthy courier to deliver the key. However, if Alice attempted to simply send a copy of her key to Bob (using a questionably trustworthy method, such as email), it is very likely that Eve would be able to gain a copy of the key while in transit. To see symmetric-key cryptography in action, see Figure 9.5.

Obviously, symmetric-key cryptography has some serious drawbacks for computer security. For instance, how do you give a secret key to someone you have never met (which is exactly what needs to happen for e-commerce)? Also, what do you do if your key is compromised or stolen? How do you get a new key to the recipient of your messages? Despite these drawbacks, however, symmetric-key cryptography *does* have a place in computer security. As it turns out, symmetric-key algorithms are the simplest, fastest cryptographic algorithms we know of. In a world built on bandwidth, this speed is a necessity.

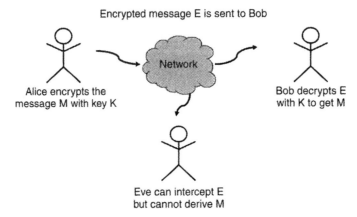

Encrypted message E is sent to Bob

Network

Alice encrypts the
message M with key K

Bob decrypts E
with K to get M

Eve can intercept E
but cannot derive M

Figure 9.5: Symmetric-key encryption example

9.4.3 Public-Key Cryptography

In the last section, we covered the oldest and most common form of cryptography. As you may have been wondering, if symmetric-key cryptography is symmetric, is there an asymmetric opposite? The answer, as you may have also guessed, is yes. *Asymmetric cryptography* is the conceptual opposite of symmetric-key cryptography. Asymmetric cryptography is usually referred to by its more common name (which is also more descriptive), *Public-key cryptography.*

Public-key cryptography is a very new method for encrypting data, developed in the 1970s as a response to the limitations of symmetric-key cryptography in an online world (recall that the Internet was in its infancy at that time). The basic idea behind the creation of public-key cryptography was to create a "puzzle" that would be very difficult to solve unless you know the trick (the key), in which case solving the puzzle would be trivial— solving the puzzle reveals the message. Using the analogy of a jigsaw puzzle, imagine that you create a blank puzzle and send the completed puzzle to your friend. Your friend writes her message on the completed puzzle, then takes the puzzle apart and returns the pieces to you (encoded with the message). If the puzzle has some trick needed to solve it, and only you know that trick, you can be assured that your friend's message arrives at your doorstep unspoiled by eavesdropping. You also have the added benefit of knowing if the message was tampered with, since you will not be able to solve the puzzle successfully if any of the pieces change. We will now look at public-key cryptography as it is implemented today.

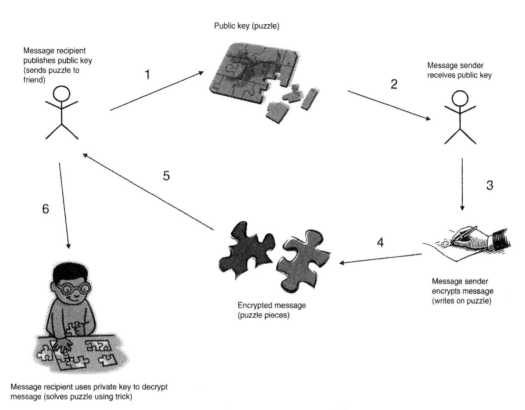

Public key (puzzle)

Message recipient
publishes public key
(sends puzzle to
friend)

1

2

Message sender
receives public key

5

3

6

4

Encrypted message
(puzzle pieces)

Message sender
encrypts message
(writes on puzzle)

Message recipient uses private key to decrypt
message (solves puzzle using trick)

Figure 9.6: Jigsaw puzzle analogy for public-key encryption

Public-key algorithms use different keys for both encryption and decryption (hence the asymmetry), and one of these keys is typically referred to as the *public-key*, since this key is usually published in some public place for anyone to access. This may at first seem counterintuitive, why would you want to *publish* your keys? Wouldn't that mean anyone can access my secret data? As it turns out, you are only publishing *half* of your total key—the part used to encrypt data (recall the puzzle example—you publish the puzzle, but keep the trick to solving it to yourself). We will illustrate public-key cryptography using Alice, Bob, and Eve once again.

Alice wants to send another message to Bob, but they have not made arrangements to exchange a secret key in order to do symmetric-key cryptography to protect the message. Fortunately, both Alice and Bob have a public-key algorithm at their disposal. Alice

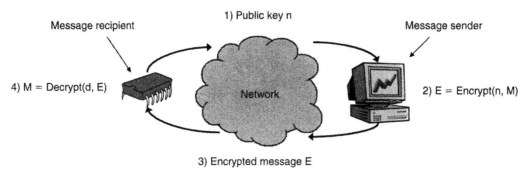

Figure 9.7: Public-key encryption

can then request that Bob publish his public-key on a reliable website (the site *must* be reliable; we will look at this problem further when we discuss *Certificate Authorities* later on). Alice then downloads Bob's key and uses it to encrypt her message. After the encryption, no one can read the message but Bob. Not even Alice can decrypt her own message. The idea behind this is that Bob keeps half of his key secret (his *private* key), the part that can decrypt information encrypted using the public-key. Eve, who has been watching the transaction, has complete access to Bob's public-key, but is helpless if she tries to read Alice's message, since she does not have Bob's private key. Alice's message is therefore delivered safely to Bob. For an illustration of public-key cryptography in action, see Figure 9.7. Further on in the next section, we will revisit public-key cryptography and see how some algorithms can be used in a novel way to prove who sent a message when we look at authentication techniques.

9.5 Data Integrity and Authentication

Cryptography is a useful tool in securing a system, but is just that, a tool, and there are many other tools available for security. To enhance cryptography in a practical system, and sometimes even replace it, we can use mechanisms and methods that protect data integrity and authenticate entities. Sometimes, one mechanism solves both of these problems. The first mechanism we will look at is what is called a *cryptographic hash* or *message digest algorithm*. Following that, we will discuss the inverse of public-key encryption; it turns out that reversing public-key algorithms (using the decryption key to encrypt a message, and using the encryption key to decrypt) allows us to do a form of authentication that is very useful in practice. We will also discuss methods and mechanisms of providing trust, such as digital signatures and certificates, and Public-Key Infrastructure (PKI).

9.5.1 Message Digests

At a very basic level, a message digest algorithm is simply a hash function: enter some arbitrary data of arbitrary size, and the hash algorithm spits out a fixed-size number that is relatively unique for the input given (note that we use the phrase *relatively unique*, as it is impossible to create a perfect hash for arbitrary input—a perfect hash can only be created if we can restrict the input to known values). What makes a hash function into a message digest is a level of guarantee that if two input datum are different (even by a single *bit*), then there is a predictably small possibility of a hash collision (those two messages generating the same hash). This property is quite important for message digests, and this will become apparent when we look at how message digests are used.

So it is great and wonderful that we have this algorithm that can take any size message and turn it into a much smaller number—so what? Remember that there is a *very small* probability of two messages generating the *same* hash value, and the chances of those two messages both containing legitimate data is even smaller. Using this knowledge, if we can provide a hash value for a message, then anyone can take the message, hash it on his or her machine, and verify that the locally generated hash and the provided hash match up. If they don't, then the message has been altered, either accidentally (some transmission error where data was lost or corrupted), or intentionally (by a malicious attacker). If used appropriately, the message digest mechanism gives us a fairly strong guarantee that a message has not been altered.

The guarantee given by hash algorithms is not perfect, however. Recently, there have been some advances in the mathematical analyses of both of the most commonly used algorithms, MD5 and SHA-1. For most intents and purposes, MD5 is considered insecure, and SHA-1 is not as secure as previously thought.[5] However, the attacks on these algorithms require a lot of computing power and dedicated attackers. This may

---

[5]Several academic papers have shown various weaknesses in both MD5 and SHA-1. There is even source code that demonstrates a hash collision in MD5. For the most part, however, these attacks remain mostly academic.

be an issue for banking transactions or other information that has a long lifetime, but for many applications, the level of security still provided by these algorithms may be sufficient. The caveat here is that if you choose an algorithm now that is known to be flawed, in the near future it is likely that, with advances in mathematics and computer technology, the algorithm will be completely useless. The problem faced by everyone right now is that there are no readily available replacements for MD5 or SHA-1. Most applications today, however, were built with some safeguards in the anticipation of these types of compromises. For example, the Transport Layer Security protocol (TLS—the IETF[6] standard for SSL—the Secure Sockets Layer protocol) uses an algorithm called HMAC, which wraps the hashing algorithm with some additional steps that allow the hashing algorithm to have some mathematically provable properties. This additional layer of security present in some current applications should help keep those applications fairly safe until suitable replacements are found for MD5 and SHA-1. The attacks against MD5 and SHA-1 also illustrate the benefits of allowing the public access to the workings of the algorithms, since these attacks likely would not have been found and exposed. We now know that these algorithms are flawed and can take additional precautions, whereas had the algorithms been proprietary, the attacks could have been discovered by an evildoer and we would never have known. It is even possible, albeit unlikely, that the faults were discovered earlier by the bad guys, but now we know what to look for, thus mitigating the potential effects of the attacks.

We have seen how hash algorithms are useful for protecting the integrity of data, but there is another common use for these handy algorithms, and that is *authentication*. Authentication is the ability to verify the correctness of some data with respect to a certain entity. For example, when someone is "carded" when trying to purchase a beer, the bartender is attempting to authenticate the customer's claim that he or she is actually old enough to purchase alcohol. The authentication comes in the form of a driver's license. That license has certain guarantees associated with it that allow the bartender to authenticate the customer's claim. In this example, the authentication comes from the government-issued identification, which is difficult to counterfeit, and is backed by certain laws and organizations—see Figure 9.8. In the world of the Internet, it is not so simple, since we do not often see who it is that we are actually communicating with

---

[6]Internet Engineering Task Force—the organization that oversees many of the publicly available standards for networking and security related to the Internet (www.ietf.org).

Figure 9.8: Driver's license security

(and in many cases, the "who" is simply a server tucked in a back room somewhere!). A driver's license does you no good if you are Amazon.com and you want to prove who you are to an online shopper. This is where a novel use of public-key cryptography becomes a very valuable tool.

It turns out that if we reverse some public-key algorithms, using the private key to *encrypt* instead of decrypt, we can use a trusted version of the corresponding key to verify that a message came from the private key's owner (assuming, of course, that the owner has not lost control of that key). The reason for this is the same property that makes public-key cryptography possible in the first place. If we use a private key to encrypt some data, then only the associated public-key will be able to decrypt that data correctly. If the recipient of a message possesses a known good, trusted copy of the sender's public-key, and if the message decrypts correctly using that key, the recipient can be almost certain that the message came from person who provided the public-key in the first place. See Figure 9.9 for an example. Using this mechanism, we can provide a level of trust in network communications, as long as we have a secure way to gain access to trusted copies of public-keys. We will discuss current approaches to this problem further on, but first we will look at public-key authentication in practice.

9.5.2 Digital Signatures

As we mentioned before, public-key cryptography is horribly inefficient. For this reason, the authentication method we just looked at would not be practical if every message had to be encrypted using the public-key algorithm. In order to make public-key authentication practical, the *digital signature* was invented. A digital signature is simply a hash of the data to be sent (using one of the message digest algorithms) encrypted using the public-key authentication method. By encrypting only the fixed-size message hash, we remove the inefficiency of the public-key algorithm and we can efficiently

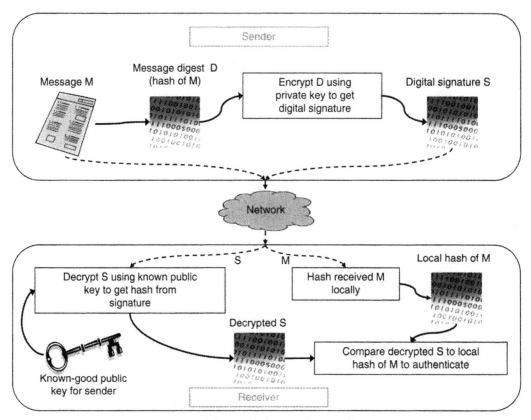

Figure 9.9: Public-key authentication

authenticate any arbitrary amount of data. Digital signatures are not foolproof, however, since they rely on hashing algorithms that may have weaknesses, the public-key must be trusted, and the private key must always remain private. In practice, however, the digital signature does provide some level of security, and in fact, forms the basis of trust for most of the Internet and e-commerce. In the next section, we will look at how digital signatures are used in practice by many protocols and applications.

9.5.3 Digital Certificates

So now we have all sorts of nifty mechanisms for encrypting data, protecting the integrity of data, and authenticating the source of that data, but how do these all work together to provide security for our applications? The most common use of the digital signature for authentication is a part of a *digital certificate*. A digital certificate consists of three primary sections: information about the owner (such as real or company name, Internet address, and physical address), the owner's public-key, and the digital signature of that data (including the public-key) created using the owner's private key. Digital certificates are typically encoded using a language called ASN.1 (Abstract Syntax Notation), developed by the telecommunications industry in the late 1970s, and more specifically, a subset of that language called Distinguished Encoding Rules (DER). This language was designed to be flexible, allowing for any number of extensions to the certificate format, which are created and used by some users of digital certificates to provide additional information, such as which web browsers should be able to accept the certificate. The public-key is stored using a base-64 encoding.

A digital certificate is provided by sending an application at the start of a secure communication. The receiving application parses the certificate, decrypts the digital signature, hashes the information, and compares it to the decrypted signature hash. If the hashes do not match, then the certificate has been corrupted in transit or is a counterfeit. If the hashes do match, then the application does a second check against a field in the data section of the certificate called the *Common Name* (*CN*). The common name represents the Internet address (url or IP address) of the sending application. If the common name and the address of the sending application do not match, then the receiving application can signal a warning to the user. Finally, a valid date range can be assigned to the certificate by the owner that will tell the recipient whether the certificate should be trusted at a certain time. If the current date (today) is between the initial date and the expiration date indicated in the certificate, then the certificate is considered valid (this does not supersede the other checks). Note that all these checks are up to the receiving application; the recipient of a digital certificate can choose to ignore all checks, or only perform one or two, but the guarantee of security is obviously not as strong if any failed checks are ignored. The digital certificate provides a useful mechanism for authentication, assuming that the public-key in the certificate can be trusted. This leaves us with a problem, however. If we only get the public-key as part of the digital certificate, how can we be certain that the person who sent the certificate is who they say they are? It turns out that this is a very difficult problem to solve, for many reasons. We will look at those reasons and the most common solution to the problem in use today, the *Public-Key Infrastructure* (*PKI*).

9.5.4 Public-Key Infrastructures

In the last section, we looked at digital certificates as a practical method for providing trust and authentication over a computer network. We also identified a problem with simply sending certificates back and forth. There is no easy way to be sure that a certificate actually belongs to the person or company detailed in the certificate. The only guarantees that we can glean from a single certificate are (1) if the digital signature matches, then the owner of the private key created the certificate and it has not been tampered with or corrupted, (2) if the address of the sender and the common name match up, then the certificate was probably created by the owner of that address (although the address can be spoofed, leading to other problems), and (3) if the current date falls within the valid date range, the certificate is not invalid. Notice the problem here? There is no way to tell for sure that the information provided on the certificate is authentic, only that the certificate was created by the owner of the private key, the data was not altered in transit, and the certificate cannot be ruled invalid based on the current date. To this day, there is no agreement on how to best handle this problem, but there are solutions in place that provide some guarantee of trust. Without them, there would be no e-commerce, and possibly no Internet (at least as we know it).

The most common solution in place today is the *Public-Key Infrastructure*, or PKI. A PKI is not a single entity or solution, but rather the idea that a known, trusted, third-party source can provide trust to anyone who needs it. To illustrate the concept, I use the analogy of a driver's license. In common, everyday life, we are occasionally called upon to provide some proof that we are who we say we are. Whether we are withdrawing money from a bank account or purchasing alcohol, we must provide proof in order to protect ourselves and others from fraud. Our typical physical proof is our driver's license. The license has some security features built in, such as a common format, identifying information (including a photograph of the licensee), and anti-copy protection (some licenses have difficult-to-copy holograms built in). Using these features, the person requiring the proof can be fairly certain of the identity of the license holder. The reason that this trust can be extended is the fact that the license was issued by a third party, in this case, the government, that is inherently trusted (some would say that we cannot trust the government at all, but if we did not, there would be no society, and we should all become hermits!). The government, in this case, extends its inherent trust to the licensee, and the anti-copying features of the physical license back up that trust.

In the electronic world, there is effectively no way to tell the difference between an original document and a forged copy. The only way that we can be sure of the validity

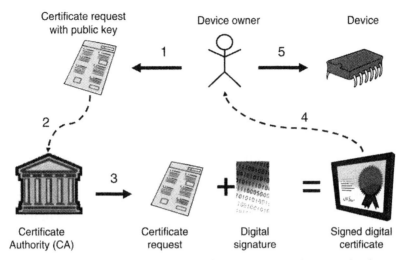

Figure 9.10: Digital signature signing using a certificate authority

of an electronic document is through some type of cryptographic mechanism—generally speaking, this is a digital certificate (assuming that a cryptographically secure channel has not been previously established). Digital certificates can be generated by anyone, but a few companies provide "signing" services where a certificate holder pays for a digital signature to be generated for that certificate. The companies that provide the signing services are generally referred to as "Certificate Authorities" (CA), and they have a private key that is used to sign customer certificates. This private key is associated with what is known as a "root" certificate, which is the basis of trust in a PKI hierarchy. The root certificate is often provided in web browsers or on secure Internet sites. The basic idea is that a root certificate can be loaded from a known trusted site, providing a fairly high level of assurance that the certificate is valid. As long as a CA keeps the private key hidden, and provides the public-key in the root certificate to the public, the PKI infrastructure works. Next we will look at how a root certificate can be used to verify an unknown but signed certificate—see Figure 9.10.

To verify a digital certificate, the recipient of that certificate must have access to the root certificate used to sign the certificate in question. The user can then decrypt the digital signature using the root certificate's public-key, and verify the hash of the remaining certificate data against the decrypted signature. Assuming that the CA keeps the corresponding private key hidden from the public, then the user can be fairly certain that the unknown certificate has been verified by the "trusted" third-party CA. As long as the CA is trusted by the user, then verification via this mechanism extends that trust to the unknown party.

Obviously, the trust in the CA is the most important link in a PKI chain. If you do not trust the company doing the signing, then there is no guarantee that a signed certificate has any validity whatsoever. However, CA's rely on trust, so their reputation, and hence their profits, are directly tied to the verification of certificate holders. Through traditional and physical means, a CA will usually follow up on the identity of a certificate holder before providing an authorized digital signature. The caveat here is that the CA provides only a guarantee that the certificate matches the person (or entity) providing it, not that the person is inherently trustworthy. It is completely up to the recipient of the certificate doing the verification to decide if the provider is trustworthy.

Certificate Authorities can also extend trust to other companies or entities that provide signing services under the umbrella of the root CA's trust. These companies are called "intermediate Certificate Authorities." An intermediate CA has its own root certificate that is actually signed by the root CA. Through this hierarchy, trust can be extended from the root CA to the intermediate CA, and finally to the end user. This hierarchy of extended trust is typically referred to as a "certificate chain," since the authentication forms a chain from the root CA to the end-user certificates. This chain is precisely like the governmental hierarchy from the analogy above, where the root CA is like the government, the intermediate CA is like the Department of Motor Vehicles, and the end-user certificate is like the driver's license. For better or worse, however, a CA is not an elected body, but rather a company that has established itself in the industry as a trusted signing entity. The foremost example of a root CA operating at the "governmental" level is Verisign. A quick look in any web browser at the built-in certificates will show a large number of Verisign certificates, or certificates from intermediate CA's that are covered under the Verisign PKI—see Figure 9.11 for a comparison with our previous driver's license example.

PKI is definitely not the only way to provide a network of trust for digital documents, but it has become the de facto standard because of the perceived trustworthiness in paying for authentication. One of the major drawbacks of PKI is that a single company or small group of companies controls the entirety of trust on the Internet, creating both a bottleneck and a single point of failure. Some security experts are of the opinion that

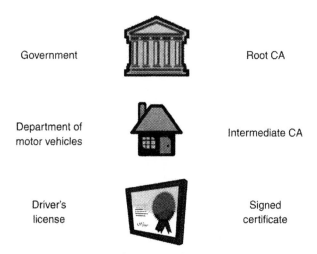

Government   Root CA

Department of
motor vehicles   Intermediate CA

Driver's
license   Signed
certificate

Figure 9.11: Department of Motor Vehicles vs. Certificate Authority

PKI is inherently flawed for this and other reasons, and an alternative is needed. One such alternative is peer networking. A person establishes trust with someone they know, and then trusts documents sent by that person. Once a certain level of trust is achieved, then that trusted peer can vouch for the validity of documents provided by people or entities they know and trust, even if the recipient does not know the sender. By keeping the number of "hops" between a sender and a recipient short, the trust can be fairly easily maintained—without a central body providing that trust. The advantages of this are fairly clear, each person controls what is trusted, and there is no single point of failure. The problem with adopting such a scheme, however, is that establishing peer networks of trust takes time, and there is a problem if the sender of a document is not connected to the recipient's network. In any case, PKI has some definite problems, and we will likely see some improvements or a replacement as the Internet continues to permeate our lives.

9.6 Recommended Reading

Practical Cryptography—Bruce Schneier

Applied Cryptography—Bruce Schneier

Computer Security: Art and Science—Matt Bishop

CHAPTER 10

Standardizing Security

Timothy Stapko

Network and Internet security are dominated by security protocols. A veritable sea of cryptic abbreviations and acronyms can confuse and confound anyone new to the discipline of security. We will look at some specifics here, covering such protocols as SSL, SSH, and IPSEC. These protocols will be introduced here, and we will cover their histories.

As you may already know, the Internet works on the same principles as your typical local area network (LAN), just at a much larger scale. For this reason, we can look at network and Internet security as being essentially the same. Similar problems plague both small networks and the Internet. Some of these problems are due to uncontrollable factors, such as weather and the unpredictability of end-users. The problems may be concerted efforts from malicious individuals or organizations that have something to gain from stealing information or making use of the network difficult. However, both purposeful and unintended failures in the system compromise the security of your application.

Almost every communications protocol has some type of built-in security, but it is up to the application developer whether or not to use those options. This obviously leads to security problems, since lazy or hurried engineers are likely to skip robust security features in an effort to get the product to market. Though this could be a problem for the aforementioned protocols, we will now look at some of the various protocols designed explicitly for secure communications. Some of them, such as the Secure Sockets Layer (SSL) are simply enhancements to existing, less-secure protocols (such as TCP). Others are stand-alone protocols that perform a specific service for the application—such as in the case of the Secure Shell protocol, or SSH, which provides a secure form of remote console (similar to Telnet).

Before we get started, we should point out a general feature of most secure communications protocols. In order to be generally useful, a secure protocol should not be limited to only supporting connections with known systems, but should allow for anonymous yet secure communications with *any* system. The reason for this is that it is extremely impractical to set up predefined trusted networks for some applications. Sometimes this is necessary, as when setting up a *virtual private network* (VPN), but most applications generally use some type of authentication, such as a password, to trust the remote machine. For this reason, most dedicated protocols utilize some form of public-key encryption, which allows for the sharing of cryptographic keys with unknown systems. The problem with using public-key encryption, as was mentioned earlier, is that the algorithms are extremely slow—sometimes thousands of times slower than their simpler, symmetric counterparts. In order to deal with this problem, many of the protocols utilizing public-key cryptography use a hybrid approach that uses the public-key algorithm to securely exchange keys for a faster symmetric algorithm.

For now, we will take a quick tour of some dedicated security protocols to introduce the reader to the technology and learn of the security options available to application designers.

10.1 Protocol Madness

The IT industry and the Internet are dominated by protocols. As was seen previously, there are many different protocols that play various roles in allowing two devices to communicate via a network. From the lowest levels of the hardware layer, to high level abstractions, protocols are everywhere. Logically, it follows that security is a concern for all these protocols, and even the security itself is defined by other protocols. Before we get into specific protocols, however, we are going to look at the security protocol and what it means, in general, for your application.

10.2 Standardizing Security—A Brief History

Security has long been a concern for computer applications developers, as there have been hackers wanting to compromise systems from the earliest days of computer networks. Indeed, the first networks were used by corporations and the military, both targets for espionage and both have information they want kept secret. In the beginning, security, like the machines being protected, was crude and often fatally flawed in execution. Not until there were those with the motives and cunning to push the boundaries of what well-behaved machines were supposed to do was there any great desire to mathematically ensure that a system was safe. However, with the growing threat of hackers in the 1970s it quickly became apparent that computer security needed some rigorous treatment if anything was ever to be secure at all.

There was a lot of work in the 1960s on security, but it was mostly unpublished. In the 1970s academics began to collect information on security and began analyzing the mathematical foundations of those ideas. They found various holes in the assumed security of preexisting systems and began a rigorous mathematical discipline to produce algorithms that would provide security and stand up to mathematical analysis. They abandoned the existing idea of "security by obfuscation," which is essentially the practice relying on the hiding of security algorithms from attackers. This turns out to be a very insecure practice, as there are many ways to attack the algorithms used for security without actually knowing the algorithm. That being said, keeping the algorithm secret can help with security—at least on a superficial level. It is now generally assumed by security experts that allowing rigorous public scrutiny of algorithms is better than keeping the algorithms secret, since there is a higher chance of finding problems in the algorithms before an attacker finds them and uses them for malicious purposes.

Besides abandoning traditional security ideas, another major development in the 1970s was the development of a new type of cryptography. Since ancient times, there had been only one basic type of security—the shared-key, or symmetric, cryptography where both parties must have knowledge of a secret key shared ahead of the communications. This was OK for some applications, such as military communications, but it presented some problems with computer networks, since there was no effective way to share the key without compromising security or making the whole network idea much less useful (if you can just hand the key to someone, why not just hand them the data?). Some pioneers thought about how they could protect data without having to share a secret ahead of time. The basic idea was to develop algorithms that have a two-part key, with one half being public to share, and the other half being secret to provide the security. Public-key security was thus born.

There were a couple of different public-key algorithms developed, but one was basically universally adopted due to a couple of unique properties. RSA, named for its inventors Rivest, Shamir, and Adelman, was reversible, unlike the other primary algorithm of the day, Diffie-Hellman (also named for its inventors). This symmetry allowed for two different but complimentary applications. RSA could be used for basic public-key cryptography, using the public-key to send an encrypted message to the owner of its matching private key. The advantage was that with RSA, the private key allowed the owner to encrypt a message that could then be decrypted with the public-key, thus proving that the message was sent from the owner of the private key that matched the public-key. This allowed for authentication over a network—if Alice knows Bob's public-key belongs to Bob, then she can decrypt an identifier sent by Bob and she will know that only he could have sent it, authenticating its validity (assuming that Bob's key is still secret). With RSA and other public-key algorithms, the fledgling Internet finally had a mechanism that could make network communications safe and efficient.

As the Internet gained more users in the 1980s, primarily corporations and academic institutions, many different protocols were developed, many competing for adoption. However, since the Internet was used not by the public, but by large organizations, there were few standards developed—each organization just used their own security. Services provided by the network, like email and file transfer, just were not secured because either no one saw the need, or the performance penalty was unacceptable.

It was not until the early 1990s that security really began to gain a mainstream foothold. The invention of the World Wide Web led to an explosion of users, and it became more and more evident that security was needed badly. Companies like Microsoft and Netscape (you may recall that Netscape Navigator and Microsoft Internet Explorer became the two predominant web browsers in the mid 1990s) tried to increase the use of their products through the introduction of generic protocols that could be used to secure any and all Web traffic. In the end, it was Netscape who won, with their implementation of the Secure Sockets Layer protocol, or SSL. SSL caught on quickly, and finally settled at version 3, which is still in widespread use today (version 1 was never released to the public and version 2 was found to be fatally flawed and nearly completely insecure). SSL was finally standardized by the Internet Engineering Task Force (IETF), one of the standards organizations for the Internet, as Transport Layer Security, or TLS. TLS was introduced in the late 1990s and is available in essentially all Web browsers and servers. The existence of SSL was one of the major factors that made the e-commerce explosion happen.

Today, security is needed more than ever. More and more, everyday items are being networked. Ordinary items are finding new uses through network interfacing, and this trend will only continue to grow. With everything connected to the global Internet, it is obvious we need to protect all that information. Fortunately, there are ways to adapt security mechanisms and protocols to resource-constrained systems so that even the most modest devices can have some modicum of security. The rest of this chapter will focus on security protocols and mechanisms, and how we can adapt them to work in resource-constrained environments.

10.3 Standardized Security in Practice

As a result of the efforts of researchers and investment in the infrastructure, we today have a large number of standard algorithms, protocols, and models to choose from when securing an application. Indeed, as the Internet is dominated by standards, so too is the realm of cryptography and security. Obviously with the pace of technological innovation today, likely all of these protocols and algorithms will be obsolete within the next decade or so, but you do not want to choose anything that is on its way to being replaced or removed.

10.3.1 Cryptographic Algorithms

The state of security standards today is pretty exciting. New algorithms are being developed, and even new types of encryption are being created, like elliptic-curve encryption and quantum encryption. There are many options to choose from and it can be difficult to choose a particular algorithm or protocol.

Several venerable algorithms are being made obsolete, such as the Data Encryption Standard (DES), which was developed in the 1970s but has recently been showing its age. Originally, DES used a 54-bit key, and at the time, it was thought that 54 bits was large enough to not be broken using a brute-force attack in any reasonable amount of time. However, the advances in computing power in the 1980s were not predicted,[1] and by the 1990s 54-bit DES was broken using a distributed application. Today, a 54-bit key is easily broken. One answer to the aging DES was a variant called 3-DES ("triple-DES"), where the key actually consists of three 54-bit DES keys strung together for a total effective key size of 162 bits. This provides a fairly effective level of encryption, but the general algorithm is slow, and other algorithms were needed for higher performance. In older applications where speed is not as much of an issue, 3DES is still widely used.

---

[1] Moore's law predicted in the 1960s that the density and performance of integrated circuits would double roughly every 18 months or so, but apparently none of the DES designers thought the prediction would hold true for so long.

A lot of cryptographic algorithms were developed in the 1990s, many of which were compiled into Bruce Schneier's classic *Applied Cryptography*. One algorithm that has remained relatively popular due to its blazing performance is Rivest Cipher 4, or RC4 (Rivest is the "R" in RSA). Originally a trade secret of RSA Security (the company founded by the creators of RSA), RC4 was leaked to the Internet and has since become a very popular cipher, since it is extremely easy to implement, and it is exceedingly fast. However, RC4 has not been through a thorough mathematical analysis (as DES was) and it is a stream cipher (which means there are some caveats in using it), so though it seems to be secure, there is some doubt that accompanies the algorithm. These doubts have not prevented RC4 from being built into virtually every web browser as part of the SSL protocol, and a large number of secure Internet transactions are performed using 128-bit RC4.

With DES becoming obsolete and algorithms like RC4 not having the rigorous treatment to give everyone confidence, there was a big push to create a new standard algorithm. To create interest in developing a new standard, the National Institute of Standards and Technology (NIST) created a competition in the late 1990s that called for submissions of algorithms that would all undergo thorough public and internal analysis, with a committee to choose the final algorithm. A number of entries from top cryptographers were considered for what would become the Advanced Encryption Standard, or AES. After much public debate and scrutiny, the NIST committee chose an algorithm called Rijndael, named for its creators (the name was actually a combination of the author's names). Now commonly known as AES, Rijndael is the standard encryption method used by the United States government and numerous other organizations. Essentially every security protocol today supports AES (except for those created before AES—and those protocols likely use a variant of DES).

In contrast to the variety of algorithms described above, which are all symmetric-key algorithms, there are relatively few public-key algorithms in use today. This is primarily due to the dominance of one particular public-key algorithm, RSA. Named for its inventors (Ron Rivest, Adi Shamir, and Leonard Adleman), RSA is so simple to implement and so elegant in its execution that most other contenders are just not attractive enough to bother using. RSA was a patented algorithm, and for a number of years was available only under license from RSA Security, the company created by Rivest, Shamir, and Adleman to govern not only RSA but a number of other security algorithms, protocols, and services. However, the patent for RSA recently expired, and with no barriers remaining it has become the primary public-key algorithm in use today. Despite RSA's dominance of the field, however, there are some other algorithms that are available, such as Diffie-Hellman (one of the first public-key algorithms), PGP (Pretty Good Privacy, used for email encryption, among other things), and DSS (Digital Signature Standard, an open protocol that was designed for digital signatures). Due to the slow performance of public-key operations, all these algorithms are usually relegated to specific uses where the public-key encryption properties are explicitly needed; otherwise a faster symmetric algorithm is usually chosen.

10.3.2 Cryptography in Practice

Now let's take a look at a couple of algorithms a little more in depth. Due to their widespread use, we will focus on four common algorithms, DES/3DES, RC4, AES, and RSA. Please note that the descriptions of these algorithms in the following sections are derived from publicly available specifications and from Bruce Schneier's excellent reference, *Applied Cryptography*. Later on, in the case studies, we will look at the implementation of these algorithms and how we can apply them in our own applications. For now, we will just take a look at the algorithms, where they come from, why they are popular, and what they can and should be used for.

10.3.2.1 DES

The Data Encryption Standard was developed in the early 1970s by a team from IBM with some help from the National Security Association (NSA). Many of the details of the implementation remained secret for many years, as the NSA did not want to reveal what techniques it had up its sleeves. The algorithm IBM had created was originally called Lucifer and was modified and tweaked by the IBM team and the NSA as a response to a call from the NIST precursor for a government standard encryption scheme. The public response was enthusiastic but underwhelming (no one was really studying cryptography at the time), so the design from the IBM team was a shoe-in. DES was originally intended to be implemented in hardware and was designed around the technology of the time. However, the published specification contained enough information for software implementations to be designed once general-purpose computers became more accessible. DES by itself is a fairly quick algorithm, but its standard 56-bit key is too small to be practical anymore (a 56-bit key can be broken using brute force methods over a distributed network in very little time). Triple-DES, or 3DES, uses 3 56-bit keys to achieve a higher level of security, and is more standard today. Despite its age and suspicious involvement of the NSA in its design, DES has stood the test of time and still remains a viable cryptographic algorithm.

The implementation of DES has some small challenges in software, as a couple of operations are trivial to implement in hardware but require some additional work to implement programmatically. The math behind DES is heavy-duty (you can read all about it in *Applied Cryptography*), but the implementation is straightforward, since most of the tricky math is actually wrapped up in a collection of constants that are used essentially to randomize the data being encrypted. These constants are put into what are called *substitution boxes*, or s-boxes, and *permutation boxes* (p-boxes). The real security of DES resides in the s-boxes and in the key—we aren't going to worry about why DES is secure (plenty of books and papers have been written on the subject)—for our purposes we can just assume that it is.

Unfortunately, the use of DES is discouraged by pretty much everyone for new development and designs, so its utility is a bit limited. In fact, original DES should not be used at all, since it is considered too weak (see the explanation earlier in this chapter), so 3DES is really the only option. Despite the fact you shouldn't use DES anymore, it is still implemented in a wide array of applications, and if you are attempting to interface with a legacy product, it may require DES, so we include it for this reason.

10.3.2.2 RC4

RC4 is probably the simplest algorithm in existence for securing information. The entire operation is essentially an XOR and a swap in a 256-entry table. The operation is completely reversible (the same operation is used both to encrypt and decrypt), and extremely fast. RC4 was originally a trade secret of RSA Security, and was invented by Ron Rivest (the "R" in RSA). The introduction of RC4 to the public is rather unique as well. Someone managed to reverse-engineer the algorithm and this person then posted the source code on the Internet that was completely compatible with the proprietary implementation! Oddly enough, the public implementation was different enough from RSA's implementation that they could not claim it was stolen. Since the algorithm was not sanctioned by RSA or Rivest, the public version of RC4 was referred to as *Alleged* RC4 (ARC4) for many years. The algorithms are exactly the same though, and as a result, there is no effective difference between the two.

RC4's unique origins and its simple implementation have made it quite popular, and essentially every web browser and SSL implementation supports it. It can be implemented in just a few lines of C code, and is blazingly fast. The real downside to RC4 is that it has not been as rigorously studied as other algorithms, and this makes many security experts a little uncomfortable. Beyond that, the fact that RC4 is a stream cipher has some inherent drawbacks to implementation. Specifically, if an RC4 key is used twice to decrypt different messages, then the encrypted messages can be combined to retrieve the plaintext message without knowing the key. However, as long as you never reuse the same key (there are a lot of keys in 128 bits), then you should be OK (see below).

Other than the issues we mentioned, there are no known attacks on RC4 itself as of this writing, and it seems to be an OK cipher (the attacks we talk about are actually targeted at poor implementations of RC4 in a system). It is a good algorithm if you want a reasonable level of security for many applications, but be advised that the jury is still out, and there is always the possibility of a breakthrough that could make it obsolete. There are also a number of drawbacks to using stream ciphers that, if not accounted for, can lead to a compromise of the security.

One major caveat to point out with RC4 is that it is a stream cipher (as opposed to a block cipher like AES or DES). A major problem in using stream ciphers (or using block ciphers in a stream-cipher mode) is what is called a substitution attack. If the attacker knows the format and portions of the plaintext message, he can anticipate where parts of the message will be and actually modify the encrypted message without knowing the key. For example, it would be fairly trivial to change the word "Canada" to "Israel" in a message (note that the words are the same number of characters so the message would still make sense) if the attacker knew that the word "Canada" was in the message and knew its position within the message. This is especially problematic for banking or database transactions, since they may be automated and can be expected to follow a particular format. Protocols that utilize stream ciphers effectively use a checksum or hash of the message included in the encrypted payload to check for any tampering.

Another problem with stream ciphers is that you must never use the same key twice. The problem here is a mathematical property of stream ciphers—if an attacker has two different encrypted messages that use the exact same key then he can retrieve the message rather trivially. The solution to this problem is to use what is called an *initialization vector,* which is essentially a one-time random string of data that is combined with the secret key to produce a relatively unique key for transferring data (SSL and other protocols that use stream ciphers do this). The initialization vector must be large enough to ensure that the probability of using the same vector twice is extremely low.[2] This was one of the issues with WEP (Wired Equivalent Privacy), a broken Wi-Fi security protocol which we will discuss more later), since it used sufficiently short initialization vectors to increase the probability of a match to a level easily broken by a computer. Roughly once in a few thousand connections there would be two connections that had the same initialization vector, allowing for the aforementioned attack to work.

RC4 (and other stream ciphers) can be used safely if the appropriate measures are taken to account for the caveats mentioned above. However, it is important to keep up on the state of cryptography to be sure no new attacks are found. If you really need security that is standard and backed by a lot of experts, you should look to AES instead—it isn't as simple or as fast as RC4 (it's close though), but it is a standard used by the US government, so at least your application will be in the same boat as everyone else.

10.3.2.3 AES

As was mentioned previously, the Rijndael algorithm was chosen in the early 2000s by NIST to be the new government standard symmetric encryption algorithm, otherwise known as the Advanced Encryption Standard.[3] Intended as a replacement for DES, AES utilizes much of the knowledge gained in the 30 years between its development and the introduction of DES as a standard. AES is also designed to take advantage of modern computer architectures as well, so it is a little less cumbersome than its predecessor. The AES algorithm is a fairly simple design that lends itself to high performance and elegance of implementation. Also, one of the stipulations of the AES competition was that the algorithm should be designed to be easily implemented in hardware, so AES has some properties that make hardware acceleration a real possibility.

---

[2]The so-called "birthday paradox" illustrates (somewhat counterintuitively) that the probability of two arbitrary numbers in a random set being the same is much higher than pure randomness would seem to dictate. This is the basis of a birthday party trick, since if there are more than 20 people in a room, the probability of two people having the same birthday is close to 50% (you just can't pick which 2).

[3]AES actually defines various modes of operation, which makes it more of a protocol than an algorithm. However, the AES moniker is usually used to refer to the basic Rijndael algorithm, which is how we will refer to it throughout this book.

AES can be implemented completely in software or completely in hardware. More recently, a hybrid approach has been taken, moving some of the more expensive operations into hardware while leaving the rest of the implementation in software. This new approach has several distinct advantages. The hardware, not being dedicated specifically to AES, can be more general and less expensive. Having the hardware acceleration, however, makes the software smaller and faster. Several companies provide various levels of AES acceleration, even going so far as to add specific AES instructions into a processor so that no external hardware devices are needed. Generally speaking, if you need a symmetric encryption algorithm for an application, you should use AES. It is the safest choice because it is the standard used by the US government and it is backed by some of the leading experts in cryptography. If AES is ever broken, it won't be your fault (and it wouldn't matter anyway, since everyone else would be in *big* trouble too).

10.3.2.4 RSA

The three algorithms we have looked at so far are all symmetric-key algorithms that are very useful if you have a good way to exchange keys with the person or machine with which you want to communicate. In some cases, a physical transfer of the key may be possible and appropriate, but more often, an electronic means of key delivery is more applicable. Today, in most cases, this method is RSA. Developed and patented by Ronald Rivest, Adi Shamir, and Leonard Adleman in 1978, RSA is the most well-known and probably most useful public-key algorithm. One of the most useful properties of RSA is that it can be used both for the basic public-key operation (I send you my public-key so you can encrypt a message to send back to me), and for authentication (I encrypt a message with my private key which you can verify came from me using my known public-key). This property makes RSA especially useful for protocols that utilize both a public-key operation and authentication, like SSL.

RSA is an elegantly simple algorithm with some extremely complex math behind it. Essentially, RSA consists of a function that utilizes some unique properties of large prime numbers and modular mathematics. The key generation for RSA involves selecting two very large prime numbers and multiplying them together. The trick is that if you know the prime factors used to generate the key then the RSA encryption function is simple to reverse (thereby decrypting an encrypted message). If you don't know those factors, then you have to find them by factoring a REALLY large number into its component prime factors, a process that takes an extremely long time using today's math and technology. It is possible that we may discover a fast way to factor large numbers in the future, and this would be a very big problem, since it would render RSA completely useless. This prime factoring can be considered similar to the brute-force search required for the naïve attack on symmetric algorithms. However, factoring a number is inherently easier than searching through all possible symmetric keys represented by a number of the same size. It is for this reason that symmetric keys are often 128 bits long, but a similarly-secure RSA key will be 1024 or 2048 bits long. Unfortunately, these large numbers are hard to deal with, even on a computer, so RSA is significantly slower than any of the symmetric algorithms we have looked at.

RSA is too slow to be generally useful, but since it does have specific useful properties, it is usually used along with a faster symmetric algorithm, usually to exchange symmetric keys.

RSA should be used with some type of hardware acceleration if at all possible. Fortunately, like AES, there are a number of products available that either implement RSA entirely in hardware or provide some type of acceleration for software implementations. The primary part of RSA that benefits from acceleration is the modular math—literally millions of operations are done in a typical RSA operation and any improvement in their speed will result in a significant improvement in performance.

RSA is an extremely useful algorithm that is employed in thousands of applications. We will look more at how RSA is implemented, and what we can do to alleviate the performance bottleneck, when we look at SSL.

10.4 Cryptography and Protocols

Most of the time, cryptographic algorithms are not used on their own, but rather as part of a complete security system or protocol. Indeed, as was mentioned earlier, RSA is pretty much always used with some symmetric algorithm as a key exchange mechanism. Security protocols are designed to address specific problems in communications. Many protocols are designed to be used for a particular application, such as the Secure Shell (SSH) protocol, which is designed to provide a remote text-based console, like Telnet but secure.[4] Some protocols are designed to be a general-purpose solution, encrypting everything that is sent between two machines on a network. Examples of this type of protocol include SSL and IPSEC.

---

[4]SSH is actually now a more general purpose collection of secure utilities, but its original intent was as a secure replacement for Telnet.

When choosing a protocol for an application, you have to look at not only the features that the protocol provides, but also how the protocol has proven itself in the field. SSL version 2 (the first publicly available version, implemented as part of the Netscape web browser) seemed to be secure, but was later shown to be fatally flawed. The replacement, SSL version 3, has been in use for almost a decade now, and seems to work pretty well. You also need to look at who designed the protocol—was it "design by committee" or were there some security experts and cryptographers involved? A recent example of why you need to research a protocol before using it is the case of the Wired-Equivalent Protocol (WEP), used by the Wi-Fi protocol suite to provide basic security for wireless transmissions. The protocol was designed by a committee that did not include the appropriate experts, and once the protocol went public, it did not take very long for some real experts to show that the protocol was fatally flawed. Having learned their lesson, the committee used some real experts and cryptographers to implement the replacement, called WPA.

Implementing a security protocol can be a bit tricky, since there are a lot of places where you can go wrong. Probably the best defense against improperly implementing a security protocol is to strictly follow good software engineering practices. You should also look into hiring a real security or cryptography expert as a consultant—an expert will know exactly where the weak points of an implementation are and help you to fix them. The other problem is that a security breach is unrelated to the protocol, residing in another part of the system entirely. This means you can easily fall into the trap of believing that your system is secure since you used a secure protocol, but neglecting the rest of the application can make all your efforts with the protocol meaningless.

10.5 Other Security Protocols

To finish out the chapter, we will look at some of the most popular alternative protocols. Some of these protocols and algorithms are more specialized than the general-purpose SSL, which can be both an advantage and an issue. On one hand, the more specialized a protocol or standard is, the more likely it is that it can be implemented in a small footprint. On the other hand, the specialized nature of the protocols leaves less room for flexibility, but for some applications, this may not be an issue. The following is intended to be a brief overview of the protocols, rather than an in-depth analysis, so we can get to the discussion of the Secure Sockets Layer.

10.5.1 SSH

Sharing many similarities to a specialized version of SSL, the Secure Shell (SSH) protocol provides a secure Telnet-style console interface, as well as several other remote-access services. This is a highly useful tool for creating secure remote configuration interfaces, since it is geared toward remote shell functionality. Due to the similarity in abbreviations, SSH is often confused with SSL, but they are separate protocols developed by separate organizations (the Netscape Corporation created SSL, and SSH was developed by a Finnish researcher). The two protocols do share many similarities, such as the use of public-key encryption to exchange symmetric-key encryption keys and the inclusion of both authentication and encryption mechanisms. The current version of SSH is an updated variant of the original that provides more security and has been standardized by the IETF in RFC 4251.

The SSH protocol is an important security tool and as an implementation that is provided as a standard utility in many operating systems. Also, like SSL, there is an open-source implementation available (www.openssh.com). For more information on SSH, check out RFC 4251 and Open SSH.

10.5.2 IPSEC

IPSEC is a protocol designed to work at the IP layer, and is widely used for implementing Virtual Private Networks. IPSEC is a security framework similar to SSL, but with a bit larger scope. The IPSEC protocol, described in RFC 2401, consists of various security mechanisms and policies that are used to build different kinds of secure tunnels between devices. The protocol is quite broad, and implementing it for an embedded system would represent many challenges. We mention IPSEC because it is an important component of network security, but SSL (or TLS) can be used for creating similar secure networking tunnels. SSL has the advantage of being a simpler protocol that is built on top of the underlying network stack (usually between the application and a TCP layer), whereas IPSEC is implemented lower in the stack, at the IP layer as seen in Figure 10.1.

10.5.3 Other Protocols

Finally, to close out our discussion on security protocols, we will briefly mention a few other protocols and standards. There are literally hundreds of protocols available for security, many of them freely available through the IETF RFC standards. There are

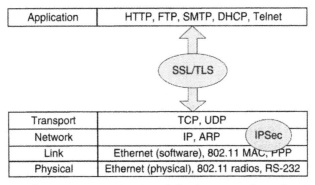

Figure 10.1: IPSEC vs. SSL in the network stack

also numerous security packages available from various vendors, and some of them are quite popular, but those products frequently use predefined and proven algorithms and mechanisms. For cryptographic algorithms, one need only look at the size of *Applied Cryptography* and glance through it to learn that those algorithms number in the thousands. In this chapter we have covered the most popular and recognizable standards, since they are proven in thousands of applications. There are many references on the more obscure algorithms, but they are generally of more academic than practical interest. A few remaining protocols that bear mentioning here are: HTTP authentication, the HTTPS protocol (HTTP over SSL), secure FTP, and PGP.

HTTP authentication is basically a simple mechanism that provides a minimum level of integrity checking. The only thing HTTP authentication does is authenticate a username and associated password for an HTTP session, but it does not do any encryption or other protection of information. HTTP authentication can use either a plaintext mode or a digest mode. In the plaintext mode, the protocol sends the username and password as plaintext over the network and checks them against a database of stored usernames. Other than maybe keeping out nosy (and stupid) kids, there is really no use for this mode. In digest mode, the authentication is a little more secure, since it utilizes a challenge-response mechanism that requires the password to be hashed before being sent. This is slightly more useful than the plaintext mode, but it still provides only a thin layer of security that really is an illusion. It may keep your kids from modifying your personal webpage, but it really doesn't fit the bill for anything else.

Despite the basic uselessness of HTTP authentication on its own, it actually does have a use when it is paired with HTTPS. The HTTPS protocol, described in RFC 2818, is simply the HTTP protocol as used over TLS (or SSL). There are a few slight differences in the way the protocol works, but those differences will not affect most applications. We will actually discuss HTTPS more in later chapters, especially when we look at SSL and client-server protocols, but we mention it here for its use with HTTP authentication. The SSL protocol provides a secure channel between two devices on a network, and it even provides some level of authentication via the use of digital certificates. However, the authentication may not be enough in some cases, as when an SSL server does not provide client certificate authentication (SSL dictates that clients perform authentication on server certificates, but the reverse operation of servers authenticating clients is optional). If you have implemented an embedded device with an HTTPS interface and it does not support client certificate authentication, then anyone can connect to the device. The channel will be secure, but the device has no way of authenticating the user on the other end, unless you implement something yourself. This type of situation is where HTTP authentication can be useful as seen in Figure 10.2. The authentication mechanism is already built into HTTP, so all you have to do is enable it for your HTTP interface. Once the SSL session

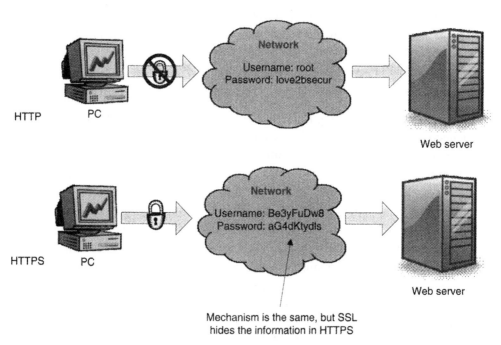

Figure 10.2: Authentication in HTTP vs. HTTPS

is established, the HTTP server will perform password authentication as normal, but in this case, the username and password will not be sent as plaintext over the network. As a result, the authentication is secure.

The last two security technologies we will mention are secure FTP (FTP over SSL) and PGP.[5] Secure FTP (not to be confused with SFTP, the Simple File Transfer Protocol), is simply the implementation of FTP over an SSL connection. It will behave in the same manner as FTP, but the information being transmitted will be encrypted. Quite different from Secure FTP, PGP is an overloaded name—it refers to a product, a company, and a technology. Originally written in the early 1990s, PGP started out as a program to secure emails using public-key cryptography (it used its own algorithm). Still in use today, PGP is provided in commercial products that provide various levels of security and authentication. The commercial products are targeted at specific platforms, but the original technology could be adapted to an embedded application. This would, of course, require you to obtain the source code for the old PGP technology, which is likely protected by copyright, but it would be an interesting project to see if it could be ported to an embedded system—encrypted email could be a useful feature in many embedded applications.

10.5.4 Wrap-Up

We have only scratched the surface of security protocols in this chapter, but it should be enough to get started. In the next chapter, we cover a single protocol in more depth, since it is so important to security in general. This protocol is used in virtually every secure Web transaction today: the Secure Sockets Layer (SSL).

---

[5]PGP is a trademark of the PGP Corporation.

CHAPTER 11

Secure Sockets Layer

Timothy Stapko

The Secure Sockets Layer is so important to network security that we have devoted this entire chapter to it. SSL is the de facto standard for secure Internet transactions. It has achieved this status by being not only secure, but being highly generic and imminently practical as well. SSL exists in the network layer between TCP and your application, providing blanket security to all data transferred over the network. The API for an SSL implementation is typically very similar to the standard network sockets API (POSIX-style). For this reason, it is simple to transform any plain TCP/IP application into a secure Internet application with very little effort.

The standard is very flexible in algorithm choice and features, so it is easy to pick and choose the features from a requirement list. Most PC-based implementations of SSL are monolithic—all features are compiled into every application. However, we will look at a more modular approach that can be used in embedded applications, allowing the developer to enable only the features that are needed, leaving the rest out of the final binary.

Aside from looking at structural procedures for making SSL work on embedded machines, we will also look at optimizations. We will look at hardware assistance as well, since SSL makes use of public-key cryptography, and some vendors have chosen to implement part or all of SSL in a hardware solution. We will also look at implementing SSL in hardware, as part of an FPGA, or as a stand-alone chip. Again, the advantage of hardware implementations is the ability to use larger keys and achieve the same or better performance. We will look at the tradeoffs between these hardware solutions, and a more flexible software implementation.

11.1 SSL History

SSL had a modest beginning as a project at Netscape Inc. in the mid-1990s to secure Web transactions for their (then-dominant) Netscape Navigator web browser. SSL version 1 was an internal project that was never released. Version 2 was publicly released and quickly became a standard way of conducting secure web transactions, eventually beating out other secure offerings from companies such as Microsoft. Unfortunately, some poor decisions in Netscape's implementation of SSL version 2 led a couple of grad students to successfully break the encryption, utilizing attacks on the random number seeds chosen (which included the system clock value—a number that is easily predicted and controlled). The attack revealed that an SSL-secured connection could be broken by a clever attacker in a manner of minutes. As a result of the compromise, SSL version 2 was deemed insecure and unfit for any real transactions. Despite this enormous shortcoming, SSL version 2 is still included in many popular web browsers, though it is now often turned off by default. You should never enable it, unless you want to try your own hand at breaking it.

In the aftermath of the SSL version 2 compromise, Netscape scrambled to provide a fixed implementation of SSL. SSL version 3 was born. At the time SSL version 3 was started, the Internet Engineering Task Force (IETF) began a project to implement a standard version of SSL that was designed to become the eventual standard. The standard, called Transport Layer Security, or TLS (a name generally disliked by the committee members but not hated as much as the other contenders), is now complete and described in RFC 2246. Due to the time it takes for a committee-designed standard to be completed, however, the Netscape version was introduced to the industry much faster than TLS, and version 3 soon became the dominant secure web protocol. TLS is structurally identical to SSL version 3, with a few minor enhancements to the way cryptography is handled, so it is sometimes referred to as SSL version 3.1. SSL version 3 to this day has not been compromised (at least as far as anyone can tell), but it is generally assumed that TLS is more secure and should be used for all new development. TLS is beginning to become the dominant standard, but SSL version 3 is still widely used and is still considered a practical way to secure web transactions. Due to the close similarity to TLS, SSL version 3 can be supported in a TLS implementation with a minimum of effort. From this point on, we will refer to TLS and SSL version 3 collectively as "SSL."

11.2 Pesky PKI

The SSL protocol provides blanket security for network communications by utilizing the advantages of both public-key cryptography and symmetric-key cryptography. As discussed earlier, there are definite benefits to public-key algorithms because we don't have to figure out how to safely exchange keys with the person (or system) with which we wish to communicate. We can just send out our public-key and in receiving keys from others, we can communicate freely and securely. Well, that's not *entirely* true, since someone could potentially provide a public-key and lie about who they really are in order to entice you into communicating sensitive information to them. This exact problem is at the heart of how SSL is typically deployed in web browsers and servers. Before we delve into the deep technical aspects of SSL implementation, we will take a short detour and look at the authentication mechanism provided by SSL, and how this mechanism is used in practice to provide a notion of trust. However, keep in mind that the way SSL is currently used does not mean it is the only method of authentication that can be used—in fact, the mechanism by which SSL certificates are distributed has nothing to do with the SSL protocol. However, SSL is most frequently paired with the distribution mechanism described below, so we will spend a little time looking at it before talking in depth about SSL.

To solve the issue of authentication (namely, the secure distribution of SSL certificates), a few companies put their reputations on the line to provide the public with signing services. Companies such as Verisign provide these services then contract with browser and server makers to provide a "known" public-key in those applications. Assuming that "In Verisign we trust" can be taken as relatively true, then we have a relatively high confidence that any certificate we receive that has been signed by Verisign is valid and correct, and has been sent by the person or organization identified by said certificate.

The whole concept of Public-Key Infrastructures (or PKI) relies on the inherent trust in companies like Verisign to do a satisfactory amount of due-diligence on each certificate they sign. Assuming that Verisign is relatively trustworthy, then this is not a terrible system. In fact, it pretty much represents the backbone of e-commerce as we know it. Any time you browse to your favorite online stores, you are effectively saying that you trust Verisign and the browser and server makers with your financial well-being. This method by which SSL certificates are deployed makes some security experts nervous, but it has proven to be highly effective in promoting "secure" transactions on the Web.

SSL makes PKI possible through the construction of the digital certificates it uses for authentication. Since signing a digital certificate with a private RSA key, for example, is able to be chained any number of times back to a common root, one company can provide that root certificate as part of standard web application implementations and extend their trust to numerous third parties. SSL is designed to use any number of root certificates for authentication, so just a few companies provide root certificates, and everyone can use the system.

To bring this discussion back into focus for the purposes of embedded systems development, let's ask the obvious question: Just how many of these "root" certificates are there, and does my embedded application need to store *all* of them? It turns out that, like almost everything we have looked at in this book, it depends on the application itself. If your application is a webserver that will be deployed on the public Internet, then all you need to do is store a single certificate that you pay to have signed. If your device is designed to connect to any number of public Internet access points, then you will probably need to store at least a dozen or two root certificates, which unfortunately can be as large as a few kilobytes each. It is quite possible that a public SSL client would need as much as a megabyte of space to store all the relevant certificates to assure the highest probability of compatibility with any unknown server. However, it is likely that most embedded applications will either be an SSL server implementation, or they will not need to connect to every server ever created. For this reason, we will look at other methods for distributing SSL certificates that are more practical for embedded systems.

11.3 PKI Alternatives

As it turns out, the PKI system used for e-commerce on the Web is not the only way to deploy SSL certificates. A lucky thing that is, too, since many embedded applications would be hard-pressed to provide the space required to store all the root certificates. PKI is simply one implementation that has been pushed by large companies for web browsers, but we have the power to do things differently. In order to provide authentication for SSL in embedded applications, we will look at a couple of different options for deploying certificates that in many cases are more applicable and practical than PKI. In Figure 11.1 we can see the relative differences in the different deployments.

The simplest path to SSL authentication is to create your own self-signed certificate. This essentially means that you will create a certificate with the appropriate public-key/ private-key pair, put the public-key in the certificate, and sign it using the private one. As it turns out, you don't have to use Verisign after all! Once you deploy your application, authentication is simply a matter of using the private key you created to authenticate the certificate being provided by your application. If your application is a web server, then you can install the certificate into the browser's root certificate cache—from that point

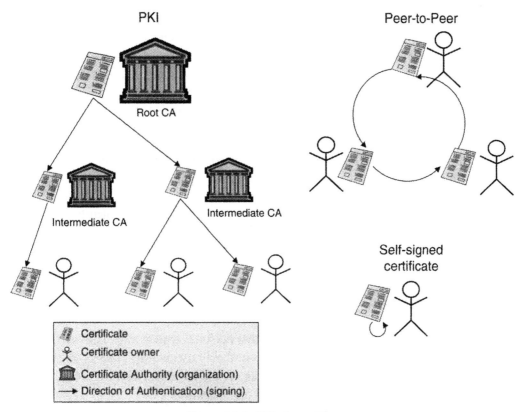

Figure 11.1: PKI alternatives

forward, the browser will always accept that certificate (assuming it hasn't been tampered with or corrupted). Alternatively, if your application is an SSL client, then your server can provide a certificate signed using the same private key, which the application can then verify directly using the single certificate. For simple applications, a self-signed certificate is the only way to go.

So we can create a self-signed certificate, but what if we have to deploy hundreds of devices, each with a unique certificate? Well, you could pay anywhere from $100 to over $1000 to get each certificate signed by a public signing company or organization, or you can do it for free and be your own CA. How does *that* work? Remember that we can self-sign a certificate, so there is nothing keeping us from signing other certificates.

All you need to do is to create a single root certificate (which simply means it is self-signed and will be used to sign other certificates), and then use that certificate's private key counterpart to sign all your other certificates. If you install your single root certificate in a web browser, then you have automatic authentication for all the certificates you create and sign using that certificate. As you might have guessed, this particular method of certificate distribution is quite useful in venues other than embedded systems, such as providing authentication on a corporate intranet, for example.

Creating a self-signed certificate and going into the business of being your own CA are both very reasonable ways to provide basic SSL authentication for your application. However, it may be that neither those methods nor PKI are suitable for what you are looking for. Fortunately, there are some other ways of distributing trust in the form of certificates, and most of them center on some concept of peer-to-peer networking.

There are various implementations of peer-to-peer authentication, and the idea hasn't caught on for e-commerce, so we will just stick to the basic ideas behind peer-to-peer authentication mechanisms. The basic idea is that of extending trust. As an example, start with someone you know and trust. Assuming that person is trustworthy, you can then assume that someone they say is trustworthy is probably trustworthy. This is fine unless that third party deceived your friend, but then again, the liar could deceive you as well. Back to the point, though—if you can extend your "network of trust" by using mutual friends to verify new people that enter the network, you can extend trust to just about anyone. Figure 11.2 shows a simple peer-to-peer PKI example.

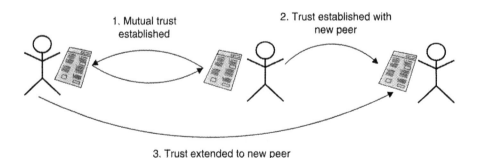

Figure 11.2: Peer-to-peer PKI

11.4 SSL under the Hood

Now that we have covered the basics of authentication in SSL, it is time to dig in and look at what makes SSL tick. To start off, we will look at the basic structure of the communications layer of the protocol, starting with the backward-compatibility handshake.

One of the issues with SSL is keeping backward-compatibility with older versions. After all, it would be better to communicate using SSL version 2 than nothing at all (well, not much better, if your information had any value, but you get the point). The other action that can occur is that two SSL implementations may actually both speak newer versions of SSL or TLS, and they can negotiate up to the most recent version they both support. This backward-compatibility feature can be optional, allowing for newer implementations to communicate directly, but the feature is required at the very beginning of the SSL handshake, so we will look at it first.

The SSL handshake begins with a network connection of some variety; in most cases the connection is a simple TCP/IP network connection (the lower-level and physical layers do not matter). SSL is essentially a TCP/IP application, and for the most part does not require any knowledge of the underlying network mechanism. That being said, SSL does rely on the network protocol being used to handle dropped packets, so a reliable protocol like TCP will work just fine for the lower level, but an unreliable protocol like UDP will not. This requirement of SSL does not reduce the security of the protocol or cause any other problems, even if the TCP connection is compromised. SSL will be able to detect any attempt to corrupt the information using the message digest authentication mechanism (discussed later in this chapter), and the encryption prevents eavesdropping. The problem with using an unreliable protocol under SSL is that SSL will flag every dropped packet as a potential attack.

Once the network connection has been established, the SSL handshake itself can begin. The SSL communication layer is similar to other network protocols in that it consists of a header containing protocol information and a body that contains the information being transmitted. In SSL, these units are called *records* (see Figure 11.3) and can be thought of as being analogous to the TCP frame and IP packet. In the handshake, the SSL records are unencrypted, since the encrypted connection, called the SSL *session*, has not yet been established. The backward-compatibility handshake begins with an unencrypted SSL record called the *Client Hello*, which is sent by the SSL client wishing to establish the

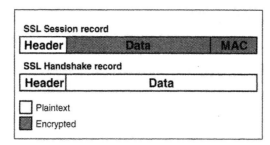

Figure 11.3: SSL record

SSL session with a particular server. By convention, the application initializing an SSL connection is always the client, and follows a specific set of actions that correspond to actions on the SSL server.

The Client Hello message kicks off the whole thing, as we see in Figure 11.4. In backward-compatibility mode, the Client Hello is actually an SSL version 2 message in an SSL version 2 record, which differs slightly from the record in later versions. Originally, SSL used some fairly simplistic methods for generating the encryption keys and verifying that the handshake was not tampered with. As a result of the simplistic nature of these methods, SSL version 2 was compromised as we mentioned earlier. In SSL version 3 and TLS, the methods were improved, resulting in a more complex handshake protocol. With the backward-compatibility mode, the initial version 2 Client Hello message is translated into a version 3 (or TLS) Client Hello, and the handshake proceeds as usual. If the client truly supports only SSL version 2, then the connection will be aborted when the server sends a Server Hello back to the client indicating that it supports only SSL version 3 or TLS.

The Server Hello is a message that is built by the server upon receiving a Client Hello. The Client Hello message contains a number of *ciphersuites* that the client supports and is willing to use. A ciphersuite is simply a collection of algorithms and some specifications for the algorithms represented by an enumeration. For SSL, the ciphersuites consist of a public-key algorithm, such as RSA, a symmetric-key algorithm such as AES (also called the *bulk cipher*), and a hashing algorithm. The Client Hello also contains a string of random data that is used as a seed for generating the *session keys* later (we will explain what session keys are and how they are generated a bit later when we get to the key-generation phase of the SSL handshake).

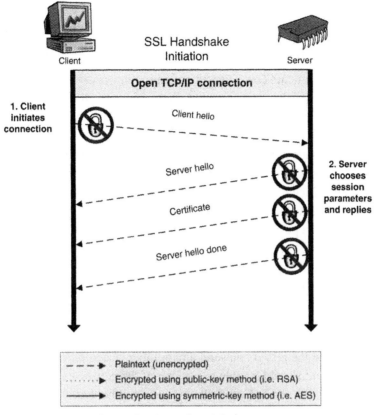

Figure 11.4: SSL handshake part 1

The server selects the ciphersuite it wants to use based upon its capabilities and the priorities set by the application designer or user for selecting cryptographic algorithms. This is the negotiation phase of the handshake—the ciphers actually used for the communication are determined by a combination of what the client offers and what the server can support. The server stores away the information from the client (such as the random data used later for seeding the session key generation) and generates a Server Hello message, which is sent back to the client. The Server Hello message contains the ciphersuite selection, a server-generated string of random data used in combination with the client random data for seeding the key generation, and some additional information to get the session started. Another thing that both the client and server do in this initial phase is to begin a hash that will eventually represent all of the messages in the entire handshake. When we get to the Finished messages at the end of the handshake, we will see how these hashes are used to determine the integrity of the handshake process.

Following the Server Hello message, the server sends a message containing its digital certificate. The SSL certificate contains authentication information about the server that the client can use to verify the authenticity of the server, as well as additional information about the server and, most importantly, the certificate contains the public-key that will be used to exchange the information used to generate the session keys used by the secure SSL tunnel once the handshake is complete and the session is established. The certificate also contains a digital signature (an encrypted hash) that the client can use to verify the information in the certificate. The digital signature is usually signed by a third-party Certificate Authority (CA) using the reverse-RSA operation (encrypting using the private key, decrypting using the public-key). The client stores a number of digital certificates from different CA entities, and uses the public-keys stored therein to decrypt incoming server certificate signatures. In this manner, a client can verify an unknown server through the concept of distributed trust. If the client trusts the CA, and the CA trusts the server, then the client can trust the server as well. As long as the CA is reputable and checks up on all the servers it signs certificates for, this system works pretty well.

SSL is designed so that the server can send any number of certificates to the client, the idea being that the client should be able to use one of the certificates to authenticate the server. For this reason, the server Certificate message can be duplicated any number of times, once for each certificate the server sends to the client. In order for the client to know that the server is done sending certificates, the server sends a final message called the "Server Hello Done" message, indicating to the client that all certificates have been sent and the next phase of the handshake can begin.

The second phase of the SSL handshake begins when the server has sent the Server Hello Done message and the client begins the key exchange phase, as seen in Figure 11.5. The first thing the client does is to parse the server's digital certificate and verify the server's authenticity. It does this by first decrypting the digital signature using a locally-stored public-key (from a CA certificate), and comparing the decrypted hash to its own hash of the server certificate's contents. If the hashes check out, the client then proceeds to check the certificates *Common Name* with the address (URL or IP address) of the server for additional verification. It also checks a field in the certificate that indicates the effective date range that the certificate is valid for. If the address does not match the Common

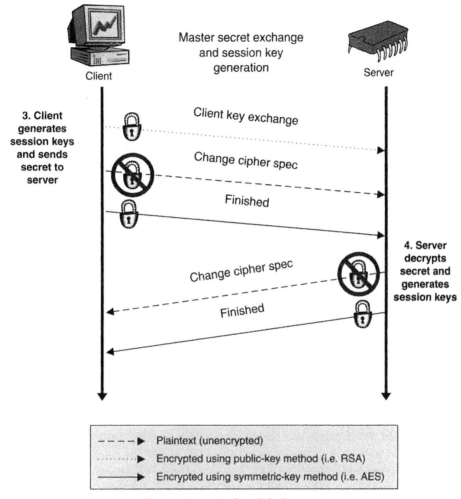

Figure 11.5: SSL handshake part 2

Name, or if the certificate has expired, the SSL connection is dropped. Optionally, and what is usually done in web browsers supporting HTTPS, the client can opt to continue the handshake even if one or all of the authentication checks fails. This usually results in a dialog box warning the user of the browser that something is amiss. If the user chooses to ignore the warnings by selecting to proceed anyway, or if the checks are all successful, the handshake proceeds.

The client uses the information from the client and server Hello messages to generate the session keys, which will be used with the bulk cipher to securely send data back and forth once the handshake is complete. The client generates a random string, called the Pre-Master Secret, which is then encrypted using the server's public-key (obtained from the server's digital certificate). The encrypted Pre-Master Secret is then sent to the server in the Client Key Exchange message. Upon receiving the message, the server can decrypt the data using its private key (the complimentary half of the public-key sent in the certificate earlier). Following the Client Key Exchange message, the client sends the last message in the handshake that is not encrypted using the session keys (which we will talk about in a moment). This message is called the Change Cipher Spec message, and it indicates to the server that all following messages will be encrypted using the session keys. The final handshake message sent by the client is called the Finished message, which is encrypted using the session keys. In a bit, we will talk about what the Finished message entails and how the server ends the handshake, but first, where do these session keys come from?

Looking back at the beginning of the handshake, we mentioned that the client and server both generate a random string of data that is communicated in the initial "hello" messages. We also said that the client generates a string of random data called the Pre-Master Secret that is used for generating session keys later. The Pre-Master Secret is small enough to be easily encrypted with the public-key algorithm being employed in the handshake, but large enough that it is not easily broken (64 bytes is a perfect size for a single 512-bit RSA-encrypted block, for example). The Pre-Master Secret and the random data sent during the initial handshake messages are used to construct the Master Secret on both the client and server—it is generated on the client as soon as the Pre-Master Secret is created, and on the server after the public-key decryption is performed (thus revealing the Pre-Master secret to the server). The Master Secret is essentially a cryptographically secure hash of the other data that is always 48 bytes in length. The Master Secret is used to generate what is called the *key material* block, essentially an arbitrary-length expansion of the Master Secret from which the session keys are derived as seen in Figure 11.6.

Generation of the key material block is done using a hash expansion algorithm that differs slightly between SSL version 3 and TLS (we will not cover what SSL version 2 does). In SSL version 3, the hash expansion is defined as part of the protocol. TLS uses a method called P_hash, which is composed of 2 hash algorithms, MD5 and SHA-1, and a hash wrapper called HMAC. HMAC, described in RFC 2104, provides a mechanism to

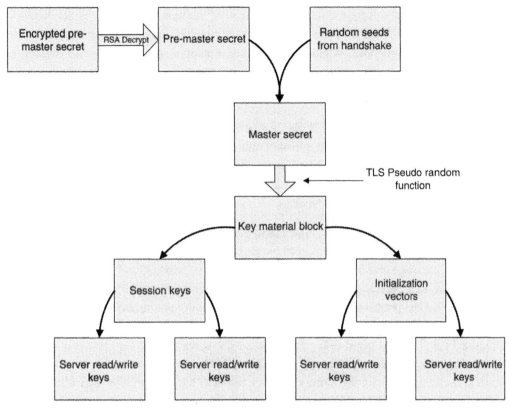

Figure 11.6: Key material block from master secret

add keying to hash functions. Essentially, both the SSL version 3 and TLS mechanisms are the same—the SSL version 3 mechanism is based on an early draft of HMAC. Both mechanisms do the same thing, which is to take the Master Secret and expand it into the arbitrary-length key material block. For typical SSL implementations, the key block is a little over 100 bytes, which requires 5 or 6 iterations of the key generation mechanism, the output of SHA-1 is 20 bytes (even with HMAC), so the algorithm strings several output strings together. The keys for HMAC (and the SSL version 3 mechanism) are part of the initial data from the Master Secret and serve to provide an additional level of security. Finally, the results of the expansion (using P_hash in TLS) are run through a final process that in TLS is called the pseudo-random function, or PRF. The result of the PRF operation is the final key material block.

The key material block represents the raw material of the session keys. Once the block is generated, the SSL/TLS protocol divides it up into pieces, each of which is used for a session key or other purpose. A general SSL session using a stream cipher will extract 2 MAC secrets, which are used to seed the integrity Message Authentication Codes (discussed below), and two session keys, one each for reading and writing. If a block cipher is used, the key block is further partitioned to produce two initialization vectors for the block cipher to use. You will notice that SSL generates two of everything, which is additional security precaution. It assumes that if one direction of communication (i.e. client to server) is broken by breaking the key, the other direction will still be relatively safe since it uses a separate key entirely. The same thought goes into the MAC secrets and the block cipher initialization vectors.

Now we go back to the handshake. The client has generated the keys and MAC secrets by the time the Pre-Master Secret is sent, so that it can immediately send its Change Cipher Spec message and begin the encrypted session by sending its Finished message to the server (the first message that is actually encrypted using the generated session keys). Upon receiving the public-key encrypted message from the client, the server generates *the same key material and session keys from the Pre-Master Secret as the client* as seen in Figure 11.7. This last point is key (pun intended)—even though the relatively small Pre-Master Secret is the only "secret" data shared by the server and client, they utilize identical methods to derive the same exact key material block, which is what allows them to communicate later. Once the server is done generating keys, it sends its own Change Cipher Spec message back to the client, indicating that it has generated its own session keys and is ready to initiate the secure session.

Finally, to close out the handshake process, the Finished messages are sent, one from the client, one from the server, as seen in Figure 11.8. The Finished message for both the client and server contains a hash of all the previous handshake messages (note that the server hashes the client's Finished message for inclusion in its own Finished hash). The message is the first to be encrypted using the newly generated session keys to avoid tampering. Both the client and server have a local hash of the handshake messages and the hash received in the other's Finished messages. Comparing the local hash to the remote hash (after the messages are decrypted) provides a final check that the handshake was not corrupted or tampered with in any way. Assuming the hashes check out, at long last the encrypted session begins.

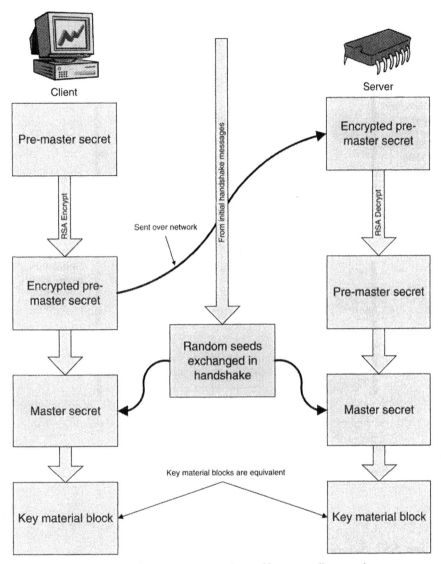

Figure 11.7: Simultaneous generation of keys on client and server

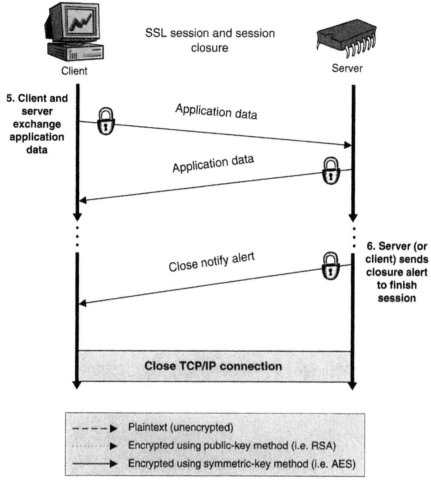

Figure 11.8: SSL handshake 3

11.5 The SSL Session

The handshake makes up the bulk of the SSL protocol, since it is the basis for all the security of the session. The session itself is fairly boring, since all that really happens is an encryption using the session keys derived during the handshake. One difference between a "regular" encrypted message and an SSL message (the SSL Record) is the use of a Message Authentication Code, or MAC, for verifying the integrity of messages.

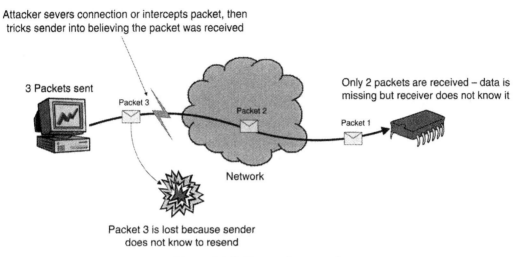

Attacker severs connection or intercepts packet, then
tricks sender into believing the packet was received

3 Packets sent

Packet 3

Packet 2

Only 2 packets are received – data is
missing but receiver does not know it

Packet 1

Network

Packet 3 is lost because sender
does not know to resend

Figure 11.9: Truncation attack

The MAC is a simple hash using an algorithm determined by the ciphersuite being used—a couple examples of hashes used are SHA-1 and MD5. The SSL MAC is slightly more than a hash of the data being sent, since it is actually seeded using the MAC secrets derived along with the session keys. The purpose of the MAC is simply to verify that the data in the SSL record was not compromised or tampered with, both things an attacker could do to disrupt communications without needing to break the actual encryption.

The session continues as long as both sides are willing to talk. At the end of the session, either the client or the server sends an *alert* to the other side to confirm that it is finished communicating. An SSL alert is simply an SSL record containing an encrypted message (again using the session keys) that usually indicates something abnormal has happened. In the case of the final alert, called close_notify, it is actually expected behavior and in fact it serves an important purpose. Without the final alert to confirm that communications are to be halted, an attacker could perform a *truncation attack* as seen in Figure 11.9, which basically means that not all of the data gets to its destination. This may not immediately seem like a major issue, but think of bank transactions—you probably wouldn't be too happy if the last few deposits you made simply disappeared since no one knew to check for them. The close_notify alert allows both the server and client to verify that all information has safely made it to its destination.

11.6 SSL in Practice

SSL is a fantastically successful protocol. Implemented in almost every web browser and web server, it provides the security backbone for the majority of Internet transactions. As the latest incarnation of SSL, TLS is being used even in other protocols as a fundamental building block for security in wireless and mobile applications. One reason for the success of SSL is the fact that it has an answer for just about every attack out there and provides multiple layers of redundant security, making it a robust and provably safe protocol. Another reason SSL is so widely implemented is that the protocol is flexible—it defines how cryptography should work together to provide security without actually defining what "cryptography" should be. Even the key generation, which uses standard hash algorithms, can be changed if needed.

CHAPTER 12

Cryptography

Timothy Stapko

In this chapter we will look at choosing and optimizing cryptographic algorithms, particularly for resource-constrained systems, such as embedded systems. We will look at various strategies for choosing algorithms for specific applications, and look at some specific algorithms as well as some strategies to avoid some of the more expensive cryptographic operations without compromising security.

First of all, we will look at whether cryptography is even necessary. Some applications can actually get away without using traditional cryptography. These applications utilize other mechanisms, such as hashing algorithms, in order to provide some assurance about data. The big advantage here is that hashing algorithms are typically many orders of magnitude faster than symmetric or public-key cryptography. One classification that will help us make the decision is whether or not we care about eavesdropping. If we only care that the data is reliable, and do not care who gets it, we can definitely avoid cryptography. We will look at these types of applications and come up with some other general rules for determining if an application requires the more expensive operations, or can get away with less.

We will look at cryptographic hashing, and discuss some optimizations and other tricks that can be used to speed up commonly used algorithms. Due to recent discoveries of weaknesses in the commonly used algorithms MD5 and SHA-1, we will also look at the future of hashing, and see if there is an alternative method, or new algorithms we can choose.

For those applications that require absolute secrecy, there are many options that will help the embedded developer meet the goals of performance, cost, and security. We will look at some specific algorithms, such as DES/3DES, which is slow and obsolete but still necessary for some applications. AES, the replacement for DES, will also be covered, and specifically, some of the hardware options available. Additionally, we will look at RC4 and other ciphers that provide a marked performance advantage over the bulkier DES and AES, but are not as provably secure.

Finally, we will cover public-key algorithms, and specifically, RSA, by far the most commonly used public-key cipher. Unfortunately, RSA and other public-key algorithms are extremely slow, requiring hardware assistance on many platforms—not just small embedded systems. However, these algorithms are essential to certain protocols (primarily SSL and SSH), so we will look at ways of handling the computationally intense public-key operations, including software tricks and hardware assistance.

12.1 Do We Need Cryptography?

One of the first steps in building a secure embedded system is to see if cryptography is actually needed. Whenever security is discussed, many engineers will immediately think of cryptography as *the* solution, when in fact, many options may exist that do not strictly require cryptography. Sure, cryptography is an important part of securing applications, but it is not a security panacea, nor is it always necessary for building a secure system. To many people, cryptography is a mysterious and confusing, almost magical area of computer science that is reserved for cloistered geniuses at top academic institutions and the US National Security Agency. The NSA is mysterious and secretive, and it is believed to employ some of the greatest cryptographers in the world, only adding to the idea that cryptography is an untouchable discipline.

There have been some attempts over the last couple of decades to bring cryptography out from behind its cloak of mystery, most notably with the publication of *Applied Cryptography* by renowned security expert Bruce Schneier. Schneier, more than any other individual, has brought cryptography into mainstream computer science by unveiling the techniques employed in most common cryptographic algorithms. That being said, a stream of inaccurate portrayals of cryptography in Hollywood and television, combined with unrelenting advertising campaigns for various security products, have only served to keep those not involved directly in computer science in the dark. For this reason, it can be difficult to explain to managers or hardware engineers what exactly cryptography is, let alone convince them that it will not solve their security problems.

Given the confusion surrounding cryptography, it can be extremely difficult for a systems engineer to determine what type of cryptography, if any, is needed. The truth is that in some circumstances, no cryptography is needed at all. For instance, if you do not care about anyone seeing the information being transmitted, but rather you are only concerned that it is not tampered with in transit, you really do not need a complete cryptographic suite to suit your needs. For example, let's take a look at publicly available stock market reports. For many investors, keeping tabs on the latest price of a stock can mean thousands or millions of dollars being made or lost. The price of stock in a public company is readily available, but if someone were able to change that price in transit during a trading session, that person would be able to wreak havoc upon unknowing investors. Obviously, there is a security risk for any system that transmits this data, but we do not need to hide it from anyone (unless your goal is to provide only your customers with the most up-to-date information, but that is another matter). Instead, we can just use a simple hash algorithm to verify the integrity of the data being transported. Not being a full-blown encryption scheme, the system's requirements can be lowered to support just the hashing (or those resources could be used for improving the performance of other parts of the system).

In order to see what type of security an application will need, we can divide applications into several distinct categories based upon the type of information each application deals with. After all, the purpose of computer security is to protect information, so this is a good place to start. The following categories should be sufficient for most applications, ordered from the lowest level of security to the highest level of security:

- No security required (may be optional)—applications such as streaming video, noncritical data monitoring, and applications without networking in a controlled environment.

- Low-level security (hashing only, plaintext data)—applications delivering publicly available information such as stock market data, data monitoring, networked applications in a controlled environment, and applications requiring an extra level of robustness without concern about eavesdropping. Another example of this would be the distribution of (usually open-source) source code or executables with an accompanying hash value for verifying integrity, usually an MD5 or SHA-1 hash of the files that is to be calculated by the end user and compared to the provided hash.

- Low-medium security (hashing plus authentication)—general industrial control, some web sites, and internal corporate network communications.

- Medium-level security (cryptography such as RC4 with authentication, small key sizes)—applications dealing with general corporate information, important data monitoring, and industrial control in uncontrolled environments.

- Medium-high security (SSL with medium key sizes, RSA and AES, VPNs)—applications dealing with e-commerce transactions, important corporate information, and critical data monitoring.

- High-level security (SSL with maximum-sized keys, VPNs, guards with guns)—applications dealing with important financial information, noncritical military communications, medical data, and critical corporate information.

- Critical-level security (physical isolation, maximum-size keys, dedicated communications systems, one-time pads, guards with really big guns)—used to secure information such as nuclear launch codes, root Certificate Authority private keys (hopefully!), critical financial data, and critical military communications.

- Absolute security (one-time pads, possibly quantum cryptography, guards with an entire military behind them)—used to secure information including Cold War communications between the United States and the Soviet Union regarding nuclear weapons, the existence of UFOs, the recipe for Coca-Cola, and the meaning of life.

The above categories are intended as a rule-of-thumb, not as strict guidelines for what level of security you will want for a particular application. Obviously, your particular application may fit into one of the above categories and require a higher level of security than indicated. You could also reduce the recommended level of security, but remember that security levels are always decreasing as hardware improves—what we consider to be "high" security today may not be very secure at all in 10 years. The prime example of this is the fact that 56-bit DES was considered nearly "unbreakable" when it was first introduced, but was broken just 20 or so years later with a concerted effort. Now, more than 10 years after that, DES is practically obsolete, and even the beefed-up triple-DES is showing its age.

If you haven't figured it out by this point, this chapter is primarily focused on applications requiring security up to medium level and some medium-high level (as described in the categories above). If your application falls into the medium-high category or higher, you will definitely want to do some serious research before jumping into implementation. In fact, you should do a lot of research if your application needs security at all. This book and others will be useful to determine what you need for your application, but you shouldn't ever take one person's or book's viewpoint as gospel. It is highly likely (pretty much guaranteed) that any single viewpoint on security will miss something important that can come back to haunt you later. That being said, we will move forward and attempt to make some sense of cryptography for systems not strictly designed for it.

12.2 Hashing—Low Security, High Performance

Ideally, we would not need any security in our applications at all, instead having the luxury to focus all resources on performance. Unfortunately, this is not reality and we have to balance security with performance. Fortunately, the security-performance tradeoff is relatively linear—higher security means some performance loss, but the balance can be adjusted fairly easily by choosing different algorithms, methods, and key sizes. For the best performance with some level of security, you can't beat the hash algorithms. As far as resources are concerned, they use far less than other types of cryptographic algorithms. Not strictly cryptography in the classic sense, hash algorithms provide integrity without protection from eavesdropping—in many applications this is all the security that is really needed.

How do we actually make hashing work for us in our applications? Probably the most common use of hashing is to provide a guarantee of integrity. The concept behind cryptographically secure hashing is that hash collisions are very difficult to create. Therefore, if you hash some information and compare it to a hash of that information provided to you by a trusted source and the hashes match, you have a fairly high certainty that the information has not changed or been tampered with in transit.

Another mechanism that employs hashing to provide a level of security is called "challenge-response," after the way the mechanism works. In a challenge-response operation, the system or user requesting the information provides a "challenge" to the sender of the requested information. The sender must then provide an appropriate "response" to the requestor, or the information is deemed invalid. There are many variants of this simple method for verifying the authenticity of information, but for our purposes, the variant we will talk about uses hashing to prove that the requestor and sender (client and server) both know some shared secret only known to those systems. The challenge is typically some cryptographically secure random number that is difficult to predict, and it is sent by the client to the server (or from the server to the client, as is the case with some HTTP applications). Once the server receives the challenge, it calculates a hash of the challenge and the secret, and sends the hash back to the client. The client also performs the same operation on its copy of the secret, so when it receives the hash it will know that the server knows the same secret.

The reason the mechanism works is because the server could not produce the correct hash value unless it knew the secret. The random number for the challenge ensures that every hash sent back to the client is different; otherwise an attacker could simply send that value to the client and there would be no security. The challenge-response hashing described here does suffer from a man-in-the-middle vulnerability, since an attacker could intercept the message in transit from the server back to the client. This attack requires the attacker to have access to the routing system of the network in order to spoof the address of the server and the client so it can intercept and retransmit the messages. However, for security at this level, it can usually be assumed that breaking the network routing will be sufficiently difficult to provide a decent level of security.

So now we know how hashing can be used to provide different types of security, how do we actually use it in practice? There are several hashing algorithms available, but the most common are MD5 and SHA-1. Unfortunately, the MD5 algorithm has been shown to have some weaknesses, and there is a general feeling of uneasiness about the algorithm. SHA-1 has been the focus of some suspicion as well, but it has not had the same negative press received by MD5. In any case, both algorithms are still heavily used in the absence of anything better. For this reason, we will look at MD5 and SHA-1 for our examples. Though there is some risk that these particular algorithms will become obsolete, the methods described here for their use in cryptographic systems can be applied to any similarly structured hash algorithms.

In practice, many embedded development tool suites provide libraries for MD5, SHA-1, or both hashing algorithms. The algorithms themselves are not too difficult to implement (they actually consist of a number of quite repetitive operations), but the convenience of having a pre-tested implementation to use is quite nice. Hashing algorithms are fairly easy to optimize as well, so it is quite likely that the provided implementations will already be fairly optimal for your target hardware.

Using hash algorithms is quite easy, since they have only three basic operations that are provided in the user API:

1. Initialization, which sets up the state data structure used to actually perform the hash.

2. Hashing, which operates on the incoming data, typically in raw bytes (may also be text).

3. Finalization, which finishes up the hash, and copies the result into an output buffer.

In this chapter, we focus primarily on C-based development tools, so the following examples of hashing in practice directly apply to the C language, but the principles can be applied to other similar languages (such as Java or C++).

The basic operation of hashing algorithms uses a buffer, in C an array of *char* type, as a workspace for the hashing. The algorithms utilize a structure that provides state information across hashing operations, allowing for the hash to be added to in multiple operations, rather than all at once in a single hashing operation. This feature is the reason that each of the hashing APIs has three distinct operations. The algorithms are set up so that they can utilize the state structure to keep the in-progress hash intact without requiring separate support from the application designer. With this setup, once the user application is ready, it can provide an output buffer for the finalization operation. This is important for networked environments where several hashes may be in progress simultaneously—the user need only keep track of the hash state structures.

So what does hashing look like in a real application? In the following C program, the user input is hashed into an output buffer (actual SHA-1 API may vary):

```
#include <sha1.h >
#include <stdio.h >
main () {

      char input_buf[128], output_buf[20];
      struct SHA1_state sha_state;
      int i, input_len;

// Initialize the state structure, requires only a reference to the struct
SHA1_init(&sha_state);

// Get user input, make sure buffer is cleared first
memset(input_buf, 0, sizeof(input_buf));
scanf("%127s", input_buf);

// Hash the input, with a length equal to the size of the user input. Note that
// the input to the SHA1_hash function can be of any length
// !!! Danger, strlen can overflow, so we terminate the buffer for safety
input_buf[127] = 0;
input_len = strlen(input_buf);

      SHA1_hash(&sha_state, input_buf, input_len);
```

```
// Finally, finalize the hash and copy it into a buffer and display
SHA1_finish(&sha_state, output_buf);
for(i = 0; i < 20; ++i) {

    printf("%X ", output_buf[i]);
}
printf("\n");

} // End program
```

Listing 12.1: Hashing with SHA-1

That's it—hashing is a very simple operation in code. Notice that we do some defensive programming when using the strlen function. Unfortunately, the C programming language does not have very good standard library support for protecting against buffer overflow.

In our little program example, if the user entered enough data to fill up the buffer to the end (more than 127 characters), we are relying on scanf to be sure that the last element of the array contains a null-terminator. In our program, the scanf "%s" type is used in the format string with the optional width format parameter, so it should not cause any issues for the call to strlen later. However, if someone was to change the scanf to some other form of input, then the assumption may be violated. For this reason, we add an unconditional null-terminator to the end of the array to be sure that strlen will terminate appropriately.

The hashing example above illustrates the use of a cryptographic algorithm to protect data, but it also highlights the fact that anything in the program can become a security issue. The use of standard C library functions such as strlen can lead to unexpected and unintentional behavior. Sometimes this behavior goes unnoticed; sometimes it leads to a crash. All it takes is one malicious attacker to find the flaw in your program and exploit it somehow. It may not be that the attacker gains access to the whole bank, but shutting down a few hundred automated teller machines could do a lot of financial damage. All that the attacker needs is for you, the developer, to stop paying attention. The example has a trick or two that help to keep the program safe, such as terminating the buffer (a simple operation that could be easily overlooked), but what if the algorithms themselves were the problem. In the next section we will look at some recent developments with two major hash algorithms (MD5 and SHA-1) that cast doubt on their continued utility.

12.2.1 Is Hashing Considered Dangerous?

In the past few years, cryptography has come into a lot of visibility, and the old faithful algorithms that have served us well for years are now being put to the test. The vast amount of information on the public Internet that needs to be protected has lead to a virtual stampede of corporations, governments, organizations, and individuals studying the security mechanisms that form the foundation of that protection. People on both sides of the law (and with varying levels of ethics) are racing to discover flaws in the most commonly used algorithms. After all, boatloads of money can be made if a security breach is discovered. For the "good" guys, the rewards are recognition and (hopefully) prompt fixing of the issue. The "bad" guys profit in numerous different ways. The end result is always the same, however: If an algorithm is broken, it usually means it's useless from that point on.

This insane virtual arms race has revealed that it is extremely hard to develop secure cryptographic algorithms (it's easy to write broken cryptographic algorithms), and it appears that hashing may be among the most difficult. The two major hash algorithms in use today (notably by SSL and TLS) are MD5 and SHA-1. At the time of the writing of this text, MD5 is considered "mostly broken" and SHA-1 is "sorta broken." What does that mean? Well, there are various ways a hash algorithm could be broken from a cryptographic standpoint. Some of these are:

- Take two arbitrary but different messages and hash them. If you can easily calculate a hash value that is the same for these different messages (a "hash collision"), then the algorithm is somewhat broken, and potentially seriously broken.

- Given a hash value, compute an arbitrary message to hash to that value. If this is easy, then the algorithm is a little more broken, since this starts to get into the area where the flaw can be exploited.

- Generate a meaningful message that generates a hash collision with another meaningful message. If this is easy, then the algorithm is mostly broken, and it is highly likely it provides no security whatsoever. If this is true for an algorithm, it is very easy to fool someone into accepting incorrect information (or worse, damaging information such as a virus or Trojan horse).

Each of the above levels of compromise is based on the idea that performing these operations on the algorithm is "hard" (infeasible given current technology and future technology for at least a few years). They all feed into one another as well, so if you can find an arbitrary hash collision, it is often easier to discover the other attacks.

Unfortunately, both MD5 and SHA-1 have been discovered to have vulnerabilities. For MD5 there are several demonstrations of ways to generate different meaningful messages that generate the same MD5 hash value. Granted, these operations are currently somewhat contrived and generally a little tricky, but it is only a matter of time until someone figures out how to do it fast and easy. Generally speaking, we don't need to rush out and pull MD5 out of all our applications, but if it isn't too difficult to do so, it is recommended. MD5 should not be used in new applications whenever possible.

The MD5 algorithm is fairly broken, but fortunately for us (and the rest of the world), SHA-1 is not as broken (yet). Researchers have discovered something akin to the first vulnerability (the arbitrary hash collision) in SHA-1, but as yet, there does not seem to be a way to translate that vulnerability into a security breach (as seems to be the case with MD5). Possibly by the time you read this, however, both SHA-1 and MD5 will be considered obsolete and will be replaced by something new (or at least in the process of being replaced).

SHA-1 and MD5, albeit broken, are still in heavy use today and will continue to be for some time. They are so integrated into the security mechanisms that we have come to rely on that it will take years to pull them all out and replace them. Even then, many legacy systems may still require them in some capacity. This scenario obviously assumes there is a decent replacement. There are some contenders in the same family as SHA-1, but if that algorithm fails, it may be hard to tell if the SHA-1 vulnerabilities translate to its brethren.

One ray of hope, however, is that there may be another competition to develop a new cryptographic algorithm as was done with AES. Only this time, the target would be a hashing algorithm to replace the aging and ailing MD5 and the slightly less damaged SHA-1. Only time will tell what ends up happening with this. Heck, we might see a quantum computer implemented in the next few years that could make all of our current strategies obsolete overnight. We still need something in the meantime, however, and it worked for AES, so it may work for this too.

12.3 To Optimize or Not to Optimize...

So far in this chapter we have focused on the choice of a single class of algorithms, hashes, as a way to get the most performance out of our security. While hashes are extremely useful for a wide variety of applications, they really cannot provide the same level of data protection that a "true" cryptographic algorithm, such as AES, can provide. One of the problems with hashes is that they produce a fixed-size output for arbitrary-length data. It doesn't take much thought to realize that if the message is larger than the size of the hash (and maybe even if it is smaller), some information is lost in processing. While hashes can work to give you a guarantee (large or small) that the data is intact and represents the original information, they cannot be used (at least directly) to encode the data such that it is extremely difficult for an attacker to get at it, but also can be decoded back into the original information given the proper "key." Hashes are, by nature, one-way operations, there is no way to build the original data from a hash value (in general, anyway—you may be able to guess at it if the message was small enough). To be able to effectively communicate information securely, it is vital that the data remains intact, albeit obfuscated, in the encrypted message. For this we need "true" cryptography as is found with symmetric and asymmetric encryption algorithms.

We talked about encryption algorithms and what they do in previous chapters, so we will not go into too much detail about symmetric and asymmetric cryptographic algorithms here, but the important things to remember are the essential properties of the different classes of algorithms. For your reference, we summarize the useful properties (and consequences of using) each of the classes of algorithms:

- Hashes—fast, efficient algorithms generally useful for verifying the integrity of data but provide no means to otherwise protect information.

- Symmetric—fast (relatively slow compared to hashes in general though), general-purpose encryption algorithms. The problem is that the keys need to be shared between receiver and sender somehow (and that sharing obviously cannot be done using a symmetric algorithm—keys must be shared physically or using another method).

- Asymmetric—slow, special-purpose algorithms that provide the useful ability to facilitate communications without having to share secret keys (public-keys obviously need to be shared, but that's generally pretty easy).

The rest of this chapter will cover the basics for optimizing cryptography for embedded applications. We covered hashing above because those algorithms are already very fast and can be used to provide a small level of security (assuming you don't care about eavesdroppers, anyway). The other cryptographic algorithms (symmetric and asymmetric) are much slower in general, and their relative performance may affect embedded applications significantly. With hashes out of the way, we can focus on the slower algorithms and just what we can do to make them a little faster.

12.3.1 Optimization Guidelines: What NOT to Optimize

Before we start any discussion about optimizing cryptographic algorithms, we need to cover a very important issue: what NOT to optimize. This is an extremely important concept, so it will be a dominant theme throughout this chapter. The reason there are parts of these algorithms that cannot be optimized is that, in order to do their job correctly, they require some significant processing. If you go into optimization with your virtual machete blindly, you are more likely than not to remove something important. Optimization is not to be taken lightly, so the rule to remember is: "if you aren't sure about it, don't do it." A corollary to that rule is to always check with an expert if security is truly important—if you are just interested in keeping curious teenagers out of your system, then it may not be as important as, say, protecting credit card numbers.

So what are these things that should never be touched? Primarily, we want to be sure that any optimization or performance tweaking does not affect the proper functioning of the algorithm itself. Fortunately, this is usually easier to find than it sounds. Almost all cryptographic algorithms have test vectors that can be found to check basic functionality of implementations (be suspicious of algorithms that aren't easily tested). You can also use another implementation (Open Source software is always a good place to start since it is free and you can look at source code—just be sure to heed all the legalities inherent in using it). If you have broken the algorithm in your efforts to make it faster, it usually shows up quickly—one of the properties of cryptographic algorithms is that they are usually either broken or not, there are few cases where an implementation is "partially functional." The problems in implementations are usually with the handling of data once it has been decrypted (this is where buffer overflow and other issues come into play). The one case to watch for is if you are implementing the algorithm on both ends of the communication channel—you may break the algorithm, but since you are implementing both sides in the same fashion, you may not notice. It is always a good idea to check your algorithm against a known good (if you cannot, then it is a good sign you need to rethink what you are doing).

Another important rule for optimizing cryptography is that you should never try to tweak the algorithm itself. For one example, the AES algorithm uses a number of "rounds" to produce a secure encoding. A "round" is generally a sequence of actions that can be repeated any number of times, with the property that each additional round further increases the security of the resulting encoding (the concept of rounds is found in some symmetric-key algorithms such as DES and AES, but is missing in others, such as RC4). A typical AES mode such as 128-bit (key size) may use, for example, 18 rounds to encrypt a message. If you were to adjust your implementation to use fewer rounds, AES would still function and it would run faster. This may at first seem like a good idea, but you have compromised the security of your implementation—in fact, there are known attacks on AES when the number of rounds is reduced. Round reduction is one of the first things cryptanalysts look at when trying to break an algorithm, so you can be sure it is a bad idea to do it purposefully. Fortunately, if your implementation has to be compatible with other implementations, reducing the number of rounds will break the basic functionality of the algorithm. Again, the situation to watch for is when you are implementing both ends.

The example of round reduction in AES (or DES or any number of other algorithms) is just a specific case that can be used to demonstrate the more general rule. Don't try to change the design of a cryptographic algorithm (unless you are, in fact, a cryptographer). If the algorithm says that a certain property must be maintained or a certain feature must be present, it is there for a reason. That reason may be "the designer felt like it," but unless you know what you are doing, it is likely a bad idea to second-guess a cryptographer.

The third rule basically follows from the first two: cryptographic primitives need to be properly implemented or you might as well not even use cryptography. So what is a cryptographic primitive? Cryptographic primitives are the random number generators, entropy sources, and basic memory or math operations that are required by the cryptographic algorithms. For example, the Pseudo-Random Number Generator (PRNG) functions that generate random numbers from some seed value are extremely important to the security of your cryptography. You can have the strongest cryptography available, at even ridiculous levels (1024-bit AES? Why not? It may eventually be broken but for now it's pretty good), but if you always use the same predictable key value the implementation can be trivial to break. Case in point, the original implementation of the Secure Sockets Layer (SSL version 2) in the Netscape web browser was broken by a couple of grad students in a few hours. Did they find a way to attack RSA and DES? No, they simply looked at the random numbers that feed into the key generation in SSL and realized that the random numbers were being seeded with a very predictable source—the system date and time. Once they knew this, they could easily reduce the amount of work needed to break the encryption since the keys being used would always fall within a relatively predictable range. As a result of this, SSL version 2 actually became almost as insecure as a plaintext communication, and sent Netscape back to the drawing board (the fact that SSL version 2 is still in use today by some systems is baffling). The moral of the story is that you should always be sure that everything used by your cryptography is appropriately secure. Random numbers better be as close to unpredictable true randomness as you can get them—a physical source is usually best. For one solution, a combination of network traffic and external interrupts works fairly well. It isn't completely random, but it gets close enough for many applications (network traffic can be manipulated by an attacker to be more predictable, thereby making it less than an ideal source of randomness). Even better would be something based upon fluid dynamics or quantum physics. Unfortunately, however, the use of lava lamps for generating true random numbers was patented by Silicon Graphics in the 1990's.[1] If you do need truly random numbers, some manufacturers produce entropy-generating hardware, usually based on interference in electronic circuits. The author is a software engineer, so the principles behind this are a bit of a mystery, but if a physicist says it's random, who can argue (besides another physicist)?

Now that we have expounded on the merits of having true randomness to make your system more secure, we will briefly cover the other parts of the third rule, such as making sure your PRNG is implemented correctly and your memory routines aren't doing something stupid. For the PRNG, it must have the property of not decreasing the entropy provided by the random seed. You could have a purely random seed, but if your PRNG always produced the same 10 numbers, it wouldn't be very effective. When implementing a PRNG, the same basic rules apply as when implementing a cryptographic algorithm. Don't change anything unless you really know what you are doing.

The final part of the third rule is to make sure that your memory and math operations are behaving properly. A perfect example of this is the use of a secure, on-chip memory to store a cryptographic key. If the chip is properly implemented, then that key should never leave that internal memory. As a programmer, however, you may have access to the key so that you can use it for your cryptographic operations. If you implement a copy routine

---

[1]Patent #5,732,138: "Method for seeding a pseudo-random number generator with a cryptographic hash of a digitization of a chaotic system."

that moves the key out into the primary system memory and leaves it there, you have essentially defeated the purpose of the internal memory. If you need to copy that key into system memory, make sure that you do it quickly and clear that memory as soon as you are done. It's not ideal (someone could access that memory while the key is there), but as long as there is a small window of opportunity, it is unlikely that an attacker would be looking and be able to extract the key.

To summarize our discussion of the basic rules of what not to optimize, we provide a brief description of the rules:

1. The implementation should not affect the basic functionality of the algorithm.

2. You should never try to optimize a cryptographic algorithm by reducing rounds or changing properties of the algorithm's design.

3. Obey basic cryptographic rules such as using an appropriate random source.

These rules are of course very generic, but should serve to protect you from most of the common mistakes that can be made when attempting to optimize cryptographic algorithms.

We have so far spent most of our time covering the basic ideas about what not to optimize, without providing much in the way of what actually *can* be optimized. The next section will cover some of the basics, providing a basis for any implementation effort. However, as we discuss the possibilities for optimization, keep the ideas from the above discussion in mind, since the line between what can be optimized and what should not be optimized can often be quite blurry. Most importantly, if you are not comfortable with the optimization (or even simply unsure), don't do it!

12.3.2 Optimization Guidelines: What Can We Optimize?

The basic rules of optimization apply to cryptography as they do in all other computer engineering disciplines—in other words, there are no rules! Unfortunately, optimization is somewhat cryptic (it is similar to cryptography in many ways) and often is considered "black magic" or "voodoo" by many engineers. While this is the case for some of the more esoteric optimization techniques (usually used on platforms employing some of the more advanced performance-enhancing features such as multistage pipelining, multithreading, instruction-level parallelism, and out-of-order execution), the optimizations we will discuss here are far more mundane, but can be extremely useful for certain architectures, especially those seen in the low-cost embedded hardware we are focusing on. For discussions on general optimization techniques, there are many excellent books written on the subject, but we are going to focus on simple tricks that are easy to spot and easy to implement.

The first thing to think about when looking to optimize a cryptographic algorithm is the math employed by the algorithm. Most cryptographic algorithms are very math-heavy, and utilize a number of mathematical primitives that may or may not be implemented in the most efficient manner possible. Two of the most common operations are the modulus and XOR, and though the latter is usually quite fast, the modulus can be quite expensive—especially if the hardware does not natively support division operations. The real trick in optimizing individual operations is to know what the compiler you are using will do to optimize the code, as well as what the underlying hardware is capable of. In some cases, it may be fine to simply let the compiler do the work for you, as would likely be the case with a multiple-thousand-dollar optimizing compiler from a company specializing in development tools, like something from Green Hills. However, with some of the less-expensive options, such as the GNU tools (gcc), C compilers for the Microchip PIC, or Dynamic C for the Rabbit microprocessors, you may need to help the compiler out a little by doing some creative manipulation of the C code or even resorting to hand-coding in assembly language. If you are used to working with embedded hardware, assembly should not be too unfamiliar, but even if it is, there are usually an ample number of samples that can provide some insight into the types of things you can do. When dealing with limited resources, concerns about portability of the code take a back seat to efficiency and code size.

So what can we do to speed up these operations? If you are looking at an XOR operation, something implemented in hardware on most (all?) platforms, then there probably is not much you can do. If the data type is larger than the basic unit supported by the hardware (such as a 32-bit value on a machine with 16-bit integers), then look at what the compiler does. It is likely that the compiled code will be as good as anything you could write by hand. However, if the operation that you are looking at is a division or modulus operation, it may be (and is likely) that your compiler will utilize a more generic version of the operation than can be implemented in your particular situation. One trick employed by programmers who know the limitations of their tools is to look for powers of 2. Since we are dealing with binary machines (try finding a non-binary machine in production), anything that can be decomposed into powers of 2 (which is essentially everything) can benefit from some reworking of the code to utilize the inherent binary capabilities of the underlying hardware. Though this is by no means the only type of math trick we can use, it can be quite powerful when properly executed.

As an example of the power-of-2 type of optimization, let's take a look at one of the most common operations in cryptography—the modulus. A staple of cryptography, the modulus is used in algorithms from the simple RC4 to the more complex RSA. For our example, we will look at an operation that is specifically used in RC4, a modulus on an index into the internal RC4 lookup table:

```
char index;
int data;
...
// Derive new index from data
index = data % 256;
```

Listing 12.2: Power-of-2 optimization example part 1

In the example, notice that the table size is a fixed 256 bytes. Conveniently (or by design), this happens to be a nice power of two (two to the eighth power, to be exact). This allows us to write the index calculation in a manner that may not be provided by the compiler itself (granted, this is a simplistic example using a constant that a good compiler should be able to do something with, but it is easy to explain). There are various ways to redo the expression, but we will look at an optimized solution utilizing a bit-mask. As it happens, a modulus by 256 will result in an integer in the range of 0 to 255, or FF in hexadecimal. Using what we know about binary, we can rewrite the modulus operation using FF as a mask:

```
char index;
int data;
...
// Derive new index from data using bit-mask in place of modulus
index = data & 0xFF;
```

Listing 12.3: Power-of-2 optimization example part 2

The trick works because the remainder of any division by 256 will be represented by the lowest 8 bits (byte) of the result. This convenient property means that we do not care what is represented by the upper portions of the result since the result of the division will never have any bits set beyond the eighth bit (an astute reader will notice that we could simply assign the integer to the character and rely on the inherent truncation that occurs when assigning to a character in C, but it is usually a bad idea to rely on the language like that). As a result of the observation above, we can omit the division operation entirely. In an algorithm like RC4 where this operation or something similar is performed on *every byte* of the data being processed, the performance improvement can be staggering. In many cases, a function call to a complicated division routine will be replaced by one or two instructions, at a savings of potentially large numbers of clock cycles.

Generally speaking, anything that you can optimize that is related to the repetitive and mathematical nature of cryptography will be a boon for developing a secure but efficient application. The modulus example above is illustrative of a specialized trick that happens to work for a few different algorithms. Unrolling loops and other generic optimization techniques can also be quite effective, but as with anything there is usually a tradeoff between efficiency and code size (which can be a big deal).

If you are implementing a cryptographic algorithm from a specification, you can make some design decisions about how the algorithm is implemented, but remember to make sure that none of the optimizations cuts any corners that may lead to a breach of the integrity of the security provided by the algorithm. If you are using a prewritten library or source code, you may be limited to the options for optimization provided by the original author. When we look at building a secure application using a Microchip PIC processor, we actually utilize an open-source implementation of AES ported to the PIC. The implementation actually allows us to choose between a compact and slow variant and a larger, faster version, controlled with macro pre-processing. On the PIC, code space is plentiful, but RAM is limited.

We can optimize certain algorithms to get a little more speed out, but unfortunately, there is only so much that we can optimize. In many cases, pure software optimization may be enough, but some applications require better performance than can be provided. One obvious choice is to move the critical operations into hardware, but that may not be an option. In the next section we will look at some performance differences between different algorithms and how algorithm choice may give your application the performance it needs.

12.4 Choosing Cryptographic Algorithms

In the last section we discussed the potential for optimizing algorithms, which can be done, but sometimes may not result in the type of performance required. As was mentioned, you can always move the critical cryptographic operations into hardware, but this may not always be possible, especially considering the additional expense and the fact the design is locked into a particular scheme for encryption. In fact, this was exactly the problem with the Wired-Equivalent Privacy (WEP) protocol originally implemented for the 802.11 wireless network protocols. Many 802.11 applications utilized specialized hardware to speed up WEP, but once WEP was discovered to have some serious security flaws, it was impractical to update all that hardware. The solution (WPA) was implemented to utilize the existing WEP hardware, but this limited the capabilities of the new protocol and took a rather significant effort to implement.

Fortunately, there is a way to get the performance you need for your application without some of the drawbacks of a hardware-based solution. The answer is to design algorithms with a natural performance advantage into your application. Obviously, this will not always work since you may have external requirements that dictate the use of particular algorithms. However, if you do have the choice, you can sometimes trade security for performance. There are enough algorithms available that you actually can choose from a relatively wide array of performance characteristics. Unfortunately, this can also be a tradeoff, since additional security often translates to more repetitions or more complex math. The RC4 cipher algorithm, for example, is extremely fast and simple to implement. Unfortunately, RC4 is usually considered less secure than more rigorous algorithms like AES. It must be noted, however, that there are no known practical attacks on RC4 as long as it is used properly (remember the caveats to using a stream cipher). The WEP protocol used RC4, and this is often cited as a reason why the protocol failed, but the problem was in how WEP generated keys in combination with RC4.[2] If you are careful, you can still utilize RC4 to provide both high performance and a relatively acceptable level of security. In fact, SSL implementations use RC4 more commonly than most other symmetric algorithms (see Chapter 4)—looking at how SSL uses RC4 and generates keys would be a good starting point to implementing an appropriate RC4 solution.

---

[2]From an RSA Laboratories technical note; RSA Security Response to Weaknesses in Key Scheduling Algorithm of RC4 (www.rsa.com)

The other common symmetric algorithms include AES and DES. DES is essentially obsolete, replaced by AES, but it may still be required for communication with some legacy applications. DES is a dinosaur and can be quite slow when implemented in software, since its original design was optimized for implementing in hardware in the 1970s. Its original replacement, 3DES ("triple-DES") is simply a method of running the algorithm over the same data three separate times, each with a different piece of the key. The key was essentially 3 DES keys strung together, and each operation used a third of the larger key. Since 3DES was the same algorithm run three times, it can be expected to be three times as slow as plain DES, making its performance characteristics even worse. 3DES, like DES, is essentially obsolete, though its significantly larger key size provides some additional security.

Compared to DES and 3DES, AES is a much more modern algorithm, and it was even designed with performance on limited-resource platforms in mind (part of the AES competition rules stated that the algorithm should be easy to implement with reasonable performance on all machines from tiny 8-bitters to supercomputers). At this point in time, choosing AES is generally a good idea for any application, since it is a standard, commonly used algorithm that provides some of the best security currently available. AES is also scalable to different key sizes and the number of rounds can be increased to increase the relative security provided. The reason AES can be implemented efficiently is due to the fact that the algorithm is essentially modular, and can be implemented in different ways that take advantage of the underlying hardware. That being said, AES is still significantly slower than RC4, but for compatibility and security, it really cannot be beat for applications requiring symmetric key encryption.

AES, RC4, and other symmetric-key algorithms are fine algorithms as long as you can arrange to share a key either before the application is deployed or somehow distribute keys afterwards. For many applications, especially embedded applications, distributing a key after deployment can be extremely difficult or simply infeasible. Presharing a key is not often an ideal situation either, since the keys are either difficult to manage (which device has which key?) or if the keys are all the same, there is a serious security hazard (if one device is compromised, they all are). With these inherent difficulties in key management for symmetric cryptography, it is sometimes necessary to employ an asymmetric (public-key) algorithm in order to have the ability to handle key management without having to pre-share symmetric keys. In some cases, it may even be useful to have the public-key algorithm handle all data transfers. An example of the use of

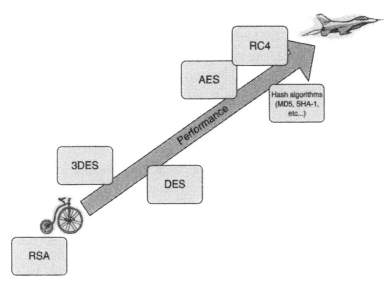

Figure 12.1: Relative performance of common cryptographic algorithms

public-key cryptography for all data transfers would be a system where the amount of data to be sent was very small (tens of bytes) and sent very infrequently, such as an irrigation controller that usually runs in an automated fashion, but occasionally needs to be overridden. If code space is extremely limited (such as on a PIC), and speed is not a factor, implementing a single cryptographic algorithm may be desirable, and an algorithm like RSA can be implemented in a small amount of code. Generally speaking, however, public-key algorithms are so slow and cumbersome in execution that they are rarely (if ever) used as the sole encryption method. More often, they are paired with more efficient symmetric-key algorithms in protocols like SSL or SSH.

Public-key algorithms are usually rolled up into a protocol like SSL, but that does not mean your application has to implement a complete security protocol. However, if you need public-key cryptography for your application and you aren't sure about needing a full protocol, you probably should opt for the full protocol, since it will be more generally useful and may save headaches later (especially for compatibility with other applications). If you absolutely must have public-key algorithm support and cannot implement a full protocol like SSL, then it should be fairly obvious how to proceed. The public-key algorithm is used for distributing the key and the symmetric-key algorithm handles the bulk of the communications encryption.

The astute reader may have noticed that the public-key algorithm does not completely remove the requirement of pre-sharing keys, since each device may require a private key. It would be possible to make each device a "client," which would allow a central machine to send out public-keys to each device (the central system has the associated private key), but this may not be feasible for many applications. However, even if each device requires a private key, the only problem is at deployment when each device is assigned its own key. Once deployed, the device can simply provide its public-key to the client at any time, thus obviating the need for any key management. For more information on how public-key algorithms are used with symmetric-key algorithms in practice, refer to the sections on SSL—the SSL protocol is a good example of the pairing of algorithms to provide the benefits of each type.

We have so far discussed the basics of how public-key algorithms are used, but the real issue is their performance. The problem is that the nature of public-key algorithms requires far more computationally intense mathematical operations than their symmetric cousins. The reason for this is the fact that the security they provide is based upon the difficulty of factoring large numbers. Modern computers are good at doing factoring by using brute-force methods on small numbers, so the numbers employed by public-key algorithms must be large enough to ensure that there will be little possibility of someone being able to factor them quickly using a computer. The result of this requirement is a much larger key size than symmetric-key algorithms for an equivalent amount of security. As an example, 512-bit RSA is considered to be about the same level of security as a typical 64-bit symmetric operation. This order-of-magnitude difference in key size translates to a similar order-of-magnitude difference in performance. The reason for this is that part of the RSA key is actually used as an exponent to derive an even larger number. Now, it should be fairly obvious that if you were to apply a large exponent to another large number, the size of the number would grow extremely rapidly. However, RSA works on the principles of modular arithmetic, so the result is truncated by the modulus (part of the public-key) after each multiplication. Even though we can store the truncated results, all this multiplication takes an extremely long time and even being able to do millions of operations a second, the result is still a significant slowdown. Without hardware assistance, a 512-bit RSA operation can take 30 seconds or more on an embedded microcontroller running at 50 MHz.

The RSA operation for encryption and decryption is relatively simple as compared to other cryptographic algorithms, but the magic of RSA lies in the keys. Key generation in RSA is less simple than the encryption operations, but it reveals the security of the algorithm. Each RSA key is derived from a couple of large prime numbers, usually denoted as *p* and *q*. The public modulus *n* is the product of p and q, and this is where the security comes in[3]—it is generally considered to be impossible to determine the prime factors of very large numbers with any method other than brute-force. Given a sufficiently large number, even a modern computer cannot calculate the prime factors of that number in a reasonable amount of time, which is usually defined in terms of years of computing power. A secure algorithm typically implies thousands or millions of years (or more) of computing power is required for a brute force attack. The trick to RSA is that the private key (*d*) is derived from p and q such that exponentiation with modulus n will result in the retrieval of the plaintext message from the encrypted message (which was calculated by applying the public exponent *e* to the original message). All of the mathematics here basically leads to the result that RSA is slow, and something needs to be done if it is going to be utilized on a lower-performance system. As we mentioned, hardware assistance is really the only way to really speed up the algorithm, but there is a method that utilizes a property of the modular math used by RSA—it is based on an ancient concept called the Chinese Remainder Theorem, or CRT. CRT basically divides the RSA operation up using the prime factors p and q used to derive the public and private keys. Instead of calculating a full private exponent from p and q, the private key is divided amongst several CRT factors that allow the algorithm to be divided up into smaller operations. This doesn't translate into much performance gain unless it is implemented on a parallel processor system, but any gain can be useful for a relatively slow, inexpensive embedded CPU.

Using CRT to speed up RSA is pretty much the only known software method for optimizing RSA—it is simply a CPU-cycle-eating algorithm. For this reason, hardware optimization is really the only choice. This was recognized early on, and in the early 1980s there was even work to design a chip dedicated to RSA operations. Today's PCs are now fast enough to run RSA entirely in software at the same level of performance as the original hardware-based solutions. However, given that a large number of modern embedded processors possess similar resources and performance characteristics to computers in the

---

[3]This description of the RSA algorithm is derived from "PKCS #1 v2.1: RSA Cryptography Standard (June 14, 2002)," from RSA Security Inc. Public-Key Cryptography Standards (PKCS). www.rsa.com.

1980s, we still have to contend with a performance-stealing algorithm. RSA is the primary reason there are few implementations of SSL for inexpensive embedded platforms. The few that do support SSL stacks typically utilize some form of hardware assistance.

12.5 Tailoring Security for Your Application

Before we take a look at hardware-based security optimization, we will spend a little time talking about tailoring security to your application. Generally speaking, SSL/TLS is going to be your best bet for a general-purpose protocol, since it provides a combination of authentication, privacy, and integrity checking. For some applications, however, the security requirements do not include all three security features provided by SSL. For example, you may only need integrity checking, but not privacy (you don't care that attackers can read what is going across the network, but you don't want them to tamper with it either). In this case, hashing alone will probably be sufficient to provide the security you need. In other cases, authentication may be all that is needed, or some combination of privacy and integrity checking. However, if you think about the security provided by SSL, you will see that the three security properties it provides are intertwined, without some privacy in the form of encrypted messages, authentication can be difficult, and without authentication, integrity checking really only helps to prevent corruption of data, not directed attacks. For this reason, you will see that the properties embodied by the SSL and TLS protocols are inherent in any good secure application.

When implementing your own security solution around a cryptographic algorithm, instead of using a publicly available protocol, you should remember the lessons provided by SSL and look to implement privacy, authentication, and integrity checking as a sound engineering practice, even if it does not seem that they are all needed.

One last thing to consider when tailoring security to your particular application is the network in which the device will be deployed. If your application needs to communicate with existing systems, then you will need to know what protocols are provided on those systems and what you will need to implement. If your application is completely under your control, then implementing your own protocol may save a lot of space. Be careful, though, since you will be walking down a dangerous path. If you need a high level of security, rolling your own protocol is almost never a good idea.

CHAPTER 13

Managing Access

John Rittinghouse
James F. Ransome

Even the most secure of systems is vulnerable to compromise if anyone can just walk in, pick up the computer, and walk out with it. Physical prevention measures must be used in conjunction with information security measures to create a total solution. Herein, we cover the essential elements every security administrator needs to know about access control and management of passwords.

13.1 Access Control

According to the Information Systems Security Association (ISSA) [1], *"access control is the collection of mechanisms for limiting, controlling, and monitoring system access to certain items of information, or to certain features based on a user's identity and their membership in various predefined groups."* In this section, we explore the major building blocks that constitute the field of access control as it applies to organizational entities and the information systems these entities are trying to protect from compromise situations.

13.1.1 Purpose of Access Control

Why should we have access control? Access control is necessary for several good reasons. Information that is proprietary to a business may need to be kept confidential, so there is a *confidentiality* issue that provides purpose to having access controls. The information that an organization keeps confidential also needs to be protected from tampering or misuse. The organization must ensure the integrity of this data for it to be useful. Having internal *data integrity* also provides purpose to having access controls. When employees of the organization show up for work, it is important that they have access to the data they need to perform their jobs. The data must be available to the employees for work to continue, or the organization becomes crippled and loses money. It is essential that *data availability* be maintained. Access controls provide yet another purpose in maintaining a reasonable level of assurance that the data is available and usable to the organization. Therefore, the answer to the question asked at the beginning of this paragraph is that there are three very good reasons for having access controls:

- Confidentiality
- Data integrity
- Data availability

13.1.2 Access Control Entities

In any discussion of access control, some common elements need to be understood by all parties. These elements compose a common body of terminology so that everyone working on security access issues is talking about the same thing. For our purposes, there are four primary elements we will discuss: (1) the Subject, an active user or process that requests access to a resource; (2) the Object, which is a resource that contains information (can be interchanged with the word "*resource*"); (3) the Domain, which is a set of objects that the subject can access; and (4) Groups, collections of subjects and objects that are categorized into groups based on their shared characteristics (e.g., membership in a company department, sharing a common job title).

13.1.3 Fundamental Concepts of Access Control

There are three concepts basic to implementation of access control in any organization. These concepts are establishment of a security policy, accountability, and assurance. We discuss each of these concepts in the following sections.

13.1.3.1 Establishment of a Security Policy

Security policy for an organization consists of the development and maintenance of a set of directives that publicly state the overall goals of an organization and recommend prescribed actions for various situations that an organization's information systems and personnel may encounter. Policy is fundamental to enabling continuity of operations. When something happens and the *one person who knows the answer* is on vacation, what is to be done? When policies are in place, administrators know what to do.

13.1.3.2 Accountability

For any information systems that process sensitive data or maintain privacy information, the organization must ensure that procedures are in place to maintain individual accountability for user actions on that system and also for their use of that sensitive data. There have been cases in industry where individuals who were employees of an organization committed criminal acts (e.g., theft of credit card data, theft of personal information for resale to mailing lists, theft of software or data for resale on eBay) that compromised the integrity of the information system. Such criminal actions cause huge problems for organizations, ranging from embarrassment to legal action. When these criminals have been caught, it has been because procedures were in place to ensure their accountability for their actions with the data. These procedures could be in the form of log files, audit trails for actions taken within an application, or even keystroke monitoring.

13.1.3.3 Assurance

As discussed previously, information systems must be able to guarantee correct and accurate interpretation of security policy. For example, if sensitive data exists on Machine A and that machine has been reviewed, inspected, and cleared for processing data of that particular level of sensitivity, when Joe takes the data from Machine A and copies it to his laptop to work on when traveling on the airplane, that data has most likely become compromised unless Joe's laptop has been reviewed, inspected, and cleared for processing that particular level of data sensitivity. If his machine has not been cleared, there is no assurance that the data has *not* been compromised. The policies in place at Joe's organization must be known to Joe in order to be effective, and they must be enforced in order to remain effective.

13.1.4 Access Control Criteria

When implementing security access controls, five common criteria are used to determine whether access is to be granted or denied: location, identity, time, transaction, and role (LITTR). Location refers to the physical or logical place where the user attempts access. Identity refers to the process that is used to uniquely identify an individual or program in a system. Time parameters can be control factors that are used to control resource use (e.g., contractors are not allowed access to system resources after 8 P.M. Monday through Friday, and not at all on weekends). Transaction criteria are program checks that can be performed to protect information from unauthorized use, such as validating whether a database query against payroll records that is coming from a user identified as belonging to the HR department is valid. Finally, a role defines which computer-related functions can be performed by a properly identified user with an exclusive set of privileges specific to that role. All of these criteria are implemented to varying degrees across the depth and breadth of a security plan. The policies and procedures used by an organization to make the plan effective determine the interplay among these criteria.

13.1.5 Access Control Models

When an organization begins to implement access control procedures, an administrator can choose to implement one of the following three basic models: (1) mandatory, (2) discretionary, and (3) nondiscretionary. Each has its particular strengths and weaknesses, and the implementor must decide which model is most appropriate for the given environment or situation. It is important to point out that most operating, network, and application systems security software in use today provide administrators with the capability to perform data categorization, discretionary access control, identity-based access control, user-discretionary access control, and nondiscretionary access control. This section provides an overview of each type of access control model. Armed with this information, an implementor of access controls will be able to make better decisions about which model is most appropriate for specific purposes.

13.1.5.1 Mandatory Access Control Model

Mandatory access control occurs when both the resource owner and the system grant access based on a resource *security label*. A security label is a designation assigned to a resource [2] (such as a file). According to the *NIST Handbook*,

> Security labels are used for various purposes, including controlling access, specifying protective measures, or indicating additional handling instructions. In many implementations, once this designator has been set, it cannot be changed (except perhaps under carefully controlled conditions that are subject to auditing). When used for Access Control, labels are also assigned to user sessions. Users are permitted to initiate sessions with specific labels only. For example, a file bearing the label "Organization Proprietary Information" would not be accessible (readable) except during user sessions with the corresponding label. Moreover, only a restricted set of users would be able to initiate such sessions. The labels

of the session and those of the files accessed during the session are used, in turn, to label output from the session. This ensures that information is uniformly protected throughout its life on the system.

Security labels are a very strong form of access control. Because they are costly and difficult to administer, security labels are best suited for information systems that have very strict security requirements (such as that used by government, financial, and research and development [R&D] organizations that handle classified information or information whose loss would severely or critically degrade the financial viability of the organization). Security labels are an excellent means of consistent enforcement of access restrictions; however, their administration and highly inflexible characteristics can be a significant deterrent to their use.

Security labels cannot generally be changed because they are permanently linked to specific information. For this reason, user-accessible data cannot be disclosed as a result of a user copying information and changing the access rights on a file in an attempt to make that information more accessible than the document owner originally intended. This feature eliminates most types of human errors and malicious software problems that compromise data. The drawback to using security labels is that sometimes the very feature that protects user data also prevents legitimate use of some information. As an example, it is impossible to cut and paste information from documents with different access levels assigned to their respective labels.

13.1.5.2 Data Categorization

One method used to ease the burden necessary for administration of security labeling is categorizing data by similar protection requirements (data categorization). As an example, a label could be developed specifically for "Company Proprietary Data." This label would mark information that can be disclosed only to the organization's employees. Another label, "General Release Data," could be used to mark information that is available to anyone.

When considering the implementation of mandatory access controls with security labels, one must decide between using a rule-based approach, where access is granted based on resource rules, or using an administratively directed approach, where access is granted by an administrator who oversees the resources. Using a rule-based approach is most often preferred because members of a group can be granted access simply by validating their membership in that group. Access levels are assigned at a group level, so all members of the group share a minimum level of access. All files that are created or edited by any one of the members of that group are equally accessible to any other member because the security labels that are instituted have all members of the group sharing equal access to the group resources. Trust is extended to the membership as a whole simply because membership in the group without having proper access *would not be allowed.* This approach is less administratively intensive than using the approach where an administrator manually oversees resources, granting or withdrawing access on an individual case-by-case basis. There are some instances where this approach is preferable, however. Consider a scenario in which only a few members need access to extremely sensitive information. The owner of this information may choose to manually oversee security label application simply to maintain a personal level of control over access to highly sensitive materials.

13.1.5.3 Discretionary Access Control Model

According to a document published in 1987 by the National Computer Security Center, [3] *discretionary access control* is defined as "a means of restricting access to objects based on the identity of subjects and/or groups to which they belong. The controls are discretionary in the sense that a subject with a certain access permission is capable of passing that permission (perhaps indirectly) on to any other subject."

Discretionary access controls restrict a user's access to resources on the system. The user may also be restricted to a subset of the possible *access types* available for those protected resources. Access types are the operations a user is allowed to perform on a particular resource (e.g., read, write, execute). Typically, for each resource, a particular user or group of users has the authority to distribute and revoke access to that resource. Users may grant or rescind access to the resources they control based on "need to know" or "job essential" or some other criteria. Discretionary access control mechanisms grant or deny access based entirely on the identities of users and resources. This is known as *identity-based discretionary access control.*

Knowing the identity of the users is key to discretionary access control. This concept is relatively straightforward in that an *access control matrix* contains the names of users on the rows and the names of resources on the columns. Each entry in the matrix represents an access type held by that user to that resource. Determining access rights is a simple process of looking up a user in the matrix row and traversing the resource columns to find out what rights are allowed for a given resource.

A variant of this is *user-directed discretionary access control.* Here, an end user can grant or deny access to particular resources based on restrictions he or she determines, irrespective of corporate policy, management guidance, and so on. With the ability to inject the human factor into this equation, you can surmise that the level of protection for an organization depends on the specific actions of those individuals tasked to protect information. One drawback to the discretionary access control model is that it is both administratively intense and highly dependent on user behavior for success in protecting resources. This has led to the creation of *hybrid access control* implementations, which grant or deny access based on both an identity-based model and user-directed controls.

13.1.5.4 Nondiscretionary Access Control Model

This access control model removes a user's *discretionary ability* and implements mechanisms whereby resource access is granted or denied based on policies and control objectives. There are three common variants of this approach: (1) role based, where access is based on a user's responsibilities; (2) task based, where access is based on a user's job duties; and (3) lattice based, where access is based on a framework of security labels consisting of a resource label that holds a security classification and a user label that contains security clearance information. The most common of these approaches is *role-based access control* (RBAC). The basic concept of RBAC is that users are assigned to roles, permissions are assigned to roles, and users acquire permissions by being members of roles. David Ferraiolo of the National Institute of Standards drafted the *"Proposed Standard for Role Based Access Controls,"* [4] which states:

> Core RBAC includes requirements that user-role and permission-role assignment can be many-to-many. Thus the same user can be assigned to many roles and a single role can have many users. Similarly, for permissions, a single permission can be assigned to many roles and a single role can be assigned to many permissions. Core RBAC includes requirements for user-role review whereby the roles assigned to a specific user can be determined as well as users assigned to a specific role. A similar requirement for permission-role review is imposed as an advanced review function. Core RBAC also includes the concept of user sessions, which allows selective activation and deactivation of roles.

As an example, Joe is an accountant and serves as the manager of payroll operations at ABC Company. His role in the company as manager of payroll would, in RBAC, allow Joe to see all materials necessary for successful conduct of payroll operations. He is also a member of the whole Accounting group at ABC Company. In that role, as a member of Accounting, he is given access to all of the general accounting resources that are made available to the accounting group, but he does not have access to specific files that belong to the Accounts Payable, Accounts Receivable, or Expense Processing teams. If the Expense Processing team decided to make an internal document available to the general Accounting group, then Joe would be able to see that document because of his membership in the Accounting group.

The difference between role-based and task-based access is subtle, but distinct. The previous scenario was built around Joe's area of responsibility and his membership in a group. In the task-based access control scenario, Joe would see only documents in Accounting that were determined by company work flow procedures to be necessary for Joe to see to successfully manage payroll operations. Based on Joe's current job duties, it is not "job necessary" for Joe to see what is produced by the Expense department, Accounts Payable, or Accounts Receivable, *even if* Joe is a member of the Accounting group. For many companies, this subtle distinction is more trouble than it is worth when the RBAC model can be more easily implemented with the newer computing platforms of today.

In the lattice-based model, Joe's access is based on a framework of security labels. The documents Joe would need to perform his job would have to have their resource label checked to see what security classification (e.g., "general release" or "company proprietary") that resource has, and a user label that contains security clearance information would be checked to ensure that Joe is entitled or "cleared" to access that company proprietary-level information. In a government scenario, working with classified material, this model is much more prevalent than it is in industry. Substitute the words *"unclassified," "confidential," "secret,"* or *"top secret"* for the words *"company proprietary"* or *"general release,"* and you get the idea how a government access control system works.

13.1.6 Uses of Access Control

There are seven general uses for access controls:

1. *Corrective*, to remedy acts that have already occurred

2. *Detective*, to investigate an act that has already occurred

3. *Deterrent*, to discourage an act before it occurs

4. *Recovery*, to restore a resource to a state of operation before when an act occurred

5. *Management*, to dictate the policies, procedures, and accountability to control system use

6. *Operational*, to set personnel procedures to protect the system

7. *Technical*, to use software and hardware controls to automate system protection

Ideally, *management* policies and procedures would dictate *operational* activities that implement *technical* solutions that *deter* unauthorized access and, when that fails, *detect* such access in a manner that allows for rapid *recovery* using *corrective* actions. There, I said it. As simplistic as that sentence sounds, it embodies the very essence of the many uses of access control in an organization. Why make it more complicated?

13.1.7 Access Control Administration Models

13.1.7.1 Centralized Administration Model

The *centralized administration model* is based on the designation of a single office location or single individual as the responsible party tasked with setting proper access controls. The advantage to using this approach is that it enforces strict controls and uniformity of access. This is because the ability to make changes to access settings resides with very few individuals in a centralized administration model. When an organization's information-processing needs change, personnel with access to that information can have their access modified, but only through the centralized location. Most of the time, these types of requests require an approval by the appropriate manager before such changes are made. Each user's account can be centrally monitored, and closing all accesses for any user can be easily accomplished if that individual leaves the organization. Because the process is managed by a few centralized resources in an organization, enforcing standard, consistent procedures is fairly easy.

The most obvious drawback to a centralized model approach is the time it takes to make changes when they must be coordinated and approved before being made. Sometimes, when there are many people in an organization, these requests can become backlogged. Most of the time, however, the trade-off between having strict enforcement and standardized processes is worth enduring the hassle of going through a little bureaucracy to get something done. An example of a centralized access model would be the use of a Remote Authentication Dial-in User Service (RADIUS) server, which is a centralized server used for a single point of network authentication for all users needing access to the network resources. Another example would be a Terminal Access Controller Access Control System (TACACS) server, which is a centralized database of accounts that are used for granting authorization for data requests against a data store or subsystem (e.g., a company-owned CRM product).

13.1.7.2 Decentralized Administration Model

Using the *decentralized administration model*, all access is controlled by the specific document or file originator. This allows control to remain with those who are responsible for the information. The belief is that these people are best suited to determine who needs access to a document and what type of access they need; however, there is a great opportunity to suffer the consequences of a lack of consistency among document originators regarding procedures and criteria that are used to make user access decisions. Also, with the decentralized administration model, it is difficult to get a composite view of all user accesses on the system at any given time.

These inconsistencies can create an environment where different application or data owners may inadvertently implement access combinations that create conflicts of interest or jeopardize the organization's best interests. Another disadvantage is that the decentralized administration model needs to be used in conjunction with other procedures to ensure that accesses are properly terminated when an individual leaves the company or is moved to another team within the organization. An example of common use of the decentralized administration model is a domain in which all file shares are accessible in read-only mode, but each file share owner would determine if a user could perform write or execute activities in the file share. In a domain with a few hundred file shares, this lack of uniformity and standardization quickly becomes apparent.

13.1.7.3 Hybrid Administration Model

The *hybrid administration model* combines the centralized and decentralized administration models into a single approach. An example would be use of a RADIUS server (centralized login/authentication) for gaining basic access to the network and having resources distributed across the network, where each domain on the network is controlled by a different administrator. This is a typical corporate model where the central administration part is responsible for the broadest and most basic of accesses, that of gaining entry to the network, and the decentralized part is where the system owners and their users (the creators of the files) specify the types of access implemented for those files under their control. The main disadvantage to a hybrid approach is the haggle over what should and should not be centralized.

13.1.8 Access Control Mechanisms

Many mechanisms have been developed to provide internal and external access controls, and they vary significantly in terms of precision, sophistication, and cost. These methods are not mutually exclusive and are often employed in combination. Managers need to analyze their organization's protection requirements to select the most appropriate, cost-effective logical access controls. Logical access controls are differentiated into internal and external access controls. Internal access controls are a logical means of separating what defined users (or user groups) can or cannot do with system resources.

13.1.8.1 Internal Access Controls

We cover four methods of internal access control in this section: passwords, encryption, *access control lists* (ACLs), and constrained user interfaces. Each of these four methods of internal access is discussed next.

13.1.8.2 Passwords

Passwords are most often associated with user authentication; however, they are also used to protect data and applications on many systems, including PCs. For instance, an accounting application may require a password to access certain financial data or to invoke a restricted application. The use of passwords as a means of access control can result in a proliferation of passwords that can reduce overall security. Password-based access control is often inexpensive because it is already included in a large variety of applications; however, users may find it difficult to remember additional application passwords, which, if written down or poorly chosen, can lead to their compromise. Password-based access controls for PC applications are often easy to circumvent if the user has access to the operating system (and knowledge of what to do).

13.1.8.3 Encryption

Another mechanism that can be used for logical access control is encryption. Encrypted information can only be decrypted by those possessing the appropriate cryptographic key.

This method is especially useful if strong physical access controls cannot be provided, such as for laptops or floppy diskettes. Thus, for example, if information is encrypted on a laptop computer, and the laptop is stolen, the information cannot be accessed. Although encryption can provide strong access control, it is accompanied by the need for strong key management. Use of encryption may also affect availability. For example, lost or stolen keys or read/write errors may prevent the decryption of the information.

13.1.8.4 Access Control Lists

ACLs refer to a matrix of users (often represented as rows in the matrix that include groups, machines, and processes) who have been given permission to use a particular system resource, and the types of access they have been permitted (usually represented in the matrix as columns). ACLs can vary widely. Also, more advanced ACLs can be used to explicitly deny access to a particular individual or group. With more advanced ACLs, access can be determined at the discretion of the policy maker (and implemented by the security administrator) or individual user, depending on how the controls are technically implemented.

13.1.8.5 Elementary ACLs

The following brief discussion of ACLs is excerpted from the NIST Special Publication 800-113. [5] Elementary ACLs (e.g., permission bits) are a widely available means of providing access control on multiuser systems. Elementary ACLs are based on the concept of owner, group, and world permissions (see Figure 13.1). These preset groups are used to define all permissions (typically chosen from read, write, execute, and delete access modes) for all resources in this scheme. It usually consists of a short, predefined list of the access rights each entity has to files or other system resources. The owner is usually the file creator, although in some cases, ownership of resources may be automatically assigned to project administrators, regardless of the creator's identity. File owners often have all privileges for their resources. In addition to the privileges assigned to the owner, each resource is associated with a named group of users. Users who are members of the group can be granted modes of access that are distinct from nonmembers, who belong to the rest of the "world," which includes all of the system's users. User groups may be arranged according to departments, projects, or other ways appropriate for the particular organization.

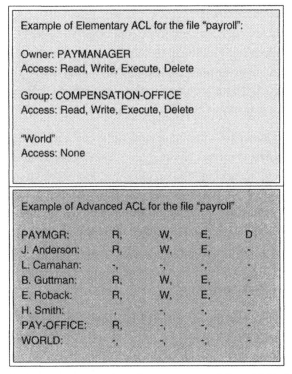

Figure 13.1: Examples of access control lists (ACLs)

13.1.8.6 Advanced ACLs

Advanced ACLs provide a form of access control based on a logical registry. They do, however, provide finer precision in control. Advanced ACLs can be very useful in many complex information-sharing situations. They provide a great deal of flexibility in implementing system-specific policy and allow for customization to meet the security requirements of functional managers. Their flexibility also makes them more of a challenge to manage, however. The rules for determining access in the face of apparently conflicting ACL entries are not uniform across all implementations and can be confusing to security administrators. When such systems are introduced, they should be coupled with training to ensure their correct use.

13.1.8.7 Constrained User Interfaces

Interfaces that restrict users' access to specific functions by never allowing them to request the use of information, functions, or other specific system resources for which they do not have access are known as *constrained user interfaces*. They are often used in conjunction with ACLs. There are three major types of constrained user interfaces: menus, database views, and physically constrained user interfaces:

- *Menu-driven systems* are a common constrained user interface, where different users are provided different menus on the same system. Constrained user interfaces can provide a form of access control that closely models how an organization operates. Many systems allow administrators to restrict users' ability to use the operating system or application system directly. Users can only execute commands that are provided by the administrator, typically in the form of a menu. Another means of restricting users is through restricted shells, which limit the system commands the user can invoke. The use of menus and shells can often make the system easier to use and can help reduce errors.

- *Database views* is a mechanism for restricting user access to data contained in a database. It may be necessary to allow a user to access a database, but that user may not need access to all of the data in the database (e.g., not all fields of a record nor all records in the database). Views can be used to enforce complex access requirements that are often needed in database situations, such as those based on the content of a field. For example, consider the situation where clerks maintain personnel records in a database. Clerks are assigned a range of clients based on last name (e.g., A–C, D–G). Instead of granting a user access to all records, the view can grant the user access to the record based on the first letter of the last name field.

- *Physically constrained user interfaces* can also limit users' abilities. A common example is an ATM, which provides only a limited number of physical buttons to select options; no alphabetic keyboard is usually present.

13.1.8.8 External Access Controls

External access controls consist of a variety of methods used to manage interactions between a system and external users, systems, and services. External access controls employ many methods, sometimes including a separate physical device placed between the system being protected and a network. Examples include Port Protection Devices (PPDs), secure gateways, and host-based authentication.

13.1.8.9 Port Protection Devices

These devices are physically connected to a communications port on a host computer. A PPD authorizes all access to the port to which it is attached. This is done before and independently of the computer's access control functions. A PPD can be a separate device in the communications stream, or it may be incorporated into a communications device (e.g., a modem). PPDs typically require a separate authenticator, such as a password, in order to access the communications port.

13.1.8.10 Secure Gateways/Firewalls

Often called *firewalls*, secure gateways block and/or filter access between two networks. They are most often employed between a private network and a larger, more public network such as the Internet. Secure gateways allow network users to connect to external networks, and simultaneously they prevent malicious hackers from compromising the internal systems. Some secure gateways allow all traffic to pass through *except* for specific traffic that has known or suspected vulnerabilities or security problems, such as remote log-in services. Other secure gateways are set up to disallow all traffic *except* for specific types, such as e-mail. Some secure gateways can make access-control decisions based on the location of the requester. Several technical approaches and mechanisms are used to support secure gateways.

13.1.8.11 Types of Secure Gateways

There are various types of secure gateways on the market today. These include packet filtering (or screening) routers, proxy hosts, bastion hosts, dual-homed gateways, and screened-host gateways. Because these secure gateways provide security to an organization by restricting services or traffic that passes through their control mechanisms, they can greatly affect system usage in the organization. This fact reemphasizes the need to establish security policy so management can decide how the organization will balance its operational needs against the security costs incurred.

Secure gateways benefit an organization by helping to reduce internal system security overhead. This is because they allow an organization to concentrate security efforts on a few machines instead of on all machines. Secure gateways allow for centralization of services. They provide a central point for services such as advanced authentication, e-mail, or public dissemination of information. This can reduce system overhead and improve service in an organization.

13.1.8.12 Host-Based Authentication

The Network File System (NFS) is an example of a host-based authentication system. It allows a server to make resources available to specific machines. Host-based authentication grants access based on the *identity of the host* originating the request rather than authenticating the identity of the user. Many network applications in use today employ host-based authentication mechanisms in order to determine whether access is allowed to a given resource. Such host-based authentication schemes are not invulnerable to attack. Under certain circumstances, it is fairly easy for a hacker to masquerade as a legitimate host and fool the system into granting access. Security measures used to protect against the misuse of some host-based authentication systems are often available, but require special steps or additional configuration actions that need to be taken before they can be used. An example would be enabling DES encryption when using remote procedure calls.

13.1.9 Techniques Used to Bypass Access Controls

In the realm of security, the use of common terms enables all parties to understand exactly what is meant when discussing security issues. When talking about attacks, four terms are quite common: *vulnerability, threat, risk,* and *exposure.* A vulnerability is a flaw or weakness that may allow harm to an information system. A threat is an activity with the potential for causing harm to an information system. Risk is defined as a combination of the chance that threat will occur and the severity of its impact. Exposure is a specific instance of weakness that could cause losses to occur from a threat event.

Hackers use several common methods to bypass access controls and gain unauthorized access to information, principally, brute force, Denial of Service (DoS), social engineering, and spoofing. The brute force method consists of a persistent series of attacks, often trying multiple approaches, in an attempt to break into a computer system. A DoS occurs when someone attempts to overload a system through an online connection in order to force it to shut down. Social engineering occurs when someone employs deception techniques against organizational personnel in order to gain unauthorized access. *This is the most common method of attack known.* Finally, spoofing is when a hacker is masquerading an ID in order to gain unauthorized access to a system.

13.2 Password Management

When granting access to a computer system, such access can be restricted by means of controls based on various kinds of identification and authorization techniques. Identification is a two-step function: (1) identifying the user, and (2) authenticating (validate) the user's identity. The most basic systems rely on passwords only. These techniques do provide some measure of protection against the casual browsing of information, but they rarely stop a determined criminal. A computer password is much like a key to a computer. Allowing several people to use the same password is like allowing everyone to use the same key. More sophisticated systems today use SmartCards and/or biometric evaluation techniques in combination with password usage to increase the difficulty of circumventing password protections.

The password methodology is built on the premise that something you know could be compromised by someone getting unauthorized access to the password. A system built on something you "know" (i.e., a password) combined with something you "possess" (i.e., a SmartCard) is a much stronger system. The combination of knowing and possessing, combined with "being" (i.e., biometrics), provides an even stronger layer of protection. Without having all three elements, even if someone could obtain your password, it is useless without the card and the right biometrics (e.g., fingerprint, retinal scan).

13.2.1 SmartCards

In general, there are two categories of SmartCards: magnetic strip cards and chip cards. As its name suggests, the magnetic strip card has a magnetic strip containing some encoded confidential information destined to be used in combination with the cardholder's personal code or password. The chip card uses a built-in microchip instead of a magnetic strip. The simplest type of chip card contains a memory chip containing information, but has no processing capability. The more effective type of chip card is the "smart" card, which contains a microchip with both memory to store some information and a processor to process it. Hence, the term *SmartCard*. Such cards are often used in combination with cryptographic techniques to provide even stronger protection.

13.2.2 Biometric Systems

Biometric systems use specific personal characteristics (biometrics) of an individual (e.g., a fingerprint, a voiceprint, keystroke characteristics, or the "pattern" of the retina).

Biometric systems are still considered an expensive solution for the most part, and as a result of the cost, they are not yet in common use today. Even these sophisticated techniques are not infallible. The adage that *"if someone wants something bad enough, he will find a way to break in and take it"* still holds true.

13.2.3 Characteristics of Good Passwords

Passwords should be issued to an individual and kept confidential. They should not be shared with anyone. When a temporary user needs access to a system, it is usually fairly simple to add him or her to the list of authorized users. Once the temporary user has finished his or her work, the user ID must be deleted from the system. All passwords should be distinctly different from the user ID and, ideally, they should be alphanumeric and at least six characters in length. Administrators should require that passwords be changed regularly, at least every 30 days. It is possible to warn users automatically when their password expires. To ensure that users enter a new password, they should be restricted in their ability to enter the system after the expiration date, although they may be allowed a limited number of grace-period logins.

Passwords must be properly managed. This entails using a password history list that maintains a list of all of the passwords that have been used in the past 6 to 12 months. New passwords should be checked against the list and not accepted if they have already been used. It is good security practice for administrators to make a list of frequently used forbidden passwords, such as names, product brands, and other words that are easy to guess and therefore not suitable as passwords. This list will be used in the same way as the history list. Only the system manager should be able to change the password history and forbidden lists.

In modern computing environments, most operating systems conform to these standards and generate passwords automatically. Passwords should be removed immediately if an employee leaves the organization or gives his or her notice of leaving. Finally, it is important to note that extreme care should be taken with the password used by network and system administrators for remote maintenance. Standard passwords that are often used to get access to different systems, for maintenance purposes, should always be avoided.

13.2.4 Password Cracking

Security experts across industry, government, and academia cite weak passwords as one of the most critical Internet security threats. Although many administrators recognize the danger of passwords based on common family or pet names, sexual positions, and so on, far fewer administrators recognize that even the most savvy users expose networks to risk as a result of the use of inadequate passwords. Data gathered and reported at one of the largest technology companies in the world, [6] where internal security policy required that passwords exceed eight characters, mix cases, and include numbers or symbols, revealed the following startling data:

- L0phtCrack obtained 18% of the user passwords in only 10 minutes.

- Within 48 hours, *90% of all the passwords were recovered* using L0phtCrack running on a very modest Pentium II/300 system.

- Administrator and most Domain Admin passwords were also cracked.

Password cracking refers to the act of attempting penetration of a network, system, or resource with or without using tools to unlock a resource secured with a password. Crack-resistant passwords are achievable and practical, but password auditing is the only sure way to identify user accounts with weak passwords. The L0phtCrack Software (now called LC4, described as follows) offers this capability.

13.2.4.1 Windows NT L0phtCrack (LC4)

LC4 is the latest version of the password auditing and recovery application, L0phtCrack. LC4 provides two critical capabilities to Windows network administrators:

1. It helps system administrators secure Windows-authenticated networks through comprehensive auditing of Windows NT and Windows 2000 user account passwords.

2. It recovers Windows user account passwords to streamline migration of users to another authentication system or to access accounts whose passwords are lost.

LC4 supports a wide variety of audit approaches. It can retrieve encrypted passwords from stand-alone Windows NT and 2000 workstations, networked servers, primary domain controllers, or Active Directories, with or without SYSKEY installed. The software is capable of sniffing encrypted passwords from the challenge/response exchanged when one machine authenticates to another over the network. This software allows administrators to match the rigor of their password audit to their particular needs by choosing from three different types of cracking methods: dictionary, hybrid, and brute force analysis. These methods are discussed in the next section. Finally, using a distributed processing approach, LC4 provides administrators with the capability to perform time-consuming audits by breaking them into parts that can be run simultaneously on multiple machines.

13.2.4.2 Password Cracking for Self-Defense

Using a tool such as LC4 internally enables an organization's password auditor to get a quantitative comparison of password strength. This is done by reviewing LC4's report on the time required to crack each password. A Hide feature even allows administrators the option to know whether a password was cracked without knowing what the password was. Password results can be exported to a tab-delimited file for sorting, formatting or further manipulation in applications such as Microsoft Excel. LC4 makes password auditing accessible to less-experienced password auditors by using an optional Wizard that walks new users through the process of configuring and running their password audit, letting them choose from preconfigured configurations. As mentioned previously, when performing the cracking process, three cracking methods (dictionary, hybrid, and brute force analysis) are used. In his Web-based article [7] *"Hacking Techniques: Introduction to Password Cracking,"* Rob Shimonski provides an excellent description of these three methods, as follows:

1. *Dictionary attack.* A simple dictionary attack is by far the fastest way to break into a machine. A dictionary file (a text file full of dictionary words) is loaded into a cracking application (such as L0phtCrack), which is run against user accounts located by the application. Because most passwords are simplistic, running a dictionary attack is often sufficient to do the job.

2. *Hybrid attack.* Another well-known form of attack is the hybrid attack. A hybrid attack will add numbers or symbols to the filename to successfully crack a password. Many people change their passwords by simply adding a number to the end of their current password. The pattern usually takes this form: first month

password is "cat"; second month password is "cat1"; third month password is "cat2"; and so on.

3. *Brute force attack.* A brute force attack is the most comprehensive form of attack, though it may often take a long time to work depending on the complexity of the password. Some brute force attacks can take a week depending on the complexity of the password. L0phtcrack can also be used in a brute force attack.

13.2.4.3 Unix "Crack"

Crack is a password-guessing program that is designed to quickly locate insecurities in Unix password files by scanning the contents of a password file, looking for users who have misguidedly chosen a weak login password. This program checks Unix operating system user passwords for guessable values. It works by encrypting a list of the most likely passwords and checking to see if the result matches any of the system users' encrypted passwords. It is surprisingly effective. The most recent version of Crack is Version 5.0.

Crack v5.0 is a relatively smart program. It comes preconfigured to expect a variety of crypt() algorithms to be available for cracking in any particular environment. Specifically, it supports "libdes" as shipped, Michael Glad's "UFC" in either of its incarnations (as "ufc" and as GNU's stdlib crypt), and it supports whatever crypt() algorithm is in your standard C library. Crack v5.0 takes an approach where the word guesser sits between two software interfaces:

1. Standard Password Format (SPF)
2. External Library Crypt Interface Definition (ELCID)

When Crack is invoked, it first translates whatever password file is presented to it into SPF; this is achieved by invoking a utility program called "*xxx*2spf." The SPF input is then filtered to remove data that has been cracked previously, is sorted, and then passed to the cracker, which starts generating guesses and tries them through the ELCID interface, which contains a certain amount of flexibility to support salt collisions (which are detected by the SPF translator) and parallel or vector computation.

13.2.4.4 John the Ripper

John the Ripper is a password cracker. Its primary purpose is to detect weak Unix passwords. It has been tested with many Unix-based operating systems and has proven to be very effective at cracking passwords. Ports of this software product to DOS and Windows environments also exist. To run John the Ripper, you must supply it with some password files and optionally specify a cracking mode. Cracked passwords will be printed to the terminal and saved in a file called */user_homedirectory/john.pot*. John the Ripper is designed to be both powerful and fast. It combines several cracking modes in one program and is fully configurable for your particular needs. John the Ripper is available for several different platforms, which enables you to use the same cracker everywhere. Out of the box, John the Ripper supports the following ciphertext formats:

- Standard and double-length DES-based format

- BSDI's extended DES-based format

- MD5-based format (FreeBSD among others)

- OpenBSD's Blowfish-based format

With just one extra command, John the Ripper can crack AFS passwords and WinNT LM hashes. Unlike other crackers, John does not use a crypt(3)-style routine. Instead, it has its own highly optimized modules for different ciphertext formats and architectures. Some of the algorithms used could not be implemented in a crypt(3)-style routine because they require a more powerful interface (bitslice DES is an example of such an algorithm).

13.2.5 Password Attack Countermeasures

One recommendation for self-defense against password cracking is to perform frequent recurring audits of passwords. It is also often a good idea to physically review workstations to see if passwords are placed on sticky notes or hidden under a keyboard, tacked on a bulletin board, and so on. You should set up dummy accounts and remove the administrator account. The administrator account is sometimes left as bait for tracking someone who has been detected attempting to use it. Finally, set local security policy to use strong passwords and change them frequently.

Endnotes

1. Information Systems Security Association, Inc. CISSP Review Course 2002, Domain 1, "Access Control Systems and Methodology." PowerPoint presentation given on August 10, 1999, slide 3.

2. U.S. Department of Commerce. Special Publication 800-12, "An Introduction to Computer Security: The NIST Handbook," undated, Chapter 17, p. 204.

3. National Computer Security Center publication NCSC-TG-003, "A Guide to Understanding Discretionary Access Control in Trusted Systems," September 30, 1987.

4. Ferraiolo, D. F. et al. "Proposed NIST Standard for Role-Based Access Control," Gaithersburg, MD: NIST, 20013.

5. *See* note 2.

6. Data obtained from public Web site of @stake, Inc., at http://www.atstake.com/research/lc/index.html.

7. Shimonski, Rob. "Hacking Techniques: Introduction to Password Cracking," July 2002, http://www.-106.ibm.com/developerworks/security/library/s-crack.

CHAPTER 14

Security and the Law

John Rittinghouse
James F. Ransome

With the rash of cyberincidents that have taken a huge financial toll on governments and businesses over the last decade, legislators began to see that laws needed to be enacted to control the Wild West environment that existed in cyberspace. Laws have been enacted to protect privacy, infrastructure, people, companies, and just about anything that uses a computer or any form of computer technology. In this chapter, we discuss the most significant of those laws and how they affect corporate operations.

14.1 The 1996 National Information Infrastructure Protection Act

In 1996, when this law was passed, legislators were presented with some startling statistics. For example, the Computer Emergency and Response Team (CERT) at Carnegie-Mellon University in Pittsburgh, Pennsylvania, reported a 498 percent increase in the number of computer intrusions and a 702 percent rise in the number of sites affected with such intrusions in the three-year period from 1991 through 1994. [1] During 1994, approximately 40,000 Internet computers were attacked in 2,460 incidents. Similarly, the FBI's National Computer Crime Squad opened more than 200 hacker cases from 1991 to 1994.

Before passing this law, legislators realized that there are two ways, conceptually, to address the growing computer crime problem. The first would be to comb through the entire U.S. Code, identifying and amending every statute potentially affected by the implementation of new computer and telecommunications technologies. The second approach would be to focus substantive amendments on the Computer Fraud and Abuse Act to specifically address new abuses that spring from the misuse of new technologies.

The new legislation adopted the latter approach for a host of reasons, but the net effect of this approach was set revamping of our laws to address computer-related criminal activity. The full text of the legislative analysis can be found on the Web. [2]

With these changes, the United States stepped into the forefront of rethinking how information technology crimes must be addressed—simultaneously protecting the confidentiality, integrity, and availability of data and systems. By choosing this path, the hope was to encourage other countries to adopt a similar framework, thus creating a more uniform approach to addressing computer crime in the existing global information infrastructure.

14.2 President's Executive Order on Critical Infrastructure Protection

Following the terrorist attack on the World Trade Center and the Pentagon that occurred on the morning of September 11, 2001, there was a growing realization in our government and across industry sectors that our national infrastructure was vulnerable and that we had become (almost completely) dependent on such critical elements that they needed specific protection. On October 16, 2001, President Bush issued an Executive Order [3] to ensure protection of information systems for critical infrastructure, including emergency preparedness communications, and the physical assets that support such systems.

The President's Executive Order established policy that reflects the fact that the information technology revolution has changed the way business is transacted, government operates, and national defense is conducted. Those three functions now depend (almost wholly) on an interdependent network of critical information infrastructures. The protection program authorized by this Executive Order requires continuous efforts to secure information systems for critical infrastructure. Protection of these systems is essential to the telecommunications, energy, financial services, manufacturing, water, transportation, health care, and emergency services sectors. The official statement of policy, excerpted from the Executive Order, follows:

> It is the policy of the United States to protect against disruption of the operation of information systems for critical infrastructure and thereby help to protect the people, economy, essential human and government services, and national security of the United States, and to ensure that any disruptions that occur are infrequent, of minimal duration, and manageable, and cause the least damage possible. The implementation of this policy shall include a voluntary public-private partnership, involving corporate and nongovernmental organizations.

Ten days after this Executive Order was issued, the 107th U.S. Congress passed H.R. 3162, which became Public Law 107-56, the *USA patriot Act of 2001*. [4]

14.3 The USA Patriot Act of 2001

Public Law 107-56, formally titled the *Uniting and Strengthening America by Providing Appropriate Tools Required to Intercept and Obstruct Terrorism Act of 2001* (USA Patriot Act) was enacted on October 26, 2001 (reauthorized in 2006). A result of the terrorist attack against the United States on September 11, 2001, carried out by members of Osama Bin Laden's Al Qaeda organization, this legislation made broad and sweeping changes that created a federal antiterrorism fund and directed law enforcement, military, and various government agencies to collectively develop a national network of electronic crime task forces throughout the United States. These task forces were designed to prevent, detect, and investigate various forms of electronic crimes, including potential terrorist attacks against critical infrastructure and financial payment systems.

Title II of this bill amends the federal criminal code to authorize the interception of wire, oral, and electronic communications to produce evidence of (1) specified chemical weapons or terrorism offenses and (2) computer fraud and abuse. This section of the law authorizes law enforcement and government personnel who have obtained knowledge of the contents of any wire, oral, or electronic communication or evidence derived therefrom, by authorized means, to disclose contents to such officials to the extent that such contents include foreign intelligence or counterintelligence.

Title III of this law amends existing federal law governing monetary transactions. The amended document prescribes procedural guidelines under which the Secretary of the Treasury may require domestic financial institutions and agencies to take specified measures if there are reasonable grounds for concluding that jurisdictions, financial institutions, types of accounts, or transactions operating outside or within the United States are part of a primary money-laundering concern. The intent of this section is to prevent terroristic concerns from using money-laundering techniques to fund operations that are destructive to national interests.

Title IV is targeted at tightening the control of our borders and immigration laws. In addition to waiving certain restrictions and personnel caps, it directs the Attorney General and the Secretary of State to develop a technology standard to identify visa and admissions applicants. This standard is meant to be the basis for an electronic system of law enforcement and intelligence sharing that will be made available to consular, law enforcement, intelligence, and federal border inspection personnel. Among the many provisions of Immigration Naturalization Service (INS) changes, this section of the law includes within the definition of "terrorist activity" the use of any weapon or dangerous device. The law redefines the phrase "engage in terrorist activity" to mean, in an individual capacity or as a member of an organization, to

1. Commit or to incite to commit, under circumstances indicating an intention to cause death or serious bodily injury, a terrorist activity;

2. Prepare or plan a terrorist activity;

3. Gather information on potential targets for terrorist activity;

4. Solicit funds or other things of value for a terrorist activity or a terrorist organization (with an exception for lack of knowledge);

5. Solicit any individual to engage in prohibited conduct or for terrorist organization membership (with an exception for lack of knowledge); or

6. Commit an act that the actor knows, or reasonably should know, affords material support, including a safe house, transportation, communications, funds, transfer of funds or other material financial benefit, false documentation or identification, weapons (including chemical, biological, or radiological weapons), explosives, or training for the commission of a terrorist activity; to any individual who the actor knows or reasonably should know has committed or plans to commit a terrorist activity; or to a terrorist organization (with an exception for lack of knowledge).

Title IV of this law also defines "terrorist organization" as a group

1. Designated under the Immigration and Nationality Act or by the Secretary of State; or

2. A group of two or more individuals, whether related or not, which engages in terrorist-related activities.

It also provides for the retroactive application of amendments under this act and stipulates that an alien shall not be considered inadmissible or deportable because of a relationship to an organization that was not designated as a terrorist organization before enactment of this act. A provision is included to account for situations when the Secretary of State may have identified an organization as a threat and has deemed it necessary to formally designate that organization as a "terroristic organization." This law directs the Secretary of State to notify specified congressional leaders seven days before formally making such a designation.

Title V, "Removing Obstacles to Investigating Terrorism," authorizes the Attorney General to pay rewards from available funds pursuant to public advertisements for assistance to the Department of Justice (DOJ) to combat terrorism and defend the nation against terrorist acts, in accordance with procedures and regulations established or issued by the Attorney General, subject to specified conditions, including a prohibition against any such reward of $250,000 or more from being made or offered without the personal approval of either the Attorney General or the President.

Title VII, "Increased Information Sharing for Critical Infrastructure Protection," amends the Omnibus Crime Control and Safe Streets Act of 1968 to extend Bureau of Justice Assistance regional information-sharing system grants to systems that enhance the investigation and prosecution abilities of participating federal, state, and local law enforcement agencies in addressing multijurisdictional terrorist conspiracies and activities. It also revised the Victims of Crime Act of 1984 with provisions regarding the allocation of funds for compensation and assistance, location of compensable crime, and the relationship of crime victim compensation to means-tested federal benefit programs and to the September 11th victim compensation fund. It established an antiterrorism emergency reserve in the Victims of Crime Fund.

Title VIII, "Strengthening the Criminal Laws Against Terrorism," amends the federal criminal code to prohibit specific terrorist acts or otherwise destructive, disruptive, or violent acts against mass transportation vehicles, ferries, providers, employees, passengers, or operating systems. It amends the federal criminal code to:

1. Revise the definition of "international terrorism" to include activities that appear to be intended to affect the conduct of government by mass destruction; and

2. Define "domestic terrorism" as activities that occur primarily within U.S. jurisdiction, that involve criminal acts dangerous to human life, and that appear to be intended to intimidate or coerce a civilian population, to influence government policy by intimidation or coercion, or to affect government conduct by mass destruction, assassination, or kidnapping.

The specific issue of information sharing that came up in many discussions of the "talking heads" around the Washington, D.C. area after the September 11th attack is addressed in

Title IX, "Improved Intelligence." Herein, amendments to the National Security Act of 1947 require the Director of Central Intelligence to establish requirements and priorities for foreign intelligence collected under the Foreign Intelligence Surveillance Act of 1978 and to provide assistance to the Attorney General to ensure that information derived from electronic surveillance or physical searches is disseminated for efficient and effective foreign intelligence purposes. It also requires the inclusion of international terrorist activities within the scope of foreign intelligence under such act.

Part of this section expresses the sense of Congress that officers and employees of the intelligence community should establish and maintain intelligence relationships to acquire information on terrorists and terrorist organizations. The law requires the Attorney General or the head of any other federal department or agency with law enforcement responsibilities to expeditiously disclose to the Director of Central Intelligence any foreign intelligence acquired in the course of a criminal investigation.

By now, it should be abundantly clear that the 107th U.S. Congress viewed the threat of terroristic activities as a huge security concern. Steps taken to close loopholes in money transaction processes, immigration and border control changes, and the hundreds of other specifics found in Public Law 107-56 reflect the determination of a nation victimized by terrorism to prevent recurrences using any means necessary and available. Citizens of the United States rallied around a cause as they have few other times in history, and the will of the people was reflected in these congressional actions.

14.4 The Homeland Security Act of 2002

Nine months after the attack on the World Trade Center and the Pentagon, President Bush proposed creation of a cabinet-level Department of Homeland Security, which was formed to unite essential agencies to work more closely together. The affected agencies consisted of the Coast Guard, the Border Patrol, the Customs Service, immigration officials, the Transportation Security Administration, and the Federal Emergency Management Agency. Employees of the Department of Homeland Security would be charged with completing four primary tasks:

1. To control our borders and prevent terrorists and explosives from entering our country

2. To work with state and local authorities to respond quickly and effectively to emergencies

3. To bring together our best scientists to develop technologies that detect biological, chemical, and nuclear weapons and to discover the drugs and treatments to best protect our citizens

4. To review intelligence and law enforcement information from all agencies of government, and produce a single daily picture of threats against our homeland, with analysts responsible for imagining the worst and planning to counter it

On November 25, 2002, President Bush signed the *Homeland Security Act of 2002* into law. The act restructures and strengthens the executive branch of the federal government to better meet the threat to our homeland posed by terrorism. In establishing a new Department of Homeland Security, the act created a federal department whose primary mission will be to help prevent, protect against, and respond to acts of terrorism on our soil. The creation of this new cabinet-level department was a historic event in American history, and it will have long-lasting repercussions on the global community. For security professionals, it adds yet another dimension to the complexity of securing infrastructure from malcontents.

14.5 Changes to Existing Laws

Since the tragic events of September 11, 2001, the U.S. Congress has enacted legislation in the *USA Patriot Act* that has strengthened or amended many of the laws relating to computer crime and electronic evidence. In this section, we review some of the more important changes that have been made to the laws [5] in the United States. In the final sections of this chapter, we discuss the topics of investigations and ethics.

14.5.1 Authority to Intercept Voice Communications

Under previous law, investigators could not obtain a wiretap order to intercept wire communications (those involving the human voice) for violations of the *Computer Fraud and Abuse Act* (18 U.S.C. § 1030). For example, in several investigations, hackers have stolen teleconferencing services from a telephone company and used this mode of communication to plan and execute hacking attacks. The new amendment changed this by adding felony violations of the *Fraud and Abuse Act* to the list of offenses for which a wiretap could be obtained.

14.5.2 Obtaining Voice-Mail and Other Stored Voice Communications

The *Electronic Communications Privacy Act* (ECPA) governed law enforcement access to stored electronic communications (such as e-mail), but not stored wire communications (such as voice-mail). Instead, the wiretap statute governed such access because the legal definition of "wire communication" included stored communications, requiring law enforcement to use a wiretap order (rather than a search warrant) to obtain unopened voice communications. Thus, law enforcement authorities were forced to use a wiretap order to obtain voice communications stored with a third-party provider, but they could use a search warrant if that same information were stored on an answering machine inside a criminal's home. This system created an unnecessary burden for criminal investigations. Stored voice communications possess few of the sensitivities associated with real-time interception of telephones, making the extremely burdensome process of obtaining a wiretap order unreasonable.

Moreover, the statutory framework mainly envisions a world in which technology-mediated voice communications (such as telephone calls) are conceptually distinct from nonvoice communications (such as faxes, pager messages, and e-mail). To the limited extent that Congress acknowledged that data and voice might coexist in a single transaction, it did not anticipate the convergence of these two kinds of communications that is typical of today's telecommunications networks. With the advent of Multipurpose Internet Mail Extensions (MIME) and similar features, an e-mail may include one or more attachments consisting of any type of data, including voice recordings. As a result, a law enforcement officer seeking to obtain a suspect's unopened e-mail from an Internet Service Provider (ISP) by means of a search warrant had no way of knowing whether the inbox messages include voice attachments (i.e., wire communications), which could not be compelled using a search warrant. This situation necessitated changes to the existing wiretap procedures.

14.5.3 Changes to Wiretapping Procedures

An amendment was written that altered the way in which the wiretap statute and the ECPA apply to stored voice communications. The amendment deleted "electronic storage" of wire communications from the definition of "wire communication" and inserted language to ensure that stored wire communications are covered under the same rules as stored electronic communications. Thus, law enforcement can now obtain such communications using the procedures set out in Section 2703 (such as a search warrant) rather than those in the wiretap statute (such as a wiretap order).

14.5.4 Scope of Subpoenas for Electronic Evidence

The government must use a subpoena to compel a limited class of information, such as the customer's name, address, length of service, and means of payment under existing law. Before the amendments enacted with the *USA Patriot Act*, however, the list of records investigators could obtain with a subpoena did not include certain records (such as credit card number or other form of payment for the communication service) relevant to determining a customer's true identity. In many cases, users register with ISPs using false names. In order to hold these individuals responsible for criminal acts committed online, the method of payment is an essential means of determining true identity. Moreover, many of the definitions used within were technology-specific, relating primarily to telephone communications. For example, the list included "local and long distance telephone toll billing records," but did not include parallel terms for communications on computer networks, such as "records of session times and durations." Similarly, the previous list allowed the government to use a subpoena to obtain the customer's "telephone number or other subscriber number or identity," but did not define what that phrase meant in the context of Internet communications.

Amendments to existing law expanded the narrow list of records that law enforcement authorities could obtain with a subpoena. The new law includes "records of session times and durations," as well as "any temporarily assigned network address." In the Internet context, such records include the Internet Protocol (IP) address assigned by the provider to the customer or subscriber for a particular session, as well as the remote IP address from which a customer connects to the provider. Obtaining such records will make the process of identifying computer criminals and tracing their Internet communications faster and easier.

Moreover, the amendments clarify that investigators may use a subpoena to obtain the "means and source of payment" that a customer uses to pay for his or her account with a communications provider, "including any credit card or bank account number." In addition to being generally helpful, this information will prove particularly valuable in identifying users of Internet services where a company does not verify its users' biographical information.

14.5.5 Clarifying the Scope of the Cable Act

Previously, the law contained several different sets of rules regarding privacy protection of communications and their disclosure to law enforcement, one governing cable service,

[6] one applying to the use of telephone service and Internet access, [7] and one called the pen register and trap and trace statute. [8] Before the amendments enacted, the *Cable Act* set out an extremely restrictive system of rules governing law enforcement access to most records possessed by a cable company. For example, the *Cable Act* did not allow the use of subpoenas or even search warrants to obtain such records. Instead, the cable company had to provide prior notice to the customer (even if he or she were the target of the investigation), and the government had to allow the customer to appear in court with an attorney and then justify to the court the investigative need to obtain the records. The court could then order disclosure of the records only if it found by "clear and convincing evidence"—a standard greater than probable cause or even a preponderance of the evidence—that the subscriber was "reasonably suspected" of engaging in criminal activity. This procedure was completely unworkable for virtually any criminal investigation.

The restrictive nature of the *Cable Act* caused grave difficulties in criminal investigations because today, unlike in 1984 when Congress passed the act, many cable companies offer not only traditional cable programming services, but also Internet access and telephone service. In recent years, some cable companies have refused to accept subpoenas and court orders pursuant to the pen/trap statute and the ECPA, noting the seeming inconsistency of these statutes with the *Cable Act's* harsh restrictions. Treating identical records differently depending on the technology used to access the Internet made little sense. Moreover, these complications at times delayed or even ended important investigations.

When this restrictive legislation was amended in the *USA Patriot Act*, Congress clarified the matter, stating that the ECPA, the wiretap statute, and the pen/trap and trace statute all govern disclosures by cable companies that relate to the provision of communication services such as telephone and Internet service. The amendment preserves the act's primacy with respect to records revealing what ordinary cable television programming a customer chooses to purchase, such as particular premium channels or pay-per-view shows. Thus, in a case where a customer receives both Internet access and conventional cable television service from a single cable provider, a government entity can use legal process under the ECPA to compel the provider to disclose only those customer records relating to Internet service, but could not compel the cable company to disclose those records relating to viewer television usage of premium channels, adult channels, and so on.

14.5.6 *Emergency Disclosures by Communications Providers*

Previous law relating to voluntary disclosures by communication service providers was inadequate for law enforcement purposes in two respects. First, it contained no special provision allowing communications providers to disclose customer records or communications in emergencies. If, for example, an ISP independently learned that one of its customers was part of a conspiracy to commit an imminent terrorist attack, prompt disclosure of the account information to law enforcement could save lives. Because providing this information did not fall within one of the statutory exceptions, however, an ISP making such a disclosure could be sued in civil courts. Second, before the *USA Patriot Act*, the law did not expressly permit a provider to voluntarily disclose noncontent records (such as a subscriber's login records) to law enforcement for purposes of self-protection, even though providers could disclose the content of communications for this reason. Moreover, as a practical matter, communications service providers must have the right to disclose to law enforcement the facts surrounding attacks on their systems. For example, when an ISP's customer hacks into the ISP's network, gains complete control over an e-mail server, and reads or modifies the e-mail of other customers, the provider must have the legal ability to report the complete details of the crime to law enforcement.

The *USA Patriot Act* corrected both of these inadequacies. The law was changed to permit, but not require, a service provider to disclose to law enforcement either content or noncontent customer records in emergencies involving an immediate risk of death or serious physical injury to any person. This voluntary disclosure, however, does not create an affirmative obligation to review customer communications in search of such imminent dangers. The amendment here also changed the ECPA to allow providers to disclose information to protect their rights and property.

14.5.7 *Pen Register and Trap and Trace Statute*

The pen register and trap and trace statute (the pen/trap statute) governs the prospective collection of noncontent traffic information associated with communications, such as the phone numbers dialed by a particular telephone. Section 216 of the *USA Patriot Act* updates the pen/trap statute in three important ways: (1) the amendments clarify that law enforcement may use pen/trap orders to trace communications on the Internet and other computer networks; (2) pen/trap orders issued by federal courts now have a nationwide effect; and (3) law enforcement authorities must file a special report with the court whenever they use a pen/trap order to install their own monitoring device on computers belonging to a public provider.

14.5.8 Intercepting Communications of Computer Trespassers

Under prior law, the wiretap statute allowed computer owners to monitor the activity on their machines to protect their rights and property. This changed when Section 217 of the *USA Patriot Act* was enacted. It was unclear whether computer owners could obtain the assistance of law enforcement in conducting such monitoring. This lack of clarity prevented law enforcement from assisting victims to take the natural and reasonable steps in their own defense that would be entirely legal in the physical world. In the physical world, burglary victims may invite the police into their homes to help them catch burglars in the act of committing their crimes. The wiretap statute should not block investigators from responding to similar requests in the computer context simply because the means of committing the burglary happen to fall within the definition of a "wire or electronic communication" according to the wiretap statute.

Because providers often lack the expertise, equipment, or financial resources required to monitor attacks themselves, they commonly have no effective way to exercise their rights to protect themselves from unauthorized attackers. This anomaly in the law created, as one commentator has noted, a "bizarre result," in which a "computer hacker's undeserved statutory privacy right trumps the legitimate privacy rights of the hacker's victims." To correct this problem, the amendments in Section 217 of the *USA Patriot Act* allow victims of computer attacks to authorize persons "acting under color of law" to monitor trespassers on their computer systems. Also added was a provision in which law enforcement may intercept the communications of a computer trespasser transmitted to, through, or from a protected computer. Before monitoring can occur, however, four requirements must be met:

1. The owner or operator of the protected computer must authorize the interception of the trespasser's communications.

2. The person who intercepts the communication must be lawfully engaged in an ongoing investigation. Both criminal and intelligence investigations qualify, but the authority to intercept ceases at the conclusion of the investigation.

3. The person acting under color of law must have reasonable grounds to believe that the contents of the communication to be intercepted will be relevant to the ongoing investigation.

4. Investigators may intercept only the communications sent or received by trespassers. Thus, this section would only apply where the configuration of the computer system allows the interception of communications to and from the trespasser and not the interception of nonconsenting users authorized to use the computer.

The *USA Patriot Act* created a definition of a "computer trespasser." Such trespassers include any person who accesses a protected computer without authorization. In addition, the definition explicitly excludes any person "known by the owner or operator of the protected computer to have an existing contractual relationship with the owner or operator for access to all or part of the computer." For example, certain Internet service providers do not allow their customers to send bulk unsolicited e-mails (or "spam"). Customers who send spam would be in violation of the provider's terms of service, but would not qualify as trespassers because they are authorized users and because they have an existing contractual relationship with the provider.

14.5.9 Nationwide Search Warrants for E-Mail

Previous law required the government to use a search warrant to compel a communications or Internet service provider to disclose unopened e-mail less than six months old. Rule 41 of the Federal Rules of Criminal Procedure required that the "property" (the e-mails) to be obtained must be "within the district" of jurisdiction of the issuing court. For this reason, some courts had declined to issue warrants for e-mail located in other districts. Unfortunately, this refusal placed an enormous administrative burden on districts where major ISPs are located, such as the Eastern District of Virginia and the Northern District of California, even though these districts had no relationship with the criminal acts being investigated. In addition, requiring investigators to obtain warrants in distant jurisdictions slowed time-sensitive investigations.

The amendment added in the *USA Patriot Act* has changed this situation in order to allow investigators to use warrants to compel records outside of the district in which the court is located, just as they use federal grand jury subpoenas and orders. This change enables courts with jurisdiction over investigations to compel evidence directly, without requiring the intervention of agents, prosecutors, and judges in the districts where major ISPs are located.

14.5.10 Deterrence and Prevention of Cyberterrorism

Several changes were made in Section 814 of the *USA Patriot Act* that improve the Computer Fraud and Abuse Act. This section increases penalties for hackers who damage protected computers (from a maximum of 10 years to a maximum of 20 years). It clarifies the *mens rea* required for such offenses to make explicit that a hacker need only intend damage, not necessarily inflict a particular type of damage. It also adds a new offense for damaging computers used for national security or criminal justice purposes, and expands the coverage of the statute to include computers in foreign countries so long as there is an effect on U.S. interstate or foreign commerce. It now counts state convictions as prior offenses for the purpose of recidivist sentencing enhancements, and it allows losses to several computers from a hacker's course of conduct to be aggregated for purposes of meeting the $5,000 jurisdictional threshold. We discuss the most significant of these changes in the following sections.

14.5.10.1 Raising Maximum Penalty for Hackers

Under previous law, first-time offenders could be punished by no more than 5 years' imprisonment, whereas repeat offenders could receive up to 10 years. Certain offenders, however, can cause such severe damage to protected computers that this five-year maximum did not adequately take into account the seriousness of their crimes. For example, David Smith pled guilty to releasing the Melissa virus that damaged thousands of computers across the Internet. Although Smith agreed, as part of his plea, that his conduct caused more than $80 million worth of loss (the maximum dollar figure contained in the Sentencing Guidelines), experts estimate that the real loss was as much as 10 times that amount. Had the new laws been in effect at the time of Smith's sentencing, he would most likely have received a much harsher sentence.

14.5.10.2 Eliminating Mandatory Minimum Sentences

Previous law set a mandatory sentencing guideline of a minimum of six months' imprisonment for any violation of the *Computer Fraud and Abuse Act* , as well as for accessing a protected computer with the intent to defraud. Under new amendments in the *USA Patriot Act*, the maximum penalty for violations for damaging a protected computer increased to 10 years for first offenders and 20 years for repeat offenders. Congress chose, however, to eliminate all mandatory minimum guidelines sentencing for Section 1030 (*Computer Fraud and Abuse Act*) violations.

14.5.10.3 *Hacker's Intent Versus Degree of Damages*

Under previous law, an offender had to "intentionally [cause] damage without authorization." Section 1030 of the *Computer Fraud and Abuse Act* defined "damage" as impairment to the integrity or availability of data, a program, a system, or information that met the following criteria:

1. Caused loss of at least $5,000;

2. Modified or impairs medical treatment;

3. Caused physical injury; or

4. Threatened public health or safety.

The question arose, however, whether an offender must intend the $5,000 loss or other special harm, or whether a violation occurs if the person only intends to damage the computer, which in fact ends up causing the $5,000 loss or harming the individuals. Congress never intended that the language contained in the definition of "damage" would create additional elements of proof of the actor's mental state. Moreover, in most cases, it would be almost impossible to prove this additional intent. Now, under new law, hackers need only intend to cause damage, not inflict a particular consequence or degree of damage. The new law defines "damage" to mean "any impairment to the integrity or availability of data, a program, a system or information." Under this clarified structure, in order for the government to prove a violation, it must show that the actor caused damage to a protected computer and that the actor's conduct caused either loss exceeding $5,000, impairment of medical records, harm to a person, or threat to public safety.

14.5.10.4 *Aggregating Damage Caused by a Hacker*

Previous law was unclear about whether the government could aggregate the loss resulting from damage an individual caused to different protected computers in seeking to meet the jurisdictional threshold of $5,000 in loss. For example, an individual could unlawfully access five computers on a network on 10 different dates—as part of a related course of conduct—but cause only $1,000 loss to each computer during each intrusion.

If previous law were interpreted not to allow aggregation, then that person would not have committed a federal crime at all because he or she had not caused more than $5,000 worth of damage to any particular computer. Under the new law, the government may now aggregate "loss resulting from a related course of conduct affecting one or more other protected computers" that occurs within a one-year period in proving the $5,000 jurisdictional threshold for damaging a protected computer.

14.5.10.5 Damaging Computers Used For National Security or Criminal Justice Purposes

Previously, the *Computer Fraud and Abuse Act* contained no special provisions that would enhance punishment for hackers who damage computers used in furtherance of the administration of justice, national defense, or national security. Thus federal investigators and prosecutors did not have jurisdiction over efforts to damage criminal justice and military computers where the attack did not cause more than $5,000 in loss (or meet one of the other special requirements). Yet these systems serve critical functions and merit felony prosecutions even where the damage is relatively slight. Furthermore, an attack on computers used in the national defense that occur during periods of active military engagement are particularly serious—even if they do not cause extensive damage or disrupt the war-fighting capabilities of the military—because they divert time and attention away from the military's proper objectives. Similarly, disruption of court computer systems and data could seriously impair the integrity of the criminal justice system. Under new provisions, a hacker violates federal law by damaging a computer "used by or for a government entity in furtherance of the administration of justice, national defense, or national security," even if that damage does not result in provable loss greater than $5,000.

14.5.10.6 "Protected Computer" And Computers in Foreign Countries

Before the law was changed, "protected computer" was defined as a computer used by the federal government or a financial institution, or one "which is used in interstate or foreign commerce." The definition did not explicitly include computers outside of the United States. Because of the interdependency and availability of global computer networks, hackers from within the United States are increasingly targeting systems located entirely outside of this country. The old statute did not explicitly allow for prosecution of such hackers. In addition, individuals in foreign countries frequently route communications through the United States, even as they hack from one foreign country to another. In such cases, their hope may be that the lack of any U.S. victim would either prevent or discourage U.S. law enforcement agencies from assisting in any foreign investigation or prosecution.

The *USA Patriot Act* amends the definition of "protected computer" to make clear that this term includes computers outside of the United States so long as they affect "interstate or foreign commerce or communication of the United States." By clarifying the fact that a domestic offense exists, the United States can now use speedier domestic procedures to join in international hacker investigations. Because these crimes often involve investigators and victims in more than one country, fostering international law enforcement cooperation is essential. In addition, the amendment creates the option of prosecuting such criminals in the United States. Because the United States is urging other countries to ensure that they can vindicate the interests of U.S. victims for computer crimes that originate in their nations, this provision will allow the United States to reciprocate in kind.

14.5.10.7 Counting State Convictions as "Prior Offenses"

Under previous law, the court at sentencing could, of course, consider the offender's prior convictions for state computer crime offenses. State convictions, however, did not trigger the recidivist sentencing provisions of the *Computer Fraud and Abuse Act*, which double the maximum penalties available under the statute.

The new law alters the definition of "conviction" so that it includes convictions for serious computer hacking crimes under state law (i.e., state felonies where an element of the offense is "unauthorized access, or exceeding authorized access, to a computer").

14.5.10.8 Definition of Loss

Calculating "loss" is important when the government seeks to prove that an individual caused more than $5,000 loss in order to meet the jurisdictional requirements found in the *Computer Fraud and Abuse Act*. Yet existing law had no definition of "loss." The only court to address the scope of the definition of "loss" adopted an inclusive reading of what costs the government may include. In *United States v. Middleton*, 231 F.3d 1207, 1210–11 (9th Cir. 2000), the court held that the definition of loss includes a wide range of harms typically suffered by the victims of computer crimes, including the costs of responding to the offense, conducting a damage assessment, restoring the system and data to their condition before the offense, and any lost revenue or costs incurred because of interruption of service. In the new law, the definition used in the *Middleton* case was adopted.

14.5.10.9 Development of Cybersecurity Forensic Capabilities

The *USA Patriot Act* requires the U.S. Attorney General to establish such regional computer forensic laboratories as he or she considers appropriate and to provide support for existing computer forensic laboratories to enable them to provide certain forensic and training capabilities. The provision also authorizes spending money to support those laboratories.

14.6 Investigations

During the conduct of any investigation following a bona fide incident, a specific sequence of events is recommended. This sequence of events should generally be followed as a matter of good practice for all incidents unless special circumstances warrant intervention by law enforcement personnel. This section provides an overview of the process taken when an investigation is needed. The sequence of events for investigations is as follows:

- Investigating the report

- Determining what crime was committed

- Informing senior management

- Determining the crime status

- Identifying company elements involved

- Reviewing security/audit policies and procedures

- Determining the need for law enforcement

- Protecting the chain of evidence

- Assisting law enforcement as necessary

- Prosecuting the case

14.7 Ethics

Internet RFC 1087 [9], "Ethics and the Internet," may have been the first document to address ethical behavior for access to and use of the Internet. It stated that such access and use is a privilege that should be treated as such by all users. An excerpt from the RFC follows:

> The IAB strongly endorses the view of the Division Advisory Panel of the National Science Foundation Division of Network, Communications Research

and Infrastructure which, in paraphrase, characterized as unethical and unacceptable any activity which purposely: (a) seeks to gain unauthorized access to the resources of the Internet, (b) disrupts the intended use of the Internet, (c) wastes resources (people, capacity, computer) through such actions, (d) destroys the integrity of computer-based information, and/or (e) compromises the privacy of users. The Internet exists in the general research milieu. Portions of it continue to be used to support research and experimentation on networking. Because experimentation on the Internet has the potential to affect all of its components and users, researchers have the responsibility to exercise great caution in the conduct of their work. Negligence in the conduct of Internet-wide experiments is both irresponsible and unacceptable. The IAB plans to take whatever actions it can, in concert with Federal agencies and other interested parties, to identify and to set up technical and procedural mechanisms to make the Internet more resistant to disruption. Such security, however, may be extremely expensive and may be counterproductive if it inhibits the free flow of information which makes the Internet so valuable. In the final analysis, the health and well-being of the Internet is the responsibility of its users who must, uniformly, guard against abuses which disrupt the system and threaten its long-term viability.

Endnotes

1. CERT Coordination Center (1994). CERT Coordination Center Web document, at http://www.cert.org. See also CERT Annual Report to ARPA for further information.

2. U.S. Department of Justice. "The National Information Infrastructure Protection Act of 1996 Legislative Analysis," at http://www.usdoj.gov/criminal/cybercrime.

3. United States of America Executive Order, issued October 16, 2001, by President George W. Bush, at http://www.whitehouse.gov.

4. Public Law 107-56, electronic document available from the Library of Congress, H.R. 3162, http://www.thomas/loc.gov.

5. More information on these changes can be found at the U.S. Department of Justice Website, at http://www.usdogj.gov.

6. Known as the *Cable Act*, 47 U.S.C. § 551.

7. Known as the wiretap statute, 18 U.S.C. § 2510 et seq.; ECPA, 18 U.S.C. § 2701 et seq.

8. Known as the pen/trap statute, 18 U.S.C. § 3121 et seq.

9. RFC 1087, "Ethics and the Internet," Internet Activities Board, January 1989, at http://www.ietf.org.

CHAPTER 15

Intrusion Process

John Rittinghouse
James F. Ransome

The scope of this chapter is to understand the technical capabilities and limitations of a potential unauthorized intruder in order to make sure that your own security measures can withstand a hacker's attempt to breach them. It is important to know not only your own tools and techniques, but also those of the potential adversary, in order to better protect against them. This is the first part of an age-old strategy used by the famous Chinese strategist, Sun Tzu, who said, *"If you know the enemy and know yourself, you need not fear the result of a hundred battles. If you know yourself but not the enemy, for every victory gained you will also suffer a defeat. If you know neither the enemy nor yourself, you will succumb in every battle."* [1]

15.1 Profiling To Select a Target or Gather Information

Target profiling is a term that describes the process of choosing a target for hacking and doing subsequent research on that specific target. Because the Internet has made public (and sometimes private) information very easy to access, once a target has been identified, it is usually a trivial matter to search for and uncover great amounts of information related to that target. Professional hackers normally carry out target profiling; it is not something done by the casual browser. These professional hackers usually choose their targets because the target has or is perceived to have some value to the hacker.

Conducting such high-level acts of intrusion generally means the profiler has a unique and specific set of tools, adequate time to carry out the task, and a strong desire to acquire whatever the target possesses. Hackers use many types of networking tools to gather sensitive information from unsecured wireless networks. These specialized tools include discovery tools, packet analyzers, application layer analyzers, network utility applications, and share enumerators. As IT security professionals become more attuned to these types of attacks and deploy better security architectures for their networks, many of these tools will become obsolete. Social engineering is typically the hacker's next-best approach.

15.2 Social Engineering

Social engineering is a term used to describe the act of convincing someone to give you something that they should not. Successful social engineering attacks occur because the target might be ignorant of the organization's information security policies or intimidated by an intruder's knowledge, expertise, or attitude. Social engineering is reported to be one of the most dangerous yet successful methods of hacking into any IT infrastructure. Social engineering has the potential of rendering even the most sophisticated security solution useless. Some favorite targets for social engineering attacks are the help desk, on-site contractors, and employees.

The help desk should be trained to know exactly which pieces of information related to the wireless network should not be given out without proper authorization or without following specific processes put in place by security policy. Items that should be marked for exclusion include the Service Set Identifier (SSID) of access points, WEP key(s), physical locations of Access Points (APs) and bridges, usernames and passwords for network access and services, and passwords and SNMP strings for infrastructure equipment. For example, if the process of defeating a WEP has stumped the hacker, he or she may try to trick an employee into providing this information. Once the correct WEP key has been obtained by the hacker, he or she will plug that key into his or her computer and use the various tools described previously to capture sensitive data in real time, just as if there was no security.

Two common tactics are often used when attempts at social engineering against help desk personnel are implemented: (1) forceful, yet professional language and (2) playing dumb. Both approaches have the same effect—obtaining the requested information. Social engineers understand that help desk employees do not wish to have their supervisors and managers brought into a discussion when their assigned customers are not happy with the service they are receiving. Social engineers also know that some people are just inept at handling conflict, and some people are easily intimidated by anyone with an authoritative voice. Playing dumb is also a favorite tactic of social engineers. The help desk personnel are often distracted and disarmed by the "dumb caller," which causes them to stop paying attention to rigid security protocols when they assume the person they are speaking with knows very little to begin with.

IT contractors can be especially good targets for social engineers. They are brought onto a job with very little training in security and may not realize the value of information they are helpfully providing the authoritative caller on the other end of the phone. How could they know the authoritative voice on the other end of the phone is a hacker anyway? Remember, most contractors are knowledgeable of the inner workings and details of just about all network resources on a site because they are often on that site to design and/or repair the very network they built. In wanting to be helpful to their customer, contractors often give out too much information to people who are not authorized to have such information.

Wireless technology is still very new to many organizations. Employees who have not been properly educated about wireless security may not realize the dangers a wireless network can pose to the organization. Nontechnical employees who use a wireless network should be trained to know that their computers can be attacked at work, at home, or on any public wireless network. Social engineers take advantage of all of these weaknesses, and they even fabricate elaborate stories to fool almost anyone who is not specifically trained to recognize these types of attacks.

15.3 Searching Publicly Available Resources

Today, it is possible to find out information on almost any conceivable topic simply by searching the Internet. This new information tool can be used to find out about almost anything, including personal information about individuals, proprietary information about corporations, and even network security information that is not intended to be made public. If public information exists, it can most likely be found on the Internet with little effort. Nefarious individuals can find out who you are, the names of your family members, where you live and work, if your residence or workplace has a WLAN, and what wireless security solutions are used by your employer. As an example, many people now post their resumes on the Internet. From an individual's resume, you may be able to determine if someone has WLAN proficiency. It is logical to therefore assume that this individual may have a WLAN set up at his or her home. This individual is also more likely than not to be connected to his or her employer's corporate network. It is also likely that sensitive corporate data is exposed on the user's laptop or desktop computer.

Because this individual connects to a laptop using the WLAN at home, he or she is susceptible to hacking attacks. All of this is probable because a small amount of personal information found on the Internet exposed critical data that led to a security breach on a corporate network. What is even worse is the fact that this data could be pilfered and the victim is likely never going to know it happened!

Many Web sites provide maps of public wireless Internet access. If someone is easily able to use a corporation's WLAN as a public access point, that WLAN may even be one of the sites listed on these maps. For instance, http://www.NetStumbler.com offers a unique mapping [2] of all reported WLANs NetStumbler has found.

15.4 War-Driving, -Walking, -Flying, and -Chalking

War-driving is the common term for unauthorized or covert wireless network reconnaissance. WLAN utilities (sniffers) are now using airborne tactics, detecting hundreds of WLAN access points from private planes cruising at altitudes between 1,500 and 2,500 feet. Recently, a Perth, Australia–based "war flier" reportedly managed to pick up e-mails and Internet Relay Chat (IRC) conversations from an altitude of 1,500 feet [3].

WLAN war drivers routinely cruise target areas in cars that are equipped with laptops. The laptops are commonly equipped with a Wireless Network Interface Card (WNIC), an external high-gain antenna, and often even with a global positioning system (GPS) receiver. The wireless LAN card and GPS receiver feed data into freely available software such as NetStumbler or Kismet, both of which detect access points and SSIDs, which are correlated to their GPS-reported locations. War-driving gets a hacker even one step closer to the actual network through a practice known as *war-walking*. This has been made possible through a software variant of Net-Stumbler made especially for the PocketPC called MiniStumbler.

The term *war-driving* is a derivation of the term *war-dialing*, which was originally used to describe the exploits of a teenage hacker portrayed in the movie *War Games* (1983), where the teenager has his computer set up to randomly dial hundreds of phone numbers seeking those that connect to modems. In the movie, the teenager eventually taps into a nuclear command and control system. Since 1983, when the movie was released, there have been several well-publicized instances of hackers breaking into government facilities. None of these break-ins have, however, resulted in the compromise of nuclear codes—that is where Hollywood ends and reality begins.

Recently, a hobbyist WLAN sniffer, alias Delta Farce, who claimed to be a member of the San Diego Wireless Users Group, purportedly conducted a *war-flying* tour of much of San Diego County in a private plane at altitudes ranging between 1,500 and 2,500 feet. According to his or her claims, Delta Farce detected 437 access points during the flight. These exploits were posted on the Ars Technica Web site [4]. Delta Farce reported that NetStumbler software had indicated that only 23 percent of the access points detected during the trip had even the simplest form of security, Wired Equivalent Privacy (WEP), enabled. The trip also showed that the range of 802.11b WLAN signals, which radiate in the 2.4-GHz unlicensed frequency band, is far greater than what manufacturers report. Delta Farce said he was able to detect wireless access points at an altitude of 2,500 feet, or about five to eight times the 300- to 500-foot range of WLANs used in a warehouse or office.

The legality of such exploits depends on where and what is done. There are federal and state laws against network intrusion and also against intercepting communications between two or more parties. Through the use of NetStumbler, Kismet, Airopeek (a spectrum analyzer), or a variety of other tools, virtually anyone can drive through a city or neighborhood and easily locate wireless networks. Once a WLAN is located, these tools will show the SSID, whether WEP is being used, the manufacturer of the equipment, IP subnet information, and the channel the network is using. Once this information is obtained, an adversary can either associate and use DHCP or make an educated guess of the actual static IP network address. At the very least, the unauthorized access will result in free Internet access. By using simple auditing tools, an adversary can now scan the network for other devices or use a VPN connection from the gateway into a corporate network. Using the trace route utility on any Windows computer can quickly give more information on a particular network connection. A trace route displays and resolves the name of all the hops between your computer and that of another host (e.g., a Web server). Trace routing provides the attacker a way to find out where he is logically located on the Internet once connected to a WLAN.

Sometimes an adversary will have a little help from war-chalkers who have already mapped out the potential target. *War-chalking* refers to the practice and development of a language of signs used to mark sidewalks or buildings located near an accessible wireless network with chalk, notifying other war-drivers that a wireless network is nearby and providing specialized clues about the structure of the network. Such clues include knowing if the network is open or closed, whether WEP is enabled or not, the speed of the Internet connection, the azimuth and distance of the access point from the mark, and so on. Most of the symbols can be found on the Web at http://www.warchalking.org. The term *war-chalking* originated from the practice of war-driving and is essentially a way for hackers to help other hackers. If your network has been war-chalked, you can bet it has been hacked or, at the very least, simply borrowed for free wireless Internet access.

15.4.1 WLAN Audit and Discovery Tools

Hackers exploit vulnerabilities discovered when using various auditing tools. WLAN auditing tools are the weapon of choice for exploiting WLAN networks. Multipurpose tools that can be used for auditing and hacking into a WLAN are described in the following sections. Many protocol analysis and site survey tools are focused on finding WiFi compliant Digital Sequence Spread Spectrum (DSSS) networks. It should not be assumed that an unauthorized user is only going to use WiFi equipment to conduct reconnaissance and penetrate a network.

15.4.1.1 NetStumbler

One of the most popular discovery tools is a free Windows-based software utility called NetStumbler. This tool is usually installed on a laptop computer. War-drivers, war-walkers, war-flyers, and war-chalkers commonly use NetStumbler to locate and interrogate WLANs. NetStumbler's popularity stems from its ease of use and wide support of a variety of network interface cards (NICs). Other networking tools can be used to gain unauthorized access to a WLAN. Once NetStumbler finds an access point, it displays the MAC Address, SSID, access point name, channel, vendor, security (WEP on or off), signal strength, and GPS coordinates if a GPS device is attached to the laptop. Adversaries use NetStumbler output to find access points lacking security or configured with manufacturer's default settings. Although WEP has exploitable vulnerabilities, a time investment is required to break WEP, and unless the adversary has specifically targeted your facility, he or she will normally take the path of least resistance and go after the more easily accessible open networks that are found everywhere.

15.4.1.2 MiniStumbler

MiniStumbler has the same functionality as NetStumbler but is designed to run on the PocketPC platform. It can operate from a very small platform, which makes it popular for use in war-walking. The ability to war-drive a wireless network with a handheld device placed in one's pocket makes MiniStumbler a valuable addition to any adversary's war chest.

15.4.1.3 Kismet

Kismet is an 802.11 wireless network sniffer. As described on the Kismet Web site [5], it differs from a normal network sniffer (such as Ethereal or tcpdump) because it separates and identifies different wireless networks in the area. Kismet works with any 802.11b wireless card capable of reporting raw packets (rfmon support), which includes any prism2-based card (e.g., Linksys, D-Link, RangeLAN), Cisco Aironet cards, and Orinoco-based cards. Kismet supports the WSP100 802.11b remote sensor by Network Chemistry and is able to monitor 802.11a networks with cards that use the ar5k chipset. Kismet runs on the Linux operating system and has similar functionality to NetStumbler, but with a few additional features. Kismet's basic feature set includes the following:

- Airsnort-compatible logging

- Channel hopping

- Cisco product detection via CDP

- Cross-platform support (handheld Linux and BSD)

- Detection of default access point configurations

- Detection of NetStumbler clients

- Ethereal/tcpdump compatible file logging

- Graphical mapping of data (gpsmap)

- Grouping and custom naming of SSIDs

- Hidden SSID decloaking

- IP blocking protection

- Manufacturer identification

- Multiple packet source

- Multiplexing of multiple capture sources

- Runtime decoding of WEP packets

- Support for multiple clients viewing a single capture stream

15.4.1.4 AiroPeek NX

AiroPeek NX is a Windows-based wireless sniffer from WildPackets. It has the capability to capture and decode packets simultaneously. Although AiroPeek can do on-the-fly decryption of WEP keys, it doesn't actually crack WEP; you must supply the valid keys [6].

15.4.1.5 Sniffer Wireless

Sniffer Wireless [7] is a Windows sniffer from Network Associates. It doesn't decode packets on the fly. You must stop sniffing before you can decode in this mode. It can decode a very large number of protocols at near-wire speeds. It also has the ability to spot rogue APs.

15.4.2 Network Discovery Tools

Management software packages such as What's Up Gold (http://www.ipswitch.com), SNMPc (http://www.castlerock.com), and Solarwinds (http://www.solarwinds.net) each contain specialized discovery tools that use the Simple Network Management Protocol (SNMP) to map their way through an enterprise. If an adversary gains access to a WLAN and steals certain SNMP strings, the attacker can then begin creating a map of the entire extended network. An insecure wireless segment that exists within an enterprise environment that has distributed WLANs can cause an otherwise secure wired network to become insecure. This is another example of the huge security risks posed by implementation and use of WLANs.

15.4.3 Networking Utilities

In order to find out what resources are available on a network, most intrusion attempts begin with a scan of the network. To gather information, the client needs to obtain a valid IP address, either through DHCP assignment or by statically assigning a valid IP address. The next logical step for the hacker is to use a network utility such as WS-Ping ProPack (http://www.ipswitch.com) or NetScan Tools professional (http://www.netscantools.com) that can perform functions such as ping sweeps (pinging every IP address in a subnet looking for active nodes), port scans for defined ports (FTP, POP3, SMTP, NETBIOS), and computer name resolution (Accounting, Human Resources, Sales, Marketing). Once these tasks are performed, more detailed probes can be accomplished with tools such as LANGuard.

Once access point scans are accomplished using NetStumbler, and ping sweeps are accomplished with networking utilities, the IP addresses of the access points can be determined by comparing the laptop's ARP cache against NetStumbler results. The ARP cache on the laptop is viewed by opening a command prompt window and typing "arp-a." This command will return the IP addresses and MAC addresses of every node detected on the network.

15.5 Exploitable WLAN Configurations

It is important that network administrators learn to properly configure administrative passwords, encryption settings, automatic network connection functions, reset functions, Ethernet Medium Access Control (MAC) Access Control Lists (ACLs), shared keys, and Simple Network Management Protocol (SNMP) agents. Doing so will help eliminate many of the vulnerabilities inherent in a vendor's out-of-the-box default configuration settings. Network administrators should configure APs in accordance with established security policies and requirements. The following is a list of vulnerabilities that can result from one of these configuration problems:

- Default passwords not updated

- WLAN encryption not set for the strongest encryption available

- No controls over the reset function

- MAC ACL functionality not in use

- Not changing the SSID from its factory default

- Not changing default cryptographic keys

- Not changing the default SNMP parameter

- Not changing the default channel

- Not using DHCP

Now, let's look at how intruders use these WLAN weaknesses to exploit an organization.

15.6 How Intruders Obtain Network Access to a WLAN

Security attacks are typically divided into two classes: *passive attacks* and *active attacks.* These two broad classes are then subdivided into other types of attacks. A passive attack is an attack in which an unauthorized party simply gains access to an asset and does not modify its content (i.e., eavesdropping). Passive attacks can be done by either simple eavesdropping or by conducting a traffic analysis (which is sometimes called traffic flow analysis). While an attacker is eavesdropping, he or she simply monitors network transmissions, evaluating packets for (sometimes) specific message content. As an example of this type of attack, suppose a person is listening to the transmissions between two workstations broadcast on a LAN or that he or she is tuning into transmissions that take place between a wireless handset and a base station. When conducting traffic analysis of this type, the attacker subtly gains intelligence by looking for patterns of traffic that occur during the communication broadcasts. A considerable amount of information is contained in message flow traffic between communicating parties.

Active attacks are attacks in which an unauthorized party makes deliberate modifications to messages, data streams, or files. It is possible to detect this type of attack, but it is often not preventable. Active attacks usually take one of four forms (or some combination of such):

1. Masquerading

2. Replay

3. Message modification

4. Denial-of-Service (DoS)

When masquerading, the attacker will successfully impersonate an authorized network user and gain that user's level of privileges. During a replay attack, the attacker monitors transmissions (passive attack) and retransmits messages as if they were sent by a legitimate user. Message modification occurs when an attacker alters legitimate messages by deleting, adding, changing, or reordering the content of the message. DoS is a condition that occurs when the attacker prevents normal use of a network.

15.6.1 WLAN Attacks

All risks known to exist with 802.11 standards–based equipment are the result of one or more of the aforementioned active or passive attack methods. These attacks generally cause a loss of proprietary information, with companies suffering legal and recovery costs and sometimes even a tarnished image as a result of publication of the attack (which is known as an *event* in the security field) that resulted in a total loss of network service. With the rapid rate of growth and adoption of 802.11b WLAN technology in many organizations attempting to capitalize on the benefits of "going wireless," there are many chances for hackers to take advantage of these known vulnerabilities when they discover that lax security practices are used by an adopter. Numerous published reports and papers have described attacks on 802.11 wireless networks and exposed risks to any organization deploying the technology. It is wise for those planning to adopt WLAN technology to find these papers and educate themselves and their staff on the risks and, moreover, to weigh these risks against the benefits of using the WLAN.

15.6.2 WEP Decryption Tools

In order to recover WEP encryption keys, WEP decryption software is used to passively monitor data transmission on a WLAN segment. When enough data has been collected, these decryption tools can compute the cryptographic key used to encrypt the data. Once this occurs, the network is totally insecure. For this to work, the decryption software must collect enough packets formed with "weak" initialization vectors. Wireless packet analyzers, such as AirSnort and WEPcrack, are common tools that are readily available on the Internet and are very popular WEP crackers. Both of these applications run in Unix-based environments.

AirSnort was one of the first tools created to automate the process of analyzing network traffic. Unfortunately, hackers quickly discovered that it is also great for breaking into wireless networks. AirSnort leverages known vulnerabilities found in the key-scheduling algorithm of RC4, which is used to form the basis of the WEP standard. The software monitors the WLAN data in a passive mode and computes encryption keys after about 100 MB of network packets have been sniffed. On a busy network, collecting this amount of data may take only three or four hours, but if traffic volume is slow, it could easily stretch out to a few days. After all of these network packets have been collected and analyzed, the cryptographic key can be determined in a matter of milliseconds. This gives the attacker access to the root key, and he or she can now read cleartext of any packet traversing the WLAN.

15.6.3 MAC Address Spoofing and Circumventing Filters

Numerous 802.11 product vendors provide capabilities for restricting access to the WLAN based on device MAC ACLs, which are stored and distributed across many APs. MAC address exploitation can be accomplished by an adversary capturing a series of wireless frame packets obtained during normal business hours at the target location. The captured frames contain all information needed to circumvent MAC filters. Using this data, the hacker is able to derive valuable information from the packet trace log, which is generated via a wireless protocol packet analyzer such as WildPackets Airopeek or Network Associates Sniffer Pro Wireless. By reviewing the BSS IDs (the MAC address of an access point) found in the packet trace, the hacker can figure out which units are access points and which are clients. Once this is known, it is a rather simple matter to deduce which SSIDs and MAC addresses are used by the connecting clients. Additionally, IP subnet information can be recorded in order to establish subsequent network connections once the hacker device is associated to the target access point. Once these data have been recorded, the hacker is in a position to gain unauthorized access to the target network.

15.6.4 Rogue AP Exploitation

Rogue APs pose huge security risks. Malicious users have been known to surreptitiously insert rogue APs into closets, under conference room tables, and in other hidden areas in buildings to gain unauthorized access to targeted networks. As long as the rogue AP location is close to WLAN users, the rogue AP can intercept wireless traffic between an authorized AP and its wireless clients without being detected. The rogue AP needs to be configured with a stronger signal than the existing AP in order to intercept client traffic. Malicious users can also gain access to a wireless network by using APs configured to allow blanket access without authorization.

15.6.5 Exploiting Confidentiality Weaknesses

Confidentiality infers that specific information is not to be made available or disclosed to unauthorized individuals, entities, or processes. Confidentiality is a fundamental security requirement for most organizations. Because of the very nature of wireless communications, confidentiality is a difficult security requirement to implement. Often, it is not possible to control the distance over which a WLAN transmission occurs. This makes traditional physical security countermeasures ineffective for WLANs. Passive eavesdropping of wireless communications is a significant risk to any organization. Because 802.11b signals can travel outside the building perimeter, hackers are often able to listen in and obtain sensitive data such as corporate proprietary information, network IDs, passwords, and network and systems configuration data. Sometimes, the hacker is even an insider who may be disgruntled. The extended range of 802.11 broadcasts enables hackers to detect transmissions from company parking lots or from positions curbside on nearby roads. This kind of attack, which is performed with a wireless network analyzer tool, or *sniffer,* is particularly easy for two reasons:

1. Confidentiality features of WLAN technology are often not even enabled.

2. Numerous vulnerabilities in the 802.11b technology security are compromised.

When an AP is connected to a network through a hub, it poses yet another risk to loss of confidentiality. Hubs generally broadcast all network traffic to *all* connected devices, which leaves hub-relayed traffic vulnerable to unauthorized monitoring. An adversary can monitor such traffic by using a laptop and wireless NIC (set to promiscuous mode) when an access point is connected to a hub instead of a switch. If the wireless AP is connected to an Ethernet hub, the hacker device monitoring broadcast traffic is able to easily pick up data that was intended for wireless clients. Consequently, organizations should consider using switches instead of hubs for connections to wireless access points.

15.6.6 Exploiting Data Integrity Weaknesses

Wireless networks face the same data integrity issues that are found in wired networks. Organizations frequently implement wireless and wired communications without adequate data encryption. As a result, data integrity can be very difficult to achieve. A determined hacker can compromise data integrity simply by deleting or modifying data in an e-mail from an account found on the wireless system. The impact of such message modification could be quite detrimental to an organization depending on the importance of the e-mail and how widespread its distribution is across the company. Existing security features of 802.11 do not provide strong message integrity. This can lead to vulnerability from other kinds of active attacks. The WEP-based integrity mechanism used in wireless networking is simply a linear Cyclical Redundancy Check (CRC). Message modification attacks are possible without implementation and use of some cryptographic checking mechanisms, such as message authentication codes and hash codes (message digests).

15.6.7 Exploiting Authentication Weaknesses of the Service Set Identifier

Two methods are defined in the 802.11b specification for validating wireless users as they attempt to gain access to a network. One method depends on cryptography. The other method consists of two types of checks used to identify a wireless client attempting to join a network. Both of these non-cryptographic approaches are considered to be identity-based verification mechanisms. When establishing a connection, the wireless station requesting access will reply to a challenge with the SSID of the wireless network—there is no true "authentication." This method is known as closed system authentication. With closed system authentication, wireless clients must respond with the actual SSID of the wireless network. That is, a client is allowed access if it responds with the correct 0- to 32-byte string identifying the BSS of the wireless network. Conversely, when using open system authentication, a client is considered authenticated if it simply responds with an empty string for the SSID—hence, the name "NULL authentication." Both of these primitive types of authentication are only identification schemes, not true authentication methods. Neither of these two schemes offers very strong security against unauthorized access. Both open and closed authentication schemes are highly vulnerable to attacks, and steps should always be taken to mitigate such risk.

It is possible for a WLAN to hide the SSID from potential intruders. Currently, a few APs have software settings used to exclude sending the SSID in order to obscure the WLAN's identity. Even with this feature, it is fairly easy for a hacker to learn the SSID of an active but hidden WLAN. The hacker will do this by sending a spoofed "disassociate" message to the AP. This message will force the wireless station to disconnect and reconnect to the WLAN. This method of forcing a hidden WLAN to reveal its SSID typically takes a hacker less than a second to execute against a station actively transmitting data.

15.6.8 Exploiting Cryptographic Weaknesses

A common cryptographic technique used for authentication is shared key authentication. It is a simple "challenge and response" scheme. The premise of this scheme is based on whether a client has knowledge of a shared secret. For example, a random challenge is generated by the access point and sent to the wireless client. The wireless client uses a cryptographic key (a.k.a., a WEP key), which is shared with the AP to encrypt the issued challenge and return the encrypted result to the AP. The AP then decrypts the encrypted challenge that was computed by the client. The AP will only allow access if the decrypted value is the same as the value issued during the challenge transmittal. The RC4 stream cipher algorithm is used to compute the encrypted and decrypted values. This authentication method is considered a rudimentary cryptographic technique. It does not provide mutual authentication. The client does not authenticate the AP. There is no assurance that a client is communicating with a legitimate AP as opposed to communicating with a rogue AP. Challenge-response schemes are considered to be a very weak form of security. Because of this weakness, challenge-response schemes are vulnerable to many types of attack, such as the man-in-the-middle attack.

15.7 Password Gathering and Cracking Software

Weak passwords are considered among the most serious of security threats in the networking environment. Security administrators have long suffered the effects of poor password administration, but they have learned over the last few years that a strong password policy in an organization can save them many hours of work in the long run. With the advent of WLANs, it was quickly discovered that passwords travel across unsecured networks from client to server all the time. Once LANs were thought to be very secure, but now, with the advent of WLANs, both network administrators and hackers have discovered that networking systems using passwords passed in cleartext across wired or wireless mediums are absolutely insecure. As a result of this discovery, password encryption has become a must. Security mechanisms such as Kerberos implement such strong encryption. Two well-known security auditing tools are used by both administrators and hackers to view cleartext passwords, namely *WinSniffer* and *Ettercap,* discussed as follows.

15.7.1 WinSniffer

WinSniffer is a utility capable of capturing SMTP, POP3, IMAP, FTP, HTTP, ICQ, Telnet, and NNTP usernames and passwords in a wired/wire-less blended networking environment. WinSniffer is a Windows-based utility. It is usually on a laptop dedicated to use for auditing wireless networks. In a switched network environment, WinSniffer captures passwords from clients or servers. WinSniffer can also be used to capture passwords saved in applications when users have forgotten them. WinSniffer can be used by an adversary to monitor users checking e-mail over an unencrypted WLAN segment. With this tool, the attacker could easily pick up a user's e-mail login information and determine which domain the user accesses when checking mail. The information obtained in this manner provides the attacker full and unrestricted access to the unwitting user's e-mail account.

Hotspots (a.k.a. public access wireless networks) are commonly found in airports or in metropolitan areas. They are some of the most vulnerable areas for user or peer-to-peer attacks. Victims who are unfamiliar with security vulnerabilities in these hotspots are easy prey. Mobile users should be trained on just how easy it is to obtain login information from a peer-to-peer attack. Often, such users check their e-mail or access a corporate network from a hotspot and in the process can unwittingly give access to their accounts to hackers. Once a hacker has obtained a valid login to the victim's corporate account, they often try to obtain further access into the corporate network using the victim's credentials in order to locate more sensitive corporate information.

15.7.2 Ettercap

Ettercap is a multipurpose sniffer/interceptor/logger for switched use on a LAN. Ettercap supports almost every major operating system platform and can be downloaded from Sourceforge [8]. Ettercap can gather data in a switched network environment. This capability exceeds the abilities of most audit tools, making ettercap a quite valuable edition to the hacker's toolbox. Ettercap uses a Unix-style *ncurses* code library to create a menu-driven user interface that is considered very user friendly for beginner-level users. Some of the better known features available in Ettercap are character injection into an established connection, SSH1 support, HTTPS support, remote traffic via GRE tunnels, PPTP brokering, plug-in support, a password collector, packet filtering and packet rejection, OS fingerprinting, a connection killer, passive LAN scanning, poison checking, and binding of sniffed data to a local port.

15.7.3 L0phtCrack

Operating systems commonly implement password authentication and encryption at the application layer. Microsoft Windows file sharing and NetLogon processes are examples of this. The challenge and response mechanism used by Microsoft over the years has changed from LM (weak security) to NTLM (medium-level security) to NTLMv2 (strong security). Before release of NTLMv2, tools such as L0phtcrack could easily crack these hashes in a matter of minutes. It is also important to properly configure your Windows operating system to use NTLMv2 and not to use the weaker versions. Proper administration of patches and service packs is not enough. To properly secure a network to use NTLMv2, much of this process must be accomplished manually [9]. LC4 is the latest version of the password auditing and recovery application L0phtCrack. According to the L0phtcrack Web site [10], LC4 provides two critical capabilities to Windows network administrators:

1. It helps systems administrators secure Windows-authenticated networks through comprehensive auditing of Windows NT and Windows 2000 user account passwords.

2. It recovers Windows user account passwords to streamline migration of users to another authentication system or to access accounts whose passwords are lost.

LC4 supports a wide variety of audit approaches. It can retrieve encrypted passwords from stand-alone Windows NT, 2000, and XP workstations, networked servers, primary domain controllers, or Active Directories, with or without Syskey installed. The software is capable of sniffing encrypted passwords from the challenge-response exchanged when one machine authenticates to another over the network. This software allows administrators to match the rigor of their password audit to their particular needs by choosing from three different types of cracking methods: dictionary, hybrid, and brute force analysis. Finally, using a distributed processing approach, LC4 provides administrators the ability to perform time-consuming audits by breaking them into parts that can be run simultaneously on multiple machines.

Once the intruder has captured the targeted password hashes, the hashes are imported into LC4's engine, and the dictionary attack automatically ensues. If the dictionary attack is unsuccessful, a brute force attack is automatically initiated. The processor power of the computer doing the audit will determine how fast the hash can be broken. L0phtCrack has many modes for capturing password hashes and dumping password repositories. One mode allows for "sniffing" in a shared medium (such as wireless), while another goes directly after the Windows Security Access Manager (SAM).

Windows 2000 service pack 3 introduced support for a feature called "SysKey" (short for System Key). This feature, first seen in Windows NT, is invoked using the *syskey. exe* executable. It encrypts the SAM so well that even L0phtCrack cannot extract passwords from it. L0phtCrack can notify an auditor that a SAM has been encrypted so the auditor need not waste time attempting to extract an uncrackable password. L0phtCrack is one of the preferred tools in a hacker's arsenal. The hacker is most likely going to use L0phtcrack in an attempt to gain access to a network. Once a hacker obtains administrator-level account information, many other tools already discussed will become quite useful to him or her.

15.7.4 Lucent Registry Crack

Proxim Orinoco PC cards store an encrypted hash of the WEP key in the Windows registry. The Lucent Registry Crack (LRC) utility is a simple command-line tool used to decrypt these values. The problem hackers face is getting these values from another computer, especially one that has the proper WEP key for the AP that the hacker wants to attack. This task is accomplished using a remote registry connection. The attacker can make a remote registry connection using the Window's Registry Editor found on his own computer. Once the hacker is remotely connected, he or she must know where the key is located in the remote registry in order to copy and paste it into a text document on his or her computer. Once this is done, the hacker can use LRC to analyze this encrypted string and produce the WEP key. This process takes only a few seconds at most to complete. When the attacker has derived the WEP key using LRC, he or she can simply insert it into a computer to gain access to the target network. This process can be defeated when wireless end users are properly trained to implement safeguards against peer-to-peer attacks (such as installing personal firewall software or enabling IPSec policies).

15.7.5 Wireless Protocol Analyzers

Wireless protocol analyzers are used to capture, decode, and filter wireless packets in real time. Many products also support multiple frequency bands used in 802.11b and 802.11a networks. Protocol analyzers operate in RF monitor mode capturing packets as they are transmitted across the medium. Protocol analyzers make no attempt to connect or communicate with APs or other wireless peers while in this mode. There are many vendors in the protocol analyzer space, whose products include the following:

- AirMagnet

- Ethereal

- Fluke WaveRunner Wireless Tester

- Network Associates Sniffer Pro Wireless

- Network Instruments Observer

- Wildpackets Airopeek

Not all wireless packet analysis tools have identical functionality. For example, some do not offer real-time packet decoding. Some force the user to capture packets and export them to a reader utility. Some analyzers decode OSI Layer 2 through 7 protocols, whereas others decode only Layer 2 frame headers.

15.8 Share Enumerators

File sharing is a major benefit of client/server networking. A major risk in file sharing arises when a node or server is improperly configured and data are exposed to unauthorized access. Share enumerators are software programs that can scan a Windows subnet for open file shares. Open file shares are directories on a Windows network that are made available to users for public browsing. Exploiting open file shares is a method used by some Internet Trojans and viruses to transmit and infect users. Others users on the Internet may be able to view or use files on the host computer. The computer could be used for distributing files (e.g., music and video) using peer-to-peer file-sharing programs. Windows open file shares provide anyone with public or domain-level access the ability to see the share, access it, and obtain data from it. Legion 2.1 is a popular freeware program that quickly scans a Windows subnet and lists all open file shares. An auditor or hacker can use Legion to quickly determine what file shares are available for access on a network. A common open file share attack methodology is to access another computer's Windows registry remotely and redefine the properties of a file share to allow root-level access. After a system reboot, the file share still appears the same to the unsuspecting victim. When a hacker browses the share, it allows him or her to view the entire contents of the root drive. If a node on the wireless segment has open file shares, those shares are exposed to any intruder who has gained access to the wireless network. Once file shares are located on the network, even those shares whose settings are not public can be cracked or their properties can be changed to allow further access.

15.9 Using Antennas and WLAN Equipment

Tools used for auditing WLANs include antennas, wireless cards, a portable computer, and specialized software. These tools are legal, readily available, and quite affordable. In most cases, the total cost for a wireless NIC, an antenna, and a pigtail cable is less than $100. The auditing software can be obtained freely from the Internet. This means *anyone who has the desire* can usually afford the equipment necessary to eavesdrop on an organization's WLAN.

15.9.1 Antennas

Antennas come in many forms. Some are magnetically mounted to the roof of a car, and some are made from Pringles potato chip cans. War drivers often use Orinoco or Cisco pigtail cables and various connectors to locate WLANs. Both omni, strong Yagi, or patch antennas are readily available for such uses. The war driver is able to easily determine network names, WEP usage, and even GPS coordinates of the target wireless devices located. Once a wireless network is found, a directional antenna such as a Yagi can be used to focus the frequency waves (beams) and listen in at great distances. This allows a hacker to operate without trespassing on a victim's property. A directional antenna is also capable of detecting much fainter signals than an omnidirectional antenna and allows the intruder to establish a better-quality link at greater distances.

15.9.2 Wireless Cards

Three very popular NICs are used by hackers to attempt intrusions into WLANs: the Lucent Gold PC Card, the Cisco 350 PC Card, and the Symbol LA-4121 PC Card. They are inexpensive, can be easily obtained, and allow external antennas to be connected using pigtail cables/connectors. Most auditing software supports the chipsets used in these NICs, and each NIC provides site-surveying software that is useful for more than just intrusions and intrusion audits.

15.10 Denial-of-Service Attacks and Tools

A denial in network availability involves some form of DoS attack, such as jamming. Jamming occurs when a malicious user deliberately sends a signal from a wireless device in order to overwhelm legitimate wireless signals. Jamming results in a breakdown in communications because legitimate wireless signals are unable to communicate on the network. Nonmalicious users can also cause a DoS. A user, for instance, may unintentionally monopolize a wireless signal by downloading large files, effectively denying other users access to the network. There are three main types of wireless DoS attacks: RF jamming, data flooding, and hijacking. The tools required to conduct any of these attacks are inexpensive and easy to acquire, but the damage to production, service, or end-user productivity can be immense if these types of attacks are not prevented.

15.10.1 RF Jamming

Jamming a Direct Sequence Spread Spectrum (DSSS) WLAN is fairly easy to do using inexpensive tools, and such jamming activities can be conducted from relatively long ranges. Most WLANs operate at power outputs that are less than 100 mW. DSSS WLANs generally use only 22 MHz of the RF spectrum in order to transmit data. An RF generator can generate very low amounts of power (less than 1 W). It utilizes either directional or omnidirectional antennas capable of transmitting a broadcast signal over very long distances. Because these devices typically use a very small power source, it provides anyone with the ability to easily jam a WLAN.

Even though DSSS WLANs are resilient to noise interference, few can function properly when competing with an RF power source jamming with a signal up to 40 times (4 W) more powerful. This amount of generated RF signal can jam nearly any WLAN and cause a complete disruption (or denial) of service to client devices using the target access point. Another consideration is that the users connecting to the jammed access point are allowed to connect to the rogue access point set up by an intruder and configured to display the same SSID as the (hijacked) authorized access point. That is why this type of attack is called hijacking. No WLAN manufacturer makes a device called an "RF jamming device" because of the legal implications involved. A hacker knows he or she must find equipment commonly used for testing WLAN antennas, cables, connectors, and accessories. One such piece of equipment is YDI's Power Signal Generator-1 (PSG-1), which can be seen at http://www.ydi.com. It is important to know that microwave ovens, Bluetooth devices, and even certain WLAN devices can inadvertently cause a jamming situation to occur on a WLAN.

15.10.2 Data Flooding

Data flooding is the act of overwhelming an infrastructure device or computer with more data than it can process. There are three primary methods of performing a data flooding attack:

1. Pull a very large file from the Internet.

2. Pull or push a very large file from or to an internal server on the LAN.

3. Use a packet generator software package.

The packet generator software is easy for even a novice. It can push enough traffic to saturate any WLAN. A packet generation attack is more likely to make it through effective network controls than the first and second methods described previously. This type of attack is very similar to RF jamming except it uses DSSS transmissions to accomplish the same result.

One might think it would require significant amounts of data to flood a WLAN, but that is not the case. A data flooding attack does not require very much data at all. An 802.11b-compliant access point will typically saturate at about 5.5 Mbps of throughput. Sometimes, saturation occurs with even less than 5.5 Mbps of throughput because APs are half-duplex devices. A WLAN client can produce the same amount of throughput. Therefore, each client device also has the ability to saturate an AP. Such methods of saturation will effectively disable the AP, creating a DoS condition by denying a reasonable Quality of Service (QoS) to other users. Because WLAN devices use a protocol known as Carrier Sense Multiple Access/Collision Detection (CSMA/CD), all nodes attached to an AP are allocated a fractional slice of time (usually calculated in a methodology known as round-robin scheduling) to transmit; however, when a single node transmits a huge chunk of data, other nodes are essentially blocked from passing even very small bits of data because the time-slicing algorithm is, in effect, paused to await the completion of processing the largest data frame allowed before allocating a small slice of time to other device connections.

For example, a time-slicing algorithm allocates a 100 millisecond block of time to each of 10 connected devices. Device 8 sends a huge chunk of data that is broken down into the largest allowable frame size and transmitted, using 900 milliseconds to do so. Once the frame is transmitted, Device 9 gets 100 milliseconds, Device 10 gets its turn, then back to Devices 1 through 7, using up a total of one second of computer clock time before getting back to Device 8 again. Device 8 sends the next chunk, using 900 milliseconds again, and the process continues until the entire amount of data sent by Device 8 has been transmitted. In this simplistic example, it is easy to figure out that Device 8 is using 90 percent of the available time and denying equal service in 10 percent increments to the remaining nine devices.

15.10.3 Client Hijacking

Hijacking occurs when an unauthorized user takes control of an authorized user's WLAN connection. In wireless environments, hijacking is done at OSI Layer 2 when the intent is to create a DoS condition. When hijacking occurs at OSI Layer 3, the intruder is most likely attempting to initiate an attack surreptitiously. The unsuspecting victim who attempts connecting to a jammed access point is allowed to connect to a rogue access point set up by an intruder and configured to display the same SSID as the (now hijacked) authorized access point. In order to successfully accomplish the hijack operation, hackers must set up the rogue AP to replicate the authorized access point. A WLAN PC card can be configured to operate as a rogue AP. When configuring a rogue software AP, it is important for the hacker to choose a channel that does not conflict with one in use by the victim. When the jamming device is used to force users to roam for a better connection, the client devices will roam off the authorized hardware AP and onto the rogue software access point. After the Layer 2 connection has been hijacked, the next logical step in the attack process is to allow the hijacked user to establish a Layer 3 connection with the hijacker. The same Layer 3 connection can be established by running a DHCP server on the laptop serving as the AP. Windows-based products automatically renew DHCP leases whenever a Layer 2 connection is broken. This autorenew function works to the hijacker's benefit.

15.11 Rogue Devices as Exploitation Tools

Ideally, rogue APs are placed to allow an intruder to gain the highest degree of access possible into a network and establish and maintain unauthorized control over the hacked network. What follows is a discussion on AP placement in order to prevent and discover rogue devices on your network.

15.11.1 Access Points

Rogue devices are usually placed in an area to appear as if the device were designed to be there in the first place. An AP should not cause any disruption in service to the existing network. It is intended to be used surreptitiously, so adversaries are generally very cautious when placing rogue APs so they will not be noticed. If an administrator happens to be scanning the area where a rogue device is suspected, he or she will search for unencrypted data packets as a first sign that a rogue device exists. There is virtually no way to tell the difference between data packets encrypted by an intruder's WEP key and data packets encrypted by an authorized WEP key.

Rogue devices are often placed near building perimeter points, especially near a window, to optimize coverage. The intruder will attempt to place the rogue device in a part of the building that has a physically insecure perimeter so he or she can be within range of the access point and not arouse suspicion.

Intruders may use 900 MHz units instead of 2.4-GHz (802.11b) or 5 GHz (802.11a) WiFi-compliant units. Virtually no WLAN discovery tool can use the 900 MHz range. Intruders may also use FHSS technology such as Bluetooth, OpenAir, or HomeRF instead of DSSS.

Few WLAN discovery tools are even able to use FHSS equipment. Additionally, intruders often use horizontally polarized antennas in order to give the rogue device a very small RF signature when scanning devices are used to find rogue devices. Such rogues are unlikely to be detected in a scan unless the administrator is physically close to the rogue device.

15.11.2 Wireless Bridges

A rogue bridge placed within the Fresnel Zone of an existing bridge link poses a great security risk. A Fresnel Zone is the area around the visual line-of-sight that radio waves spread out into after they leave the antenna. This area must be clear or signal strength will weaken. Fresnel Zones are an area of concern for wireless transmissions using the 2.4-GHz range. The 2.4-GHz signals can pass through walls easily, but they have a tough time passing through trees because of the water content; 2.4-GHz signals are absorbed in water, so any barrier with a high water content becomes a problem. The Fresnel Zone of a wireless bridge link may span several miles and can be extremely broad. This fact makes placement of a rogue bridge much easier for an intruder. Conversely, rogue detection becomes much tougher for an administrator. A rogue bridge must be set up with a very low priority; otherwise, it will become the root bridge and be detected. Intruders tend to use high-gain directional antennas in order to ensure a consistent, high-quality connection. Locating a rogue bridge in a three-mile point-to-point bridge link lessens the chances of being discovered significantly when compared to setting up the rogue device inside a corporate office. Administrators are rarely able to detect the presence of rogue bridges.

References

Material in this section is excerpted from the *Cybersecurity Operations Handbook*, by John W. Rittinghouse and William M. Hancock, New York: Digital Press, 2003. Reprinted with permission.

http://www.anti-spy.com.

Endnotes

1. Sun-tzu S. *The Art of War*, Chapter 3. Seattle, WA: Clearbridge Publishing, 2002.

2. Consult URL http://www.netstumbler.com/nation.php for a map of wireless access points.

3. The URL for this story is http://www.computerworld.com/mobiletopics/mobile/story.html.

4. See http://www.arstechnica.com/wankerdesk/3q02/warflying-1.html.

5. See http://www.kismetwireless.net.

6. For further details, see http://www.wildpackets.com/products/airopeek.

7. For further details, see http://www.networkassociates.com/us/products/sniffer/field/snifferbasic.htm.

8. Sourceforge's URL is http://ettercap.sourceforge.net.

9. Instructions to do this can be found at http://www.technet.com.

10. See http://www.atstake.com/research/lc.

CHAPTER 16

Security Policy

George L. Stefanek

Establishing a security policy is the starting point in designing a secure network. It is essential that a set of minimum security requirements be gathered, formalized and included as the basis of your security policy. This security policy must be enforceable by your organization and will create an additional cost to running and monitoring your network. This additional cost/benefit of a security policy must be understood and embraced by your organization's management in order to enhance and maintain network and system security.

The lack of an accepted and well-thought-out security policy and guidelines document is one of the major security vulnerabilities in most companies today. This section discusses several Best Practices related to the production of such a document. The importance of a meaningful security policy cannot be over-emphasized.

16.1 Best Practice #1

Perform a threat analysis and risk analysis for your organization to determine the level of security that must be implemented.

First, identify all the threats to your network; second, determine threat categories; third, perform the risk assessment; and, fourth, recommend action. Risk assessment should be performed by constructing a "consequence" matrix vs. "likelihood" matrix[1] as shown in Figure 16.1.

---

[1] NDIA, Undersea Warfare Systems Division, "INFOSEC Considerations for Submarine Systems Study, Technical Report II: Threat/Risk Decomposition", 1998.

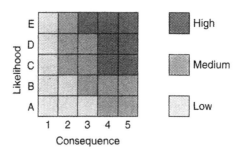

Figure 16.1: Risk rating matrix

For each threat, construct a likelihood/consequence table. The "likelihood" rating for a threat can use categories such as A = remote, B = unlikely, C = likely, D = highly likely, and E = near certainty. The "Consequence" rating of the threat should use a point system where block 1 in the table = 0–1 points, block 2 = 2–3 points, block 3 = 4–6 points, block 4 = 7–9 points, block 5 = 9–11 points. Points are assigned based upon your subjective assessment of a compromise in security as an impact on functioning of your organization and data compromise. For instance, 5 points = organization grinds to a halt, loss of mission-critical systems; 3 points = some application(s) don't function, loss of non-mission-critical systems, but the organization can function; and 0 points = no loss of function. Data compromise points may be assigned as, for example, 2 points = compromise of sensitive data to people outside the organization; 1 point = compromise of sensitive data to people internal to the organization that should not have access; and 0 point = no compromise of data. This type of matrix will immediately highlight where to emphasize security. Also, a form should be created and filled out that:

1. identifies the threat

2. gives solutions to the threat

3. shows the likelihood of the threat (i.e., A, B, etc.)

4. gives the impact of the threat on an organization's function and data

5. shows the consequences of a compromise.

16.2 Best Practice #2

Define a security policy for the entire site and use it as a guide for the network security architecture.

Define a policy that includes sections for confidentiality, integrity, availability, accountability, assurance, and enforcement, as described in the following paragraphs. The policy should address as much as possible of what is included in these sections according to risk and affordability. The general security policy described in this section was developed from DoD 5200.28 and SECNA VINST 5239.3.

Confidentiality – The system must ensure the confidentiality of sensitive information by controlling access to information, services, and equipment. Only personnel who have the proper authorization and need-to-know can have access to systems and data. The system must include features and procedures to enforce access control policies for all information, services, and equipment comprising the system.

Integrity – The system must maintain the integrity (i.e., the absence of unauthorized and undetected modification) of information and software while these are processed, stored and transferred across a network or publicly accessible transmission media. Each file or data collection in the system must have an identifiable source throughout its life cycle. Also, the system must ensure the integrity of its mission-critical equipment. Automated and/or manual safeguards must be used to detect and prevent inadvertent or malicious destruction or modification of data.

Availability – The system must protect against denial of service threats. Protection must be proportionate to the operational value of the services and the information provided. This protection must include protection against environmental threats such as loss of power and cooling.

Accountability – The system must support tracing of all security relevant events, including violations and attempted violations of security policy to the individual subsystems and/or users including external connections. The system must enforce the following rules:

1. Personnel and systems connecting to the system must be uniquely identifiable to the system and must have their identities authenticated before being granted access to sensitive information, services, or equipment.

2. Each subsystem handling sensitive or mission-critical information must maintain an audit trail of security relevant events, including attempts by individual users or interfacing subsystems to gain access through interfaces not authorized for that particular purpose. This audit trail must be tamper-resistant and always active.

Assurance – The criticality and sensitivity of the information handled, equipment and services, and the need-to-know of personnel must be identified in order to determine the applicable security requirements. The security implementations chosen must provide adequate security protection commensurate with the criticality of the data, in accordance with the security policy.

Enforcement – The security policy must be enforced throughout the life cycle of the system. All implementations of system security functions including those implemented at the subsystem level must be evaluated to ensure that they adequately enforce the requirements derived from the security policy. Each platform must be evaluated to ensure that the installed system configuration enforces the stated security policy. As a result of this evaluation, an assessment of the vulnerability can be generated. This assessment must be evaluated by the security manager or system administrator to decide if any modifications to the system must be made so that it complies with the security policy. Security best practices must be employed throughout the life cycle of a system to ensure continued compliance with the stated security policy. New system projects must have information security representatives during the planning and preliminary design stages in order to implement security into the design.

16.3 Best Practice #3

Create a plan for implementing your security policy.

Once a security policy is established, an implementation plan should be created. Incremental, staged infrastructure improvements and new hires (if any) will help management plan for expenses and create a timetable for implementation.

The implementation plan should include the following steps:

1. Defining implementation guidelines. These guidelines should specify the personnel to receive security alarms and what action is to be taken, chains of command for incident escalation, and reporting requirements.

2. Educating staff, customers, etc. about the security policy.

3. Purchasing any needed hardware/software and hiring anyneeded personnel.

4. Installing and testing equipment/software.

Part 3
Wireless Network Security

CHAPTER 17

Security in Traditional Wireless Networks

Praphul Chandra

17.1 Security in First Generation TWNs

Earlier, we discussed the Advanced Mobile Phone System (AMPS) as an example of a first-generation traditional wireless network (TWN). These networks were designed with very little security.[1] Since the AMPS radio interface was analog and since AMPS used no encryption, it was relatively simple for a radio hobbyist to intercept cellular telephone conversations with a police scanner. In the AMPS network, for the purposes of authenticating itself to the network, the mobile station sends the Electronic Serial Number (ESN) that it stores to the network. The network verifies that this is a valid ESN and then allows the subscriber access to network services. The problem with the authentication process is that the ESN is sent in clear over the air interface (obviously, since there is no encryption). This means that a radio hobbyist can not only eavesdrop on cellular telephone conversations but can also capture a valid ESN and then use it to commit cellular telephone fraud by cloning another cellular phone and making calls with it. It was the cellular fraud attack along with the concern for subscriber confidentiality that prompted cellular service providers to demand a higher level of security while designing second generation TWNs.

17.2 Security in Second Generation TWNs

One of the prominent design decisions of second generation TWNs was the move from an analog system to use of a digital system. This design decision led to a significant improvement

---

[1]To be fair to AMPS designers, they had too many other problems before security became a priority.

in the security of the system. The use of a speech coding algorithm, Gaussian Minimum Shift Keying (GMSK), digital modulation, slow frequency hopping and TDMA made casual eavesdropping by radio hobbyists significantly more difficult since it required use of much more highly specialized and expensive equipment than a simple police scanner. However, the use of a digital system is only one of the many security provisions that were designed into the second generation TWNs. In this section, we look at security in GSM networks.

Figure 17.1 shows the high level architecture of GSM, which is the most widely deployed TWN in the world today. Before dwelling on the security of the GSM network, we need to understand the service model of GSM networks. Recall from our earlier discussion of TWNs that TWNs evolved from the PSTN with the aim of extending voice communication services to mobile subscribers. Not surprisingly therefore, the GSM network designers aimed to make the GSM "as secure as the PSTN." To appreciate what this means, realize that the PSTN is an extremely controlled environment. The core PSTN network is controlled and regulated by a small group of operators worldwide. For most part the security in the PSTN is ensured by restricting physical access to the network. This is true both at the access network level and at the core network level.

Since the GSM standard evolved from the PSTN, it carries forward the security philosophy of the PSTN. The network beyond the BTS is considered a controlled environment,

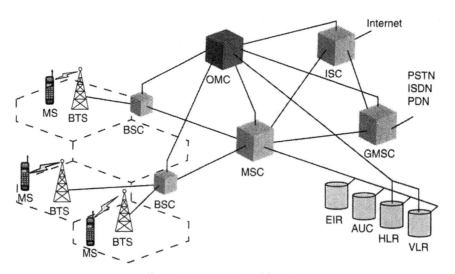

Figure 17.1: GSM architecture

since access to this part of the network is controlled by the service provider. It is only the access network (connecting the ME/MS to the BTS) that is considered a hostile operating environment. GSM security therefore aims to secure this part of the network. We look at the details of securing the access network in GSM in the next few sections.

17.2.1 Anonymity in GSM

One of the first things that a ME has to do when it switches on (or roams into) in a coverage area is to identify itself to the network requesting services from the network. Recall from our earlier discussion that the IMSI is the unique number, contained in the SIM, by which a subscriber is identified by the network for call signaling purposes. In other words, TWNs use the IMSI to route calls. It is therefore imperative for the network to know where each IMSI is at all times. This functionality wherein the network keeps track of where each IMSI (subscriber) is at any given time is known as *location management*. Even though the details of this topic are beyond the scope of this book, the basic underlying concept of location management is that each time the subscriber crosses a cell boundary,[2] the ME should inform the network about the IMSI's new location. This allows the network to route an incoming call to the correct cell.

In summary, the location update messages from the ME to the network need to carry the identity of the subscriber so that the network knows where to route an incoming call to at any given time. Combine this with the fact that the one-to-one mapping between the IMSI (telephone number) and the subscriber identity is publicly available. This means that if an eavesdropper can capture the IMSI over the air, they can determine the identity of the subscriber and their location. In simpler terms, this means that if you are using a cell phone anywhere in the world, your geographical location can be easily determined. This is not acceptable to most subscribers and therefore this "property" is treated as a security threat in TWNs.

The anonymity feature was designed to protect the subscriber against someone who knows the subscriber's IMSI from using this information to trace the location of the subscriber or to identify calls made to or from the subscriber by eavesdropping on the

---

[2]To be precise, a set of adjoining cells are grouped together to form a location area and the location updates are required only when the subscriber crosses a location area boundary. This reduces the number of messages and thus saves bandwidth.

air interface. GSM protects against subscriber traceability by using temporary mobile subscriber identity (TMSI). Unlike the IMSI which is globally unique, the TMSI has only local significance; that is, the IMSI-TMSI mapping is maintained in the VLR/MSC. When a SIM has authenticated with the network, the network allocates a TMSI to the subscriber. For all communication with the SIM, the network uses this TMSI to refer to the SIM. The use of a TMSI reduces the exposure of IMSI over the air interface to a minimum thus minimizing the probability that an eavesdropper may be able to identify and locate a subscriber.

17.2.2 Key Establishment in GSM

Once the subscriber has identified itself into the network, the next step is to prove to the network that the ME is actually who they are claiming to be: this is the authentication process. We discuss the authentication process used in GSM networks in the next section. Before we go into that discussion it is important to talk about the key establishment procedure used in GSM networks. A key establishment procedure is used to establish some sort of a secret or key between two communicating parties. This shared secret then forms the basis for securing the network.

The GSM security model uses a 128-bit preshared secret key (K_i) for securing the ME-to-BTS interface. In other words, there is no key establishment protocol in the GSM security architecture model. Instead each SIM is burnt or embedded with a unique K_i; that is, each subscriber has a unique K_i. Since this is a "shared" secret between the subscriber and the network, it is obvious that the key has to be stored somewhere in the network too. This "somewhere" is the authentication center (AuC) which is basically a database which stores the K_i of all subscribers.[3] It is this shared secret (K_i) between the SIM and the AuC that forms the basis for securing the access interface in GSM networks.

17.2.3 Authentication in GSM

When a ME first switches on, it searches for a wireless network to connect to by listening to a certain set of frequencies. When it finds a wireless network to connect to, the ME-SIM sends a sign-on message to the BTS requesting access to the network. The BTS then contacts the mobile switching center (MSC) to decide whether or not to allow

---

[3]In reality, each service provider maintains its own AuC.

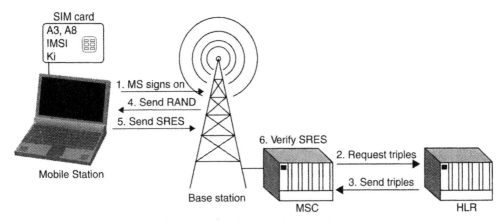

Figure 17.2: GSM authentication

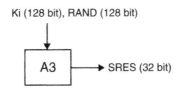

Figure 17.3: GSM SRES generation

the ME-SIM access to the network. In order to make this decision, the MSC asks the home location register (HLR) to provide it with five sets of security triplets. A security triplet consists of three numbers: RAND (a 128-bit random number), SRES (a 32-bit signed response to the RAND generated using the preshared K_i) and a session key K_c (an encryption key generated using K_i). The HLR supplies these triplets to the MSC by using the K_i from the AuC. The MSC then picks one of these five sets of triplets to use for the current "session." The RAND from this triplet is then sent to the ME (via the base station controller (BSC) and the BTS) as a challenge. The ME-SIM is then expected to generate a SRES to this RAND using the A3 algorithm and the K_i stored in its SIM (as shown in Figure 17.3). This SRES is sent back to the MSC (via the BTS and the BSC). The MSC compares the SRES received from the ME to the SRES contained in the triplet it received from the HLR. If the two match, the MSC can safely deduce that the ME has a SIM which contains a valid K_i. The MSC can therefore safely allow the ME access to the network. On the other hand, if the two SRESs do not match, the MSC would not allow the ME access to the network. The GSM authentication process is shown in Figure 17.2.

As shown in Figure 17.2, the authentication process is carried out between the SIM and the MSC. The SIM uses the preshared secret K_i that it stores and carries out the A3 and the A8 algorithms to generate the SRES and the session key K_c. It is important to note that the K_i, IMSI and the A3 and A8 algorithms are stored and implemented in the SIM. More importantly, the K_i (which forms the basis of all security in GSM networks) never leaves the SIM.[4]

In the authentication process just described, note the inherent trust relationship between the HLR and the MSC. Such an inherent trust relationship is also present between the BSC and the MSC and again between the BSC and the BTS. This brings us to a very important characteristic of the GSM security model: it aims to secure the wireless part of the GSM network only. In retrospect, this may be considered a .aw, but remember that the GSM network evolved from the PSTN network and the aim of the GSM security designers was to make the GSM network as secure as the PSTN network. Realize that access to the PSTN network was (and still is) very tightly controlled. There are only a very small number of PSTN service providers and therefore getting access to the core network is not trivial. In other words, the core PSTN network is secured by restricting physical access to the network.[5] The GSM network designers carried forward this philosophy. The core network in the GSM architecture refers to the network beyond the BSC and it is considered "secure" since it is controlled by the service provider and access to it is tightly controlled. Therefore, the aim of the GSM security designers was to secure the wireless access network only. However, there is a missing link even if we assume that the core GSM network is secured by the service provider (either by restricting physical access to the network or by other proprietary means). This missing the link is the one between the BTS and the BSC. Remember that this link is not part of the core network. Combine this with the fact that GSM does not specify how the BTS and the BSC need to be connected.[6] In practice, it is common for the BTS and the BSC to be connected by

---

[4]Compare this with first generation TWNs where the ESN was transmitted in clear over the air interface.

[5]The access network of the PSTN, on the other hand, is much easier to access as compared to the core network and therefore the PSTN security in the access network is much easier to violate.

[6]GSM just specifies the interface between the BTS and the BSC.

microwave (wireless links). GSM does not specify how to secure this link, thus making it susceptible to attacks.

There is another important characteristic of the GSM authentication process that is worth discussing. In GSM, the authenticating entity is the SIM and not the subscriber per se. In other words, the network authenticates the SIM card and not the subscriber of the SIM card. Remember that the authentication process relies on a preshared secret (K_i) between the SIM and the AuC. During the authentication process, the MSC validates that the SIM trying to access the network has a valid K_i. What happens if a ME is stolen and is used for making calls (and using other GSM services)?

GSM does have some countermeasures to protect against equipment theft. For one, the GSM core network maintains a database of all valid mobile equipment[7] on the network. This database is called the Equipment Identity Register (EIR). If a subscriber loses their ME, it is their responsibility to report it to the service provider. Before authenticating the ME into the network, the MSC also ensures that the ME that is trying to authenticate in to the network has not been compromised. Extrapolating this approach, a service provider may also maintain a list of compromised SIMs. When a SIM is reported stolen, the service provider marks the IMSI and the corresponding K_i[8] as compromised. If a compromised SIM tries to access the network, it is denied access.

Note that when the GSM authentication process completes, it has also established a security context: the session key K_c which can then be used for providing confidentiality in the network. It is the preshared secret key (K_i) between the SIM and the AuC that forms the basis of generating the session key. GSM uses the A8 algorithm to derive a session key K_c from the preshared secret key K_i as shown in Figure 17.4.

Compare Figure 17.4 with Figure 17.3. The purpose of the A8 algorithm is to derive a 64-bit session key (K_c) given the 128-bit K_i and the 128-bit RAND. On the other hand, the purpose of the A3 algorithm is to derive a 32-bit SRES given the same two inputs (the K_i and the RAND). The important thing to note here is that A3 and A8 are not algorithms per se: they are just labels (reference names) for algorithms. In other words, a service provider is free to use

---

[7]Each ME in the GSM network is uniquely identified by the international mobile equipment identity (IMEI).

[8]There is a one-to-one mapping between the IMSI and the K_i.

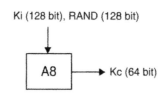

Figure 17.4: GSM K_c generation

any algorithm that it wishes to generate SRES from K_i and RAND. The GSM specification just uses the name A3 to reference such an algorithm. Similarly, the service provider is also free to use any algorithm that it wishes to generate K_c from K_i and the name A8 is just used by the specification to reference this algorithm. Most GSM implementations combine the A3 and A8 functionality and use a single algorithm to serve both the purposes. The COMP128 algorithm, which is the reference algorithm specified in the GSM specification, takes as input the 128-bit K_i and the 128-bit RAND and generates the 32-bit SRES and a 54-bit number. The 54-bit number is appended with 10-zeros to form the 64-bit session key: K_c. We will see in Section 17.2.3 how this session key is used for providing confidentiality.

GSM allows the service provider to choose an algorithm for A3 and A8 implementation while still ensuring seamless roaming among networks of different service providers. This is an important accomplishment and is achieved because even though the authentication process is carried out between the ME and the servicing MSC, the servicing MSC utilizes the HLR of the ME to authenticate the network. Indirectly therefore, it is the home network of the ME which authenticates the ME into another service provider's network. Since the A3 and A8 algorithms need to execute only at the HLR and the SIM[9] (both of which "belong to" the service provider), they can be proprietary algorithms.

One of the finer details of the authentication process in GSM is the use of five sets of security triplets that the MSC gets from the HLR. Even though only one set of triplets is required for authenticating a subscriber into the network, five sets are requested so as to improve roaming performance. Realize that a ME needs to authenticate with a MSC each time it enters its network from another service provider's network. Instead of contacting the HLR for security triplets each time a ME roams into its coverage area, the MSC gets five sets of triplets: one for the current authentication process and four for future use. This reduces the roaming/handover time and improves system performance.

---

[9]The A3 and A8 algorithms are implemented in the SIM.

17.2.4 Confidentiality in GSM

In the previous section, we saw how the GSM authentication process establishes a security context (the session key K_c) when it completes. This session key is used for providing confidentiality over the wireless (ME – BTS) interface. The algorithm used for encrypting packets over the air interface is the A5 algorithm. Unlike A3 and A8 which are just names used by the GSM standard to reference operator-specific algorithms, the A5 is actually an encryption algorithm specified by the GSM standard. The reasoning behind this design decision is the need to support seamless roaming across networks of different service providers. As we saw in Section 17.2.3, the choice of A3 and A8 could be left to the operator since the authentication process is carried out between the SIM and the service providers HLR. The process of encryption on the other hand must necessarily be carried out between the BTS and ME without involving the home network.[10] For achieving seamless roaming between different networks, it is therefore imperative that all service providers use the same encryption algorithm.

The A5 algorithm is basically a stream cipher which generates a unique key stream for every packet by using the 64-bit session key (K_c) and the sequence number of the frame as the input. Since the sequence number of each packet can be easily determined, the confidentiality of the packets depends on keeping the session key (K_c) secret. There is therefore provision in GSM to change the ciphering key K_c: thus making the system more resistant to eavesdropping. The ciphering key may be changed at regular intervals or as required by the service provider.

Once the ciphering key has been established between the SIM and the network, the encryption of signaling messages and subscriber data traffic begins as soon as the GSM network sends a ciphering mode request to the ME. Note that unlike the A3 and the A8 algorithms, the encryption algorithm A5 is implemented in the ME.

17.2.5 What's Wrong with GSM Security?

Probably the most glaring vulnerability in the GSM security architecture is that there is no provision for any integrity protection of data or messages. The GSM security architecture talks about authentication and confidentiality but not about integrity

---

[10]A packet sent from a ME may very well reach its destination without traversing through its home network (packet routing is done by the serving MSC and not the home MSC).

protection. The absence of integrity protection mechanisms means that the receiver cannot verify that a certain message was not tampered with. This opens the door for multiple variation of man-in-the-middle attacks in GSM networks.

Another important vulnerability in the GSM security architecture is the limited encryption scope. In simpler terms, GSM concentrates only on securing the ME-BTS interface. We saw in Section 17.2.1 that the reason behind this design decision lies in the evolution of GSM from the PSTN. The fact however remains that the only link which is cryptographically protected in the GSM network is the ME-BTS wireless interface. This exposes the rest of the network to attacks.[11] One of the most exposed links which is not cryptographically protected in the GSM network is the BTS-BSC interface. Since this link is not part of the "core" network and since this link is often a wireless link (microwave-based, satellite-based and so on), it becomes an attractive target for attacks.

The GSM cipher algorithms are not published along with the GSM standards. In fact, access to these algorithms is tightly controlled. This means that the algorithms are not publicly available for peer review by the security community. This has received some criticism since one of the tenets of cryptography is that the security of the system should lie not in the algorithm but rather in the keys. The thinking is that it is therefore best to let the algorithm be publicly reviewed so that the loopholes in the algorithm are discovered and published. Workarounds can then be found to close these loopholes. However, keeping the algorithms secret (like GSM does) denies this opportunity: hence the criticism. To be fair to GSM designers, the GSM specifications came out at a time when the controls on the export and use of cryptography were extremely tight and therefore not making the algorithms public was at least partly a regulatory decision.

Even the algorithm used for encryption in the ME-BTS link is no longer secure given the increasing processing power of hardware available today. Using the simplest of all attacks, the brute force attack (which works by trying to break down the security of the system by trying each one of all possible keys), the GSM encryption algorithm A5 can be compromised within a matter of hours. The primary problem is the small key length of the session key K_c. The actual length of K_c is 64 bits. However, the last 10 bits of this key are specified to be 0 thus reducing the effective key size to 54 bits. Even though this key size

---

[11]Unless the service provider explicitly secures these links.

is big enough to protect against real-time attacks (decrypting packets being transmitted in real-time), the state of the hardware available today makes it possible to record the packets between the MS and the BTS and then decode them at a later time. An important thing to note is that there are multiple A5 algorithms specified in the GSM standard. The first (and probably the strongest) A5 algorithm is the A5/1 algorithm. However, the A5/1 algorithm was too strong for export purposes and therefore the GSM standard specified other A5 variations which are named A5/x, for example the A5/2 algorithm has an effective security of only 2^{16} against brute force attacks (as opposed to A5/1 which has an effective security of 2^{54} against brute force attacks). As we know, brute-force attacks are not the most efficient attacks on the network. It is often possible to reduce the effective security of the system by exploiting loopholes in the security algorithm. For the A5 algorithm, differential cryptanalysis has been shown to reduce the effective security of the system even more. Lastly, the GSM security architecture is inflexible; in other words, it is difficult to replace the existing encryption algorithm (A5) with a more effective algorithm or to increase the length of the key used in the A5 encryption algorithm.[12] In a sense, therefore, the GSM networks are "stuck with" the A5 algorithm.

Another important vulnerability in the GSM security architecture is that it uses one-way authentication where the network verifies the identity of the subscriber (the ME, to be accurate). There is no way for the ME to verify the authenticity of the network. This allows a rogue element to masquerade as a BTS and hijack the ME. Again, to be fair to GSM security designers, at the time of the writing of the GSM standards, it was hard to imagine a false base station attack (an attacker masquerading as the GSM network) since the equipment required to launch such an attack was just too expensive. However, with the phenomenal growth in GSM networks, the cost of this equipment has gone down and the availability has gone up, thus making these attacks much more probable.

A very real attack against the GSM network is known as SIM cloning. The aim of this attack is to recover the K_i from a SIM card. Once the K_i is known, it can be used not only to listen to the calls made from this SIM but also to place calls which actually get billed to this subscriber. The SIM cloning attack is a chosen-plaintext attack which sends a list of chosen plaintexts to the SIM as challenges (RAND). The A8 algorithm generates the

---

[12]It is however possible to increase the effective size of the key from 54 bits to the actual length of 64 bits by removing the requirement of having the leading 10 bits of the key to be all zeros.

SRES to these challenges and responds back. The attacker therefore now has access to a list of chosen-plaintext, ciphertext pairs. If the algorithm used for A8 implementation is the COMP128 reference algorithm and if the RANDs are chosen appropriately, this list of pairs can be analyzed to reveal enough information to recover the K_i using differential cryptanalysis. There are many variations of the SIM cloning attack. In one approach, the attacker has physical access to the SIM card and a personal computer is used to communicate with the SIM through a smart card reader. This approach recovers the K_i in a matter of few hours.

However, it is not always possible to have physical access to the SIM. Therefore, another approach is to launch this attack wirelessly over the air interface. Even though this approach removes the requirement of having the physical access to the SIM (thus making the attack far more attractive), it introduces obstacles of its own. First, the attacker should be capable of masquerading as a rogue BTS. This means that it should be capable of generating a signal strong enough to overpower the signal of the legitimate BTS. Only if this is true would the attacker be able to communicate with the ME. One workaround is to launch this attack when the signal from the legitimate BTS is too weak (in a subway, elevator and so on) The second obstacle arises if the ME is moving. In this case there might not be enough time to collect enough chosen- plaintext, ciphertext pairs to recover the K_i because the inherent latency in the wireless interface increases the time required for each transaction. A workaround to this problem is break up the attack over a period of time. Instead of trying to get all the plaintext, ciphertext pairs in one run, the attacker gets only as many pairs as they can and stores them. They repeat this process over a period of days till they get enough data to recover the K_i.

Yet another variation of this attack attempts to have the AuC generate the SRES of given RANDs instead of using the SIM. This attack exploits the lack of security in the SS7 signaling network. Since the core signaling network is not cryptographically protected and incoming messages are not verified for authenticity, it is possible to use the AuC to generate SRESs for chosen RANDs.

A salient feature of the GSM security architecture is that it is transparent to the subscriber. However this feature sometimes becomes a loophole. There are scenarios where a service provider may choose to use null encryption (A5/0). If a ME is in such a cell, should it be allowed to connect to such a BTS or not? The current design is to allow the ME to connect to such a cell.

17.3 Security in 2.5 Generation TWNs

As we have discussed, the GSM network evolved from the PSTN network and the GSM security architecture followed the same path. GSM security architecture was designed to secure only the last hop (BTS-ME) in the network since the rest of the network was assumed to be a "secure environment" controlled and secured by the service provider. This architecture worked for voice communication and PSTN-based networks because it was relatively easy for the limited number of service providers to maintain a secure environment in the core network.

With the explosive growth in the Internet, 2G service providers upgraded their networks to 2.5G networks to provide data services to their subscribers. These data services basically consisted of connecting the ME to the Internet (that is, various web servers). The 2.5G system architecture looks like the one shown in Figure 17.5.

General Packet Radio Service (GPRS) was basically intended to provide the ME with data-connectivity to various web servers. Since data usually requires more bandwidth than voice, the GSM network achieves this by allocating multiple timeslots to an ME which is trying to access data services.[13] This has an interesting implication on the security architecture of the network. Recall that for voice calls the encryption and decryption happens at the BTS on the network side. For the A5 algorithm, this is possible because the BTS knows the ciphering key K_c and can implicitly deduce the sequence number. These are the only two inputs required to operate the A5 algorithm. In the GPRS architecture, since a ME has multiple timeslots to transmit, it is possible that multiple timeslots are allocated on channels belonging to different BTSs to connect to the network. This may happen for example during roaming as shown in Figure 17.6. This in turn means that the BTS cannot implicitly deduce the sequence number of a packet. To solve this problem, GPRS transfers the responsibility of encryption and decryption on the network side from the BTS to the SGSN. The SGSN is the equivalent of the VLR and MSC. This means that the GPRS architecture effectively prevents (protects against) eavesdropping on the backbone between the BTS and the SGSN too.

17.3.1 WAP

The GPRS protocol provides a connectivity mechanism for the ME to connect to a data network (Internet). From an OSI layer perspective, GPRS provides Layer 2 (point-to-point)

---

[13]Recall that for a voice call, GSM assigns only one timeslot to a ME.

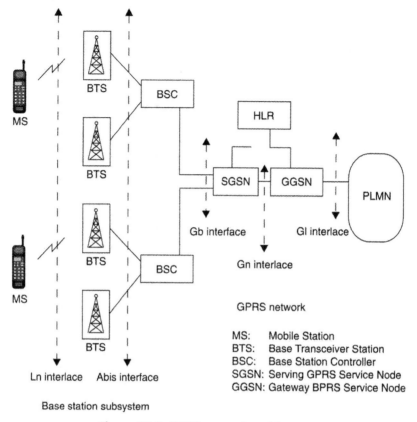

Figure 17.5: GPRS network architecture

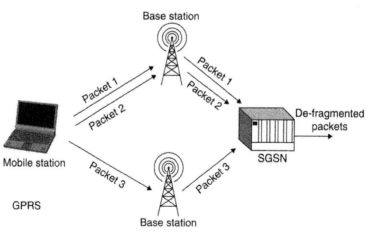

Figure 17.6: GPRS roaming

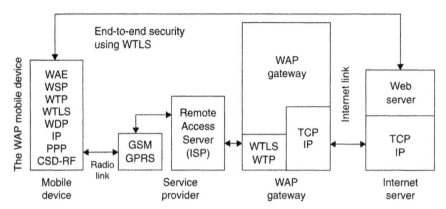

Figure 17.7: WAP—Network architecture

connectivity. What is still required is a set of higher layer protocols (see Figure 17.7). In the wired network, internet applications use the Hyper Text Transfer Protocol (HTTP) and the Hyper Text Markup Language (HTML) to access and retrieve data content (web pages, applets and so on) from web servers. Ideally, the same protocols could have been used over GPRS. This, along with an embedded browser in the ME, would have made the ME a PC-like medium to browse the internet. The problem is that we are operating in a bandwidth-constrained medium and a memory-constrained, CPU-constrained, screen-size constrained end-point (the ME). HTTP and HTML are not optimized for operating under such conditions. This is where Wireless Application Protocol (WAP) comes in.

WAP is an open specification that offers a standard method to access Internet-based content and services from wireless devices such as mobile phones and Personal Digital Assistants (PDAs). The WAP protocol stack is designed for minimizing bandwidth requirements and guaranteeing that a variety of wireless networks can run WAP applications. The information content meant for the ME is formatted suitably for the ME's small screen, and a low bandwidth, high latency environment; i.e., the Wireless Application Environment (WAE).

Figure 17.8 shows the WAP programming model. The client is the embedded browser in the ME and the server may be any regular web server. The new entity in the architecture is the WAP gateway. The embedded browser connects to the WAP gateway and makes requests for information from web servers in the form of a normal universal resource locator (URL). The gateway forwards this request to the appropriate web server and gets the information using HTTP in HTML format. Note that the gateway to the web servers

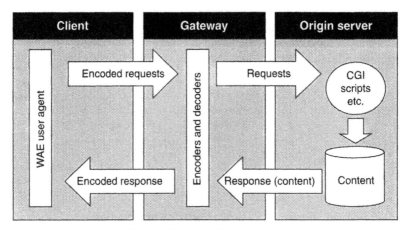

Figure 17.8: WAP—Overview

is usually a wired link, which is appropriate for using HTTP and HTML. The role of the gateway is to reformat the content from the web server suitable for transmission in a WAE and for display on a ME. The language used for creating this content is called Wireless Markup Language (WML) which is optimized for low bandwidth, high latency connections. To summarize, the WAP gateway is the translator between HTTP, HTML on the web server side and WTP, WML on the ME side.

So, together WAP and GPRS allow the ME to connect to the Internet. Since the Internet is a huge uncontrolled network of haphazardly connected (and growing) nodes, this breaks one of the biggest assumptions of the GSM security architecture—that the core network is a controlled secure environment. In this new operating environment, securing just the last link is not enough. Instead, an end-to-end security architecture is desired. This end-to-end security is achieved by the Wireless Transport Layer Security (WTLS) layer in the WAP stack.

WTLS is modeled along the lines of Secure Sockets Layer (SSL)/Transport Layer Security (TLS). The reason for designing a new protocol along the lines of TLS and not using TLS itself is optimization. First, TLS was designed to be used over a reliable transport layer (such as TCP) whereas WTLS needs to operate over an unreliable datagram transport where datagrams may be lost, duplicated or re-ordered. Second, the WTLS protocol was modified to cope with long roundtrip times and limited bandwidth availability typical of the wireless environment. Finally, WTLS has been optimized to operate with limited processing power and limited memory of the ME. Figure 17.9 shows a WTLS session.

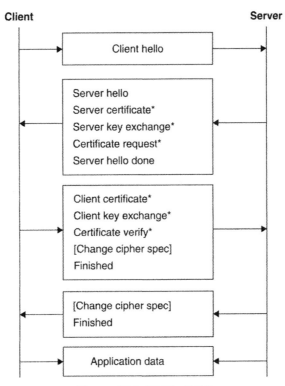

Figure 17.9: TLS in WAP

17.3.2 Code Security

For most part, network security is concerned with transactional security on the network. In other words, we talk about securing the link(s) in the network using encryption, securing access to the network using authentication and so on. The merging of the GSM network with the Internet, however, adds another dimension to the concept of network security. The ME in a GPRS network can "browse" the Internet. In the Internet architecture, web servers sometimes download applets (short programs) on to the client (ME) over the network. These applets then execute at the client (ME). Consider what happens if the applet is a malicious piece of code or it simply has a bug which can harm the ME. To protect against such attacks, it is extremely important that the applets (or other programs which are downloaded from remote sites but execute on the client) be secure.

From a client's (ME's) perspective however, it is difficult to ensure the security of the applet by examining the source code. Therefore, most applets are signed by Certificate Authorities (CAs) (See the section on Public Key Infrastructure (PKI)). Before executing the applet, the subscriber can be informed of the CA which has signed the applet. If the subscriber trusts that CA, they can allow the applet to be executed on their ME otherwise they can block the execution of the applet.

17.4 Security in 3G TWNs

The UMTS security architecture was designed using the GSM security as the starting point. The reason behind doing so was to adopt the GSM security features that have proved to be robust and redesign the features that have been found to be weak. Another reason for doing so was to ensure interoperability between GSM and Universal Mobile Telecommunications System (UMTS) elements.

17.4.1 Anonymity in UMTS

UMTS anonymity builds on the concept of TMSI introduced by GSM (see Section 17.2.1). To avoid subscriber traceability, which may lead to the compromise of subscriber identity, the subscriber should not be identified for a long period by means of the same temporary identity. To achieve this, the UMTS architecture provides provisions for encrypting any signaling or subscriber data that might reveal the subscriber's identity.

Note that we seemingly have a chicken and egg situation here. As we discussed in Section 17.2.1, one of the first things that the ME has to do is to identify itself (its IMSI) to the network. On the other hand, the TMSI allocation procedure (initiated by the VSC/MLR) should be performed after the initiation of ciphering to ensure that the TMSI (and hence the subscriber identity) is not vulnerable to eavesdropping. The problem is that ciphering cannot start unless the CK has been established between the UMTS Subscriber Identity Module (USIM) and the network and the CK cannot be established unless the network first identifies the subscriber using its IMSI. The problem is not as complicated as it appears though. Let's dig a little deeper.

Recall from Section 17.2.1 that the TMSI has only local significance; in other words, the TMSI is allocated by the VLR/MSC and the IMSI-TMSI mapping is maintained in the VLR/MSC. When the subscriber roams into the coverage area of another VLR/ MSC (hereafter referred to as VLRn), it continues to identify itself with the TMSI that it was allocated by the previous VLR/MSC (hereafter referred to as VLRo). Obviously VLRn does not recognize this TMSI, since it was allocated by VLRo. In the UMTS architecture, VLRn should request VLRo to get the IMSI corresponding to this TMSI. If VLRn cannot retrieve this information from VLRo, only then should VLRn request the subscriber to identify itself by its IMSI.

The bottom line is that most times, VLRn can determine the IMSI of a subscriber without the ME actually having to transmit the IMSI over the air interface. Instead the ME can identify itself using the TMSI that is already assigned to it. The AKA procedure can then be carried out from this point on.[14] At the completion of the AKA procedure, the CK has been established between the USIM and the network and the VLR/MSC can therefore assign a new TMSI to the ME while ensuring that it is encrypted. From this point on, the TMSI can be safely used by the network and the USIM to identify the subscriber.

Besides the IMSI and the TMSI which can cause a subscriber's identity to be compromised, there is another identity in the UMTS security architecture that can be exploited to trace a subscriber. This is the Sequence Number (SQN) which is used by the ME to authenticate the network. The reason why the SQN can also be used to identify the subscriber is that the network maintains a per-subscriber SQN which is incremented sequentially. It is therefore necessary to encrypt the SQN to protect against subscriber traceability. As explained in Section 17.4.2 the authentication process (also known as the Authentication and Key Agreement (AKA)[15] process) in UMTS establishes various keys and one of these is the anonymity key (AK). Just like the CK and IK, the AK is established or derived independently at the USIM and the VLR/MSC without ever being transmitted over the air. In other words, the AK is known only to the USIM and the VLR/MSC. The use of the AK is to protect the Sequence Number (SQN) from eavesdropping.

17.4.2 Key Establishment in UMTS

Just like GSM, there is no key-establishment protocol in UMTS and just like GSM it uses a 128-bit preshared secret key (K_i) between the UMTS Subscriber Identity Module (USIM) and the authentication center (AuC) which forms the basis for all security in UMTS.

---

[14]Or VLRn can decide to use a previously existing set of keys.

[15]Authentication and key agreement.

17.4.3 Authentication in UMTS

The authentication model in UMTS follows closely from the GSM authentication model but there is one significant difference. The authentication procedure is mutual; that is, the network authenticates the subscriber (USIM) and the subscriber (USIM) authenticates the network. This is unlike GSM where there is no provision for the subscriber to authenticate the validity of the network.

Figure 17.10a shows the authentication procedure in UMTS networks. The process starts when a subscriber (USIM) first sends a sign-on message to the base station it wants to connect to (not shown in Figure 17.10a). The base station then contacts the VLR/MSC (via the RNC) to decide whether or not to allow the USIM access to the network. In order to make this decision, the MSC asks the HLR to provide it with a set of authentication vectors. The authentication vector is the equivalent of the security triplets in GSM. The UMTS authentication vector is actually a security quintet[16] which consists of five numbers: RAND (a 128-bit random number), XRES (the 32-bit expected signed response to the RAND), CK (a 128-bit session cipher or encryption key), IK (a 128-bit integrity key) and AUTN (a 128-bit network authentication token).

When the HLR receives a request for generating authentication vectors from a MSC/VLR, it first generates a random number, RAND and a sequence number, SQN. The HLR then requests the AuC to supply the preshared secret K_i corresponding to this USIM. The RAND, SQN, K_i and the Authentication Management Field (AMF) serve as the input to the five functions (f1 – f5) and generate the security quintet. We tabulate below all the relevant entities involved in the process.

RAND: Random challenge generated by AuC.

XRES: $f2_K$(RAND): Expected subscriber RESponse as computed by AuC.

CK: $f3_K$(RAND): Cipher Key used for encrypting data and signaling messages.

IK: $f4_K$(RAND): Integrity Key.

---

[16]A GSM security triplet consists of three numbers: RAND (a 128-bit random number), SRES (a 32-bit signed response to the RAND generated using the preshared K_i) and a session key K_c (an encryption key generated using K_i).

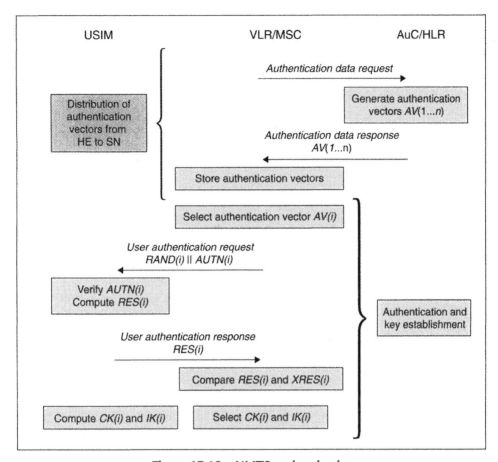

Figure 17.10a: UMTS authentication

AK: f5$_K$(RAND): Anonymity Key.

SQN: Sequence Number.

AMF: Authentication Management Field.

MAC: f1$_K$(SQN ∥ RAND ∥ AMF): Message Authentication Code.

AUTN: SQN (+) AK ∥ AMF ∥ MAC: Network Authentication Token.

Security Quintet: (RAND, XRES, CK, IK, AUTN).

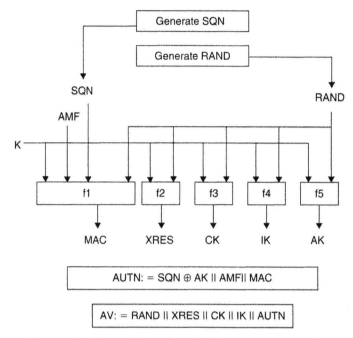

Figure 17.10b: UMTS authentication vector generation

When the VLR/MSC receives the set of authentication vectors, it selects the first[17] authentication vector and stores the rest for later use. The VLR/MSC then sends the RAND and the AUTN from the selected authentication vector to the USIM. The RAND serves as a challenge. When the USIM receives these, it carries out the procedure shown in Figure 17.11.

Note that the procedure shown in Figure 17.11 requires only three inputs: AUTN, RAND and K_i. The former two entities are received from the VLR/MSC and K_i is already stored in the USIM. Note also that AUTN basically consists of SQN (+) AK || AMF || MAC which are used as shown in Figure 17.11. Once the USIM has calculated all the entities, it verifies that the MAC received in the AUTN and the XMAC calculated by the USIM match. It also verifies that the SQN obtained from the network's message is in the correct

---

[17]Note the importance of sequence. In the GSM security architecture, the MSC/VLR selects any one of the five security triplets.

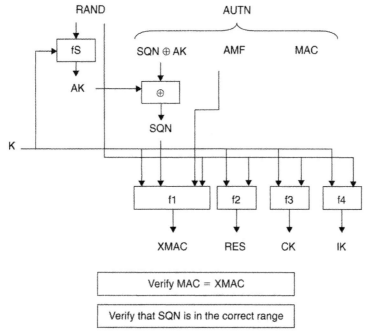

Figure 17.11: UMTS response generation at USIM

range. Once the USIM has verified these matches, the USIM has authenticated the network since a rogue network can't generate a valid SQN[18].

At this point, one half of the authentication process is now complete. What is now left is for the network to authenticate the USIM. To complete this process, the USIM sends back the RES to the network. Note that the RES is generated from the RAND using the preshared secret key that it stores, K_i and the function f2; i.e., RES = $f2_K$(RAND). The RES is therefore the response to the challenge (RAND). When the VLR/MSC receives the RES from the USIM, it compares it with the XRES in the corresponding authentication vector that it received from the HLR. If the two match, the network (VLR/MSC) has successfully authenticated the USIM and allows it to access network services.

---

[18]Recall from Section 17.4.1 that the network maintains a per-subscriber SQN which is incremented sequentially each time the ME and the network carry out the authentication process. Also, this SQN can be kept secret by using the AK.

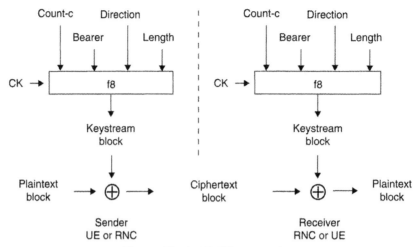

Figure 17.12: UMTS encryption

At this point, the mutual authentication process has completed and the following keys have been established between the network and the USIM: CK, IK and AK. These keys now form the basis of providing confidentiality, integrity and anonymity in the UMTS architecture.

Recall from Section 17.2.3 that GSM left the choice of the authentication protocol to the service provider. Most service providers used the COMP128 algorithm for this purpose. UMTS follows the same philosophy. It leaves the choice of the authentication protocol to the service provider but does provide an example algorithm, MILENAGE, that service providers may use.

17.4.4 Confidentiality in UMTS

The GSM encryption algorithm was A5, which used a 64-bit session key K_c. The UMTS encryption algorithm is known as KASUMI and uses a 128-bit session key CK. The KASUMI algorithm is more secure than A5 and one of the reasons for this is simply the use of longer keys for encryption. Figure 17.12 shows the encryption process used in UMTS networks.

Let's take a look at the input values used for encryption. First there is the 128-bit ciphering key (also known as encryption key), CK which has been established at the

USIM and at the VLR/MSC as a result of the authentication process. Second, there is the 32-bit COUNT-C which is a ciphering sequence number which is updated sequentially for each plaintext block. Third, there is the 5-bit BEARER which is a unique identifier for the bearer channel (the channel number which is used for carrying end-user's traffic) in use. Fourth, there is the 1-bit DIRECTION value which indicates the direction of transmission (uplink or downlink). Finally, there is the 16-bit LENGTH which indicates the length of the key-stream block. All these values are input into the f8 encryption algorithm[19] to generate a key stream. This key stream is XORed with the plaintext block to generate the ciphertext block. At the receiving end, the same process is repeated except that the generated key stream is XORed with the received ciphertext to get back the plaintext.

Besides the use of increased key lengths, there is another significant improvement in UMTS security architecture over the GSM security architecture. Recall from Section 17.2.2 that the GSM confidentiality was limited to securing the link between the ME and the BTS. This made the link between the BTS and the BSC (usually a microwave link) unsecure. The UMTS security architecture extends the encrypted interface from the BTS back to the RNC (the UMTS equivalent of BSC). Since the traffic between the USIM and RNC is encrypted, this protects not only the wireless interface between the USIM and base station but also the interface between the base station and the RNC, thus closing one of the loopholes of GSM security. One last thing to note is that the encryption in the UMTS security architecture is applied to all subscriber traffic as well as signaling messages.

17.4.5 Integrity Protection in UMTS

As we discussed in Section 17.2.5, one of the gaping loopholes in the GSM security architecture was the absence of an integrity protection mechanism. UMTS attempts to solve this problem using the integrity key IK derived using the authentication process as described in Section 17.4.2.

The UMTS integrity mechanism is shown in Figure 17.13. Let's look at the input values required for the integrity mechanism. First, there is the 128-bit integrity key, IK, which is established as a part of the UMTS authentication process. Second, there is the

---

[19]Just like in GSM, "f8" is a label for an algorithm rather than an algorithm itself. One oft-used f8 algorithm is the Kasumi algorithm.

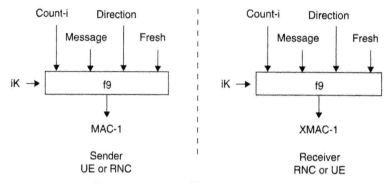

Figure 17.13: UMTS message integrity

32-bit integrity sequence number which is updated sequentially for each plaintext block that is integrity protected. Third, there is the message itself which needs to be integrity protected. Fourth, there is the DIRECTION bit (uplink or downlink). Finally, there is the 32-bit FRESH which is a per-connection nonce. All these values are input to the f9 algorithm and the output is a 32-bit MAC-I (message authentication code). This MAC is attached to the message by the sender. At the receiving end, the same process is repeated to calculate the XMAC-I. The receiver then compares the computed XMAC to the received MAC, so the receiver can deduce that the message was not tampered with. Thus we have integrity protected the message. The integrity protection mechanism just described is applied to all but a specifically excluded set of signaling messages.

Ideally, the integrity protection mechanism should be used for protecting user traffic too. However, integrity protecting each packet involves a lot of overhead in terms of processing and bandwidth, but let's take a step back and think about it: what is it that we are trying to achieve by integrity protecting user traffic? The aim is to ensure that the information content of the voice conversation is not modified; in other words, words are not inserted, deleted or modified in the conversation. However, ensuring this does not really require that each voice packet be integrity protected, since inserting, deleting or modifying a word in a conversation (without the user realizing) would affect voice samples spanning several packets.[20] Almost invariably,

---

[20]Voice Packets basically consist of samples of digitized voice. Typical voice calls over TWNs usually use a rate of 10–20 Kbps to digitize voice. Therefore a single spoken word would usually span several packets.

inserting, deleting or modifying words in a conversation without the user realizing this, would lead to a change in the number of packets. It is therefore usually sufficient to integrity protect the number of user packets in a conversation. This is the compromise that UMTS uses. The RNC monitors the sequence numbers that are used for ciphering voice packets related to each radio bearer (channel). Periodically, the RNC may send a message containing these sequence numbers (which reflect the amount of data sent and received on each active radio bearer) to the ME. Obviously, this message itself is integrity protected. On receiving this message, the ME can verify that the value received in this message matches the counter values that the ME maintains locally. This provides integrity protection of the user traffic.

17.4.6 Putting the Pieces Together

The UMTS security architecture is huge, complex and entails a lot of features. We have seen individual pieces of this security architecture in the last few sections. In this section, we try to put all the pieces together to get a complete picture.

Figure 17.14 shows the overview of the UMTS security process. The process starts when the ME first starts the Layer 2 connection (RRC layer connection) procedure. Note that the Layer 2 connection refers to the connection between the MS and the RNC. The message exchange between the ME and the VLR/MSC is a Layer 3 level connection. In any case, as part of the RRC layer connection procedure, the MS sends, among other things, its list of security capabilities: User Encryption Algorithms (UEAs), User Integrity Algorithms (UIAs) and so on to the RNC. The RNC stores these subscriber security capabilities.

When the ME sends its first Layer 3 Connection message to the VLR/MSC, it includes in this message the subscriber identity (either the TMSI or the IMSI) and Key Set Identifier (KSI). The KSI identifies the set of keys (CK, IK, and so on) that were last established between this ME and this CN domain.[21] Note that the first Layer 3 message can be any one of a set of possible initial Layer 3 messages (location update request, routing update request, attach request and so on). The important thing is that whichever Layer 3 is sent first,[22] it will carry the subscriber identity and the KSI.

Once the VLR/MSC receives the subscriber identity and the KSI, it may decide to start a new AKA procedure or it may decide to continue using the keys that it already shares

---

[21]Usually a service provider's domain.

[22]This will depend on the context in which it is sent.

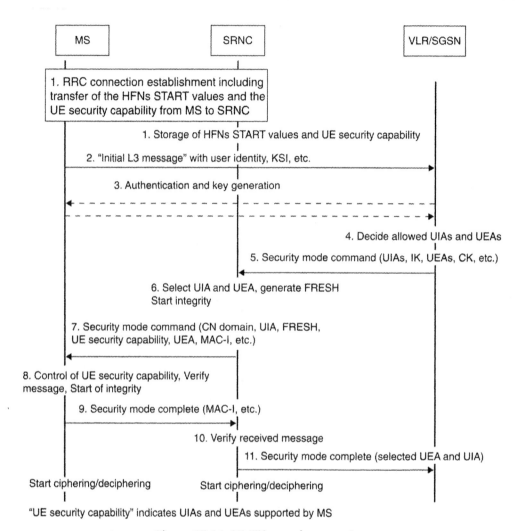

Figure 17.14: UMTS security—*overview*

with the ME from last time. This decision is left to the VLR/MSC. In other words, the service provider is free to control when (and how frequently) it authenticates the identity of the subscriber and establishes a new key set. In either case, at this point, both the ME (the USIM to be precise) and the VLR/MSC trust each other's identity and have agreed upon the set of keys (CK, IK) that they will use.

Note however that the security algorithms to be used have not been agreed on. This is the next step. Towards this end, the VLR/MSC sends a list of the encryption and integrity algorithms that this subscriber is allowed to use to the RNC. The RNC now has the set of UEAs and the UIAs that the ME wishes to use and also the set of encryption and integrity algorithms that the VLR/MSC wishes to use. The RNC takes the intersection of these two sets and determines an encryption and integrity algorithm that is most acceptable to the ME and the VLR/MSC. In other words, the RNC has now decided upon the encryption and the integrity algorithm that would be used between the ME and the VLR/MSC. It does however need to pass this information on to the ME and the VLR/MSC.

The RNC sends the selected algorithm information in the next message. Also included in this message is the full set of security capabilities that the MS had sent to the RNC during RRC connection establishment time. Sending the UEAs and UIAs that the RNC received from the MS back to the MS may seem redundant at first. Realize however, that this is the first message that is integrity protected; in other words, no message sent over the air interface before this message was integrity protected.[23] This means that the set of security capabilities that the MS had sent to the RNC as part of RRC connection establishment procedure was not integrity protected. Therefore, a malicious eavesdropper could have modified this message to change the set of security capabilities that the MS offers to the network thus forcing the MS and the network to use weaker encryption and integrity algorithms. Such an attack is known as the bidding down attack. To prevent against such an attack, the RNC returns the full set of security capabilities that the MS had sent to the RNC in an integrity protected message.

On receiving this message, the MS ensures the integrity of this message by calculating the XMAC on this message and comparing it with the MAC received in the message, thus verifying that the message has not been modified by a malicious eavesdropper. Next, the MS verifies that the set of capabilities in this message is the same as the set of capabilities that the MS had sent to the RNC during the RRC connection setup. Once the MS has verified that the set of security capabilities were not modified, it sends a security mode complete message back to the RNC. Note that this and all subsequent messages from the MS are integrity protected.

At this point the MS has authenticated the network, the network has authenticated the MS and the MS has all the information (algorithms, keys and so on) that it requires to

---

[23]But all messages from now on are integrity protected.

provide confidentiality and integrity protection. However, the VLR/MSC does not have all the information that it requires for this purpose since the encryption and the integrity protection algorithms to be used were determined by the RNC and never conveyed to the VLR/MSC. This is what is done next. The RNC sends a message specifying the encryption and the integrity protection algorithms to be used to the VLR/MSC. From this point on not only are all the messages integrity protected but also encrypted to provide confidentiality.

17.4.7 Network Domain Security

In the last few sections, we have seen how UMTS provides security in the access network. The term *access network* here refers to the connection between the ME and the VLR/MSC. As we discussed in Section 17.2, second generation TWNs were concerned only with securing the wireless access network and the core network was considered to be a secure operating environment. This security in the core network was based on the fact that the core network was accessible to only a relatively small number of well established institutions and it was therefore very difficult for an attacker to get access to this network.

However, this assumption of the core network being inherently secure is no longer valid since with the opening up of telecom regulations all over the world, the number of service providers has grown significantly. This has two important implications. One, there are a lot more institutions that now have access to the core network, thus increasing the probability that the core network may be compromised. Two, since the access networks of these service providers need to communicate with each other (to provide seamless mobility for example), there is a growing need to make this internetwork communication secure.

The ideal solution obviously would have been to secure the core network. The problem is that this was a huge task and beyond the scope of UMTS network designers. The UMTS designers therefore limited their scope to securing the mobile specific part of the network, which is known as the Mobile Application Part (MAP). To this end, UMTS specifies the MAPSEC protocol, which works at the application layer to protect MAP messages cryptographically.[24] Figure 17.15 shows how MAPSEC is used for protecting MAP messages being exchanged between two networks.

---

[24]Some MAP messages are still sent in plaintext without any protection, to avoid performance penalties.

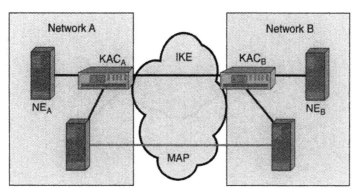

Figure 17.15: MAPSEC

The Key Administration Center (KAC) is a new entity introduced in the system architecture by MAPSEC. Each network which wishes to use MAPSEC has a KAC. The purpose of KAC in Network A is to establish a Security Association (SA) with the KAC in Network B. The term security association refers to the set of security algorithms, keys, key lifetimes and so on that the two networks will use to secure MAP messages that they exchange. To establish a SA, the KACs use the Internet Key Exchange (IKE) protocol. Once the SAs have been established, the KAC distributes this information to its Network Elements (NE). The network elements then use these SAs to protect the MAP messages. MAPSEC allows for three modes of protection: no protection, integrity protection only and integrity with confidentiality.

As you will have noticed, the design of the MAPSEC protocol is strongly influenced by the IPSec protocol. This influence comes across in the use of IKE for key establishment and the use of SAs, among other things. There is an important reason for this. The exploding growth in data networks in the last few years has led to the convergence of voice and data networks and the boundaries between these two networks are fast disappearing. 2.5G TWNs marked the integration of data networks with TWNs. 3G networks are expected to be even more closely tied to IP-based networks. This means that replacing SS7 signaling with IP-based signaling (like SIP) is extremely likely. Therefore, the UMTS network designers provided a method not only for securing MAP in SS7 networks (MAPSEC) but also for using MAP over IP-based networks which may be protected by the already well-established IPSec protocol. Keeping this

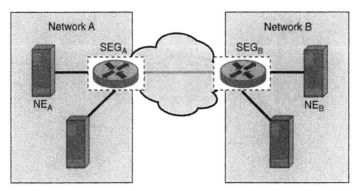

Figure 17.16: MAP over IP-based networks

convergence in mind, the UMTS network designers tried to model MAPSEC along the IPSec lines.

Figure 17.16 shows how network domain security is achieved for IP-based control messages (out of which MAP messages may just be one type of messages being protected). Note that Figure 17.16 resembles Figure 17.15 closely. The only difference seems to be that the KAC has been replaced by another entity known as the Security Gateway (SEG). Like the KAC, the SEG in Network A establishes SAs with its peer in Network B. However, unlike the KAC, the SEG does not distribute the SAs to its network elements. Instead, it maintains a database of established SAs and a database of security policies which specify how and when the SAs are to be used. When the network elements in Network A want to send control messages (like MAP messages) to their peers in Network B, they send the message to the SEG. It is the SEG which is then responsible for protecting the message in accordance with the policy using the established SA. In other words, the SEG is responsible not only for establishing the SAs but also for using them to protect control messages.

17.5 Summary

From the rudimentary ESN in first generation TWNs to the provisions for confidentiality, integration, mutual authentication, and anonymity in the UMTS networks, security in TWNs has come a long way. This improvement in security can be attributed to many factors. Probably the most important among these are the demand for security from subscribers and technological enhancements in cryptography. There is also however another important factor which has made this possible and this is Moore's Law. Remember that security always comes at a cost. Without the significant growth in processing power per square millimeter, it would probably have been impossible to provide the kind of security that TWN subscribers have come to expect. In the near future, the convergence of TWNs with the Internet will surely bring new challenges. What is important is that we be willing to admit the loopholes and then fix them.

CHAPTER 18

Wireless LAN Security

Praphul Chandra

18.1 Introduction

The 802.11 security architecture and protocol is called Wired Equivalent Privacy (WEP). It is responsible for providing authentication, confidentiality and data integrity in 802.11 networks. To understand the nomenclature, realize that 802.11 was designed as a "wireless Ethernet." The aim of the WEP designers was therefore to provide the same degree of security as is available in traditional wired (Ethernet) networks. Did they succeed in achieving this goal?

A few years back, asking that question in the wireless community was a sure-fire way of starting a huge debate. To understand the debate, realize that wired Ethernet[1] (the IEEE 802.3 standard) implements no security mechanism in hardware or software. However, wired Ethernet networks are inherently "secured" since the access to the medium (wires) which carry the data can be restricted or secured. On the other hand, in "wireless Ethernet" (the IEEE 802.11 standard) there is no provision to restrict access to the (wireless) media. So, the debate was over whether the security provided by WEP (the security mechanism specified by 802.11) was comparable to (as secure as) the security provided by restricting access to the physical medium in wired Ethernet. Since this comparison is subjective, it was difficult to answer this question. In the absence of quantitative data for comparison, the debate raged on. However, recent loopholes discovered in WEP have pretty much settled the debate, concluding that WEP fails to achieve its goals.

---

[1]We use 802.3 as a standard of comparison since it is the most widely deployed LAN standard. The analogy holds true for most other LAN standards—more or less.

In this chapter, we look at WEP, why it fails and what has and is being done to close these loopholes. It is interesting to compare the security architecture in 802.11 with the security architecture in traditional wireless networks (TWNs). Note that both TWNs and 802.11 use the wireless medium only in the access network; that is, the part of the network which connects the end-user to the network. This part of the network is also referred to as the last hop of the network. However, there are important architectural differences between TWNs and 802.11.

The aim of TWNs was to allow a wireless subscriber to communicate with any other wireless or wired subscriber anywhere in the world while supporting seamless roaming over large geographical areas. The scope of the TWNs therefore, went beyond the wireless access network and well into the wired network.

On the other hand, the aim of 802.11 is only last-hop wireless connectivity. 802.11 does not deal with end-to-end connectivity. In fact, IP-based data networks (for which 802.11 was initially designed) do not have any concept of end-to-end connectivity and each packet is independently routed. Also, the geographical coverage of the wireless access network in 802.11 is significantly less than the geographical coverage of the wireless access network in TWNs. Finally, 802.11 has only limited support for roaming. For all these reasons, the scope of 802.11 is restricted to the wireless access network only. As we go along in this chapter, it would be helpful to keep these similarities and differences in mind.

18.2 Key Establishment in 802.11

The key establishment protocol of 802.11 is very simple to describe—there is none. 802.11 relies on "preshared" keys between the mobile nodes or stations (henceforth stations (STAs)) and the *access points* (APs). It does not specify how the keys are established and assumes that this is achieved in some "out-of-band" fashion. In other words, key establishment is outside the scope of WEP.

18.2.1 What's Wrong?

As we saw earlier, key establishment is one of the toughest problems in network security. By not specifying a key establishment protocol, it seems that the 802.11 designers were side-stepping the issue. To be fair to 802.11 designers, they did a pretty good job with the standard. The widespread acceptance of this technology is a testament to this. In retrospect, security was one of the issues where the standard did have many loopholes, but then again everyone has perfect vision in hindsight. Back to our issue, the absence of any key management protocol led to multiple problems as we discuss below.

1. In the absence of any key management protocol, real life deployment of 802.11 networks ended up using manual configuration of keys into all STAs and the AP that wish to form a *basic service set* (BSS).

2. Manual intervention meant that this approach was open to manual error.

3. Most people cannot be expected to choose a "strong" key. In fact, most humans would probably choose a key which is easy to remember. A quick survey of the 802.11 networks that I had access to shows that people use keys like "abcd1234" or "12345678" or "22222222" and so on. These keys, being alphanumeric in nature, are easy to guess and do not exploit the whole key space.

4. There is no way for each STA to be assigned a unique key. Instead, all STAs and the AP are configured with the same key. As we will see in Section 18.4.4, this means that the AP has no way of uniquely identifying a STA in a secure fashion. Instead, the STAs are divided into two groups. Group One consists of stations that are allowed access to the network, and Group Two consists of all other stations (that is, STAs which are not allowed to access the network). Stations in Group One share a secret key which stations in Group Two don't know.

5. To be fair, 802.11 does allow each STA (and AP) in a BSS to be configured with four different keys. Each STA can use any one of the four keys when establishing a connection with the AP. This feature may therefore be used to divide STAs in a BSS into four groups if each group uses one of these keys. This allows the AP a little finer control over reliable STA recognition.

6. In practice, most real life deployments of 802.11 use the same key across BSSs over the whole extended service set (ESS).[2] This makes roaming easier and faster, since an ESS has many more STAs than a BSS. In terms of key usage, this means that the same key is shared by even more STAs. Besides being a security loophole to authentication (see Section 18.4.4), this higher exposure makes the key more susceptible to compromise.

---

[2]Recall that an ESS is a set of APs connected by a distribution system (like Ethernet).

18.3 Anonymity in 802.11

We saw that subscriber anonymity was a major concern in TWNs. Recall that TWNs evolved from the voice world (the PSTN). In data networks (a large percentage of which use IP as the underlying technology), subscriber anonymity is not such a major concern. To understand why this is so, we need to understand some of the underlying architectural differences between TWNs and IP-based data networks. As we saw earlier, TWNs use IMSI for call routing. The corresponding role in IP-based networks is fulfilled by the IP address. However, unlike the IMSI, the IP address is not permanently mapped to a subscriber. In other words, given the IMSI, it is trivial to determine the identity of the subscriber. However, given the IP address, it is extremely difficult to determine the identity of the subscriber. This difficulty arises because of two reasons. First, IP addresses are dynamically assigned using protocols like DHCP; in other words, the IP address assigned to a subscriber can change over time.

Second, the widespread use of Network Address Translation (NAT) adds another layer of identity protection. NAT was introduced to deal with the shortage of IP addresses.[3] It provides IP-level access between hosts at a site (local area network (LAN)) and the rest of the Internet without requiring each host at the site to have a globally unique IP address. NAT achieves this by requiring the site to have a single connection to the global Internet and at least one globally valid IP address (hereafter referred to as GIP). The address GIP is assigned to the NAT translator (also known as NAT box), which is basically a router that connects the site to the Internet. All datagrams coming into and going out of the site must pass through the NAT box. The NAT box replaces the source address in each outgoing datagram with GIP and the destination address in each incoming datagram with the private address of the correct host. From the view of any host external to the site (LAN), all datagrams come from the same GIP (the one assigned to the NAT box). There is no way for an external host to determine which of the many hosts at a site a datagram came from. Thus, the usage of NAT adds another layer of identity protection in IP networks.

18.4 Authentication in 802.11

Before we start discussing the details of authentication in 802.11 networks, recall that the concepts of authentication and access control are very closely linked. To be precise, one of the

---

[3]To be accurate, the shortage of IPv4 addresses. There are more than enough IPv6 addresses available but the deployment of IPv6 has not caught on as fast as its proponents would have liked.

primary uses of authentication is to control access to the network. Now, think of what happens when a station wants to connect to a LAN. In the wired world, this is a simple operation. The station uses a cable to plug into an Ethernet jack, and it is connected to the network. Even if the network does not explicitly authenticate the station, obtaining physical access to the network provides at least some basic access control if we assume that access to the physical medium is protected. In the wireless world, this physical-access-authentication disappears.

For a station to "connect to" or associate with a wireless local area network (WLAN), the network-joining operation becomes much more complicated. First, the station must find out which networks it currently has access to. Then, the network must authenticate the station and the station must authenticate the network. Only after this authentication is complete can the station connect to or associate with the network (via the AP). Let us go over this process in detail.

Access points (APs) in an 802.11 network periodically broadcast beacons. Beacons are management frames which announce the existence of a network. They are used by the APs to allow stations to find and identify a network. Each beacon contains a Service Set Identifier (SSID), also called the *network name*, which uniquely identifies an ESS. When an STA wants to access a network, it has two options: passive scan and active scan. In

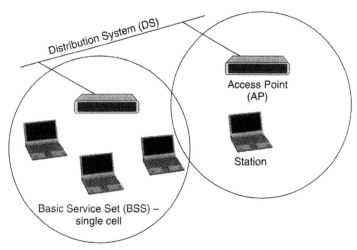

Figure 18.1: 802.11 system overview

the former case, it can scan the channels (the frequency spectrum) trying to find beacon advertisements from APs in the area. In the latter case, the station sends probe-requests (either to a particular SSID or with the SSID set to 0) over all the channels one-by-one. A particular SSID indicates that the station is looking for a particular network. If the concerned AP receives the probe, it responds with a probe-response. A SSID of 0 indicates that the station is looking to join any network it can access. All APs which receive this probe-request and which want this particular station to join their network, reply back with a probe-response. In either case, a station finds out which network(s) it can join.

Next, the station has to choose a network it wishes to join. This decision can be left to the user or the software can make this decision based on signal strengths and other criteria. Once a station has decided that it wants to join a particular network, the authentication process starts. 802.11 provides for two forms of authentication: Open System Authentication (OSA) and Shared Key Authentication (SKA). Which authentication is to be used for a particular transaction needs to be agreed upon by both the STA and the network. The STA proposes the authentication scheme it wishes to use in its authentication request message. The network may then accept or reject this proposal in its authentication response message depending on how the network administrator has set up the security requirements of the network.

18.4.1 Open System Authentication

This is the default authentication algorithm used by 802.11. Here is how it works. Any station which wants to join a network sends an authentication request to the appropriate AP. The authentication request contains the authentication algorithm that the station the wishes to use (0 in case of OSA). The AP replies back with an authentication response thus

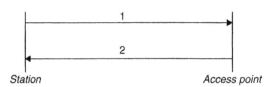

1. Authentication Request: Auth Alg = 0; Trans. Num = 1.
2. Authentication Resp.: Auth Alg = 0; Trans. Num = 2; Status = 0/*

Figure 18.2: 802.11 OSA

authenticating the station to join the network[4] if it has been configured to accept OSA as a valid authentication scheme. In other words, the AP does not do any checks on the identity of the station and allows any and all stations to join the network. OSA is exactly what its name suggests: open system authentication. The AP (network) allows any station (that wishes to join) to join the network. Using OSA therefore means using no authentication at all.

It is important to note here that the AP can enforce the use of authentication. If a station sends an authentication request requesting to use OSA, the AP may deny the station access to the network if the AP is configured to enforce SKA on all stations.

18.4.2 Shared Key Authentication

Shared Key Authentication (SKA) is based on the challenge-response system. SKA divides stations into two groups. Group One consists of stations that are allowed access to the network and Group Two consists of all other stations. Stations in Group One share a secret key which stations in Group Two don't know. By using SKA, we can ensure that only stations belonging to Group One are allowed to join the network.

Using SKA requires (1) that the station and the AP be capable of using WEP and (2) that the station and the AP have a preshared key. The second requirement means that a shared key must be distributed to all stations that are allowed to join the network before attempting authentication. How this is done is not specified in the 802.11 standard. Figure 18.3 explains how SKA works in detail.

When a station wants to join a network, it sends an authentication request to the appropriate AP which contains the authentication algorithm it wishes to use (1 in case of SKA). On receiving this request, the AP sends an authentication response back to the station. This authentication response contains a challenge-text. The challenge text is a 128-byte number generated by the pseudorandom-number-generator (also used in WEP) using the preshared secret key and a random Initialization Vector (IV). When the station receives this random number (the challenge), it encrypts the random number using WEP[5] and its own IV to generate a response to the challenge. Note that the IV that the station uses for encrypting the challenge is different from (and independent of) the IV that the

---

[4]The authentication request from the station may be denied by the AP for reasons other than authentication failure, in which case the status field will be nonzero.
[5]WEP is described in Section 18.5.

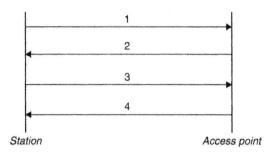

Station   Access point

1. Authentication Request: Auth Alg = 1; Trans. Num = 1.
2. Authentication Resp: Auth Alg = 1; Trans. Num = 2; Data = 128-byte random number.
3. Authentication Resp: Auth Alg = 1; Trans. Num = 3; Data = Encrypted (128-byte number rcvd in.
4. Authentication Resp: Auth Alg = 1; Trans. Num = 4; Status = 0/*.

Figure 18.3: 802.11 SKA

AP used for generating the random number. After encrypting the challenge, the station sends the encrypted challenge and the IV it used for encryption back to the AP as the response to the challenge. On receiving the response, the AP decrypts the response using the preshared keys and the IV that it receives as part of the response. The AP compares the decrypted message with the challenge it sent to the station. If these are the same, the AP concludes that the station wishing to join the network is one of the stations which knows the secret key and therefore the AP authenticates the station to join the network.

The SKA mechanism allows an AP to verify that a station is one of a select group of stations. The AP verifies this by ensuring that the station knows a secret. This secret is the preshared key. If a station does not know the key, it will not be able to respond correctly to the challenge. Thus, the strength of SKA lies in keeping the shared key a secret.

18.4.3 Authentication and Handoffs

If a station is mobile while accessing the network, it may leave the range of one AP and enter into the range of another AP. In this section we see how authentication fits in with mobility.

A STA may move inside a BSA (intra-BSA), between two BSAs (inter-BSA) or between two Extended Service Areas (ESAs) (inter-ESAs). In the intra-BSA case, the STA is static for all handoff purposes. Inter-ESA roaming requires support from higher layers (MobileIP for example) since ESAs communicate with each other at Layer 3.

It is the inter-BSA roaming that 802.11 deals with. A STA keeps track of the received signal strength (RSS) of the beacon with which it is associated. When this RSS value falls

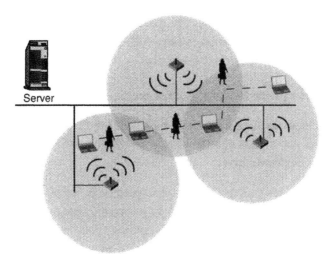

Figure 18.4: 802.11 handoffs and security

below a certain threshold, the STA starts to scan for stronger beacon signals available to it using either active or passive scanning. This procedure continues until the RSS of the current beacon returns above the threshold (in which case the STA stops scanning for alternate beacons) or until the RSS of the current beacon falls below the break-off threshold, in which case the STA decides to handoff to the strongest beacon available. When this situation is reached, the STA disconnects from its prior AP and connects to the new AP afresh (just as if had switched on in the BSA of the new AP). In fact, the association with the prior-AP is not "carried-over" or "handed-off" transparently to the new AP: the STA disconnects with the old AP and then connects with the new AP.

To connect to the new AP, the STA starts the connection procedure afresh. This means that the process of associating (and authenticating) to the new AP is the same as it is for a STA that has just powered on in this BSS. In other words, the prior-AP and the post-AP do not coordinate among themselves to achieve a handoff.[6] Analysis[7] has shown that authentication delays are the second biggest contributors to handoff times next only to

---

[6]To be accurate, the IEEE 802.11 standard does not specify how the two APs should communicate with each other. There do exist proprietary solutions by various vendors which enable inter-AP communication to improve handoff performance.

[7]An Empirical Analysis of the IEEE 802.11 MAC layer Handoff Process—Mishra et al.

channel scanning/probing time. This re-authentication delay becomes even more of a bottleneck for real time applications like voice. Although this is not exactly a security loophole, it is a "drawback" of using the security.

18.4.4 What's Wrong with 802.11 Authentication?

Authentication mechanisms suggested by 802.11 suffer from many drawbacks. As we saw, 802.11 specifies two modes of authentication—OSA and SKA. OSA provides no authentication and is irrelevant here.

SKA works on a challenge-response system as explained in Section 18.4.2. The AP expects that the challenge it sends to the STA be encrypted using an IV and the preshared key. As described in Section 18.2.1, there is no method specified in WEP for each STA to be assigned a unique key. Instead all STAs and the AP in a BSS are configured with the same key. This means that even when an AP authenticates a STA using the SKA mode, all it ensures is that the STA belongs to a group of STAs which know the preshared key. There is no way for the AP to reliably determine the exact identity of the STA that is trying to authenticate to the network and access it.[8]

To make matters worse, many 802.11 deployments share keys across APs. This increases the size of the group to which a STA can be traced. All STAs sharing a single preshared secret key also makes it very difficult to remove a STA from the allowed set of STAs, since this would involve changing (and redistributing) the shared secret key to all stations.

There is another issue with 802.11 authentication: it is one-way. Even though it provides a mechanism for the AP to authenticate the STA, it has no provision for the STA to be able to authenticate the network. This means that a rogue AP may be able to hijack the STA by establishing a session with it. This is a very plausible scenario given the plummeting cost of APs. Since the STA can never find out that it is communicating with a rogue AP, the rogue AP has access to virtually everything that the STA sends to it.

Finally, SKA is based on WEP, discussed in Section 18.5. It therefore suffers from all the drawbacks that WEP suffers from too. These drawbacks are discussed in Section 18.5.1.

---

[8]MAC addresses can be used for this purpose but they are not cryptographically protected in that it is easy to spoof a MAC address.

18.4.5 Pseudo-Authentication Schemes

Networks unwilling to use SKA (or networks willing to enhance it) may rely on other authentication schemes. One such scheme allows only stations which know the network's SSID to join the network. This is achieved by having the AP responding to a probe-request from a STA only if the probe request message contains the SSID of the network. This in effect prevents connections from STAs looking for any wild carded SSIDs. From a security perspective, the secret here is the SSID of the network. If a station knows the SSID of the network, it is allowed to join the network. Even though this is a very weak authentication mechanism, it provides some form of protection against casual eavesdroppers from accessing the network. For any serious eavesdropper (hacker), this form of authentication poses minimal challenge since the SSID of the network is often transmitted in the clear (without encryption).

Yet another authentication scheme (sometimes referred to as address filtering) uses the MAC addresses as the secret. The AP maintains a list of MAC addresses of all the STAs that are allowed to connect to the network. This table is then used for admission control into the network. Only stations with the MAC addresses specified in the table are allowed to connect to the network. When a station tries to access the network via the AP, the AP verifies that the station has a MAC address which belongs to the above mentioned list. Again, even though this scheme provides some protection, it is not a very secure authentication scheme since most wireless access cards used by stations allow the user to change their MAC address via software. Any serious eavesdropper or hacker can find out the MAC address of one of the stations which is allowed access by listening in on the transmissions being carried out by the AP and then change their own MAC address to the determined address.

18.5　Confidentiality in 802.11

WEP uses a preestablished/preshared set of keys. Figure 18.5 shows how WEP is used to encrypt an 802.11 MAC Protocol Data Unit (MPDU). Note that Layer 3 (usually IP) hands over a MAC Service Data Unit (MSDU) to the 802.11 MAC layer. The 802.11 protocol may then fragment the MSDU into multiple MPDUs if so required to use the channel efficiently.

The WEP process can be broken down into the following steps.

Step 1: Calculate the Integrity Check Value (ICV) over the length of the MPDU and append this 4-byte value to the end of the MPDU. Note that ICV is another name for Message Integrity Check (MIC). We see how this ICV value is generated in Section 18.6.

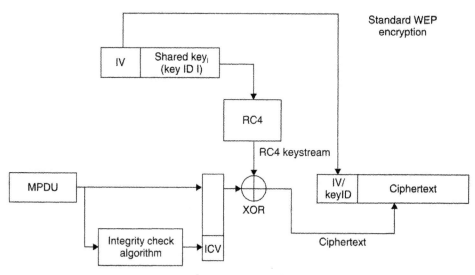

Figure 18.5: WEP

Step 2: Select a master key to be used from one of the four possible preshared secret keys. See Section 18.2.1 for the explanation of the four possible preshared secret keys.

Step 3: Select an IV and concatenate it with the master key to obtain a key seed. WEP does not specify how to select the IV. The IV selection process is left to the implementation.

Step 4: The key seed generated in Step 3 is then fed to an RC4 key-generator. The resulting RC4 key stream is then XORed with the MPDU + ICV generated in Step 1 to generate the ciphertext.

Step 5: A 4-byte header is then appended to the encrypted packet. It contains the 3-byte IV value and a 1-byte key-id specifying which one of the four preshared secret keys is being used as the master key.

The WEP process is now completed. An 802.11 header is then appended to this packet and it is ready for transmission. The format of this packet is shown in Figure 18.6.

18.5.1 What's Wrong with WEP?

WEP uses RC4 (a stream cipher) in synchronous mode for encrypting data packets. Synchronous stream ciphers require that the key generators at the two communicating

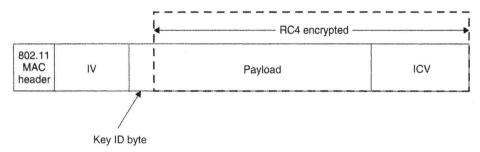

Figure 18.6: A WEP Packet

nodes must be kept synchronized by some external means because the loss of a single bit of a data stream encrypted under the cipher causes the loss of ALL data following the lost bit. In brief, this is so because data loss desynchronizes the key stream generators at the two endpoints. Since data loss is widespread in the wireless medium, a synchronous stream cipher is not the right choice. This is one of the most fundamental problems of WEP. It uses a cipher not suitable for the environment it operates in.

It is important to re-emphasize here that the problem here is not the RC4 algorithm.[9] The problem is that a stream cipher is not suitable for a wireless medium where packet loss is widespread. SSL uses RC4 at the application layer successfully because SSL (and therefore RC4) operates over TCP (a reliable data channel) that does not lose any data packets and can therefore guarantee perfect synchronization between the two end points.

The WEP designers were aware of the problem of using RC4 in a wireless environment. They realized that due to the widespread data loss in the wireless medium, using a synchronous stream cipher across 802.11 frame boundaries was not a viable option. As a solution, WEP attempted to solve the synchronization problem of stream ciphers by shifting synchronization requirement from a session to a packet. In other words, since the synchronization between the end-points is not perfect (and subject to packet loss), 802.11 changes keys for every packet. This way each packet can be encrypted or decrypted irrespective of the previous packet's loss. Compare this with SSL's use of RC4, which can afford to use a single key for a complete TCP session. In effect, since the wireless medium is prone to data loss, WEP has to use a single packet as the synchronization unit rather than a complete session. This means that WEP uses a unique key for each packet.

---

[9]Though loopholes in the RC4 algorithm have been discovered too.

Using a separate key for each packet solves the synchronization problem but introduces problems of its known. Recall that to create a per-packet key, the IV is simply concatenated with the master key. As a general rule in cryptography, the more exposure a key gets, the more it is susceptible to be compromised. Most security architectures therefore try to minimize the exposure of the master key when deriving secondary (session) keys from it. In WEP however, the derivation of the secondary (per-packet) key from the master key is too trivial (a simple concatenation) to hide the master key.

Another aspect of WEP security is that the IV which is concatenated with the master key to create the per-packet key is transmitted in cleartext with the packet too. Since the 24-bit IV is transmitted in the clear with each packet, an eavesdropper already has access to the first three bytes of the per-packet key.

The above two weaknesses make WEP susceptible to an Fluhrer-Mantin-Shamir (FMS) attack, which uses the fact that simply concatenating the IV (available in plain text) to the master key leads to the generation of a class of RC4 weak keys. The FMS attack exploits the fact that the WEP creates the per-packet key by simply concatenating the IV with the master-key. Since the first 24 bits of each per-packet key is the IV (which is available in plain text to an eavesdropper),[10] the probability of using weak keys[11] is very high. Note that the FMS attack is a weakness in the RC4 algorithm itself. However, it is the way that the per-packet keys are constructed in WEP that makes the FMS attack a much more effective attack in 802.11 networks.

The FMS attack relies on the ability of the attacker to collect multiple 802.11 packets which have been encrypted with weak keys. Limited key space (leading to key reuse) and availability of IV in plaintext which forms the first 3 bytes of the key makes the FMS attack a very real threat in WEP. This attack is made even more potent in 802.11 networks by the fact that the first 8 bytes of the encrypted data in every packet are known to be the Sub-Network Access Protocol (SNAP) header. This means that simply XORing the first 2 bytes of the encrypted pay-load with the well known SNAP header yields the first 2 bytes of the generated key-stream. In the FMS attack, if the first 2 bytes of enough

---

[10]Remember that each WEP packet carries the IV in plaintext format prepended to the encrypted packet.

[11]Use of certain key values leads to a situation where the first few bytes of the output are not all that random. Such keys are known as weak keys. The simplest example is a key value of 0.

key-streams are known then the RC4 key can be recovered. Thus, WEP is an ideal candidate for an FMS attack.

The FMS attack is a very effective attack but is by no means the only attack which can exploit WEP weaknesses. Another such attack stems from the fact that one of the most important requirements of a synchronous stream cipher (like RC4) is that the same key should not be reused *EVER*. Why is it so important to avoid key reuse in RC4? Reusing the same key means that different packets use a common key stream to produce the respective ciphertext. Consider two packets of plaintext (P1 and P2) which use the same RC4 key stream for encryption.

Since $C1 = P1 \oplus RC4(key)$

And $C2 = P2 \oplus RC4(key)$

Therefore $C1 \oplus C2 = P1 \oplus P2$

Obtaining the XOR of the two plaintexts may not seem like an incentive for an attack but when used with frequency analysis techniques it is often enough to get lots of information about the two plaintexts. More importantly, as shown above, key reuse effectively leads to the effect of the key stream canceling out! An implication of this effect is that if one of the plaintexts (say P1) is known, P2 can be calculated easily since $P2 = (P1 \oplus P2) \oplus P1$. Another implication of this effect is that if an attacker (say, Eve) gets access to the <P1, C1> pair,[12] simply XORing the two produces the key stream K. Once Eve has access to K, she can decrypt C2 to obtain P2. Realize how the basis of this attack is the reuse of the key stream, K.

Now that we know why key reuse is prohibited in RC4, we look at what 802.11 needs to achieve this. Since we need a new key for every single packet to make the network really secure, 802.11 needs a very large key space, or rather a large number of unique keys. The number of unique keys available is a function of the key length. What is the key length used in WEP? Theoretically it is 64 bits. The devil, however, is in the details. How is the 64-bit key constructed? 24 bits come from the IV and 40 bits come from the base-key. Since the 40-bit master key never changes in most 802.11 deployments,[13] we must ensure that we use different IVs for each packet in order to avoid key reuse. Since the master key

---

[12]This is not as difficult as it sounds.

[13]This weakness stems from the lack of a key-establishment or key-distribution protocol in WEP.

is fixed in length and the IV is only 24 bits long, the effective key length of WEP is 24 bits. Therefore, the key space for the RC4 is 2^N where N is the length of the IV. 802.11 specified the IV length as 24.

To put things in perspective, realize that if we have a 24 bit IV ($\rightarrow 2^{24}$ keys in the key-space), a busy base station which is sending 1500 byte-packets at 11 Mbps will exhaust all keys in the key space in $(1500*8)/(11*10^6*2^{24})$ seconds or about five hours. On the other hand, RC4 in SSL would use the same key space for 2^{24} ($=10^7$) sessions. Even if the application has 10,000 sessions per day, the key space would last for three years. In other words, an 802.11 BS using RC4 has to reuse the same key in about five hours whereas an application using SSL RC4 can avoid key reuse for about three years. This shows clearly that the fault lies not in the cipher but in the way it is being used. Going beyond an example, analysis of WEP has shown that there is a 50% chance of key reuse after 4823 packets, and there is 99% chance of collision after 12,430 packets. These are dangerous numbers for a cryptographic algorithm.

Believe it or not, it gets worse. 802.11 specifies no rules for IV selection. This in turn means that changing the IV with each packet is optional. This effectively means that 802.11 implementations may use the same key to encrypt all packets without violating the 802.11 specifications. Most implementations, however, vary from randomly generating the IV on a per-packet basis to using a counter for IV generation. WEP does specify that the IV be changed "frequently." Since this is vague, it means that an implementation which generates per-packet keys (more precisely the per-MPDU key) is 802.11-compliant and so is an implementation which re-uses the same key across MPDUs.

18.6 Data Integrity in 802.11

To ensure that a packet has not been modified in transit, 802.11 uses an Integrity Check Value (ICV) field in the packet. ICV is another name for message integrity check (MIC). The idea behind the ICV/MIC is that the receiver should be able to detect data modifications or forgeries by calculating the ICV over the received data and comparing it with the ICV attached in the message. Figure 18.7 shows the complete picture of how WEP and CRC32 work together to create the MPDU for transmission.

The underlying assumption is that if Eve modifies the data in transit, she should not be able to modify the ICV appropriately to force the receiver into accepting the packet. In

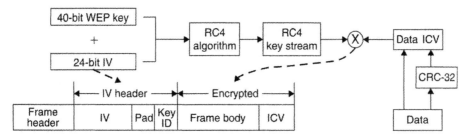

Figure 18.7: Data integrity in WEP

WEP, ICV is implemented as a Cyclic Redundancy Check-32 bits (CRC-32) checksum which breaks this assumption. The reason for this is that CRC-32 is linear and is not cryptographically computed, i.e., the calculation of the CRC-32 checksum does not use a key/shared secret. Also, this means that the CRC32 has the following interesting property:

$$CRC(X \oplus Y) = CRC(X) \oplus CRC(Y)$$

Now, if X represents the payload of the 802.11 packet over which the ICV is calculated, the ICV is CRC(X) which is appended to the packet. Consider an intruder who wishes to change the value of X to Z. To do this, they calculate $Y = X \oplus Z$. Then she captures the packet from the air-interface, XORs X with Y and then XORs the ICV with CRC(Y). Therefore, the packet changes from $\{X, CRC(X)\}$ to $\{X \oplus Y, CRC(X) \oplus CRC(Y)\}$ or simply $\{X \oplus Y, CRC(X \oplus Y)\}$. If the intruder now re-transmits the packets to the receiver, the receiver would have no way of telling that the packet was modified in transit. This means that we can change bits in the payload of the packet while preserving the integrity of the packet if we also change the corresponding bits in the ICV of the packet.

Note that an attack like the one described above works because flipping bit x in the message results in a deterministic set of bits in the CRC that must be flipped to produce the correct checksum of the modified message. This property stems from the linearity of the CRC32 algorithm.

Realize that even though the ICV is encrypted (cryptographically protected) along with the rest of the payload in the packet, it is not cryptographically computed; that is, calculating the ICV does not involve keys and cryptographic operations. Simply encrypting the ICV does not prevent an attack like the one discussed above. This is so because the flipping of a bit in the ciphertext carries through after the RC4 decryption into the plaintext because RC4(k, X $\oplus$ Y) = RC4(k, X) $\oplus$ Y and therefore:

$$RC4(k, CRC(X \oplus Y)) = RC4(k, CRC(X)) \oplus CRC(Y)$$

The problem with the message integrity mechanism specified in 802.11 is not only that it uses a linear integrity check algorithm (CRC32) but also the fact that the ICV does not protect all the information that needs to be protected from modification. Recall from Section 18.5 that the ICV is calculated over the MPDU data; in other words, the 802.11 header is not protected by the ICV. This opens the door to redirection attacks as explained below.

Consider an 802.11 BSS where an 802.11 STA (Alice) is communicating with a wired station (Bob). Since the wireless link between Alice and the access point (AP) is protected by WEP and the wired link between Bob and access point is not,[14] it is the responsibility of the AP to decrypt the WEP packets and forward them to Bob. Now, Eve captures the packets being sent from Alice to Bob over the wireless link. She then modifies the destination address to another node, say C (Charlie), in the 802.11 header and retransmits them to the AP. Since the AP does not know any better, it decrypts the packet and forwards it to Charlie. Eve, therefore, has the AP decrypt the packets and forward them to a destination address of choice.

The simplicity of this attack makes it extremely attractive. All Eve needs is a wired station connected to the AP and she can eavesdrop on the communication between Alice and Bob without needing to decrypt any packets herself. In effect, Eve uses the infrastructure itself to decrypt any packets sent from an 802.11 STA via an AP. Note that this attack does not necessarily require that one of the communicating stations be a wired station. Either Bob or Charlie (or both) could as easily be other 802.11 STAs which do not use WEP. The attack would still hold since the responsibility of decryption would still be with the AP. The bottom line is that the redirection attack is possible because the ICV is not calculated over the 802.11 header. There is an interesting security lesson here. A system can't have confidentiality without integrity, since an attacker can use the redirection attack and exploit the infrastructure to decrypt the encrypted traffic.

Another problem which stems from the weak integrity protection in WEP is the threat of a replay attack. A replay attack works by capturing 802.11 packets transmitted over the wireless interface and then replaying (retransmitting) the captured packet(s) later on

---

[14]WEP is an 802.11 standard used only on the wireless link.

with (or without) modification such that the receiving station has no way to tell that the packet it is receiving is an old (replayed) packet. To see how this attack can be exploited, consider a hypothetical scenario where Alice is an account holder, Bob is a bank and Eve is another account holder in the bank. Suppose Alice and Eve do some business and Alice needs to pay Eve $500. So, Alice connects to Bob over the network and transfers $500 from her account to Eve. Eve, however, is greedy. She knows Alice is going to transfer money. So, she captures all data going from Alice to Bob. Even though Eve does not know what the messages say, she has a pretty good guess that these messages instruct Bob to transfer $500 from Alice's account to Eve's. So, Eve waits a couple of days and replays these captured messages to Bob. This may have the effect of transferring another $500 from Alice's account to Eve's account unless Bob has some mechanism for determining that he is being replayed the messages from a previous session.

Replay attacks are usually prevented by linking the integrity protection mechanism to either timestamps and/or session sequence numbers. However, WEP does not provide for any such protection.

18.7 Loopholes in 802.11 Security

To summarize, here is the list of things that are wrong with 802.11 security:

1. 802.11 does not provide any mechanism for key establishment over an unsecure medium. This means key sharing among STAs in a BSS and sometimes across BSSs.

2. WEP uses a synchronous stream cipher over a medium, where it is difficult to ensure synchronization during a complete session.

3. To solve the previous problem, WEP uses a per-packet key by concatenating the IV directly to the preshared key to produce a key for RC4. This exposes the base key or master key to attacks like FMS.

4. Since the master key is usually manually configured and static and since the IV used in 802.11 is just 24 bits long, this results in a very limited key-space.

5. 802.11 specifies that changing the IV with each packet is optional, thus making key reuse highly probable.

6. The CRC-32 used for message integrity is linear.

7. The ICV does not protect the integrity of the 802.11 header, thus opening the door to redirection attacks.

8. There is no protection against replay attacks.

9. There is no support for a STA to authenticate the network.

Note that the limited size of the IV figures much lower in the list than one would expect. This emphasizes the fact that simply increasing the IV size would not improve WEP's security considerably. The deficiency of the WEP encapsulation design arises from attempts to adapt RC4 to an environment for which it is poorly suited.

18.8 WPA

When the loopholes in WEP, the original 802.11 security standard, had been exposed, IEEE formed a Task Group: 802.11i with the aim of improving upon the security of 802.11 networks. This group came up with the proposal of a Robust Security Network (RSN). A RSN is an 802.11 network which implements the security proposals specified by the 802.11i group and allows only RSN-capable devices to join the network, thus allowing no "holes." The term *hole* is used to refer to a non-802.11i compliant STA which by virtue of not following the 802.11i security standard could make the whole network susceptible to a variety of attacks.

Since making a transition from an existing 802.11 network to a RSN cannot always be a single-step process (we will see why in a moment), 802.11i allows for a Transitional Security Network (TSN) which allows for the existence of both RSN and WEP nodes in an 802.11 network. As the name suggests, this kind of a network is specified only as a transition point and all 802.11 networks are finally expected to move to a RSN. The terms RSN and 802.11i are sometimes used interchangeably to refer to this security specification.

The security proposal specified by the Task Group-i uses the Advanced Encryption Standard (AES) in its default mode. One obstacle in using AES is that it is not backward compatible with existing WEP hardware. This is so because AES requires the existence of a new more powerful hardware engine. This means that there is also a need for a security solution which can operate on existing hardware. This was a pressing need for vendors of 802.11 equipment. This is where the Wi-Fi alliance came into the picture.

The Wi-Fi alliance is an alliance of major 802.11 vendors formed with the aim of ensuring product interoperability. To improve the security of 802.11 networks without requiring a hardware upgrade, the Wi-Fi alliance adopted Temporal Key Integrity Protocol (TKIP) as the security standard that needs to be deployed for Wi-Fi certification. This form of security has therefore come to be known as Wi-Fi Protected Access (WPA). WPA is basically a prestandard subset of 802.11i which includes the key management and the authentication architecture (802.1X) specified in 802.11i. The biggest difference between WPA and 802l.11i (which has also come to be known as WPA2) is that instead of using AES for providing confidentiality and integrity, WPA uses TKIP and MICHAEL respectively. We look at TKIP/WPA in this section and the 802.11i/WPA2 using AES in the next section.

TKIP stands for Temporal Key Integrity Protocol. It was designed to fix WEP loopholes while operating within the constraints of existing 802.11 equipment (APs, WLAN cards and so on). To understand what we mean by the "constraints of existing 802.11 hardware," we need to dig a little deeper. Most 802.11 equipment consists of some sort of a WLAN Network Interface Card (NIC) (also known as WLAN adapter) which enables access to an 802.11 network. A WLAN NIC usually consists of a small microprocessor, some firmware, a small amount of memory and a special-purpose hardware engine. This hardware engine is dedicated to WEP implementation since software implementations of WEP are too slow. To be precise, the WEP encryption process is implemented in hardware. The hardware encryption takes the IV, the base (master) key and the plaintext data as the input and produces the encrypted output (ciphertext). One of the most severe constraints for TKIP designers was that the hardware engine cannot be changed. We see in this section how WEP loopholes were closed given these constraints.

18.8.1 Key Establishment

One of the biggest WEP loopholes is that it specifies no key-establishment protocol and relies on the concept of preshared secret keys which should be established using some out-of-band mechanism. Realize that this is a system architecture problem. In other words, solving this problem requires support from multiple components (the AP, the STA and usually also a backend authentication server) in the architecture.

One of the important realizations of the IEEE 802.11i task group was that 802.11 networks were being used in two distinct environments: the home network and the enterprise network. These two environments had distinct security requirements and different infrastructure capacities to provide security. Therefore, 802.11i specified two distinct security architectures. For the enterprise network, 802.11i specifies the use of IEEE 802.1X for key establishment and authentication. As we will see in our discussion in the next section, 802.1X requires the use of a backend authentication server. Deploying a back end authentication server is not usually feasible in a home environment. Therefore, for home deployments of 802.11, 802.11i allows the use of the "out-of-band mechanism" (read manual configuration) for key establishment.

We look at the 802.1X architecture in the next section and see how it results in the establishment of a Master Key (MK). In this section, we assume that the two communicating end-points (the STA and the AP) already share a MK which has either been configured manually at the two end-points (WEP architecture) or has been established using the authentication process (802.1X architecture). This section looks at how this MK is used in WPA.

Recall that a major loophole in WEP was the manner[15] in which this master key was used which made it vulnerable to compromise. WPA solves this problem by reducing the exposure of the master key, thus making it difficult for an attacker to discover the master key. To achieve this, WPA adds an additional layer to the key hierarchy used in WEP. Recall from Section 17.4 that WEP uses the master key for authentication and to calculate the per-packet key. In effect there is a two-tier key hierarchy in WEP: the master (preshared secret) key and the per-packet key.

WPA extends the two-tier key-hierarchy of WEP to a multitier hierarchy (See Figure 18.8). At the top level is still the master key, referred to as the Pair-wise Master Key (PMK) in WPA. The next level in the key hierarchy is the PTK which is derived from the PMK. The final level is the per-packet keys which are generated by feeding the PTK to a key-mixing function. Compared with the two-tier WEP key hierarchy, the three-tier key hierarchy of WPA avoids exposing the PMK in each packet by introducing the concept of PTK.

---

[15]The per-packet key is obtained by simply concatenating the IV with the preshared secret key. Therefore, a compromised per-packet key exposes the preshared secret key.

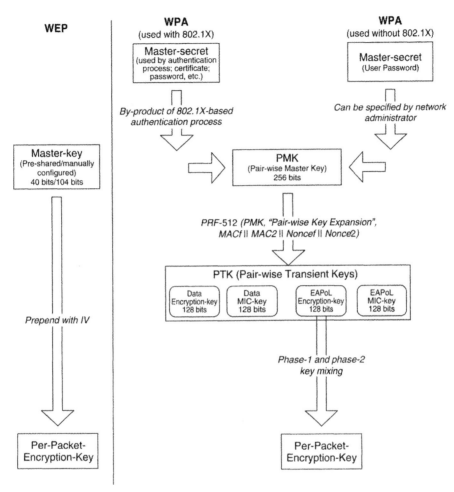

Figure 18.8: Key hierarchy in 802.11

As we saw, WPA is flexible about how the master key (PMK in WPA) is established. The PMK, therefore, may be a preshared[16] secret key (WEP-design) or a key derived from an authentication process like 802.1X.[17] WPA does require that the PMK be 256 bits (or 32 bytes) long. Since a 32-byte key is too long for humans to remember, 802.11 deployments

---

[16]As we saw, this usually means that the keys are manually configured.

[17]It is expected that most enterprise deployments of 802.11 would use 802.1X while the preshared secret key method (read manual configuration) would be used by residential users.

using preshared keys may allow the user to enter a shorter password which may then be used as a seed to generate the 32-byte key.

The next level in the key hierarchy after the PMK are the PTK. WPA uses the PMK for deriving the Pair-wise Transient Keys (PTK) which are basically session keys. The term PTK is used to refer to a set of session keys which consists of four keys, each of which is 128 bits long. These four keys are as follows: an encryption key for data, an integrity key for data, an encryption key for EAPoL messages and an integration key for EAPoL messages. Note that the term *session* here refers to the association between a STA and an AP. Every time an STA associates with an AP, it is the beginning of a new session and this results in the generation of a new PTK (set of keys) from the PMK. Since the session keys are valid only for a certain period of time, they are also referred to as temporal keys and the set of four session keys together is referred to as the Pair-wise Transient Keys (PTK). The PTK are derived from the PMK using a *pseudorandom function* (PRF). The PRFs used for derivation of PTKs (and nonces) are explicitly specified by WPA and are based on the HMAC-SHA algorithm.

> PTK = PRF-512(PMK, "Pair-wise key expansion", AP_MAC || STA_MAC || ANonce || SNonce)

Realize that to obtain the PTK from the PMK we need five input values: the PMK, the MAC addresses of the two endpoints involved in the session and one nonce each from the two endpoints. The use of the MAC addresses in the derivation of the PTK ensures that the keys are bound to sessions between the two endpoints and increases the effective key space of the overall system.

Realize that since we want to generate a different set of session keys from the same PMK for each new session,[18] we need to add another input into the key generation mechanism which changes with each session. This input is the nonce. The concept of nonce is best understood by realizing that it is short for *Number-Once*. The value of nonce is thus arbitrary except that a nonce value is never used again.[19] Basically it is a number that is used only once. In our context, a nonce is a unique number (generated randomly) that can

---

[18]If a STA disconnects from the AP and connects back with an AP at a later time, these are considered two different sessions.

[19]To be completely accurate, nonce values are generated such that the probability of the same value being generated twice is very low.

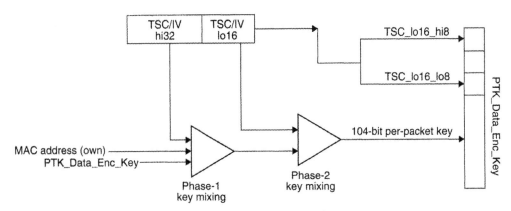

Figure 18.9: TKIP encryption

distinguish between two sessions established between a given STA and an AP at different points in time. The two nonces involved in PTK generation are generated, one each, by the two end points involved in the session; i.e., the STA (SNonce) and the AP (ANonce). WPA specifies that a nonce should be generated as follows:

$$\text{ANonce} = \text{PRF-256(Random Number, "Init Counter", AP\_MAC} \parallel \text{Time)}$$
$$\text{SNonce} = \text{PRF-256(Random Number, "Init Counter", STA\_MAC} \parallel \text{Time)}$$

The important thing to note is that the PTKs are effectively shared between the STA and the AP and are used by both the STA and the AP to protect the data/EAPoL-messages they transmit. It is therefore important that the input values required for derivation of PTK from the PMK come from *both* the STA and the AP. Note also that the key derivation process can be executed in parallel at both endpoints of the session (the STA and the AP) once the nonces and the MAC addresses have been exchanged. Thus, both the STA and the AP can derive the same PTK from the PMK simultaneously.

The next step in the key hierarchy tree is to derive per-packet keys from the PTK. WPA improves also upon this process significantly. Recall from Section 18.5 that the per-packet key was obtained by simply concatenating the IV with the master key in WEP. Instead of simply concatenating the IV with the master key, WPA uses the process shown in Figure 18.9 to obtain the per packet key. This process is known as per-packet key mixing.

In phase one, the session data encryption key is "combined" with the high order 32 bits of the IV and the MAC address. The output from this phase is "combined" with the lower order 16 bits of the IV and fed to phase two, which generates the 104-bit per-packet key. There are many important features to note in this process:

1. It assumes the use of a 48-bit IV (more of this in Section 18.8.2).

2. The size of the encryption key is still 104 bits, thus making it compatible with existing WEP hardware accelerators.

3. Since generating a per-packet key involves a hash operation which is computation intensive for the small MAC processor in existing WEP hardware, the process is split into two phases. The processing intensive part is done in phase one whereas phase two is much less computation intensive.

4. Since phase one involves the high order 32 bits of the IV, it needs to be done only when one of these bits change; that is, once in every 65,536 packets.

5. The key-mixing function makes it very hard for an eavesdropper to correlate the IV and the per-packet key used to encrypt the packet.

18.8.2 Authentication

As we said in the previous section, 802.11i specified two distinct security architectures. For the home network, 802.11i allows the manual configuration of keys just like WEP. For the enterprise network however, 802.11i specifies the use of IEEE 802.1X for key establishment and authentication. Earlier chapters explained the 802.1X architecture in detail. We just summarize the 802.1X architecture in this section.

802.1X is closely architected along the lines of EAPoL (EAP over LAN). Figure 18.10a shows the conceptual architecture of EAPoL and Figure 18.10b shows the overall system architecture of EAPoL. The controlled port is open only when the device connected to the authenticator has been authorized by 802.1x. On the other hand, the uncontrolled port provides a path for extensible authentication protocol over LAN (EAPoL) traffic *ONLY*. Figure 18.10a shows how access to even the uncontrolled port may be limited using MAC filtering.[20] This scheme is sometimes used to deter DoS attacks.

EAP specifies three network elements: the supplicant, the authenticator and the authentication server. For EAPoverLAN, the end user is the supplicant, the Layer 2

---

[20]Allowing only STAs with have a MAC address which is "registered" or "known" to the network.

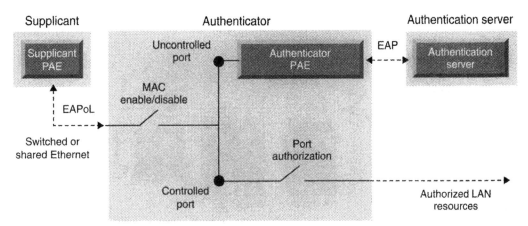

Figure 18.10a: 802.1X/EAP port model

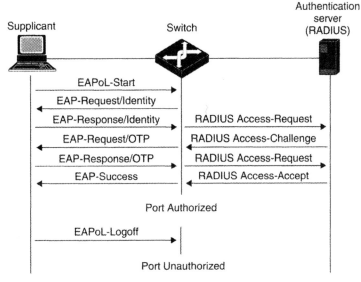

Figure 18.10b: EAPoL

(usually Ethernet) switch is the authenticator controlling access to the network using logical ports, and the access decisions are taken by the backend authentication server after carrying out the authentication process. Which authentication process to use (MD5, TLS and so on) is for the network administrator to decide.

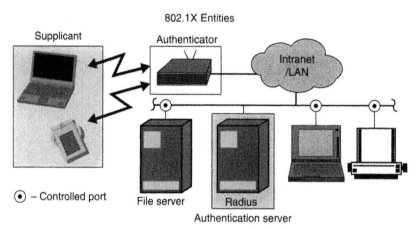

Figure 18.10c: EAP over WLAN

EAPoL can be easily adapted to be used in the 802.11 environment as shown in Figure 18.10c. The STA is the supplicant, the AP is the authenticator controlling access to the network, and there is a backend authentication server. The analogy is all the more striking if you consider that an AP is in fact just a Layer 2 switch, with a wireless and a wired interface.

There is however one interesting piece of detail that needs attention. The 802.1X architecture carries the authentication process between the supplicant (STA) and the backend authentication server.[21] This means that the master key (resulting from an authentication process like TLS) is established between the STA and backend server. However, confidentiality and integrity mechanisms in the 802.11 security architecture are implemented between the AP and the STA. This means that the session (PTK) and per packet keys (which are derived from the PMK) are needed at the STA and the AP. The STA already has the PMK and can derive the PTK and the per-packet keys. However, the AP does not yet have the PMK. Therefore, what is needed is a mechanism to get the PMK from the authentication server to the AP securely.

Recall that in the 802.1X architecture, the result of the authentication process is conveyed by the authentication server to the AP so that the AP may allow or disallow the STA access to the network. The communication protocol between the AP and the authentication server is not specified by 802.11i but is specified by WPA to be RADIUS. Most deployments of 802.11

---

[21]With the AP controlling access to the network using logical ports.

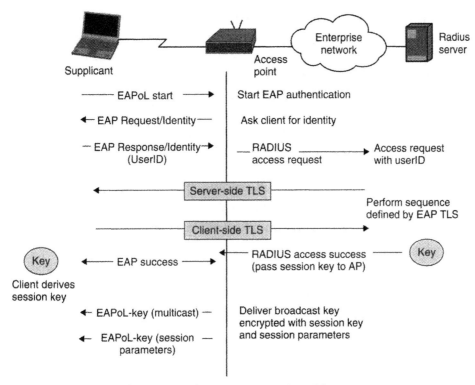

Figure 18.10d: 802.1X network architecture

would probably end up using RADIUS. The RADIUS protocol does allow for distributing the key securely from the authentication server to the AP and this is how the PMK gets to the AP.

Note that 802.1X is a framework for authentication. It does not specify the authentication protocol to be used. Therefore, it is up to the network administrator to choose the authentication protocol they want to plug in to the 802.1X architecture. One of the most often discussed authentication protocols to be used with 802.1X is TLS. Figure 18.10d summarizes how TLS can be used as an authentication protocol in a EAP over WLAN environment. The EAP-TLS protocol is well documented. It has been analyzed extensively and no significant weaknesses have been found in the protocol itself. This makes it an attractive option for security use in 802.1X. However, there is a deployment issue with this scheme.

Note that EAP-TLS relies on certificates to authenticate the network to the clients and the clients to the networks. Requiring the network (the servers) to have certificates is a common theme in most security architectures. However, the requirement that each client be issued a certificate leads to the requirement of the wide spread deployment of PKI. Since this is sometimes not a cost effective option, a few alternative protocols have been proposed: EAP-TTLS (tunneled TLS) and PEAP. Both of these protocols use certificates to authenticate the network (the server) to the client but do not use certificates to authenticate the client to the server. This means that a client no longer needs a certificate to authenticate itself to the server: instead the clients can use password-based schemes (CHAP, PAP and so on) to authenticate themselves. Both protocols divide the authentication process in two phases. In phase 1, we authenticate the network (the server) to the client using a certificate and establish a TLS tunnel between the server and the client. This secure[22] TLS channel is then used to carry out a password-based authentication protocol to authenticate the client to the network (server).

18.8.3 Confidentiality

Recall from Section 18.5.1 that the fundamental WEP loophole stems from using a stream cipher in an environment susceptible to packet loss. To work around this problem, WEP designers changed the encryption key for each packet. To generate the per-packet encryption key, the IV was concatenated with the preshared key. Since the preshared key is fixed, it is the IV which is used to make each per-packet key unique. There were multiple problems with this approach.

First, the IV size at 24 bits was too short. At 24 bits there were only 16,777,216 values before a duplicate IV value was used. Second, WEP did not specify how to select an IV for each packet.[23] Third, WEP did not even make it mandatory to vary the IV on a per-packet basis—realize that this meant WEP explicitly allowed reuse of per-packet keys. Fourth, there was no mechanism to ensure that the IV was unique on a per station basis. This made the IV collision space shared between stations, thus making a collision even more likely. Finally, simply concatenating the IV with the preshared key to obtain a per-packet key is cryptographically unsecure, making WEP vulnerable to the FMS attack. The FMS attack exploits the fact that the WEP creates the perpacket key by simply

---

[22]Secure since it protects the identity of the client during the authentication process.

[23]Implementations vary from a sequential increase starting from zero to generating a random IV for each packet.

concatenating the IV with the master-key. Since the first 24 bits of each per-packet key is the IV (which is available in plain text to an eavesdropper),[24] the probability of using weak keys[25] is very high.

First off, TKIP doubles the IV size from 24 bits to 48 bits. This results in increasing the time to key collision from a few hours to a few hundred years. Actually, the IV is increased from 24 bits to 56 bits by requiring the insertion of 32 bits between the existing WEP IV and the start of the encrypted data in the WEP packet format. However, only 48 bits of the IV are used since eight bits are reserved for discarding some known (and some yet to be discovered) weak keys.

Simply increasing the IV length will, however, not work with the existing WEP hardware accelerators. Remember that existing WEP hardware accelerators expect a 24-bit IV as an input to concatenate with a preshared key (40/104-bit) in order to generate the per-packet key (64/128-bit). This hardware cannot be upgraded to deal with a 48-bit IV and generate an 88/156-bit key. The approach, therefore, is to use per-packet key mixing as explained in Section 18.8.1. Using the per-packet key mixing function (much more complicated) instead of simply concatenating the IV to the master key to generate the per-packet key increases the effective IV size (and hence improves on WEP security) while still being compatible with existing WEP hardware.

18.8.4 Integrity

WEP used CRC-32 as an integrity check. The problem with this protocol was that it was linear. As we saw in Section 18.6, this is not a cryptographically secure integrity protocol. It does however have the merit that it is not computation intensive. What TKIP aims to do is to specify an integrity protocol which is cryptographically secure and yet not computation intensive so that it can be used on existing WEP hardware which has very little computation power. The problem is that most well known protocols used for calculating a message integrity check (MIC) have lots of multiplication operations and multiplication operations are computation intensive.

---

[24]Remember that each WEP packet carries the IV in plain text format prepended to the encrypted packet.

[25]Use of certain key values leads to a situation where the first few bytes of the output are not all that random. Such keys are known as weak keys. The simplest example is a key value of 0.

Therefore, TKIP uses a new MIC protocol—MICHAEL—which uses no multiplication operations and relies instead on shift and add operations. Since these operations require much less computation, they can be implemented on existing 802.11 hardware equipment without affecting performance.

Note that the MIC value is added to the MPDU in addition to the ICV which results from the CRC32. It is also important to realize that MICHAEL is a compromise. It does well to improve upon the linear CRC-32 integrity protocol proposed in WEP while still operating within the constraints of the limited computation power. However, it is in no way as cryptographically secure as the other standardized MIC protocols like MD5 or SHA-1. The TKIP designers knew this and hence built in countermeasures to handle cases where MICHAEL might be compromised. If a TKIP implementation detects two failed forgeries (two packets where the calculated MIC does not match the attached MIC) in one second, the STA assumes that it is under attack and as a countermeasure deletes its keys, disassociates, waits for a minute and then re-associates. Even though this may sound a little harsh, since it disrupts communication, it does avoid forgery attacks.

Another enhancement that TKIP makes in IV selection and use is to use the IV as a sequence counter. Recall that WEP did not specify how to generate a per-packet IV.[26] TKIP explicitly requires that each STA start using an IV with a value of 0 and increment the value by one for each packet that it transmits during its session[27] lifetime. This is the reason the IV can also be used as a TKIP Sequence Counter (TSC). The advantage of using the IV as a TSC is to avoid the replay attack to which WEP was susceptible.

TKIP achieves replay protection by using a unique IV with each packet that it transmits during a session. This means that in a session, each new packet coming from a certain MAC address would have a unique number.[28] If each packet from Alice had a unique number, Bob could tell when Eve was replaying old messages. WEP does not have replay protection since it cannot use the IV as a counter. Why? Because WEP does not specify how to change IV from one packet to another and as we saw earlier, it does not even specify that you need to.

---

[26]In fact, WEP did not even specify that the IV had to be changed on a per-packet basis.

[27]An 802.11 session refers to the association between a STA and an AP.

[28]At least for 900 years—that's when the IV rolls over.

18.8.5 The Overall Picture: Confidentiality + Integrity

The overall picture of providing confidentiality and message integrity in TKIP is shown in Figure 18.10e.

18.8.6 How Does WPA Fix WEP Loopholes?

In Section 18.7 we summarized the loopholes of WEP. At the beginning of Section 18.8 we said that WPA/TKIP was designed to close these loopholes while still being able to work with existing WEP hardware. In this section, Table 18.1 summarizes what WPA/TKIP achieves and how.

18.9 WPA2 (802.11i)

Recall from Section 18.8 that Wi-Fi protected access (WPA) was specified by the Wi-Fi alliance with the primary aim of enhancing the security of existing 802.11 networks by

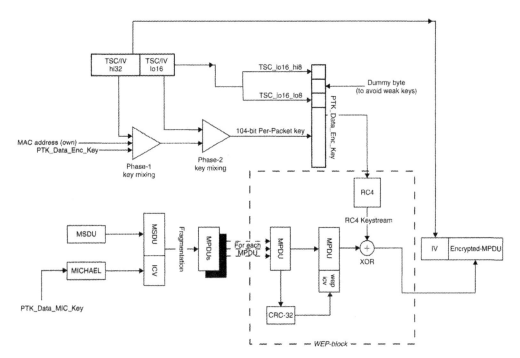

Figure 18.10e: TKIP—The complete picture

Table 18.1: WEP loopholes and WPA fixes

WEP	WPA
Relies on preshared (out-of-band) key establishment mechanisms. Usually leads to manual configuration of keys and to key sharing among STAs in a BSS (often ESS).	Recommends 802.1X for authentication and key-establishment in enterprise deployments. Also supports preshared key establishment like WEP.
Uses a synchronous stream cipher which is unsuitable for the wireless medium.	Same as WEP.
Generates per-packet key by concatenating the IV directly to the master/preshared key thus exposing the base-key/master-key to attacks like FMS.	Solves this problem by (a) introducing the concept of PTK in the key hierarchy and (b) by using a key mixing function instead of simple concatenation to generate per-packet keys. This reduces the exposure of the master key.
Static master key + Small size of IV + Method of per-packet key generation $\longrightarrow$ Extremely limited key space.	Increases the IV size to 56 bits and uses only 48 of these bits reserving 8-bits to discard weak keys. Also, use of PTK which are generated afresh for each new session increases the effective key space.
Changing the IV with each packet is optional $\longrightarrow$ key reuse highly probable.	Explicitly specifies that both the transmitter and the receiver initialize the IV to zero whenever a new set of PTK is established* and then increment it by one for each packet it sends.
Linear algorithm (CRC-32) used for message integrity $\longrightarrow$ Weak integrity protection.	Replaces the integrity check algorithm to use MICHAEL which is nonlinear. Also, specifies countermeasures for the case where MICHAEL may be violated.
ICV does not protect the integrity of the 802.11 header $\longrightarrow$ Susceptible to Redirection Attacks.	Extends the ICV computation to include the MAC source and destination address to protect against Redirection attacks.
No protection against replay attacks.	The use of IV as a sequence number provides replay protection.
No support for a STA to authenticate the network.	Use of 802.1X in enterprise deployments allows for this.

*This usually happens every time the STA associates with an AP.

designing a solution which could be deployed with a simple software (firmware) upgrade and without the need for a hardware upgrade. In other words, WPA was a stepping stone to the final solution which was being designed by the IEEE 802.11i task group. This security proposal was referred to as the Robust Security Network (RSN) and also came to be known as the 802.11i security solution. The Wi-Fi alliance integrated this solution in their proposal and called it WPA2. We look at this security proposal in this section.

18.9.1 Key Establishment

WPA was a prestandard subset of IEEE 802.11i. It adopted the key-establishment, key hierarchy and authentication recommendations of 802.11i almost completely. Since WPA2 and 802.11i standard are the same, the key-establishment process and the key hierarchy architecture in WPA and WPA2 are almost identical. There is one significant difference though. In WPA2, the same key can be used for the encryption and integrity protection of data. Therefore, there is one less key needed in WPA2. For a detailed explanation of how the key hierarchy is established see Section 18.8.1.

18.9.2 Authentication

Just like key establishment and key hierarchy, WPA had also adopted the authentication architecture specified in 802.11i completely. Therefore, the authentication architecture in WPA and WPA2 is identical. For a detailed explanation of the authentication architecture, see Section 18.8.2.

18.9.3 Confidentiality

In this section we look at the confidentiality mechanism of WPA2 (802.11i). Recall that the encryption algorithm used in WEP was RC4, a stream cipher. Some of the primary weaknesses in WEP stemmed from using a stream cipher in an environment where it was difficult to provide lossless synchronous transmission. It was for this reason that Task Group i specified the use of a block encryption algorithm when redesigning 802.11 security. Since AES was (and still is) considered the most secure block cipher, it was an obvious choice. This was a major security enhancement since the encryption algorithm lies at the heart of providing confidentiality.

Recall from earlier discussions that specifying an encryption algorithm is not enough for providing system security. What is also needed is to specify a mode of operation.

To provide confidentiality in 802.11i, AES is used in the counter mode. Counter mode actually uses a block cipher as a stream cipher, thus combining the security of a block cipher with the ease of use of a stream cipher. Figure 18.11 shows how AES counter mode works.

Using the counter mode requires a counter. The counter starts at an arbitrary but predetermined value and is incremented in a specified fashion. The simplest counter operation, for example, would start the counter with an initial value of 1 and increment it sequentially by 1 for each block. Most implementations however, derive the initial value of the counter from a nonce value that changes for each successive message. The AES cipher is then used to encrypt the counter to produce a key stream. When the original message arrives, it is broken up into 128-bit blocks and each block is XORed with the corresponding 128 bits of the generated key stream to produce the ciphertext.

Mathematically, the encryption process can be represented as $C_i = M_i (+) E_k(i)$ where i is the counter. The security of the system lies in the counter. As long as the counter value is never repeated with the same key, the system is secure. In WPA2, this is achieved by using a fresh key for every session (see Section 18.8.1).

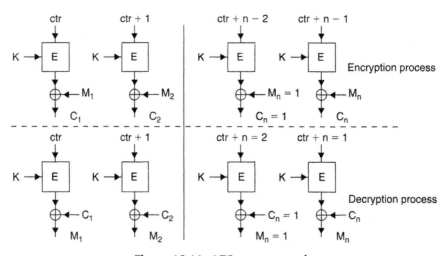

Figure 18.11: AES counter mode

To summarize, the salient features of AES in counter mode are as follows:

1. It allows a block cipher to be operated as a stream cipher.

2. The use of counter mode makes the generated key stream independent of the message, thus allowing the key stream to be generated before the message arrives.

3. Since the protocol by itself does not create any interdependency between the encryption of the various blocks in a message, the various blocks of the message can be encrypted in parallel if the hardware has a bank of AES encryption engines.

4. Since the decryption process is exactly the same as encryption,[29] each device only needs to implement the AES encryption block.

5. Since the counter mode does not require that the message be broken up into an exact number of blocks, the length of the encrypted text can be exactly the same as the length of the plain text message.

Note that the AES counter mode provides only for the confidentiality of the message and not the message integrity. We see how AES is used for providing the message integrity in the next section. Also, since the encryption and integrity protection processes are very closely tied together in WPA2/802.11i, we look at the overall picture after we have discussed the integrity process.

18.9.4 Integrity

To achieve message integrity, Task Group i extended the counter mode to include a Cipher Block Chaining (CBC)-MAC operation. This is what explains the name of the protocol: AES-CCMP where CCMP stands for Counter-mode CBC-MAC protocol. It is reproduced here in Figure 18.12 where the black boxes represent the encryption protocol (AES in our case).

As shown in the figure, CBC-MAC XORs a plaintext block with the previous cipher block before encrypting it. This ensures that any change made to any cipher text (for example by a malicious intruder) block changes the decrypted output of the last block and

---

[29]XORing the same value twice leads back to the original value.

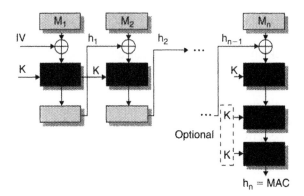

Figure 18.12: AES CBC-MAC

hence changes the residue. CBC-MAC is an established technique for message integrity. What Task Group i did was to combine the counter mode of operation with the CBC-MAC integrity protocol to create the CCMP.

18.9.5 The Overall Picture: Confidentiality + Integrity

Since a single process is used to achieve integrity and confidentiality, the same key can be used for the encryption and integrity protection of data. It is for this reason that there is one less key needed in WPA2. The complete process which combines the counter mode encryption and CBC-MAC integrity works as follows.

In WPA2, the PTK is 384 bits long. Of this, the most significant 256 bits form the EAPoL MIC key and EAPoL encryption key. The least significant 128 bits form the data key. This data key is used for both encryption and integrity protection of the data. Before the integrity protection or the encryption process starts, a CCMP header is added to the 802.11 packet before transmission. The CCMP header is eight bytes in size. Of these eight bytes, six bytes are used for carrying the Packet Number (PN) which is needed for the other (remote) end to decrypt the packet and to verify the integrity of the packet. One byte is reserved for future use and the remaining byte contains the key ID. Note that the CCMP header is prepended to the payload of the packet and is not encrypted since the remote end needs to know the PN before it starts the decryption or the verification process. The PN is a per-packet sequence number which is incremented for each packet processed.

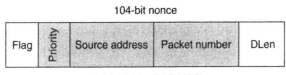

Figure 18.13: IV for AES CBC-MAC

The integrity protection starts with the generation of an Initialization Vector (IV) for the CBC-MAC process. This IV is created by the concatenation of the following entities: flag, priority, source MAC address, a PN and DLen as shown in Figure 18.13.

The flag field has a fixed value of 01011001. The priority field is reserved for future use. The source MAC address is self explanatory and the packet number (PN) is as we discussed above. Finally, the last entity DLen indicates the data length of the plaintext. Note that the total length of the IV is 128 bits and the priority, source address and the packet number fields together also form the 104-bit nonce (shaded portion of Figure 18.13) which is required in the encryption process. The 128-bit IV forms the first block which is needed to start the CBC-MAC process described in Section 18.9.4. The CBC-MAC computation is done over the 802.11 header and the MPDU payload. This means that this integrity protection scheme also protects the source and the destination MAC address, the quality of service (QoS) traffic class and the data length. Integrity protecting the header along with the MPDU payload protects against replay attacks. Note that the CBC-MAC process requires an exact number of blocks to operate on. If the length of the plaintext data cannot be divided into an exact number of blocks, the plaintext data needs to be padded for the purposes of MIC computation.

Once the MAC has been calculated and appended to the MPDU, it is now ready for encryption. It is important to re-emphasize that only the data-part and the MAC part of the packet are encrypted whereas the 802.11 header and the CCMP header are not encrypted. From Section 18.9.3, we know that the AES-counter mode encryption process requires a key and a counter. The key is derived from the PTK as we discussed. The counter is created by the concatenation of the following entities: Flag, Priority, Source MAC address, a packet number (PN) and Ctr as shown in Figure 18.14.

Comparing Figure 18.14 with Figure 18.13, we see that the IV for the integrity process and the counter for the encryption process are identical except for the last sixteen bits.

Whereas the IV has the last sixteen bits as the length of the plaintext, the counter has the last sixteen bits as Ctr. It is this Ctr which makes the counter a real "counter." The value of Ctr starts at one and counts up as the counter mode proceeds. Since the Ctr value is sixteen bits, this allows for up to 2^{16} (65,536) blocks of data in a MPDU. Given that AES uses 128-bit blocks, this means that an MPDU can be as long as 2^{23}, which is much

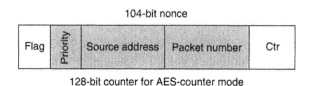

Figure 18.14: Counter for AES counter mode

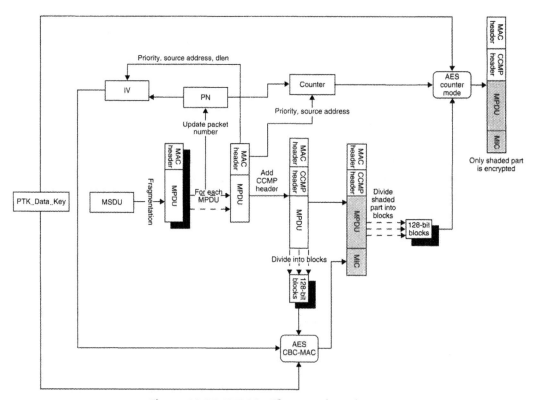

Figure 18.15: WPA2—The complete picture

more than what 802.11 allows, so the encryption process does not impose any additional restrictions on the length of the MPDU.

Even though CCMP succeeds in combining the encryption and integrity protocol in one process, it does so at some cost. First, the encryption of the various message blocks can no longer be carried out in parallel since CBC-MAC requires the output of the previous block to calculate the MAC for the current block. This slows down the protocol. Second, CBC-MAC requires the message to be broken into an exact number of blocks. This means that if the message cannot be broken into an exact number of blocks, we need to add padding bytes to it to do so. The padding technique has raised some security concerns among some cryptographers but no concrete deficiencies/attacks have been found against this protocol.

The details of the overall CCMP are shown in Figure 18.15 and finally Table 18.2 compares the WEP, WPA and WPA2 security architectures:

Table 18.2: Comparison of WEP, WPA and WPA2 security architectures

WEP	WPA	WPA2
Relies on preshared a.k.a. out-of-band key establishment mechanisms. Usually leads to manual configuration of keys and to key sharing among STAs in a BSS (often ESS).	Recommends 802.1X for authentication and key-establishment in enterprise deployments. Also supports preshared key establishment like WEP.	Same as WPA.
Uses a synchronous stream cipher which is unsuitable for the wireless medium.	Same as WEP.	Replaces a stream cipher (RC4) with a strong block cipher (AES).
Generates per-packet key by concatenating the IV directly to the master/preshared key thus exposing the basekey/master-key to attacks like FMS.	Solves this problem (a) by introducing the concept of PTK in the key hierarchy and (b) by using a key mixing function instead of simple concatenation to generate per-packet keys. This reduces the exposure of the master key.	Same as WPA.

(Continued)

Table 18.2: (Continued)

WEP	WPA	WPA2
Static master key + Small size of IV + Method of per-packet key generation → Extremely limited key space.	Increases the IV size to 56 bits and uses only 48 of these bits reserving 8-bits to discard weak keys. Also, use of PTK which are generated afresh for each new session increases the effective key space.	Same as WPA.
Changing the IV with each packet is optional → key-reuse highly probable.	Explicitly specifies that both the transmitter and the receiver initialize the IV to zero whenever a new set of PTK is established* and then increment it by one for each packet it sends.	Same as WPA.
Linear algorithm (CRC-32) used for message integrity → Weak integrity protection.	Replaces the integrity check algorithm to use MICHAEL which is nonlinear. Also, specifies countermeasures for the case where MICHAEL may be violated.	Provides for stronger integrity protection using AES-based CCMP.
ICV does not protect the integrity of the 802.11 header → Susceptible to Redirection Attacks.	Extends the ICV computation to include the MAC source and destination address to protect against Redirection attacks.	Same as WPA.
No protection against replay attacks.	The use of IV as a sequence number provides replay protection.	Same as WPA.
No support for a STA to authenticate the network.	No explicit attempt to solve this problem but the recommended use of 802.1X could be used by the STA to authenticate the network.	Same as WPA.

*This usually happens every time the STA associates with an AP.

CHAPTER 19

Security in Wireless Ad Hoc Networks

Praphul Chandra

19.1 Introduction

The term *ad hoc networks* refers to networks formed on-the-fly (ad hoc), in other words on an as-needed basis.

The term *ad hoc wireless networks* refers to those ad hoc networks which use a wireless medium for communication. Since a wired ad hoc network would be synonymous with a LAN, the term *ad hoc networks* almost always means ad hoc wireless networks and the two terms are used interchangeably throughout this text.

The term *mobile ad hoc networks* (MANETs) refers to ad hoc networks in which the nodes forming the ad hoc network are mobile. Most ad hoc networks allow their nodes to be mobile and are therefore MANETs.

In other words, these networks are formed on an as-needed basis and do not require the existence of any infrastructure. This property makes ad hoc wireless networks suitable for use in various scenarios like disaster recovery, enemy battlefields or in areas where user density is too sparse or too rare to justify the deployment of network infrastructure economically. Figure 19.1 shows some examples of ad hoc wireless networks.

The scenarios and examples shown in Figure 19.1 present a small subset of scenarios where ad hoc networks may be useful. An ad hoc network may operate in a standalone fashion or may be connected to a larger network like the Internet. Since ad hoc networks have such varied areas of use, it is instructive to classify them based on certain features.

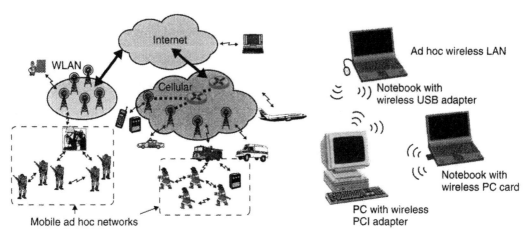

Figure 19.1: Examples of ad hoc networks

First, ad hoc networks may be classified on the basis of their geographical coverage. Therefore we have ad hoc personal area networks (PANs), ad hoc local area networks (LANs) and ad hoc wide area networks (WANs).

Second, ad hoc networks may be classified based on whether or not nodes in the network are capable of acting as routers. To understand this classification, realize that the wireless networks that we have looked at till now in this book always used the fixed, static, wired infrastructure for routing. In TWNs, call routing was achieved by dedicated routing switches of the PSTN and the core GSM network (which consisted of MSCs and GMSCs). Furthermore, since both the PSTN and the core GSM network are wired networks which are static (that is, their network topology almost never changes), it is relatively easy to proactively distribute the network topology information to the routing switches. This in turn allows each routing switch to precompute and maintain routes to other switches, thus facilitating routing. Similarly, in wireless local area networks (WLANs), packet routing is achieved by using Layer 2 switches and IP routers. Again, since these routing devices are connected by a wired infrastructure and are static, it is relatively easy to proactively distribute the network topology information to the routers and switches.

Ad hoc networks have two major limitations: (a) there are no dedicated routing devices (since there is no infrastructure available) and (b) the network topology may change rapidly and unpredictably as nodes move. In the absence of any routing infrastructure, the nodes forming the ad hoc networks themselves have to act as routers. A MANET may therefore be defined as an autonomous system of mobile routers (and associated hosts) connected by wireless links—the union of which forms an arbitrary graph. Given the central importance of routing in ad hoc networks, it is not surprising that routing forms a basis for classifying ad hoc networks into two groups: single-hop ad hoc networks and multihop ad hoc networks. Single-hop ad hoc networks are ad hoc networks where nodes do not act as routers and therefore communication is possible only between nodes which are within each other's radio frequency (RF) range. On the other hand, multihop ad hoc networks are ad hoc networks where nodes are willing to act as routers and route or forward the traffic of other nodes.

If you look at the basis for two classifications closely that—is, the geographical coverage and the routing capability of nodes—the two classifications are not completely orthogonal. Ad hoc PANs are more likely to be single hop ad hoc networks since nodes would be close enough to be within each other's RF range. On the other hand, ad hoc LANs and ad hoc WANs are more likely to require nodes to have routing capability and therefore form multihop networks.

Multihop ad hoc networks and their security is an active area of research as of the writing of this chapter. In the next section we look at some of the emerging security concepts (and areas of active research) in multihop ad hoc networks. Single-hop ad hoc networks are now being used commercially and one of the most popular single-hop ad hoc wireless standard is Bluetooth. We look more closely at Bluetooth and how it implements security in the next section.

19.2 Bluetooth

One of the most popular ad hoc standards today is Bluetooth. Some of the salient features of Bluetooth are as follows:

- Wireless ad hoc networking technology.
- Operates in the unlicensed 2.4 GHz frequency range.
- Geographical coverage limited to personal area networks (PAN).
- Point-to-point and point-to-multipoint links.
- Supports synchronous and asynchronous traffic.

- Concentrates on single-hop networks.

- Frequency hopping spread spectrum (FHSS) with Gaussian frequency shift keying (GFSK) modulation at the physical layer.

- Low power and low cost given important consideration.

- Adopted as the IEEE 802.15.1 standard for physical layer (PHY) and media access control (MAC) layers.

The Bluetooth standard limits its scope by dealing only with single-hop ad hoc networks with limited geographical coverage (PAN). In the previous sections we saw that multihop ad hoc networks present a unique set of challenges which are still an active area of research. The Bluetooth standard brings ad hoc networks to the commercial forefront by concentrating on single-hop PAN ad hoc networks. Removing the multihop feature from ad hoc networks makes things a lot simpler.

The Bluetooth Special Interest Group (SIG) was founded in 1998 with the aim of developing Bluetooth as a short-range wireless inter-connectivity standard.[1] In other words, Bluetooth deals with ad hoc networks whose geographical coverage is limited to PAN. Typical applications of Bluetooth today include connecting a wireless headset with its cell phone, interconnecting the various components (keyboard, mouse, monitor, and so on) of a PC, and so on.

Before we get into the details of Bluetooth and its security, it is important to emphasize that Bluetooth is by no means the only ad hoc network standard. Another popular ad hoc standard is 802.11 in its IBSS mode.

Since Bluetooth networks have been so commercially successful, we briefly look at Bluetooth security in this section.

19.2.1 Bluetooth Basics

A typical Bluetooth network, called the piconet, is shown in Figure 19.2. Each piconet has one master and can have up to seven slaves.[2] Therefore, there can be at most eight

---

[1]The Bluetooth standard is also being accepted as the IEEE 802.15 standard.

[2]To be precise, a piconet has one master and up to seven active slaves. There is no limit on the number of slaves in a piconet which are in "park" or "hold" state. This distinction is irrelevant from a security perspective however.

Piconet and scatternet configurations

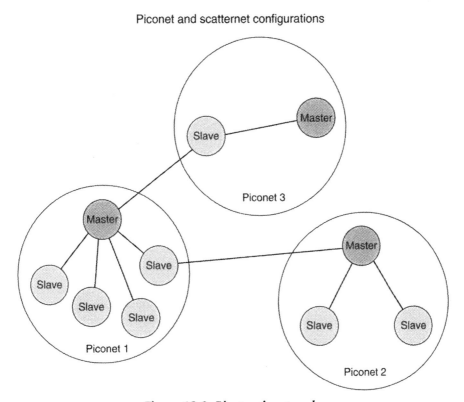

Figure 19.2: Bluetooth networks

devices in a piconet. A slave can communicate only with the master and a master can obviously communicate with any of the slaves. If two slaves wish to communicate with each other, the master should relay this traffic. In effect, therefore, we have a logical star topology in a piconet, with the master device at the center.

Comparing the piconet to a 802.11 network, the piconet is the equivalent of a BSS (though with a much smaller geographical coverage), the master device is the equivalent of the AP (except that it is not connected to any distribution system) and the slave devices are the equivalent of the Stations (STAs).

A Bluetooth device may participate in more than one piconet simultaneously, as shown in Figure 19.3. In such a scenario, it is possible for the devices in two piconets to communicate with each other by having the common node act as the bridge and

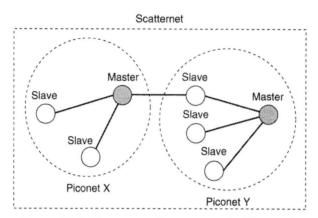

Figure 19.3: Piconets and scatternets in Bluetooth

relay the inter-piconet traffic. The two piconets are now joined together and form a scatternet. Even though scatternets are theoretically possible, they are rare in commercial deployments since they pose tough practical problems like routing and timing issues. The Bluetooth standard concentrates mostly on single-hop piconets and we limit our discussion to piconet security. Scatternets (and their security) are an active area of research and involve a lot of the security issues that we discussed earlier.

19.2.2 Security Modes

Just like IEEE 802.11 standard, the Bluetooth standard also defines Layer 1 and Layer 2 of the OSI stack to achieve communication in single-hop personal-area ad hoc networks. However, by their very nature, ad hoc networks (Bluetooth) are a much less controlled environment than WLANs (802.11). This, combined with the fact that the Bluetooth standard may be used by a wide range of applications in many different ways, makes interoperability a much bigger challenge in Bluetooth networks. To ease the problem of interoperability, the Bluetooth SIG defined application profiles. A profile defines an unambiguous description of the communication interface between two Bluetooth devices or one particular service or application.

There are basic profiles which define the fundamental procedures for Bluetooth connection and there are special profiles defined for distinct services and applications. New profiles can be built using existing profiles, thus allowing for a hierarchical profile structure as shown in Figure 19.4.

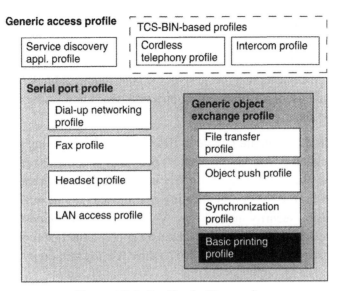

Figure 19.4: Profiles in Bluetooth

Each service or application selects the appropriate profile depending on its needs, and since each application may have different security requirements, each profile may define different security modes. The most fundamental profile is the Generic Access Profile (GAP) which defines the generic procedure related to the discovery of the Bluetooth devices and link management aspects of connection between them. The GAP defines three basic security modes of a Bluetooth device.

Before we discuss the different security modes, it is important to keep a few things in mind. First, the security mechanisms (authentication and encryption) specified by the Bluetooth standard are implemented at the link layer (Layer 2). This means that the scope of Bluetooth security is the Layer 2 level link between two nodes separated by a single hop. To be explicit, Bluetooth security does not deal with end-to-end security[3] and does not deal with application layer security. If such security mechanisms are required, they have to be arranged for outside the scope of the Bluetooth standard.

Second, all Bluetooth devices must implement an authentication procedure: that is a requirement.[4] Bluetooth devices may or may not implement encryption procedures: that

---

[3]The source and destination nodes may be more than one hop away as in a scatternet.
[4]On the other hand, implementing encryption procedures is optional.

is optional. However, just because a device implements or supports authentication and/or encryption, does not mean that this device would use these security features in a connection. What security features are used for a Bluetooth connection depends on the security modes of the master and the slave in the connection. Table 19.1 summarizes what security features are used in a Bluetooth connection depending on the security mode of the master and the slave.

Security mode 1 is the unsecured mode in Bluetooth. A device using this mode does not demand authentication or encryption from its peer at connection establishment. This means that a device in security mode 1 offers its services to all devices which wish to connect to it. However, if the communicating peer device (say B) wishes to authenticate a node which is using security mode 1 (say A), then A must respond to the challenge that B sends since A as a Bluetooth device must support authentication. Similarly, if B wishes to use encryption on its link with A, then A must turn on its encryption if it supports it.

On the other end of the spectrum is security mode 3 which is the always-on security mode in Bluetooth. A device which is in security mode 3 shall always initiate an authentication

Table 19.1: Security features of Bluetooth connection

Security Modes Master Slave	1	2	3
1	**Authentication** = No; **Encryption** = No;	**Authentication** = if the master *app.* demands it; **Encryption** = if the master *app.* demands it;	**Authentication** = Yes; **Encryption** = if the master *policy* demands it.
2	**Authentication** = if the slave *app.* demands it. **Encryption** = if the slave *app.* demands it.	**Authentication** = if the master or the slave *app.* demands it; **Encryption** = if the master or the slave *app.* demands it;	**Authentication** = Yes; **Encryption** = if the master *policy* or if the slave *app.* demands it.
3	**Authentication** = Yes. **Encryption** = if the slave *policy* demands it;	**Authentication** = Yes. **Encryption** = if the slave *policy* or if the master *app.* demands it.	**Authentication** = Yes. **Encryption** = if the master or the slave *policy* demands it;

procedure. This means that this device will not communicate with a device unless it can authenticate it. The encryption part in security mode 3 is not as simple. Recall that the Bluetooth standard does not require every device to implement encryption. If a device in security mode 3 (say A) is trying to communicate with a peer which implements encryption, then this is not an issue since the link can be secured. However, if A wishes to communicate with a peer which does not implement encryption (say, B), then we have a problem. How this situation is to be handled is left to higher layers (in other words, the security policy manager). The security manager may decide not to communicate with B, or it may decide to communicate with B without using encryption.

Security mode 2 lies in between modes 1 and 3 and has been designed to offer applications flexibility. Whether or not authentication and/or encryption is used, is left to the decision of the security policy manager. This is achieved by relinquishing control of security use to a higher layer security manager. In other words, security mode 2 works by using service level-enabled security. This mode is most useful in scenarios where multiple applications (with different security requirements) may run on a single Bluetooth device. The security manager can then co-ordinate what security policy to use depending on the application which is running.

In summary, whether or not authentication and/or encryption is used in a Bluetooth communication link depends on a lot of factors: the security needs of the application, the security capabilities of the devices, the security mode of the device and the security (access) policies. The security manager considers all these factors when deciding if (and at which stage of connection establishment) the device should start using the security procedures.

19.2.3 Key Establishment

Key establishment is probably the most complex part of Bluetooth security. A large part of this complexity is in the key hierarchy because of the large number of keys involved in the Bluetooth security process. Figure 19.5 shows a classification of most of the keys involved in the Bluetooth security process. The key hierarchy in Bluetooth varies a little bit depending on whether we are dealing with unicast communication (between two devices) or broadcast communication. For the rest of this section we assume that we are dealing with unicast communication. Finally, in Section 19.2.3.7 we will point out how broadcast communication key hierarchy is different from unicast key hierarchy.

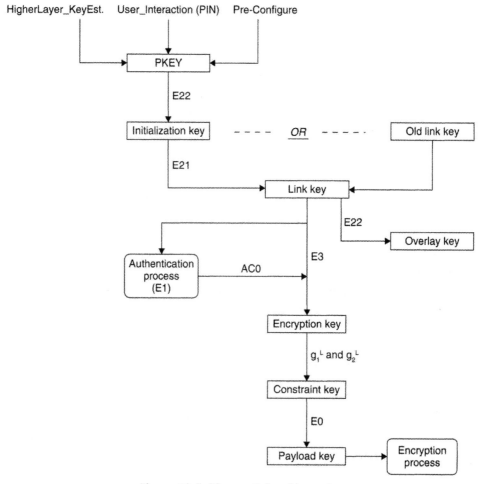

Figure 19.5: Bluetooth key hierarchy

19.2.3.1 Pass Key

At the top level is the Pass Key (PKEY). The PKEY is basically the shared secret between the two communicating devices. There are two types of PKEYs: variable PKEYs and fixed PKEYs. Variable PKEYs refer to PKEYs that can be chosen at the time of the "pairing" (the process by which two Bluetooth devices establish a shared secret that they can use for securing communication). This is usually achieved by prompting the user to enter a PKEY during the pairing procedure. Obviously, users of both devices should enter the same PKEY. On the other hand, fixed PKEYs refer to PKEYs that are preconfigured into the Bluetooth device. Again, both the communicating devices should be preconfigured with the same PKEY. Even though variable PKEYs are more secure (since the PKEY can be changed on a per-pairing basis), both variable and fixed PKEYs serve specific purposes. Consider for example, a scenario where users in a conference room wish to form a Bluetooth network using their laptops. Such a scenario is well-suited for using variable PKEYs, since each device has user interaction capabilities. On the other hand, consider the Bluetooth network between the headset and its cell phone. The Bluetooth headset must use a fixed PKEY since there is no[5] user interaction capability on the headset. The Bluetooth standard also allows the use of higher layer key-establishment protocols to generate the PKEY and pass it on to the Bluetooth stack.

Since the PKEY can come from one of many sources, instead of specifying the exact length of the PKEY, the Bluetooth standard specifies that the PKEY can be as long as 128 bits. This allows for devices prompting the user for a PKEY to enter a much smaller PKEY (or a PIN) thus making user interaction a little more convenient. However, using a smaller PKEY has other drawbacks. As we will see in the next few sections, the PKEY is the starting point for establishing the Link Key, which in turn forms the basis of all security in Bluetooth. To be precise, the PKEY is the shared secret between the two communicating endpoints that ensures the Link Key is known only to the communicating end-points. The use of a smaller PKEY means that an attack like the dictionary attack becomes much easier to launch. In this context, a dictionary attack involves calculating all the link keys that can be derived from all possible PKEYs. This list of link keys is maintained in a table and can then be used against <plaintext, ciphertext pairs>[6] to determine which link key is being used.

19.2.3.2 Initialization Key

The next level in the hierarchy is the Initialization Key (IK or K_{INIT}). The K_{INIT} is a short-lived temporary key that is used (and exists only) during the pairing process when two Bluetooth devices start communicating for the first time. The K_{INIT} is derived using the E_{22} algorithm and three inputs: PKEY, IN_RAND and L_{PKEY}. The PKEY is the pass-key that we just talked about and L_{PKEY} is the length of this pass-key in bytes. Finally, the

---

[5]Or rather hardly any.

[6]For example <AU_RAND, SRES> pairs.

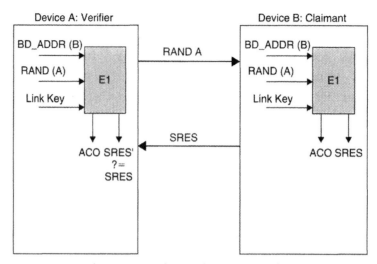

Figure 19.6: Bluetooth authentication

IN_RAND is a 128-bit random number generated by the device. The process of deriving the IK from these inputs is as follows:

$$PKEY' = PKEY + padding\ bits.$$

$$L_{PKEY}' = min(L_{PKEY} + 6, 16).$$

$$K_{INIT} = E_{22}(PKEY', IN\_RAND, L_{PKEY}')$$

The need for padding bits stems from the flexibility of allowing the PKEY to be as long as 128 bits but not specifying the exact length of the PKEY. The padding shown above is needed to ensure that PKEY is exactly 128 bits and these padding bits come from the claimant's[7] BD_ADDR (the 48-bit MAC address).

19.2.3.3 Link Key

The next level in the key hierarchy is the Link Key (LK). The link key is the shared secret established between the two Bluetooth devices when the pairing sequence ends.

---

[7]Bluetooth uses the terms claimant and verifier to refer to the two Bluetooth devices which are involved in the communication. Claimant refers to the device which is claiming an identity and verifier refers to the device which verifies this claim.

There are two types of link keys specified by the Bluetooth standard: the unit key and the combination key. The use of unit key has now been deprecated because of the security loopholes that result from its use. We therefore concentrate only on the combination link keys here. The terms *link key* and *combination key* from now on are used interchangeably throughout the text to refer to the same key. The combination/link key is derived either from the existing link key (if such a key exists) or the initialization key, K_{INIT} (if no link key exists between the two devices). We will see how the combination link key is generated but note that we talk about generating a link key from the existing link key! What does this mean?

Consider two devices which establish a combination/link key and communicate securely for a while. When the communication is terminated, what should the two devices do with the link key which they used for this session? One approach is to discard these keys. This approach requires that every time these two devices want to establish a Bluetooth session, they must generate the link key from scratch. Even though cryptographically this is a more pure approach, in the interest of efficiency Bluetooth allows devices to store the link keys that they generate in nonvolatile memory and reuse this link key for future communications with the same device. In other words, unlike the initialization key, the link key is a semi-permanent key. Therefore, each device may[8] maintain a database of <remote_device_address, link_key> pairs for the link keys it wishes to reuse. Such an approach is specially suited to devices which repeatedly connect to a small fixed set of devices for example the Bluetooth headset which usually connects to its cell phone.

Note that just because devices can reuse link keys does not mean that they should never change the link key. Periodically changing the link key is a recommended practice since, as we know by now, the more a key is used or exposed, the more it is vulnerable to compromise. It is in these scenarios that a link key is used to generate another link key. In other scenarios where two devices do not already have a link key shared between them, the K_{INIT} is used to generate the link key. In summary, the end of the pairing process in Bluetooth should lead to the establishment of a link key which the two devices can use for securing their communication. This link key (combination key) can come from three sources:

1. Use an existing link key that the two devices had established previously.

2. Use an existing link key to generate a fresh link key.

3. Use the initialization key, K_{INIT}, to generate a link key.

---

[8]The decision whether or not to store a particular link key in the database may be left to the user.

The process of generating the link key is as follows. We start with either the existing link key or the K_{INIT} depending on the context. Let us call this starting key K_{START}.[9] The most important property to remember about K_{START} is that it is shared secretly between the two communicating devices; that is, it is known only to the two communicating devices. Now, each of the communicating devices (say A and B) generate a private key using the E_{21} algorithm, their BD_ADDR and a self-generated random number (LK_RAND). Therefore,

$$K_A = E_{21}(LK\_RAND_A, BD\_ADDR_A)$$
$$K_B = E_{21}(LK\_RAND_B, BD\_ADDR_B)$$

The combination link key is simply the XOR of K_A and K_B. However, the process of establishing the link key is not yet complete since K_A is available only at A and K_B is available only at B. What is needed is a way for B to be able to generate K_A and for A to be able to generate K_B.

To be able to generate KB, A needs to know LK_RAND_B and BD_ADDR_B. Arguably, A already knows BD_ADDR_B since it is going to communicate with it. So, what is needed is a secure way to get LK_ADDR_B from B to A. Here is where K_{START} comes in. B XORs LK_ADDR_B with K_{START} and sends it to A. Since K_{START} is a shared secret known only to A and B, we are assured that this transmission of LK_ADDR_B is secure. Once A gets to know LK_ADDR_B, A can generate K_B too. Following the exact same procedure, B can generate K_A too. At this point both A and B have calculated K_A and K_B and can therefore easily calculate the combination link key as K_A XOR K_B.

Note from Figure 19.7 that after the establishment of the new combination link key K_{START} is deleted by both endpoints since the new combination link key should be used from now on.

19.2.3.4 Encryption Key

The combination link key is used in the authentication process (see Section 19.2.4) as we will see in the next section. The link key is also used for generating the ciphering key (CK or K_C) which is the next key in the key hierarchy. The K_C is derived from the link key using the E3 algorithm as follows:-

$$K_C = E_3(K, EN\_RAND, COF)$$

---

[9]Therefore $K_{START} = LK$ or K_{INIT}.

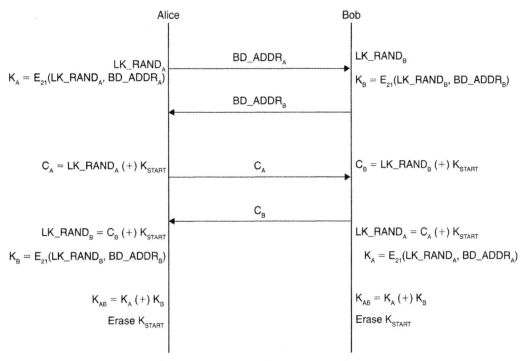

Figure 19.7: Bluetooth key establishment

The link key is denoted by K. EN_RAND refers to a 128-bit random number and COF refers to a 96-bit Ciphering Offset. The value of COF is equal to the value of the Authentication Ciphering Offset (ACO), which is derived from the authentication process. (See Section 19.2.4 for details on this.)

19.2.3.5 Constraint Key

The next step in the key hierarchy is the constraint key (K_C'), a.k.a. the constraint encryption key. The reason for the existence of this key is export restrictions. Many countries impose restrictions on the export of encryption hardware. Specifically, hardware which is capable of encrypting above certain key strengths is not exportable. For this purpose, Bluetooth put in a key strength constraining mechanism that reduces the 128-bit K_C to a 128-bit K_C' whose effective key length (key strength) can be any value less than 128 bits. The derivation of K_C' from K_C is achieved in hardware using linear feedback and feed forward registers and can be given as:

$$K_C'(x) = g_2^L(x) \{K_C[\bmod g_1^L(x)]\}$$

where L is the desired effective length and g_1 and g_2 are two polynomials specified by the Bluetooth standard.

19.2.3.6 Payload Key

Finally, the payload key (P_K) is the actual key that is used to encrypt (decrypt) Bluetooth packets. Therefore, P_K is derived from the Constraint Key K_C' using the E_0 algorithm as follows:

$$K_P = E_0(K_C', CK\_VAL, BD\_ADDR, EN\_RAND)$$

where BD_ADDR is the 48-bit Bluetooth (read MAC) address of the device, EN_RAND is a 128-bit random number and CK_VAL is the 26 bits of the current clock value (bits 1–26 of the master's native clock).

19.2.3.7 Broadcast Key Hierarchy

All our discussion regarding Bluetooth key hierarchy has assumed that the Bluetooth communication is between two devices (a master and a slave). However, the Bluetooth standard also supports broadcast communication where a master may broadcast data to all its slaves. Realize that a broadcast transmission is different from multiple unicast transmissions. In a broadcast transmission, the master sends out a message to a special broadcast address and all slaves in the piconet accept this message. In the latter case, a master sends out multiple copies of a message to each slave individually. In this section, we talk about broadcast in the former sense.

Recall from Section 19.2.3.3 that the (combination) link key in unicast communication in a piconet was generated using the addresses and random numbers from both endpoints involved in the communication. In broadcast communication, this becomes a dilemma since communication is not between two endpoints. Therefore, the link key is replaced by the use of a master key (K_{master}). The master key is derived independently by the master without involving any of the slaves. This is done using the E_{22} algorithm as follows:

$$K_{master} = E_{22}(LK\_RAND1, LK\_RAND2, 16)$$

Since the master key is derived only at the master, we need a way to communicate the K_{master} to all the slaves in the piconet securely. This is done using the overlay key, K_{ovl}. The overlay key is derived from the current link key as follows:

$$K_{overlay} = E_{22}(K, RAND3, 16)$$

Since the master and each of the slaves in a Bluetooth piconet share a link key, the overlay key can be securely established between the master and each of the slaves. This overlay key can then be used for conveying the master key to each of the slaves. Finally, the master key is used for securing broadcast communication in a piconet.

Note that unlike the link key which is a semi-permanent key (stored in nonvolatile memory for future use), the master key is a temporary key which is never stored in nonvolatile memory (and never re-used).

19.2.3.8 The Algorithms

The key hierarchy that we have discussed in the previous sections uses five algorithms: E_0, E_1, E_3, E_{21} and E_{22}. Of these five algorithms, four (E_1, E_3, E_{21} and E_{22}) are based on a block cipher and one on a stream cipher (E_0). The discussion of E_0 as a stream cipher is postponed to Section 19.2.5 where we see how Bluetooth packets are encrypted. To understand the use of the block cipher for four of these algorithms, recall that all keys in Bluetooth have a length of 128 bits. Using a 128-bit block cipher in key derivation means that we can feed one key directly as input into the block cipher to generate the key of the next level in the hierarchy. All these four algorithms use the same underlying block cipher: SAFER+ .

At the time the Bluetooth standard was being ratified, the National Institute of Standards and Technology (NIST) was considering contenders for the Advanced Encryption Standard (AES). SAFER+ was one of the strong contenders for AES which had been very thoroughly cryptanalyzed. Bluetooth therefore chose SAFER+ as the block cipher to be used for the implementation of E_1, E_3, E_{21} and E_{22}. The details of this algorithm are not discussed here for reasons of brevity and the interested reader is referred to the Bluetooth standards.

19.2.4 Authentication

The authentication process always involves two endpoints: the claimant (which claims a certain identity) and the verifier (which wishes to verify that the claimant is actually the identity it is claming to be). In the Bluetooth piconet context, the roles of claimant and verifier are orthogonal to the rule of the master and the slave. In other words, either the master or the slave can act as the verifier. Who is the verifier depends on higher layers. The application or the user who wishes to ensure the identity of the remote end (and who therefore starts the authentication process) takes on the role of the verifier. For mutual authentication, both end-points take on the role of the verifier one at a time. Figure 19.8 shows a mutual authentication process in Bluetooth.

The authentication process in Bluetooth is basically a challenge-response protocol which is carried out using the existing link key. Consider a device (claimant) which initiates communication with A claiming to be B. A wishes to verify that the claimant is in fact B. To verify this A sends the claimant a random number, AU_RAND. On receiving this, the claimant is expected to send back a signed response SRES calculated using the E_1 algorithm as follows:

$$SRES = E_1(K, AU\_RAND, BD\_ADDR_B).$$

Since the AU_RAND and BD_ADDR_B may easily be known publicly, the security lies in K, the link key. The underlying assumption of the Bluetooth authentication process is that the link key is known only to the two endpoints which established it (see Section 19.2.3.3 on how this is achieved). Since only B knows the correct K which it established with A,

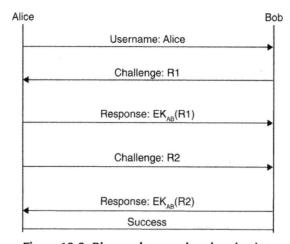

Figure 19.8: Bluetooth mutual authentication

only B would be able to generate the correct SRES. So, all A needs to verify the identity of the claimant is to ensure that the response sent back by the claimant is equal to SRES.

As Figure 19.8 shows, mutual authentication can also be carried out in Bluetooth with B now taking on the role of a verifier. B sends out a challenge to A and carries on the authentication process to verify the identity of A. The E_1 algorithm used for generating the SRES in the authentication process also produces a 96-bit Authentication Ciphering Offset (ACO) as an output. As we saw in Section 19.2.3.4, this ACO is used for generating the ciphering key, K_C. It is this ACO which "links" the authentication process to the rest of the session. In other words, the ACO serves to link the security context established by the authentication process to the rest of the session.

Note that if mutual authentication is desired, first A acts as the verifier and B as the claimant. Next, the roles are swapped with B acting as the verifier and A as the claimant. Therefore, in mutual authentication, there would be two ACOs that would be produced: one from each authentication process. In such a scenario, the standard specifies that the ACO from the latter authentication process be used for generating the encryption key. Therefore, the security context is linked to the last authentication process that is carried out.

19.2.5 Confidentiality

As we discussed in Section 19.2.3.6, the payload key P_K, which is used for encrypting outgoing messages is derived using the E_0 algorithm. The E_0 algorithm is basically a stream cipher which generates a key stream. This key stream is then XORed with the plaintext of the messages to create the ciphertext. The design of the E_0 stream cipher is not based on any existing stream cipher but is a proprietary algorithm specified by the Bluetooth standard.

Figure 19.9 shows the encryption process in Bluetooth networks. From Section 19.2.3.6, we know that P_K is derived from the constraint key K_C' using the E_0 algorithm as follows:

$$P_K = E_0(KC', CK\_VAL, BD\_ADDR, EN\_RAND)$$

where BD_ADDR is the 48-bit Bluetooth address of the device, EN_RAND is a 128-bit random number and CK_VAL is the 26 bits of the current clock value (bits 1–26 of the master's native clock). Next, this P_K is fed into the key stream generator. This key stream generator then produces a key stream which is XORed with the plaintext to produce

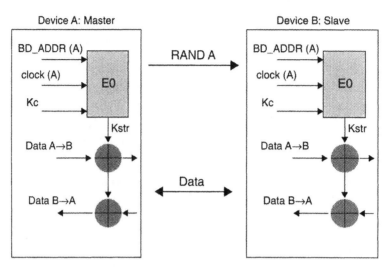

Figure 19.9: Bluetooth encryption

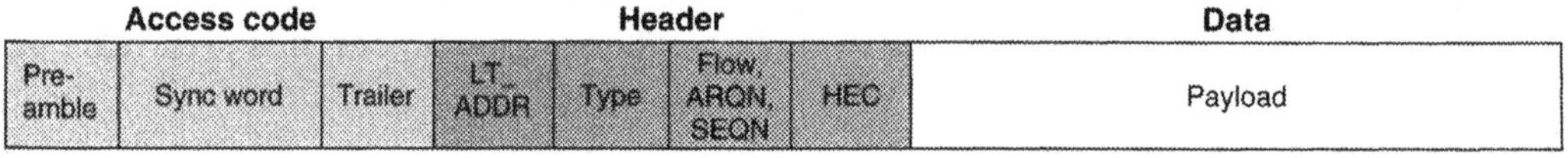

Access code			Header				Data
Pre-amble	Sync word	Trailer	LT_ADDR	Type	Flow, ARQN, SEQN	HEC	Payload

Figure 19.10: Bluetooth packet format

the ciphertext. There are a few important things to note about the encryption process in Bluetooth.

First, not all bits of the Bluetooth packet are encrypted. Figure 19.10 shows the format of a Bluetooth packet. It consists of an access code followed by a header and finally the payload. The access code is derived from the BD_ADDR of the master of the piconet and since every piconet has a unique master, the access code uniquely identifies a piconet. The access code is therefore used by the devices in a piconet to determine if a packet is for another piconet, in which case the packet is discarded.

The access code also defines where a slot boundary lies and is therefore also used by the slaves in a piconet to synchronize their clocks to the master's clock. It is therefore not a surprise that the access code in Bluetooth packets is not encrypted. Next, the header in the Bluetooth packet is also not encrypted. The reason for this is also pretty obvious when you consider that the header contains the address of the destination device. This information is obviously needed by all devices in the piconet to determine whether or not a particular packet is intended for it or not. Therefore, the bottom line is that only the payload in the Bluetooth packet is encrypted.

Second, as shown in Figure 19.9, the CRC is appended to the packet before it is encrypted. In other words, the CRC along with the payload is also encrypted.

Third, realize that using a stream cipher in a wireless medium is a security loophole as discussed in Section 18.5.1 (What's Wrong with WEP?). Just like WEP tries to overcome the drawbacks of using a stream cipher by changing the key for each packet, Bluetooth uses the same approach: the P_K is derived for each Bluetooth packet.[10] However, unlike WEP where the per-packet key is calculated simply by prepending the IV with the master key, the derivation of the per-packet key in Bluetooth is much more cryptographically involved, thus making Bluetooth encryption more secure than WEP. Let us take a closer look at Figure 19.9.

As Figure 19.9 shows, the encryption process in Bluetooth can be separated into three distinct blocks. The first block consists of deriving the payload key, P_K. We saw at the beginning of this section how this is achieved. This P_K feeds into the second block and acts as the initialization seed to a key stream generator. In other words, P_K is used to initialize the key stream generator. The key stream generated from the second block feeds into the third block which is nothing but a simple XOR operation between this key stream and the payload[11] of the Bluetooth packet.

Finally, realize that to change the P_K on a per-packet basis, we need a variable which changes on a per-packet basis. One of the inputs required for generating the payload key, P_K is CK_VAL, the lower 26 bits of the master clock. Since the lowest bit of the master clock (and hence the CK_VAL) changes every 625 microseconds, this means the value of the P_K can be derived afresh every 625 microseconds. However initializing the key stream generator with P_K takes some time. This is where guard space comes in. The Bluetooth standard specifies a guard space between the end of the payload in one packet and the start of the next packet. This guard space must be at least 259 microseconds, which is

---

[10]Multislot packets do not require a change of the payload key when passing a slot boundary within a packet.

[11]Along with the CRC.

Figure 19.11: Bluetooth stream cipher periodicity

enough time for the key stream generator to be initialized. The overall timing sequence is as shown in Figure 19.11 where "running the stream cipher" and "initializing key stream generator" slots alternate.

19.2.6 Integrity Protection

The Bluetooth standard relies on a cyclic redundancy check (CRC) to provide message integrity. Recall from Section 18.6 that WEP also used CRC for providing message integrity. We discussed the loopholes of this approach in Section 18.7 and concluded that using a linear noncryptographic integrity check mechanism like CRC leaves a lot to be desired as far as integrity protection is concerned. We also saw that just encryption does not automatically provide integrity. In other words, just because a message is encrypted does not mean that it can't be modified in-transit without the receiver's knowledge. The bottom line is that by choosing CRC, Bluetooth fails to provide any real[12] integrity protection.

19.2.7 Enhancements

As we said at the beginning of this chapter, ad hoc networking technology is still in its nascent stage. There are many issues to be resolved and security is such an issue. The Bluetooth standard is also expected to evolve as ad hoc networking technology evolves.

---

[12]That is, any strong enough.

CHAPTER 20

Implementing Basic Wireless Security

Chris Hurley

20.1 Introduction

This chapter looks at how to configure the basic security settings on the four most popular access points currently deployed for wireless home use (Linksys WAP11, BEFW11SR, and WRT54G, and the D-Link DI-624). It covers the practical steps for implementing wireless security for home systems.

There are four very simple steps that are sufficient for most home users:

1. Use a unique Service Set Identifier (SSID).

2. Disable SSID broadcast.

3. Enable Wired Equivalent Privacy (WEP) encryption.

4. Filter access by Media Access Control (MAC) address.

20.2 Enabling Security Features on a Linksys WAP11 802.11b Access Point

The Linksys Wireless Access Point (WAP) 11 802.11b access point was one of the first access points deployed by a large number of people. The WAP11 was the wireless access point many users purchased that already had a home network with a router set up. The WAP11 requires a separate router in order to allow access to any non-wireless devices, to include Internet access. This section details the minimum steps you should take to configure the WAP11 securely. All of the steps outlined in this section should be done from a computer that is connected to your wired network. This is because as you make

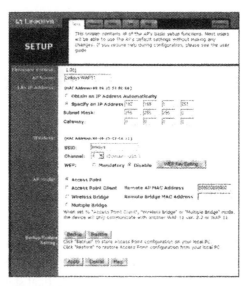

Figure 20.1: The Linksys WAP11 initial setup screen

changes the access point will need to reset. When the access point resets, you will likely lose your wireless connection momentarily.

20.2.1 Setting a Unique SSID

The first step you need to take is to set a unique SSID for your access point. When you log in to your access point, by default there is no username assigned to the WAP11 and the password is *admin*. This brings up the initial setup screen (Figure 20.1).

In the **AP Name** field, choose a name for your access point. This is NOT the SSID, but it is prudent to set this to a unique name. Many access points are named after the address of the owner or the company name, making them easier for an attacker to target. Next, replace "Linksys" with a unique SSID. This can be anything that you want, though it is not a good idea to use your address, phone number, social security number, or any other information that identifies you specifically. Figure 20.2 depicts the setup screen after a unique AP Name and SSID have been chosen.

Once these are set, click **Apply** and your changes are stored. Setting a unique SSID is a good first step for practicing security, but without taking additional steps it is relatively useless. A unique SSID will contain a combination of upper- and lowercase letters,

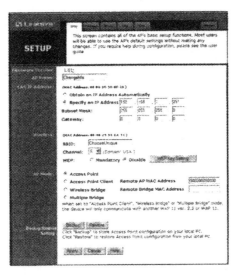

Figure 20.2: A unique AP name and SSID are set

numbers, and special characters. Additionally, using nonprintable characters will cause some WarDriving applications to crash or identify the SSID incorrectly.

20.2.2 Disabling SSID Broadcast

After you have set a unique SSID on your access point, the next step is to disable the SSID broadcast. By default, access points transmit a beacon to let wireless users know that they are there. Active scanners such as NetStumbler rely on this beacon to find access points. By disabling the SSID broadcast, you have effectively placed your access point in *stealth*, also known as *cloaked*, mode.

To disable the SSID broadcast, first click the **Advanced** tab on the initial setup screen. This will take you to the screen shown in Figure 20.3.

Next, click the **Wireless** tab to bring up the advanced wireless settings, as shown in Figure 20.4.

Select the **Disable** radio button, and then click **Apply** to save your settings. Passive scanners, such as Kismet and AirSnort, have the ability to detect cloaked access points, but disabling SSID broadcast is one more step toward an effective overall security posture.

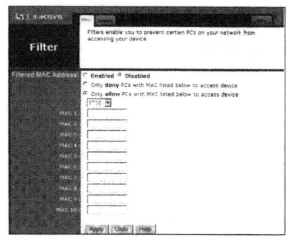

Figure 20.3: The advanced settings

Figure 20.4: The advanced wireless settings

20.2.3 Enabling WEP

After you have set a unique SSID and disabled SSID broadcast, the next step is to enable WEP encryption. The flaws associated with WEP have been widely publicized and discussed. Inasmuch, because it is possible to crack WEP keys, you should not rely on WEP alone, but use WEP as a part of your overall security posture.

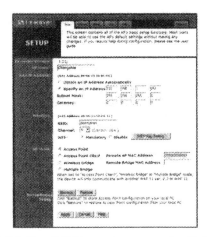

Figure 20.5: Making WEP mandatory on the Linksys WAP11

Although WEP is flawed, actually cracking the WEP key is not a simple process on a home network, for two primary reasons.

1. The amount of traffic that must be generated in order to successfully crack the WEP key.

2. Vendors have taken steps to eliminate or reduce the number of Weak Initialization Vectors (IVs) that are transmitted.

It usually requires at least 1200 Weak IVs to be collected before a WEP key is cracked. On a home network it can take days, weeks, or even months to generate enough traffic to capture that many Weak IVs. It is highly unlikely that an attacker will invest that amount of time into attacking a simple home network, especially when there are so many networks that don't have WEP enabled.

Many vendors have also developed firmware upgrades that reduce or eliminate the number of Weak IVs that are generated. This further increases the amount of time it takes to successfully crack a WEP key.

To enable WEP on the Linksys WAP11, on the main setup screen select the **Mandatory** radio button, as shown in Figure 20.5.

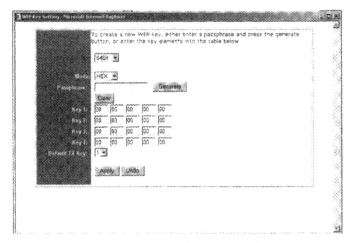

Figure 20.6: The WEP key setting window

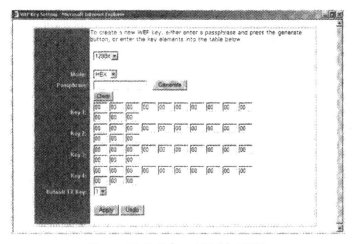

Figure 20.7: Select 128-bit WEP

Next, click **WEP Key Setting** to open the WEP Key Setting window, as shown in Figure 20.6.

In the **WEP Key Setting** window, change **64Bit** to **128Bit** in the drop-down box (as shown in Figure 20.7) to require 128-bit WEP keys. As the number of bits implies, 128-bit WEP provides a stronger, harder to crack key than 64-bit.

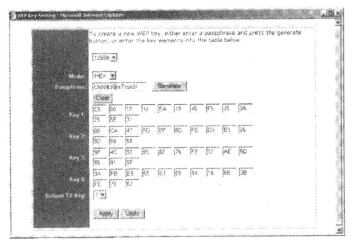

Figure 20.8: Generating WEP keys

Leave the Mode set to HEX and choose a strong passphrase to generate your keys. A strong passphrase consists of a combination of upper- and lowercase letters, numbers, and special characters. Once chosen, enter your passphrase in the **Passphrase** text box and click **Generate**. This will create four WEP keys. See Figure 20.8.

Since four keys are generated, you need to decide which one your client should use. Set the **Default TX Key** to the number (1–4) that you want to use on your network. Once you have generated your WEP keys and chosen the key to transmit, click **Apply** to save your settings.

Information on configuring your client software is provided later in the "Configuring Security Features on Wireless Clients" section of this chapter.

20.3 Filtering by Media Access Control (MAC) Address

Once you have set a unique SSID, disabled SSID broadcast, and required the use of WEP encryption, you should take at least one more step: filtering by Media Access Control (MAC) address. To enable MAC address filtering on the Linksys WAP11, from the main setup screen click the Advanced tab to display the **Advanced** wireless settings (Figure 20.9). Click the **Enabled** radio button to enable MAC address filtering. Next, select the radio

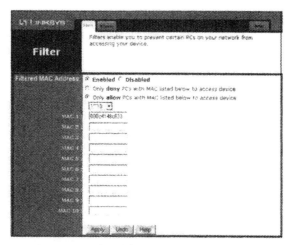

Figure 20.9: Enable MAC address filtering

button for **Only Allow PCs With MAC Listed Below To Access Device**. Finally, in the text boxes labeled **MAC 1** thru **MAC 10**, list the MAC addresses of any wireless clients that are allowed to access your wireless network. Click **Apply** to save and enable your settings. Instructions for finding the MAC address of your card are provided in the "Tools & Traps" sidebar in this chapter.

Tools & Traps...

Finding the Media Access Control (MAC) Address of Wireless Cards

Finding the Media Access Control (MAC) address of your wireless card is a simple process. The easiest way is to look at the back of the card itself, as every wireless card has a label on the back that provides information like: the FCC ID, the encryption standard that is supported, and the MAC address. Figure 20.10 shows this label.

Windows 2000 and XP users can find the MAC address using the *ipconfig/all* command.

Figure 20.10: Finding the MAC address on the card label

Figure 20.11: Using *ipconfig/all* in Windows to determine the MAC address

The Physical Address highlighted in Figure 20.11 is the MAC address for the wireless card. Linux users can determine the MAC address of their card using *ifconfig* <*interface*> command.

The highlighted HWaddr shown in Figure 20.12 is the MAC address of the wireless card.

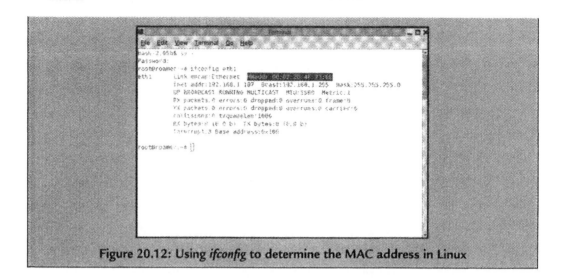

Figure 20.12: Using *ifconfig* to determine the MAC address in Linux

20.4 Enabling Security Features on a Linksys BEFW11 SR 802.11b Access Point/Router

The most popular 802.11b access point currently deployed is the Linksys BEFW11SR Access Point/Router. The BEFW11SR gained popularity because it combined both a wireless access point and a wired router into one device. This section details the minimum steps you should take to configure the BEFW11SR securely. All of the steps outlined in this section should be done from a computer that is connected to your wired network. This is because, as you make changes, the access point will need to reset. When the access point resets, you will likely lose your wireless connection momentarily.

20.4.1 Setting a Unique SSID

The first step you need to take is to assign a unique SSID to your access point. When you log in, you'll find, by default, that BEFW11SR has no username and the password is *admin*. This brings up the initial setup screen (see Figure 20.13).

Next, enter a unique **SSID** in the SSID textbox, as shown in Figure 20.14.

Finally, click the **Apply** button to save your new SSID.

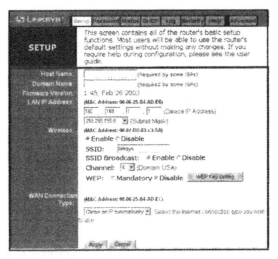

Figure 20.13: The Linksys BEFW11SR initial setup screen

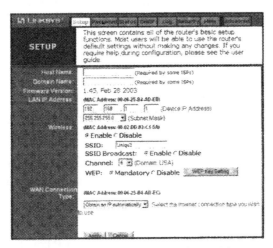

Figure 20.14: Entering a unique SSID

20.4.2 Disabling SSID Broadcast

After you have assigned a unique SSID to your BEFW11SR, you should disable SSID broadcast. Disabling SSID broadcast prevents active wireless scanners like NetStumbler from finding your access point.

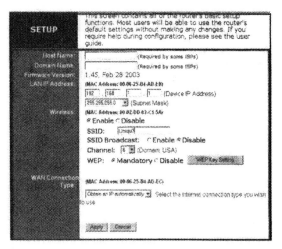

Figure 20.15: Disable SSID broadcast on the Linksys BEFW11SR

From the initial setup screen, select the **Disable** radio button next to SSID Broadcast. Next, click the **Apply** button to save your settings and disable SSID broadcast, as shown in Figure 20.15.

20.4.3 Enabling WEP

Once you have set a unique SSID and disabled SSID broadcast, the next security measure you should enable is WEP encryption. On the initial setup screen, select the **Mandatory** radio button next to WEP, as shown in Figure 20.16.

Next, click the WEP Key Setting button to open the WEP Key Setting window. Enter a strong passphrase in the **Passphrase** textbox and click the **Generate** button. This generates a WEP key for your network, as shown in Figure 20.17.

Next, click the **Apply** button to save your settings. You also need to configure any wireless clients that use the BEFWllSR with the same WEP key. Instructions for setting up wireless clients are presented in the "Configuring Security Features on Wireless Clients" section later in this chapter.

20.4.4 Filtering by Media Access Control (MAC) Address

After you have set a unique SSID, disabled SSID broadcast, and enabled WEP, your next step is to filter by Media Access Control (MAC) address. Filtering by MAC address

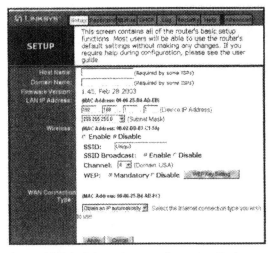

Figure 20.16: Select the mandatory radio button

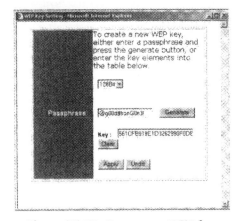

Figure 20.17: Generate a WEP key

allows only wireless clients with pre-determined MAC addresses to connect to the access point.

From the initial setup screen, click the **Advanced** tab. This brings up the advanced settings window. Next, click the **Wireless** tab to open the advanced wireless settings window, as shown in Figure 20.18.

Select the **Enable** radio button next to Station MAC Filter, as shown in Figure 20.19.

Figure 20.18: The advanced wireless settings window

Figure 20.19: Enable Station MAC filter

Then, click the **Edit MAC Filter Setting** button to open the **Wireless Group MAC Table** window, shown in Figure 20.20.

Next, enter the MAC addresses of the clients that are allowed to access your wireless network in the **MAC Address** textboxes and select the **Filter** checkbox next to each of them, as shown in Figure 20.21.

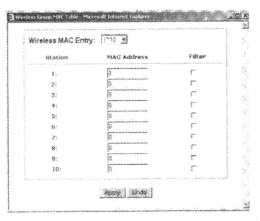

Figure 20.20: The wireless group MAC table window

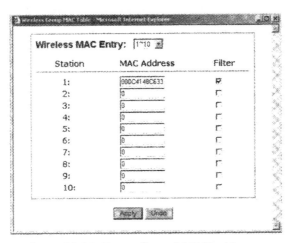

Figure 20.21: Enter allowed MAC addresses

Finally, click the **Apply** button on the **Wireless Group MAC Table** window, the
Advanced Wireless Settings window, and the initial setup screen to save your settings.

20.5 Enabling Security Features on a Linksys WRT54G 802.1 lb/g Access Point/Router

The most popular 802.11g device is the Linksys WR.T54G 802.11b/g Access Point/
Router. The WRT54G gained popularity in 2003 as 802.11g devices became more

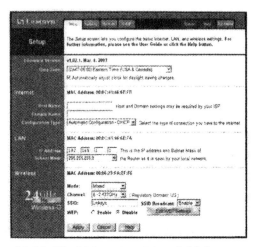

Figure 20.22: The Linksys WRT54G initial setup screen

common and affordable. 802.11g devices operate on the 2.4 GHz band, like 802.11b, but offer speeds up to 54 megabits per second (Mbps). Additionally, 802.11 g devices are compatible with 802.11 b cards. This section details the minimum steps you should take to configure the WRT54G securely. All of the steps outlined in this section should be done from a computer that is connected to your wired network.

20.5.1 Setting a Unique SSID

The first security measure you should enable on the Linksys WRT54G is setting a Unique SSID. When you log in to the WRT54G, by default, there is no username required and the password is *admin*. This brings up the initial setup screen (see Figure 20.22).

In the **SSID** textbox, enter a unique SSID, as shown in Figure 20.23.

Then, click the **Apply** button to save your settings.

20.5.2 Disabling SSID Broadcast

After you have set a unique SSID, disable the SSID broadcast. From the initial setup screen, select **Disable** from the **SSID Broadcast** drop-down box, as shown in Figure 20.24.

Then, click the **Apply** button to save your settings and disable SSID broadcast.

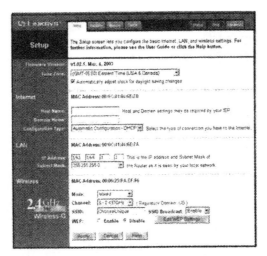

Figure 20.23: Setting a unique SSID on the WRT54G

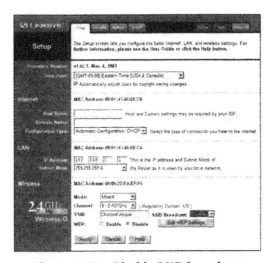

Figure 20.24: Disable SSID broadcast

20.5.3 Enabling WEP

Once you have set a unique SSID and disabled SSID broadcast, you need to require the use of 128-bit WEP encryption. From the initial setup screen, choose the **Enable** radio button next to WEP to require WEP encryption, as shown in Figure 20.25.

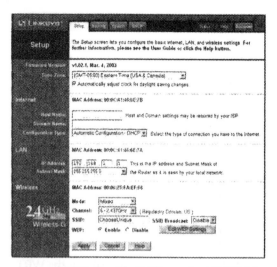

Figure 20.25: Enable WEP on the WRT54G

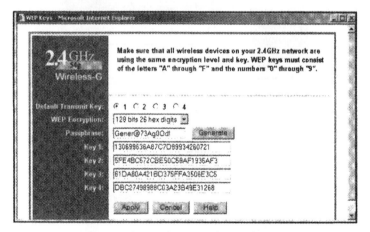

Figure 20.26: The WEP keys window

Next, click the **Edit WEP Settings** button, opening the **WEP Keys** window. Select **128 bits 26 hex digits** from the **WEP Encryption** drop-down box to require 128-bit WEE Type a strong passphrase in the **Passphrase** textbox. This is the passphrase that will be used as the basis for generating WEP keys. Click the **Generate** button to generate four WEP keys, as shown in Figure 20.26.

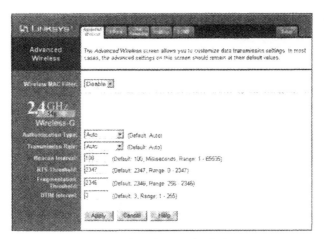

Figure 20.27: The advanced wireless screen

Next, select the key (1–4) that you will initially use by choosing the appropriate radio button next to **Default Transmit Key**. Finally, click **Apply** on both the **WEP Keys** window and the initial setup screen to save your settings.

20.5.4 Filtering by Media Access Control (MAC) Address

After you have set a unique SSID, disabled SSID broadcast, and enabled WEP encryption, you need to filter access to the WRT54G by MAC address.

First, from the initial setup screen click the **Advanced** tab to display the **Advanced Wireless** screen (see Figure 20.27).

Next, select **Enable** from the **Wireless MAC Filter** drop-down box. This will reveal the MAC filter options, as shown in Figure 20.28.

Choose the **Permit Only PCs Listed Below To Access The Wireless Network** radio button, and click the **Edit MAC Filter List** button to display the **MAC Address Filter List** window (see Figure 20.29).

Enter the MAC addresses of wireless clients that are allowed to access your wireless network in the provided textboxes and then click **Apply**, as shown in Figure 20.30.

Finally, click **Apply** in the **Advanced Wireless** window to save your settings and enable filtering by MAC address.

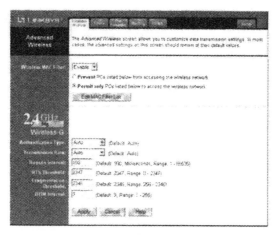

Figure 20.28: The wireless MAC filter options

Figure 20.29: The MAC address filter list window

20.6 Enabling Security Features on a D-Link DI-624 AirPlus 2.4 GHz Xtreme G Wireless Router with 4-Port Switch

Although Linksys has a sizable share of the home access point market, D-Link also has a large market share. D-Link products are also sold at most big computer and electronics stores such as Best Buy and CompUSA. This section details the steps you need to take to enable the security features on the D-Link 624 AirPlus 2.4 GHz Xtreme G Wireless

Figure 20.30: Enter allowed MAC addresses

Router with 4-Port Switch. The DI-624 is an 802.11g access point with a built-in router and switch similar in function to the Linksys WRT54G.

20.6.1 Setting a Unique SSID

The first security measure to enable on the D-Link DI-624 is setting a unique SSID.

First, you need to log into the access point. Point your browser to 192.168.0.1. Use the username **admin** with a blank password to access the initial setup screen (see Figure 20.31).

Next, click the **Wireless** button on the left side of the screen to bring up the **Wireless** Settings screen, as shown in Figure 20.32.

In the **SSID** textbox, enter a unique SSID, as shown in Figure 20.33, and click **Apply** to save and enable the new SSID.

20.6.2 Enabling Wired Equivalent Privacy

After you have set a unique SSID, you will need to enable 128-bit WEP encryption.

First, choose the **Enabled** radio button next to WEP, as shown in Figure 20.34.

Next, choose **128Bit** from the **WEP Encryption** drop-down box, as shown in Figure 20.35.

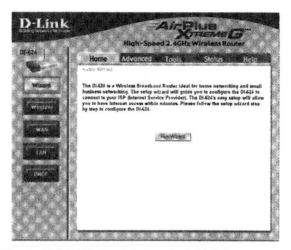

Figure 20.31: The D-Link DI-624 initial setup screen

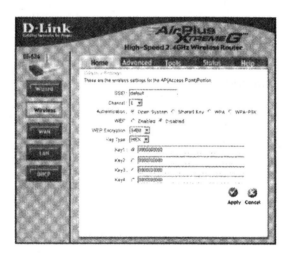

Figure 20.32: The wireless settings screen

Then, you will need to assign a 26-character hexadecimal number to at least Key1 (see Figure 20.36). A 26-digit hexadecimal number can contain the letters A–F and the numbers 0–9.

Finally, after you have assigned your WEP keys, click **Apply** to save your settings. Any wireless clients that connect to the DI-624 must be configured to use this WEP key.

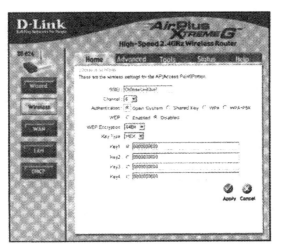

Figure 20.33: Set a unique SSID

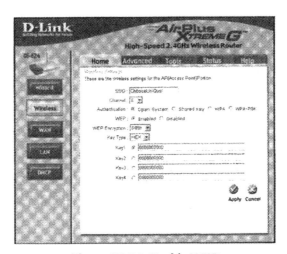

Figure 20.34: Enable WEP

20.6.3 Filtering by Media Access Control (MAC) Address

After you have set a unique SSID and enabled 128-bit WEP encryption, you should filter access to the wireless network by MAC address.

First, click the **Advanced** tab, as shown in Figure 20.37.

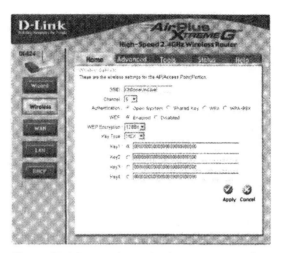

Figure 20.35: Require 128-Bit WEP encryption

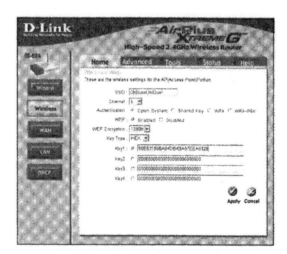

Figure 20.36: Assign WEP keys

Next, click the **Filters** button on the left side of the screen, as shown in Figure 20.38.

Then choose the **MAC Filters** radio button. This makes the MAC Filtering options visible, as shown in Figure 20.39.

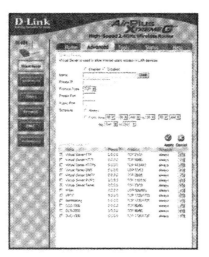

Figure 20.37: The advanced options screen

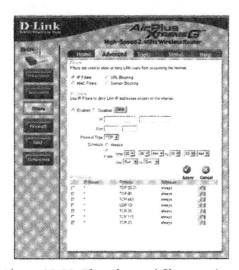

Figure 20.38: The advanced filters options

Finally, select the **Only Allow Computers With MAC Address Listed Below To Access The Network** radio button and enter the MAC address of each client card that is allowed to access the network. You must also enter a descriptive name, of your choice, for each client in the **Name** textbox (see Figure 20.40). Note that you must click **Apply** after each MAC address entered.

Figure 20.39: The MAC filtering options

Figure 20.40: Filter by MAC address

20.6.4 Disabling SSID Broadcast

After you have set a unique SSID, enabled 128-bit WEP, and filtered access by MAC address, you need to disable SSID broadcast.

From the **Advanced Features** screen, click the **Performance** button, as shown in Figure 20.41.

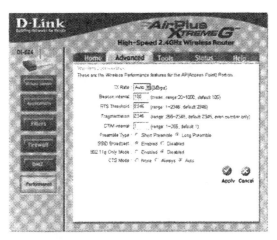

Figure 20.41: The advanced performance options

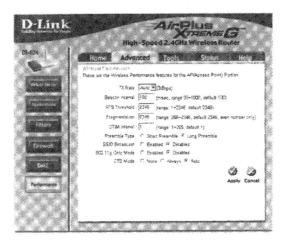

Figure 20.42: Disabling SSID broadcast

Select the **Disabled** radio button next to SSID Broadcast and click **Apply** to save your settings, as shown in Figure 20.42.

20.7 Configuring Security Features on Wireless Clients

After you have configured your access points to utilize security features, you will then need to configure each wireless client to work with your access points. This means

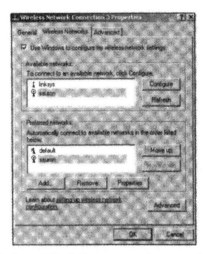

Figure 20.43: The Windows XP wireless network connection properties window

that each wireless client needs to be configured with the correct SSID and have the appropriate WEP key entered and selected.

20.7.1 Configuring Windows XP Clients

For Windows XP Clients, double-click the **Wireless Network Connections** icon on the taskbar to bring up the **Wireless Network Connection Properties** window (see Figure 20.43).

Next, click the **Add** button to add a new preferred network. Enter the **SSID** of your network in the SSID textbox, and un-check **The Key Is Provided For Me Automatically**. Then, enter the WEP key for your wireless network in the **Network Key** and **Confirm Network Key** textboxes, as shown in Figure 20.44.

Finally, choose the appropriate key index (1–4) and click **OK**. If your settings are correct, you will be connected to your access point.

20.7.2 Configuring Windows 2000 Clients

Unlike Windows XP, with Windows 2000 you will need to configure the wireless client software that came with your wireless card. The examples shown in this section are for

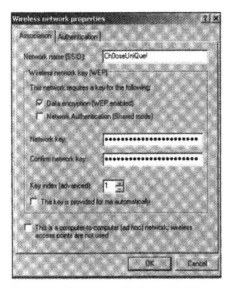

Figure 20.44: Configuring Windows XP clients for use

the ORINOCO client manager. The exact steps for other client managers may differ, but the basic idea is the same.

1. Create a wireless profile for your access point.

2. Enter the SSID of your access point.

3. Enter the WEP key for your wireless network.

To create a wireless profile for your access point, double-click the **client manager** icon on the taskbar to open the Client Manager. Choose **Actions | Add/Edit Configuration Profile** to open the **Add/Edit Configuration Profile** window, as shown in Figure 20.45. Select the radio button next to a blank profile and enter the name of the profile.

Next, click the **Edit Profile** button. On the **Basic** screen, enter the SSID of your access point in the **Network Name** textbox. Then, click the **Encryption** tab, as shown in Figure 20.46. Choose the **Use Hexadecimal** radio button and enter your WEP keys (1–4).

Finally, click **OK** to save your settings. Your wireless client is now configured for use with your access point.

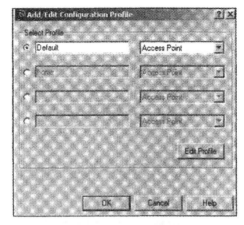

Figure 20.45: The configuration profiles

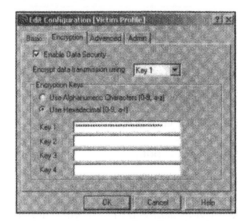

Figure 20.46: Entering the WEP key

20.7.3 Configuring Linux Clients

Configuring wireless clients for use with your network after security features have been enabled is a simple two-step process.

1. Edit the /etc/pcmcia/wirless.opts.

2. Restart PCMCIA services or reboot.

Figure 20.47: Editing the wireless.opts file

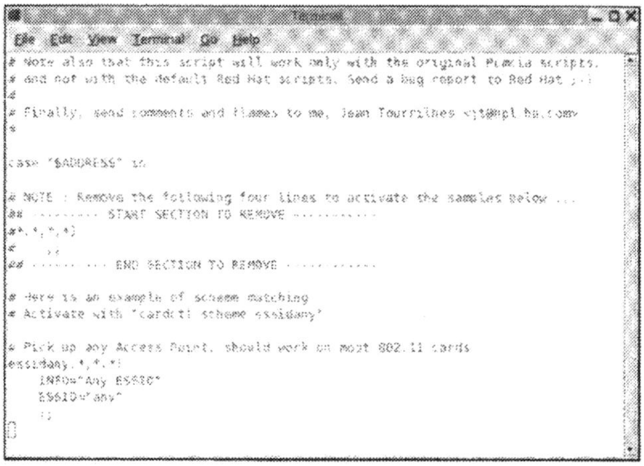

Figure 20.48: Commenting lines out of the wireless.opts file

First, you need to edit the /etc/pcmcia/wireless.opts file (see Figure 20.47).

Ensure that the four lines starting with the line that reads **START SECTION TO REMOVE** have been commented out by placing a pound sign (#) in front of each, as shown in Figure 20.48.

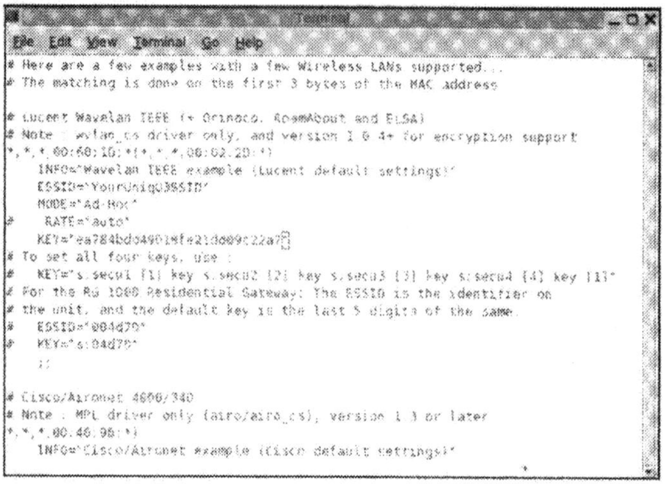

Figure 20.49: Entering the SSID and WEP key

Next, find the appropriate section for your wireless card. Enter the SSID your access point uses in the **ESSID** section. Then, enter the WEP key for your wireless network in the **KEY** section, as shown in Figure 20.49, and save your changes.

Before your settings take effect, you need to restart the Personal Computer Memory Card International Association (PCMCIA) services. The method for restarting PCMCIA services varies and depends on the Linux distribution you are using. In Slackware, Linux PCMCIA services are restarted by issuing the following command:

root@roamer: ~# /etc/rc.d/rc/pcmcia restart

If you are unsure how to restart PCMCIA services on your distribution, you can reboot. When the system restarts, your settings will take effect.

NOTES FROM THE UNDERGROUND...

Enabling Security Features on the Xbox

Many Xbox owners like to take advantage of the Xbox Live feature. Xbox Live allows gamers to connect their Xbox to the Internet and play selected games against online

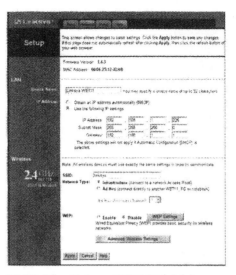

Figure 20.50: The Linksys WET 11 initial setup screen

opponents. Since the Xbox is often connected to a TV that isn't necessarily in the same room with most of the household computer equipment, wireless networking is a natural choice for this connection.

There are several wireless bridges (like the Linksys WET Wireless Ethernet Bridge) available that will connect the Xbox to a home network. These devices must be configured to use the wireless network's security features.

First, log in to the WET 11. By default, the WET 11 is configured to use the IP address 192.168.1.251 (see Figure 20.50).

Enter the SSID for your wireless network in the SSID textbox, and then select the Enable radio button next to WEP (see Figure 20.51).

Click WEP Settings button to open the Shared Keys window (see Figure 20.52). Select 128 bit 26 hex digits from the drop-down box and then enter the WEP keys that your wireless network uses. The WEP keys can be entered in either of two ways:

- Generate the keys using the same passphrase used to generate the keys on your access point.

- Manually enter the WEP keys that your access point uses.

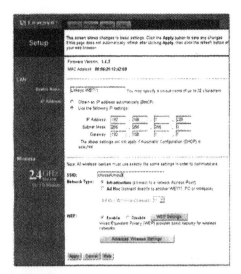

Figure 20.51: Set the SSID and enable WEP

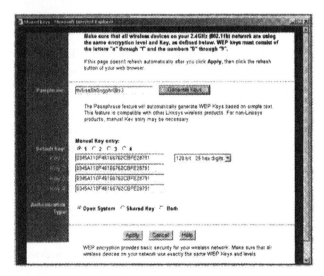

Figure 20.52: Enter the WEP keys

Next, click the Apply button on the Shared Keys window and the initial setup to save your settings. Finally, add the MAC address of your WET 11 to your allowed MAC address list on your access point.

20.8 Summary

Regardless of the particular brand or model of access point that you choose for your home network, by default it is set up very insecurely. At a minimum, four basic security measures should be enabled:

1. Change the default Service Set Identifier (SSID).

2. Disable the SSID broadcast.

3. Enable 128-Bit Wired Equivalent Privacy (WEP) encryption.

4. Filter by Media Access Control (MAC) address.

These steps are enough in most cases to protect low-traffic wireless networks used at home. It is important to use all four of these measures in order to deploy your wireless network using a layered security approach. While it is true that no network is completely secure, enabling these security measures on your wireless network reduces the risk of becoming an easy target and increases the likelihood that a prospective attacker will move on to a wireless network that does not employ the same precautions.

20.9 Solutions Fast Track

20.9.1 Enabling Security Features on a Linksys WAP11 802.11b AP, Linksys BEFW11SR 802.11b AP/Router, WRT54G 802.11b/g AP/Router, and D-Link DI-624 AirPlus 2.4 GHz Xtreme G Wireless Router with 4-Port Switch

These have been consolidated because they are the recommendations for securing any AP/router and are not specific to a particular hardware.

- Assigning a unique SSID to your wireless network is the first security measure that you should take. Any attacker with a "default" configuration profile is able to associate with an access point that has a default SSID. While assigning a unique SSID in and of itself doesn't offer much protection, it is one layer in your -wireless defense.

- Many attackers use active wireless scanners to discover target wireless networks. Active scanners rely on the access point beacon to locate it. This beacon broadcasts the SSID to any device that requests it. Disabling SSID broadcast makes your access point "invisible" to active scanners. Because your access point can still be discovered by passive wireless scanners, this step should be used in conjunction with other security measures.

- Wired Equivalent Privacy (WEP) encryption, at a minimum, should be used on your home wireless network. Although there are tools available that make it possible to crack WEP, the amount of traffic that needs to be generated make it unlikely an attacker will take the time to do so on a home, or low-traffic, network. Adequate security for these networks is provided by 128-bit WEP.

- Filtering by Media Access Control (MAC) address allows only wireless cards that you specifically designate to access your wireless network. Again, it is possible to spoof MAC addresses, therefore you shouldn't rely on MAC address filtering exclusively. It should be part of your overall security posture.

- Each of the four security steps presented in this chapter can be defeated. Fortunately, for most home users they do provide adequate security for a wireless network. By enacting a four-layer security posture on your wireless network, you have made it more difficult for an attacker to gain access to your network. Because the likelihood of a strong "return" on the attacker's time investment would be low, he is likely to move on to an easier target. Don't allow your wireless network to be a target of convenience.

20.9.2 Configuring Security Features on Wireless Clients

Windows XP clients are configured using the Wireless Connection Properties and the Windows XP Wireless Client Manager. To associate with your access point once the security features have been enabled, your access point must be added as a Preferred Network. You need to enter the SSID and the WEP key during the configuration process.

Windows 2000 does not have a built-in wireless client manager like Windows XP. You need to enter the SSID and WEP key into a profile in the client manager software that shipped with your wireless card.

Linux users need to configure the /etc/pcmcia/wireless.opts file in order to access a wireless network with the security features enabled. The SSID and WEP key need to be entered in the appropriate section of the /etc/pcmcia/wireless.opts file. Restarting PCMCIA services or rebooting allows these settings to take effect.

CHAPTER 21

Implementing Advanced Wireless Security

Chris Hurley

21.1 Introduction

The practical security measures discussed in Chapter 20, "Basic Wireless Network Security," are, as a general rule, sufficient for most home wireless users. Corporate users, however, should not rely on basic security measures alone to protect their wireless networks. In this chapter, we discuss some of the more advanced ways to reduce the risk of your wireless network being compromised.

In this chapter, you will also learn about different methods of secondary authentication such as Virtual Private Networks (VPNs) and Remote Authentication and Dial-In User Service (RADIUS). A secondary authentication mechanism requires that, in addition to association with a wireless access point, a user must authenticate (that is, log in) using some other means. But, first, we will look at the replacement for Wired Equivalent Privacy (WEP): Wi-Fi Protected Access (WPA).

21.2 Implementing Wi-Fi Protected Access (WPA)

Wi-Fi Protected Access (WPA) is designed to provide wireless users with an encryption mechanism that is not susceptible to the vulnerabilities of Wired Equivalent Privacy (WEP). Most 802.11g access points either ship with the option to use WPA or a firmware upgrade can be downloaded from the access point manufacturer.

Before enabling WPA, you should ensure that your wireless card has WPA drivers. As with access points, you often need to update the card's drivers, firmware, or both in order to take advantage of WPA. This section details how to set up WPA encryption on two access points: the D-Link DI-624 and the Linksys WRV54G. You will also learn how to configure your wireless client to use WPA.

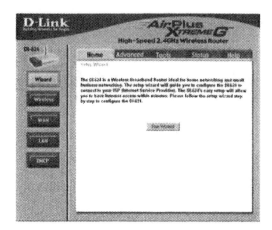

Figure 21.1: The DI-624 initial configuration screen

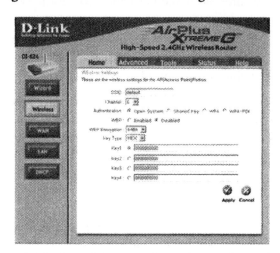

Figure 21.2: The Wireless Configuration Options window

21.2.1 Configuring the D-Link DI-624 AirPlus 2.4 GHz Xtreme G Wireless Router with 4-Port Switch

The D-Link DI-624 ships with WPA capability. This means that no firmware upgrade is necessary and you can start using WPA as soon as the DI-624 comes out of the box. First, you need to log into the DI-624 from a wired connection. Then, point your browser to 192.168.0.1 and supply the username *admin* with a blank password when prompted. This opens the initial configuration screen, as seen in Figure 21.1.

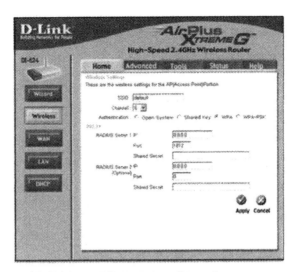

Figure 21.3: The WPA configuration screen

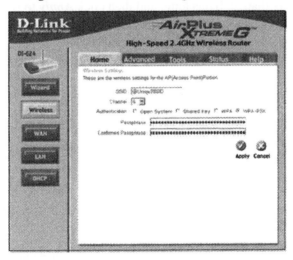

Figure 21.4: The WPA-PSK configuration screen

Next, click the *Wireless* button on the left to open the wireless configuration options window, as shown in Figure 21.2.

Next, choose either the *WPA* or *WPA-PSK* Authentication options. The WPA option requires a RADIUS server, whereas WPA-PSK (Pre Shared Key) sets a passphrase that must also be entered in the client WPA configuration settings. See Figures 21.3 and 21.4.

Damage & Defense

Known WPA-PSK Vulnerability

WPA-PSK utilizes a 256-bit pre-shared key or a passphrase that can vary in length from 8 to 63 bytes. Short passphrase-based keys (less than 20 bytes) are vulnerable to the offline dictionary attack. The pre-shared key that is used to set up the WPA encryption can be captured during the initial communication between the access point and the client card. Once an attacker has captured the pre-shared key, he can use that to essentially "guess" the WPA key using the same concepts used in any password dictionary attack. In theory, this type of dictionary attack takes less time and effort than attacking WER Choosing a passphrase that is more than 20 bytes mitigates this vulnerability.

Enter either your RADIUS server information and Shared Secret for WPA or a strong passphrase that is more than 20 bytes long, and then click *Apply* to save your settings and enable WPA.

21.2.2 Configuring the Linksys WRV54G VPN Broadband Router

The Linksys WRV54G VPN-Broadband Router may require a firmware upgrade to allow WPA capability. Firmware version 2.10 or later is required for WPA functionality on the WR.V54G. To enable WPA, you need to log in to the WRV54G, as shown in Figure 21.5. Point your browser to the IP address of the WRV54G. By default, this is 192.168.1.1. There is no username required and the default password is *admin*.

Next, click the *Wireless* tab to display the Wireless Network Settings, as seen in Figure 21.6.

Then, choose the *Wireless Security* option to display the Wireless Security settings, as seen in Figure 21.7.

The *Security Mode* drop-down box displays the four modes of security available on the WRV54G:

- WPA Pre-Shared Key
- WPA Radius
- RADIUS
- WEP

Figure 21.5: The Linksys WRV54G initial configuration screen

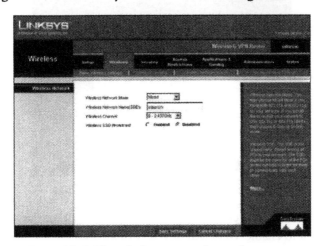

Figure 21.6: The wireless networks settings screen

WPA RADIUS requires a RADIUS server, as shown in Figure 21.8. WPA Pre-Shared Key (Figure 21.9) allows you to enter a strong pre-shared key. All wireless clients must also be configured to use the WPA pre-shared key in order to authenticate to the wireless network.

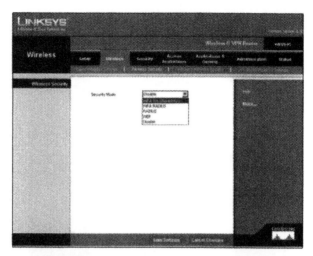

Figure 21.7: The wireless security settings

Figure 21.8: The WPA RADIUS settings

Finally, enter the RADIUS server IP address and shared secret, or the pre-shared key and choose *Save Settings* to enable WPA support.

21.2.3 Configuring Windows XP Wireless Clients for WPA

In order to take advantage of WPA, you must configure your wireless client. To allow Windows XP to work with WPA you must first install the Microsoft Update for Microsoft Windows XP (KB826942). This patch enables WPA compatibility in Windows XP. After

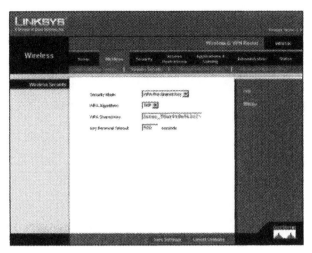

Figure 21.9: The WPA Pre-Shared Key settings

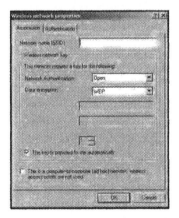

Figure 21.10: The Connection Properties window

installing KB826942, double-click the *Wireless Network Connection* icon on the toolbar. This opens the Wireless Network Connection Properties window, as seen in Figure 21.10. If you have a profile for your access point already set up, select it and click *Properties*. Otherwise, select Add under the Preferred Networks. The connection properties window will open.

Next, enter the SSID for your access point in the *Network Name* textbox, as shown in Figure 21.11. Then, choose the type of encryption you configured your access point to use—WPA or WPA-PSK—and then the encryption standard: WEP, Temporal Key

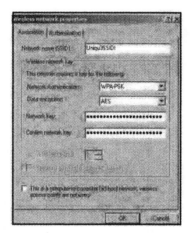

Figure 21.11: WPA Client Settings

Integrity Protocol (TKIP), or Advanced Encryption Standard (AES). Finally, input the pre-shared key configured on your access point into the *Network key* and *Confirm network key* textboxes.

Your client setup is now complete and you can utilize your wireless network with WPA security.

NOTES FROM THE UNDERGROUND

WPA In Linux with Linuxant DriverLoader

In order to utilize WPA, you must have drivers for your wireless card that support it. Most 802.11g cards either have a WPA capable driver when they are purchased, or one can be downloaded. The problem is that these drivers are for Windows. Linux users have not been able to enjoy the benefits of WPA because there are very few card manufacturers that have released WPA drivers for Linux. Linuxant has offered a solution to this problem for many cards: DriverLoader.

DriverLoader allows you to use the Windows driver for cards based on the Atheros, Broadcom, Cisco, Intel Centrino, Prism, Realtek, and Texas Instruments chipsets in Linux. DriverLoader also supports WPA. It is available for a free trial from the Linuxant web site (www.linuxant.com/driverloader/) and a permanent license can be purchased for $19.95 at the time of this writing.

21.3 Implementing a Wireless Gateway with Reef Edge Dolphin

The first solution we'll examine is freeware. Free is always a good thing, especially in the IT industry! Reef Edge (www.reefedge.com) produces several commercial products for use in securing wireless networks, including Connect Manager. Dolphin is a somewhat scaled-down version of Connect Manager that still provides the same basic features, but is flee. If you need to add an unlimited number of users, or add new user groups, you should investigate Connect Manager. The Dolphin FAQ (http://techzone.reefedge. com/dolphin/index/faq.page) provides more information on the limitations of Dolphin in comparison to Connect Manager.

Dolphin runs a hardened version of Linux and, once installed, acts almost the same as any other network appliance. The chief difference is that console and Telnet logins are not supported; all access is via the Secure Socket Layer (SSL) secured Web interface. An aging piece of Intel 586 hardware can be quickly and easily transformed into a secure wireless gateway, providing access control from the wireless network to the wired network, which we demonstrate in this chapter. Dolphin is a noncommercial product and not to be used in large implementations, but it does provide an ideal (and affordable) solution for Small Office/ Home Office (SOHO) applications and serves as an excellent test bed for administrators who want to get their feet wet with wireless without opening their networks to security breaches. If you find that Dolphin is to your liking, you might want to consider contacting Reef Edge to purchase Connect Manager or an edge controller. An edge controller is a second, or satellite, machine that can be set up to support your wireless network. You will be able to easily move up to these solutions with the knowledge you gain by configuring and using Dolphin.

> **NOTE**
>
> SSL was developed in 1996 by Netscape Communications to enable secure transmission of information over the Internet between the client end (Web browsers) and Web servers. SSL operates between the application and transport layers and requires no actions on the part of the user. It is not a transparent protocol that can be used with any application layer protocol; instead, it works only with those application layer protocols for which it has been explicitly implemented. Common transport layer protocols that make use of SSL include: HyperText Transfer Protocol (HTTP), Simple Mail Transfer Protocol (SMTP), and Network News Transfer Protocol (NNTP).

SSL provides the three tenants of Public Key Infrastructure (PKI) security to users:

- *Authentication* Ensures that the message being received is from the individual claiming to send it.

- *Confidentiality* Ensures that the message cannot be read by anyone other than the intended recipient.

- *Integrity* Ensures that the message is authentic and has not been altered in any way since leaving the sender.

Dolphin provides some robust features that are typically found in very expensive hardware-based solutions, including secure authentication, IPSec security, and session roaming across subnets. Users authenticate to the Dolphin server over the WLAN using SSL-secured communications and then are granted access to the wired network. Dolphin supports two groups, users and guests, and you can control the access and quality of service of each group as follows:

- *Users* Trusted users who can use IPSec to secure their connection and access all resources.

- *Guests* Unknown users who are not allowed to use IPSec to secure their communications and have access control restrictions in place.

Finally, Dolphin supports encrypted wireless network usage through IPSec tunnels. Through the creation of IPSec VPN tunnels, users can pass data with a higher level of security (encryption) than WEP provides.

To begin working with Dolphin, you need to register for the Reef Edge TechZone at http://techzone.reefedge.com. Once this is done, you will be able to download the CD-ROM ISO image and bootable diskette image files from the Reef Edge download page. The server that you are using for Dolphin must meet the following minimum specifications:

- Pentium CPU (586) or later

- Pentium CPU (586) or later

- 64 MB IDE hard drive as the first boot IDE device

- IDE CD-ROM

- Diskette drive if the CD-ROM drive being used is not E1 Torrito compliant (see www.area51partners.com/ftles/eltorito.pdf for more information on this specification).

- Two Peripheral Component Interconnect (PCI) network adapters from the following list of compatible network adapters:

  - 3Corn 3c59x family (not 3c905x)
  - National Semiconductor 8390 family
  - Intel EtherExpress 100
  - NE2000/pci
  - PCNet32
  - Tulip family

The Dolphin implementation is depicted in Figure 21.12.

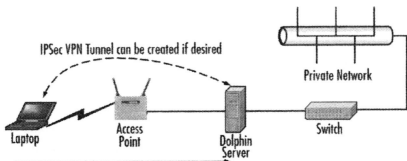

IPSec VPN Tunnel can be created if desired

Private Network

Laptop   Access Point   Dolphin Server   Switch

1. Client attempts to access network, but is stopped by Dolphin
2. Dolphin checks internal database and grants access if user is authenticated.
3. Client can now access network resources.

Figure 21.12: Dolphin provides gateway services for the wireless network

21.3.1 Installing Dolphin

Once you've gathered all the required items, you can begin installing Dolphin on your server. To do so, perform these steps:

1. Create the CD-ROM from the ISO image. If required, create the bootable diskette from the floppy disk image.

2. Connect a keyboard, mouse, and monitor to the Dolphin server.

3. Power on the Dolphin server and place the Dolphin CD-ROM in the CD-ROM drive. If your computer is not capable of booting directly from the CD, you will also need to use the boot diskette.

4. Select OK when prompted to start the installation.

5. Accept the EULA when prompted.

6. Acknowledge, when prompted, that installing Dolphin will erase the contents of the first physical disk.

7. Restart the Dolphin server as prompted after the installation has been completed. After the restart, you will see a long series of dots followed by this message:

    ```
    System Ready. IP address: 192.168.0.1/255.255.255.0.
    ```

    This value represents the wired side of the Dolphin server and can be changed later if you desire by completing the steps in the "Configuring Dolphin" section of this chapter.

8. Determine which network adapter is which on the Dolphin server. Configure the network adapter on your management station (depicted in Figure 21.15) with the IP address of 10.10.10.10 and a subnet mask of 255.255.255.0, as shown in Figure 21.13.

9. Connect directly using a crossover cable between your management station and one of the network adapters on the Dolphin server, ping the Dolphin server with an IP address of 10.10.10.1. If you receive an echo reply, as shown in Figure 21.14, you have located the wireless side of the Dolphin server. If you don't get an echo reply, make the connection to the other network adapter on the Dolphin server. Attempt to ping the other network adapter on the Dolphin server with the IP address of 10.10.10.1 to verify connectivity. The wired side of the Dolphin server intially has the IP address of 192.168.0.1 with a subnet mask of 255.255.255.0, as mentioned in Step 7. You can, however, change the IP

10 . 10 . 10 . 10
255 . 255 . 255 . 0
OK
Cancel

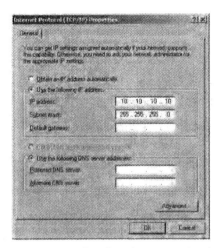

Figure 21.13: Configuring the network adapter

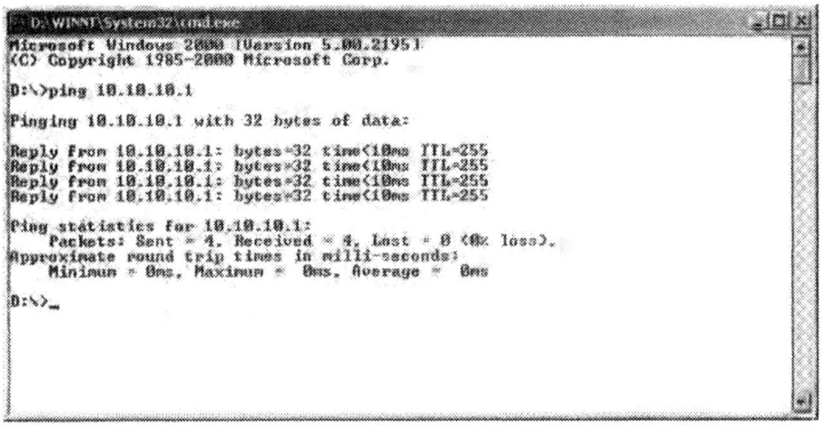

D:\WINNT\System32\cmd.exe
Microsoft Windows 2000 [Version 5.00.2195]
(C) Copyright 1985-2000 Microsoft Corp.
D:\>ping 10.10.10.1
Pinging 10.10.10.1 with 32 bytes of data:
Reply from 10.10.10.1: bytes=32 time<10ms TTL=255
Reply from 10.10.10.1: bytes=32 time<10ms TTL=255
Reply from 10.10.10.1: bytes=32 time<10ms TTL=255
Reply from 10.10.10.1: bytes=32 time<10ms TTL=255
Ping statistics for 10.10.10.1:
Packets: Sent = 4, Received = 4, Lost = 0 (0% loss),
Approximate round trip times in milli-seconds:
Minimum = 0ms, Maximum = 0ms, Average = 0ms
D:\>_

Figure 21.14: Finding the wireless side of the Dolphin server

addresses and subnet masks of both the wireless and wired side of the Dolphin server if you so desire, as discussed in the next section, "Configuring Dolphin."

10. Configure your management station with an IP address in the 192.168.0.x range, such as 192.168.0.180, and connect it to the wired side (192.168.0.1) of the Dolphin server, preferably through a switch, but you can use a crossover cable to make a direct connection.

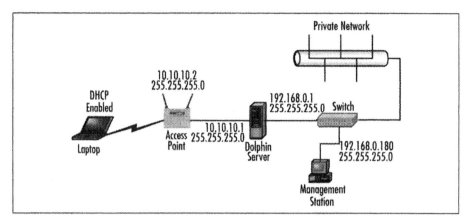

Figure 21.15: Making the Dolphin connections

11. Configure your wireless client for Dynamic Host Configuration Protocol (DHCP) so that it can receive an IP address and DNS server information from the Dolphin server. (You can change the DHCP values passed out later in this procedure.)

12. Connect the AP to the wireless side of the Dolphin server (10.10.10.1). Ensure that the AP and the wireless side of the Dolphin server are configured correctly, with IP addresses on the same subnet. You should now have an arrangement like the one shown in Figure 21.15.

13. Force the wireless client to renew its DHCP lease and check to see that it looks something like the one shown in Figure 21.16.

14. Ping the wireless side of the Dolphin server, from the wireless client, at 10.10.10.1 to verify connectivity.

15. Ping the wireless side of the Dolphin server again, from the wireless client, using the DNS name mobile.domain.

16. Attempt to access resources on the wired network from the wireless client. Acknowledge the SSL connection if prompted to do so (although you won't actually see any SSL-secured pages until you attempt to log in at the next step).

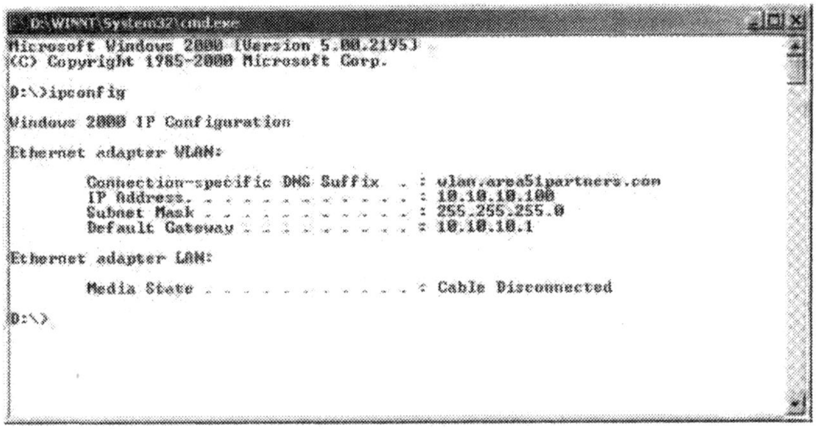

Figure 21.16: Verifying the DHCP lease

Figure 21.17: Connecting to the Dolphin server

If you see the Web page in Figure 21.17, congratulate yourself—your Dolphin installation is operating properly!

17. Log in from the page shown in Figure 21.18 using the username temp and the password temp. If login is successful, you will see the page shown in Figure 21.19. Notice that the IPSec key shown at the bottom of the page is actually your shared key that you would use to create IPSec connections.

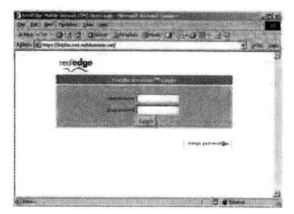

Figure 21.18: Logging into the Dolphin web page

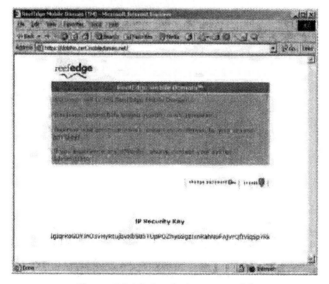

Figure 21.19: Login is successful

NOTE

Although these are the default credentials to enter your Dolphin system, it is critical that you change them once you are done with the initial configuration. After you have created your first user account, Dolphin will delete the temp account for you automatically to ensure that no one compromises the server or gains unauthorized network access.

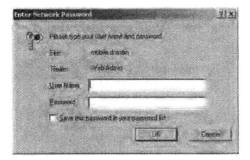

Figure 21.20: Logging in to the administrative interface

18. Log in to the Dolphin Web management interface by entering *https://mobile.domain/ admin*into your browser. You will be prompted to log in, as shown in Figure 21.20.

NOTE

If you don't have a crossover cable, you can use a switch or hub and two standard straight-through cables. Simply connect the Dolphin server to the management station through the switch or hub. Ensure that you are using the uplink port on the switch or hub and, if required by your hardware, ensure that the uplink port is selected for uplink via regular use. Also make sure that you are on the right network segment with correct IP addressing configured.

21.3.2 Configuring Dolphin

Your Dolphin server is now installed and operable on your wireless network. You now need to perform some configuration and management tasks before your server is ready to be placed into production. You need to add users to the Dolphin database who will be allowed to gain access to the wireless network. (Dolphin does not support RADIUS and thus must use a local user database.) In addition, you might want to change the IP addresses and subnets assigned to the Dolphin server network adapters. The following steps walk you through the process of configuring some of these options:

1. Log in to the Dolphin server by completing Steps 17–18 of the previous procedure.

2. Click the *Wired LAN* link from the menu on the left side of the window. This provides you with the capability to change the wired-side properties, as shown in

Figure 21.21: Changing the wired-side network properties

Figure 21.21. In most cases, you'll need to change the wired-side IP address from the default configuration of 192.168.0.1 because this is typically reserved for use by the default gateway. Be sure to enter the default gateway and DNS server IP addresses as well as to enable wireless network clients to access network resources. After making your changes, click the *Save* button. (Note that you will have to restart the Dolphin server to commit the changes to the running configuration. You can, however, make all your changes and then restart the server.)

3. Click the *Wireless LAN* link to configure the wireless-side properties, as shown in Figure 21.22. You can change all these properties as you see fit. By default, the Dolphin server is configured with the domain name reefedge.com and DHCP address range of 10.10.10.10–10.10.10.253. After making your changes, click the *Save* button. If you want to configure the quality of service that wireless clients receive, click the *Wireless LAN Bandwidth* link to configure your values, as shown in Figure 21.23. After making your changes, click the *Save* button.

4. Create a list of authorized users for Dolphin—that is, a listing of users who can authenticate to Dolphin and then be granted wireless network access. Click the *Add New User* link to open the User Management page shown in Figure 21.24.

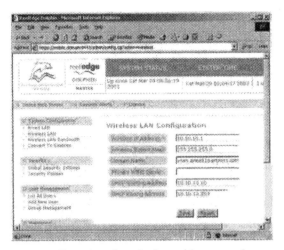

Figure 21.22: Changing the wireless-side network properties

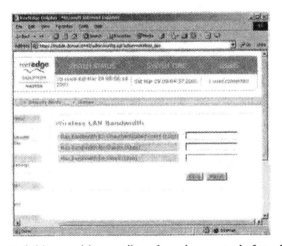

Figure 21.23: Dolphin provides quality of service controls for wireless clients

Note that you can only choose between the users group and the guests group—
Dolphin does not support creating custom groups (a limitation due to its freeware
status). After supplying the required information, click *Save*. After creating your
first Dolphin user, the "temp" account will be deleted for security reasons. If you
want to configure additional security policies, click the *Security Policies* link to
open the Security Policies For User Groups page, shown in Figure 21.25. This

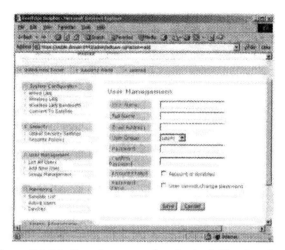

Figure 21.24: Creating users for the Dolphin Database

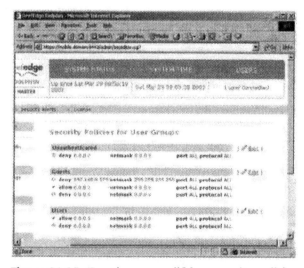

Figure 21.25: Creating or modifying security policies

page allows you to configure the equivalent of a firewall rule set for your Dolphin server.

5. Change the administrative password to restart your Dolphin server. To do this, scroll the page all the way to the bottom and click the *Admin Password* link. Click *Save* after making your change (see Figure 21.26).

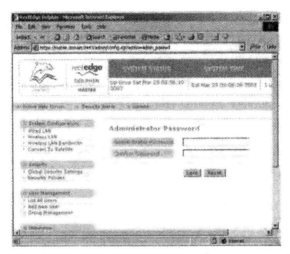

Figure 21.26: Changing the administrator password

6. Restart your Dolphin server. After Dolphin has completed loading, you will see the familiar series of dots, this time followed by the new wired-side IP address that you have configured.

21.3.3 Improving the User Experience

Should you not want authorized users to need to use the Web interface to Dolphin to authenticate, you can equip them with a small utility that is available from Reef Edge, and can be used to perform regular and IPSec-secured logins/logouts. The process to install and use this utility is outlined here:

1. Download the *Active TCL* package from Active State at www.activestate.com/ Products/Download/Download.plex?id=ActiveTCL and install it onto your wireless client computer.

2. Download the *TCL TLS 1.4* package from Reef Edge's download page. Create a folder called *tlsl.4* in the lib directory of the Active TCL installation path and extract the contents of the TLS 1.4 archive into this folder.

3. Download the *dolphin_status.tcl* file, also located at the Reef Edge download page.

4. Place the dolphin_status.tcl file in a convenient location on the client computer. Once Active TCL has been installed, the dolphin_status.tcl file will act as an executable and can be double-clicked to open.

5. Execute the *dolphin_status.tcl* file to get the login prompt shown in Figure 21.27. You have the option of creating an IPSec tunnel at this time as well. The tcl file will create a configuration ftle named dolphin in the same directory it is located in.

Tools and Traps

Using Enterprise Wireless Gateways

Don't think of Dolphin as a full-featured Enterprise Wireless Gateway (EWG). However, you should consider it a wireless gateway. For a full featured EWG, you might want to consider one of the more capable and robust (and more expensive) solutions offered from one of the following vendors:

- *Bluesocket* www.bluesocket.com

- *Columbitech* www.columbitech.com

- *Reef Edge* www.reefedge.com

- *Sputnik* www.sputnik.com

- *Vernier Networks* www.verniernetworks.com

- *Viator Networks* www.viatornetworks.com

These solutions offer the same features as Dolphin—authentication and VPN support—but they also provide many other options, such as RADIUS server support, hot failover support, and multiple protocol support (such as WAP, 3G, and 802.11). The EWG market is still in a great deal of flux as vendors try to refine their products. That does not mean, however, that you cannot create very secure solutions using today's technology. A word of caution, though: You should expect to find bugs and other errors with most of these solutions because the technology is still so new. Caveat emptor.

Figure 21.27: Using the Dolphin_status.tcl file to log in

21.3.4 Dolphin Review

As you've seen in this chapter, the Dolphin product provides a very inexpensive solution for small wireless environments. It is very lightweight and has mammal hardware requirements; you most likely have an old PC stuffed in a storage room that could be turned into a dedicated wireless gateway by installing the Dolphin application on it.

On the up side, Dolphin is easy to use and configure, is inexpensive, and provides a relatively good amount of security for smaller organizations. In addition, Dolphin supports the creation of IPSec-secured VPN tunnels between the wireless clients and the Dolphin server. On the down side, Dolphin is limited in the number of users it can support as well as the number of groups you can create to classify users. Dolphin also does not provide for the use of an external RADIUS server. These limitations, however, are clearly stated by Reef Edge because Dolphin is not intended for commercial usage. If you have a small home or office wireless network that needs to be secured by an access-granting device, Dolphin might be an ideal choice for you.

Now that we've spent some time looking at the freeware Dolphin product, let's step up the discussion and examine some more robust (and more costly) solutions that you might implement to secure a larger wireless network necessitating control over user access in a larger enterprise environment.

21.4 Implementing a VPN on a Linksys WRV54G VPN Broadband Router

The Linksys WRV54G is an access point/router combination that Linksys designed for the small office, or home user that desires a higher level of security than WEP or WPA can provide. The W-RV54G offers all of the security features of other access points, but also provides the capability of setting up an IPSec VPN tunnel. A VPN tunnel allows two points to establish an encrypted session using a selected protocol. Other protocols can then be transmitted through this tunnel. A basic example of this is a Secure Shell (SSH) tunnel. A firewall can be configured to allow only SSH traffic (port 22) inbound. The client can then tunnel other traffic, such as HTTP (port 80) through the established SSH tunnel. This both encrypts the HTTP traffic, and removes the requirement to allow port 80 traffic through the firewall. Additionally, because some form of authentication (passphrase, key exchange, or both) is required to establish the initial SSH tunnel, additional user level access controls are in place.

This section describes the process of setting up an IPSec tunnel to utilize the VPN features on the WRKV54G. First, we discuss the steps that must be taken on Windows 2000 or XP clients to prepare for VPN access. Then, the configuration steps that are required on the WRKV54G are detailed.

21.4.1 Preparing Windows 2000 or XP Computers for Use with the WRV54G

There are four steps that you need to take to configure your Windows 2000 or XP computer to establish a VPN tunnel with the WRV54G.

1. Create an IPSec policy.

2. Build two filter lists.

3. Establish the tunnel rules.

4. Assign the IPSec policy to the computer.

21.4.1.1 Creating an IPSec Policy

Click *Start | Run* and type *secpol.msc* in the *Open* textbox to open the Local Security Settings screen, as seen in Figure 21.28.

Right-click *IP Security Policies on Local Computer* and select *Create IP Security Policy* to open the IP Security Policy Wizard. Click *Next* on the IP Security Policy Wizard window.

Enter a name for your security policy in the *Name* textbox (as shown in Figure 21.29) and click *Next*.

Remove the checkbox next to *Activate the default response rule*, as shown in Figure 21.30, and click *Next*.

Finally, make sure that the *Edit properties* checkbox is selected, as shown in Figure 21.31, and click *Finish*.

21.4.1.2 Building Filter Lists

Selecting the Edit properties checkbox before finishing the IP Security Policy Wizard opens the Properties window for your new security policy (Figure 21.32).

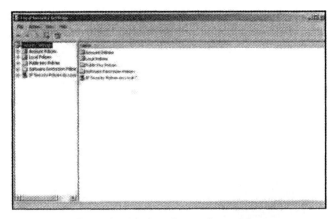

Figure 21.28: Local security settings

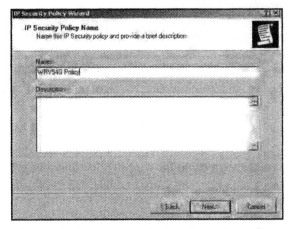

Figure 21.29: Naming the local security policy

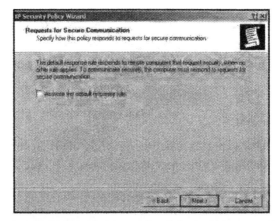

Figure 21.30: Deactivate the default response rule

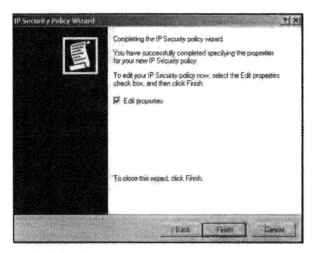

Figure 21.31: Completing the local policy creation

Deselect the *Use Add Wizard* checkbox and click *Add* to open the New Rule Properties window. By default, this window opens on the IP Filter List tab. Click *Add* again to open the *IP Filter List* window. Enter a name for the filter, as shown in Figure 21.33. Deselect the *Use Add Wizard* checkbox and click *Add*.

The Filter Properties window opens on the Addressing tab. Choose *My IP Address* in the Source Address field and *A specific IP Subnet* in the Destination Address field. In the *IP*

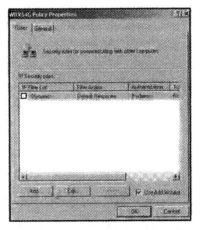

Figure 21.32: The Policy Properties

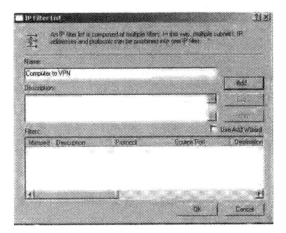

Figure 21.33: The IP Filter List window

Address field enter *192.168.1.0*. This represents all addresses in the range 192.168.1.1–192.168.1.255. If you are using a different range, make sure to adjust this accordingly. Enter the Subnet Mask for your network in the *Subnet Mask* field (see Figure 21.34). By default, this is 255.255.255.0. Click the *OK* button to close this window.

Next, click *OK* in Windows XP or *Close* in Windows 2000. This filter is used for communication from your computer to the router.

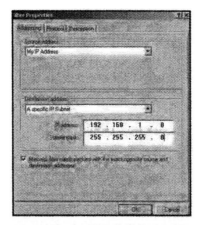

Figure 21.34: The IP filter settings

Figure 21.35: Creating the second filter

You will then need to create a filter for communication from the router to your computer. In the *New Rule Properties* window, highlight the rule you just created, as shown in Figure 21.35, and click *Add*.

This opens the *IP Filter List* window. Enter a name for the new filter in the *Name* textbox and click Add. On the Addressing tab, choose *A specific IP Subnet* in the *Source Address* field. In the *IP Address* field, enter *192.168.1.0*. This represents all addresses in the range 192.168.1.1–192.168.1.255. If you are using a different range, you will need to adjust

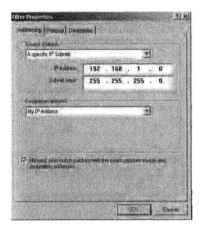

Figure 21.36: The Filter Properties window

this accordingly. Enter the subnet mask for your network in the *Subnet Mask* field. By default, this is 255.255.255.0. Choose *My IP Address* in the *Destination Address* field (Figure 21.36).

Click the *OK* button to close this window. Next, click *OK* in Windows XP or *Close* in Windows 2000. This filter is used for communication from the router to your computer.

21.4.1.3 Establishing the Tunnel Rules

The rules that are employed by the tunnels must be set up in order to properly filter traffic through the VPN tunnel. First, select the tunnel you created for communication from your computer to the router and then click the *Filter Action* tab. Next, select the *Require Security* radio button and click *Edit* to open the Require Security Properties window, as shown in Figure 21.37.

Ensure that the *Negotiate security* radio button is selected. Then, deselect *Accept unsecured communication, but always respond using IPSec* and select *Session key perfect forward security (PFS)*, as shown in Figure 21.38.

Click *OK* to return to the *New Rule Properties* window. Select the *Authentication Methods* tab and click *Edit* to open the *Edit Authentication Method Properties* window. Choose the *Use this string (preshared key)* radio button and enter the pre-shared key in

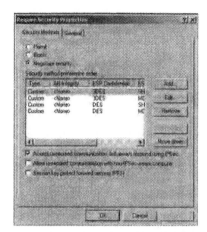

Figure 21.37: The require security properties window

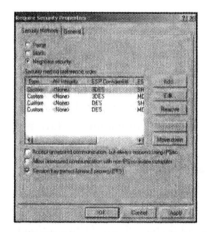

Figure 21.38: The security methods options

the textbox (Figure 21.39). This can be a combination of up to 24 letters and numbers, but special characters are not allowed. Make sure that you remember this key as it will be used later when the router is configured.

Next, click the *OK* button in Windows XP or the Close button in Windows 2000.

Select the ***Tunnel Setting*** tab on the New Rule Properties window. Select *The tunnel endpoint is specified by this IP address* and enter the external IP address of the WRV54G, as shown in Figure 21.40. This is the IP address your router uses to communicate with the Internet.

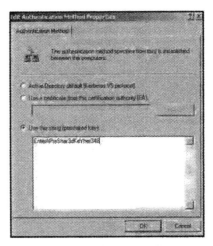

Figure 21.39: Entering the pre-shared key

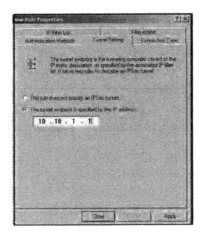

Figure 21.40: The Tunnel setting tab

Next, click the *Connection* Type tab (as shown in Figure 21.41). Select *All network connections* if you want this rule to apply to both Internet and *local area network (LAN)* connections. Choose Local area network (LAN) if you want this tunnel to apply only to connections made from the local network. Choose *Remote access* if you want this rule to apply only to connections made from the Internet.

After you have selected the type of network connections that the rule applies to, click *Close*.

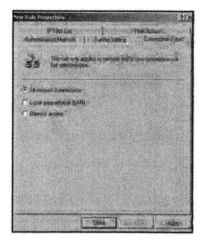

Figure 21.41: Select the connection type

Another filter rule must be created to allow communication from the router to your computer. To create this rule, repeat the steps outline in this section, but enter the IP address of your computer as the Tunnel Endpoint instead of the IP address of the router.

21.4.1.4 Assigning the Security Policy

Finally, you must assign your new security policy to the local computer. In the *Local Security Settings* window, right-click the new policy that you have just created and select *Assign*, as shown in Figure 21.42.

Your computer is now configured to communicate over a VPN tunnel.

21.4.2 Enabling the VPN on the Linksys WRV54G

Now that your computer is configured to communicate over an IPSec VPN tunnel, you must configure the WRV54G to communicate with your computer. Using your web browser, type the IP Address of the WRV54G into your address bar. This is 192.168.1.1, by default. You will be prompted for your username and password.

From the setup screen, select *Security | VPN* to display the VPN settings, as seen in Figure 21.43.

Select the *Enabled* radio button for VPN Tunnel. Choose a name for this tunnel and enter it in the *Tunnel Name* textbox. Next, enter the IP address and netmask of the local

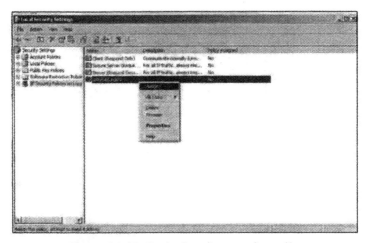

Figure 21.42: Assigning the security policy

Figure 21.43: The WRV54G VPN settings

network in the *IP Address* and *Mask* fields for the Local Secure Group. Use 192.168.1.0 to allow all *IP* addresses between 192.168.1.1–192.168.1.255.

Enter the IP address and netmask of the computer you just configured in the *IP Address* and *Mask* fields for the Remote Secure Group. Next, choose *3DES* from the *Encryption*

Figure 21.44: The completed VPN settings.

drop-down box. This requires the use of Triple Data Encryption Standard encryption. Choose *SHA1* from the *Authentication* drop-down box.

Choose *Auto(IKE)* as the *Key Exchange Method* and select the *Enabled* radio button for *PFS*. This enables the use of Internet Key Exchange (IKE) and Perfect Forward Secrecy (PFS).

Finally, select the radio button next to *Pre-Shared Key* and enter the same pre-shared key you entered on your computer while setting it up.

Once you have entered these settings, your VPN setup screen should look similar to Figure 21.44.

Click *Save Settings* to save your settings and establish a VPN tunnel between the WRV54G and your computer.

21.5 Implementing RADIUS with Cisco LEAP

The use of RADIUS servers to authenticate network users is a longstanding practice. Using RADIUS server dynamic per-user, per-session WEP keys combined with IV randomization is a fairly new practice. Another new addition is Cisco's proprietary offering (now being used by many third-party vendors), Lightweight Extensible Authentication Protocol (LEAP).

LEAP is one of approximately 30 different variations of the Extensible Authentication Protocol (EAP). Other variants include Extensible Authentication Protocol-Message Digest Algorithm 5 (EAP-MD5), Extensible Authentication Protocol-Transport Layer Security (EAP-TLS), Extensible Authentication Protocol-Tunneled TLS (EAP-TTLS), and Protected Extenstible Authentication Protocol (PEAP). EAP allows other security products (such as LEAP) to be used to provide additional security to Point-to-Point Protocol (PPP) links through the use of special Application Programming Interfaces (APIs) that are built into operating systems and, in the case of the Cisco Aironet hardware, hardware device firmware.

LEAP (also known as EAP-Cisco Wireless) uses dynamically generated WEP keys, 802.1x port access controls, and mutual authentication to overcome the problems inherent in WEP. 802.1x is an access control protocol that operates at the port level between any authentication method (LEAP in this case) and the rest of the network. 802.1x does not provide authentication to users; rather, it translates messages from the selected authentication method into the correct frame format being used on the network. In the case of our example, the correct frame format is 802.11, but 802.1x can also be used on 802.3 (Ethernet) and 802.5 (Token Ring) networks, to name a few. When you use 802.1x, the choice of the authentication method and key management method are controlled by the specific EAP authentication being used (LEAP in this case).

NOTE

RADIUS is defined by Requests for Comments (RFC) 2865. The behavior of RADIUS with EAP authentication is defined in RFC 2869. RFC can be searched and viewed online at www.rfc-editor.org. 802.1x is defined by the IEEE in the document located at http://standards.ieee.org/getieee802/download/802.1X-2001.pdf.

LEAP creates a per-user, per-session dynamic WEP key that is tied to the network logon, thereby addressing the limitations of static WEP keys. Since authentication is performed against a back-end RADIUS database, administrative overhead is minimal after initial installation and configuration.

21.5.1 LEAP Features

Through the use of dynamically generated WEP keys, LEAP enhances the basic security of WEP. This feature significantly decreases the predictability of the WEP key through the use of a WEP key-cracking utility by another user. In addition, the WEP keys that are generated can be tied to the specific user session and, if desired, to the network login as well. Through the use of Cisco (or other third-party components that support LEAP) hardware from end to end, you can provide a robust and scalable security solution that silently increases network security not only by authenticating users but also by encrypting wireless network traffic without the use of a VPN tunnel. (You can, however, opt to add the additional network overhead and implement a VPN tunnel as well to further secure the communications.)

Cisco LEAP provides the following security enhancements:

- *Mutual authentication.* Mutual authentication is performed between the client and the RADIUS server, as well as between the AP and the RADIUS server. By using mutual authentication between the components involved, you prevent the introduction of both rogue APs and RADIUS servers. Furthermore, you provide a solid authentication method to control whom can and cannot gain access to the wireless network segment (and thus the wired network behind it). All communications carried out between the AP and the RADIUS server are done using a secure channel, further reducing any possibility of eavesdropping or spoofing.

- *Secure-key derivation.* A preconfigured shared-secret secure key is used to construct responses to mutual authentication challenges. It is put through an irreversible one-way hash that makes recovery or replay impossible and is useful for one time only at the start of the authentication process.

- *Dynamic WEP keys.* Dynamic per-user, per-session, WEP keys are created to easily allow administrators to quickly move away from statically configured WEP keys, thus significantly increasing security. The single largest security vulnerability of a properly secured wireless network (using standard 802.11 b security measures) is the usage of static WEP keys that are subject to discovery through special software. In addition, maintaining static WEP keys in an enterprise environment is an extremely time-consuming and error-prone process. In using LEAP, the session-specific WEP keys that are created are unique to that specific user and are not used by any other user. In addition, the broadcast WEP key (which is statically configured in the AP) is encrypted using the session key before being delivered to the client. Since each session key is unique to the user and can be tied to a network login, LEAP also completely eliminates common vulnerabilities due to lost or stolen network adapters and devices.

- *Reauthentication policies.* Policies can be set that force users to reauthenticate more often to the RADIUS server and thus receive fresh session keys. This can further reduce the window for network attacks as the WEP keys are rotated even more frequently.

- *Initialization vector changes.* The IV is incremented on a per-packet basis, so hackers cannot find a predetermined, predictable sequence to exploit. The capability to change the IV with every packet, combined with the dynamic keying and reauthentication, greatly increases security and makes it that much more difficult for an attacker to gain access to your wireless network.

21.5.2 Building a LEAP Solution

To put together a LEAP with RADIUS solution, you need the following components:

- A Cisco Aironet AP that supports LEAP. Currently, this includes the 350, 1100, and 1200 models. The 350 is the oldest of the bunch and offers the least amount of configurability. The 1100 is the newest and runs IOS, offering both Command Line Interface (CLI)- and Graphical User Interface (GUI)-based management and configuration.

- A Cisco Aironet 350 network adapter.

- The most up-to-date network adapter driver, firmware, and Aironet Client Utility (ACU). You can download this driver using the Aironet Wireless Software Selector on the Cisco Web site at www.cisco.com/pcgi-bin/Software/WLAN/wlplanner.cgi.

- A RADIUS server application that supports LEAP. For our purposes, we use Funk Software's (www.funk.com) Steel Belted Radius/Enterprise Edition.

As shown earlier, our LEAP solution will look (basically) like the diagram shown in Figure 21.45.

Tools and Traps

Nothing in Life Is Perfect...

LEAP has two potential weaknesses that you need to be aware of.

The first weakness is that the EAP RADIUS packet transmitted between the AP and the RADIUS server is sent in cleartext. This packet contains the shared secret used to perform mutual authentication between these two devices. The reality of this weakness, however, is that you can mitigate its potential effects by having good network authentication policies for your wired network. Thus, an attacker would have to plug directly into a switch sitting between the AP and the RADIUS server and use a special network sniffing capable of sniffing over a switched network, such as dsniff.

The second weakness of LEAP is that the username is transmitted in cleartext between the wireless client and the AP. This opens the door to the possibility of a dictionary attack. Note that the password is encrypted using MS-CHAPv1. Your defense against a dictionary attack on your LEAP user's passwords is to implement a solid login policy for your network. For example, if you are using Active Directory and performing network authentication against it using domain user accounts, you could require strong passwords through the Password Policy options and account lockout through the Account Lockout Policy options.

For more information on configuring Active Directory for enhanced security, see *MCSE/MCSA Implementing and Administering Security in a Windows 2000 Network: Study Guide and DVD Training System* (Exam 70-214) by Will Schmied (Syngress Publishing 2003, ISBN 1931836841).

21.5.3 Installing and Configuring Steel Belted RADIUS

To get started with your LEAP/RADIUS solution, you first need to install and configure the RADIUS server of your choosing. As previously stated, we'll use Steel Belted RADIUS (SBR) for this purpose because it integrates tightly with Cisco LEAP. Perform the following steps to get SBR installed and configured for LEAP:

1. Download the SBR installation package from the Funk Web site (www.funk. com). You can download it for a 30-day trial if you are not ready to purchase it.

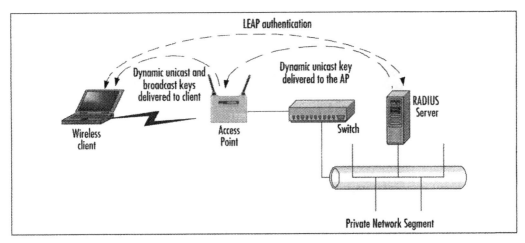

Figure 21.45: The Cisco LEAP and RADIUS solution

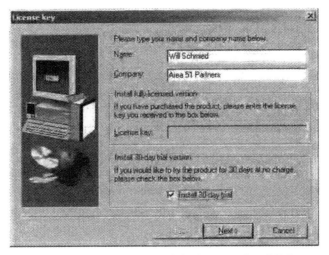

Figure 21.46: SBR has a "Try it before you buy it" feature

2. Provide your name, your organization's name, and your product key, as shown in Figure 21.46. Note that you can opt to exercise the 30-day trial if you desire. Click *Next* to continue.

3. Select the *SBR Enterprise Edition* option on the next page, and click *Next* to continue.

Figure 21.47: Choosing the installation options and location

4. Click Yes to accept the EULA.

5. Click Next to start the setup routine.

6. Select your installation location. Note that you will want to install both the Radius Admin Program and the Radius Server, shown in Figure 21.47. Click *Next* to continue.

7. Continue with the installation routine to complete the installation process.

8. Ensure that the *Yes, launch Radius Administrator* option is selected, and click *Finish*. Once the installation has completed, the Admin application opens. Select the *Local* option and click the *Connect* button. If the display you see is something like that shown in Figure 21.48, you've successfully installed SBR.

9. Close the *SBR Admin* application to begin the configuration of SBR for LEAP.

10. Navigate to the SBR. installation directory and open the Service folder. Locate and open the *eap.ini* file for editing. For this example, we use native RADIUS authentication, meaning that users will be authenticating directly against the SBR RADIUS database. (You can, optionally, configure SBR for Windows domain authentication, as discussed later in this chapter.) Under the

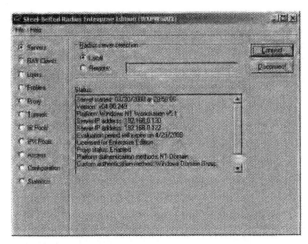

Figure 21.48: Launching the admin application

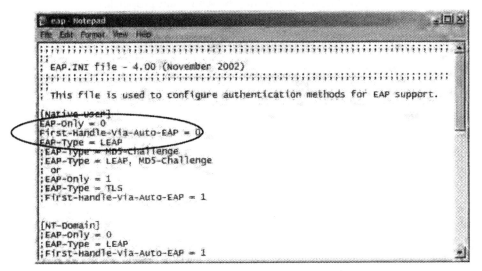

Figure 21.49: Configuring SBR for LEAP

[Native-User] heading, remove the semicolon from the first three items to enable LEAP. Save and close the *eap.ini* file (see Figure 21.49).

11. Restart the Steel Belted Radius service to force it to reload the eap.ini file from the **Services** console located in the *Administrative Tools* folder. Launch the *SBR Admin* application and connect to the local server.

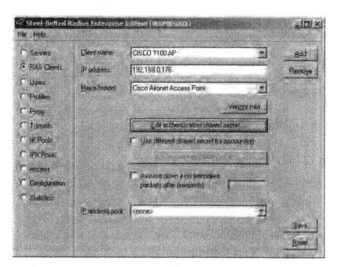

Figure 21.50: Configuring the RAS client properties

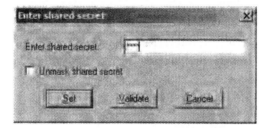

Figure 21.51: Entering the shared secret

12.   Click the *RAS Clients* option to configure SBR for the Cisco Access Point, as shown in Figure 21.50. Note that the AP is the RAS client since it is performing authentication on behalf of the wireless network client. Click the *Add* button to create a new client, and click *OK* to confirm it. Next, specify the client IP address (the IP address assigned to the AP) and the type of client (Cisco Aironet Access Point).

13.   Click the *Edit* authentication shared secret button. This will enter the shared secret to be used for the AP and SBR server to authenticate each other, as shown in Figure 21.51. After entering your shared secret, click the Set button to confirm it. (Remember your shared secret; you will need it again when you

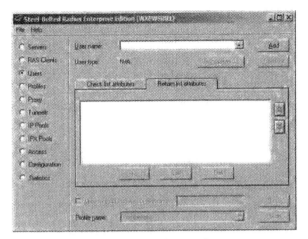

Figure 21.52: Creating native users

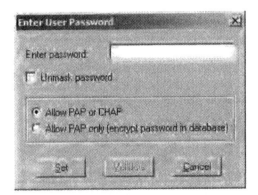

Figure 21.53: Entering the user password

configure the AP later.) Click the *Save* button on the RAS Clients page to confirm the client details.

14. Click the *Users* option to create native users (users internal to the SBR server), as shown in Figure 21.52. Click the *Add* button to add a new username. After entering the username, click *OK* to confirm it.

15. Click the *Set password* button to open the Enter User Password dialog box shown in Figure 21.53. You need to leave the *Allow PAP or CHAP* option selected because LEAP actually makes use of an MS CHAPv1 derivative. After setting the

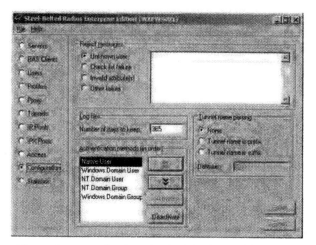

Figure 21.54: Selecting the authentication methods

password, click the Set button to confirm it. Click the Save button on the Users page to confirm the user.

16. Click the *Configuration* option to set the authentication methods (and their order) to be used, as shown in Figure 21.54. Since we are using *native users*, ensure that the Native User option is placed first in the list. Click *Save* to confirm the change if required.

21.5.4 Configuring LEAP

Once you've gotten your RADIUS server installed and configured, the hard work is behind you. All that is left now is to configure LEAP on the AP and client network adapter. To configure LEAP on the AP, perform the following steps. (Note that the exact screen will vary among the 350, 1100, and 1200 APs—the end configuration is the same, however. For this discussion, a Cisco Aironet 1100 AP is used with all configurations performed via the Web interface instead of the CLI.)

1. Log in to your AP via the Web interface.

2. Configure your network SSID and enable EAP authentication, as shown in Figure 21.55. Save your settings to the AP after configuring them.

3. Enter a 128-bit broadcast WEP key, as shown in Figure 21.56. Save your settings to the AP after configuring them.

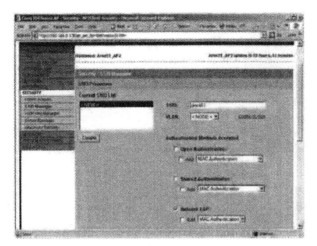

Figure 21.55: Enabling EAP authentication

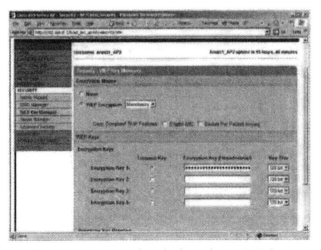

Figure 21.56: Entering the broadcast WEP key

4. Configure your RADIUS server IP address and shared-secret key information, as shown in Figure 21.57. In addition, you need to ensure that the *EAP Authentication* option is selected. Save your settings to the AP after configuring them. If you want to enable a reauthentication policy, you can do so from the Advanced Security-EAP Authentication page shown in Figure 21.58. The default option is Disable Reauthentication. Save your settings to the AP after configuring them.

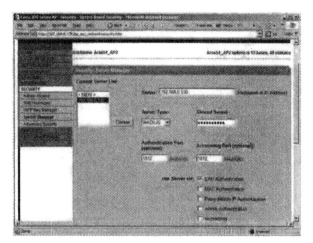

Figure 21.57: Configuring the RADIUS server information

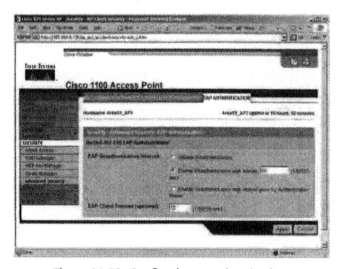

Figure 21.58: Configuring reauthentication.

To enable the wireless client for LEAP, first ensure that it is using the most recent firmware and drivers. Once you've got the most up-to-date files, proceed as follows to get the client configured and authenticated using LEAP:

1. Launch the Cisco Aironet Client Utility (ACU), shown in Figure 21.59. Notice that the ACU reports that the network adapter is not associated with the AP.

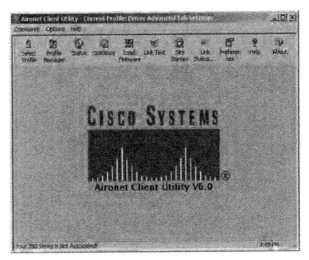

Figure 21.59: Using the Cisco ACU

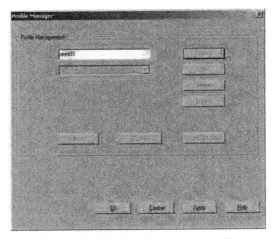

Figure 21.60: Creating a new profile

This is normal at this point because the AP is configured to require LEAP authentication.

2. Click the *Profile Manager* button to create a new profile, as shown in Figure 21.60. Click the *Add* button to enter the new profile name, and then click the *OK* button to begin configuring the profile.

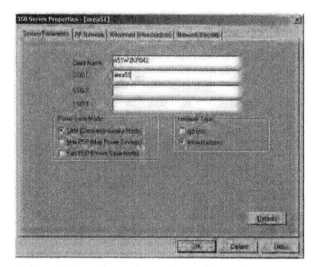

Figure 21.61: Configuring the SSID for the profile

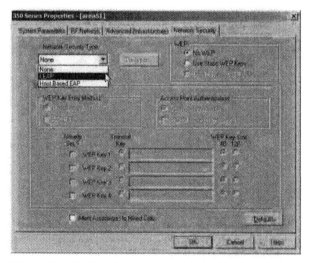

Figure 21.62: Configuring the authentication method

3. Enter the correct SSID for your network (as previously configured for the AP) on the *System Parameters* tab, shown in Figure 21.61.

4. Switch to the *Network Security* tab and select *LEAP* from the drop-down list, as shown in Figure 21.62. After selecting LEAP, click the *Configure* button.

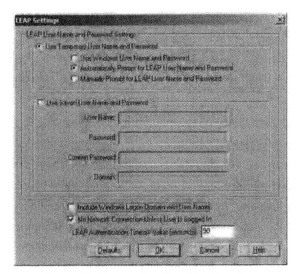

Figure 21.63: Configuring LEAP options

5. Ensure that the *Use Temporary User Name and Password* option is selected with the *Automatically Prompt for LEAP User Name and Password* suboption on the *LEAP Settings* page. Remove the *check* mark from the *Include Windows Logon Domain with User Name* option because we are using Native mode authentication in this example. Click *OK* after making your configuration (see Figure 21.63).

6. Click *OK* twice more and you will be prompted with the LEAP login dialog box shown in Figure 21.64. Enter your details and click *OK*. If you look at the SBR Admin application on the Statistics page, you can see successful and failed authentications. Notice that the statistics shown in Figure 21.65 represent clients that are being forced to reauthenticate to the RADIUS server fairly often.

21.5.5 Windows Active Directory Domain Authentication with LEAP and RADIUS

In the preceding sections, we only looked at creating native users in Steel Belted Radius. As mentioned, however, you can create AD domain users and authenticate directly against Active Directory. This offers many advantages, such as preventing dictionary attacks by enforcing account lockout policies. If you want to use domain user

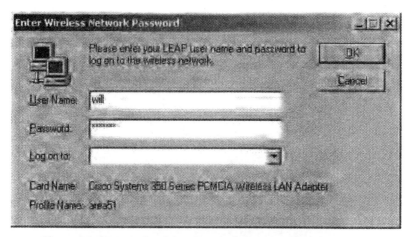

Figure 21.64: Logging into the wireless network using LEAP

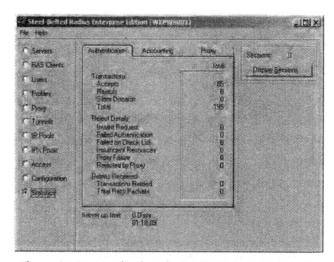

Figure 21.65: Monitoring the RADIUS server statistics

accounts for LEAP authentication, you need only perform the following additions and modifications to the procedures we outlined earlier in this chapter:

1. Make modifications to the eap.ini file, as shown in Figure 21.66. Under the *[NT-Domain]* heading, remove the semicolon from the first three items to enable LEAP. Save and close the eap.ini file.

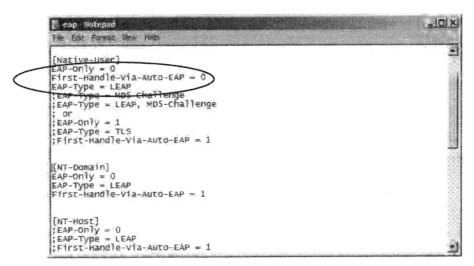

Figure 21.66: Modifying the eap.ini file for domain authentication

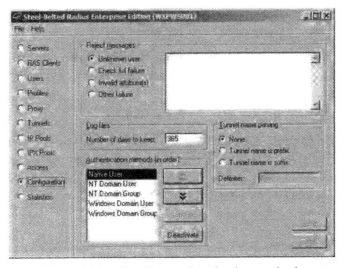

Figure 21.67: Checking authentication methods

2. Restart the Steel Belted Radius service to force it to reload the eap.ini file from the *Services* console located in the *Administrative Tools* folder, Launch the *SBR Admin* application and connect to the local server. On the *Configuration* page, shown in Figure 21.67, ensure that the *NT Domain User* and *NT Domain Group*

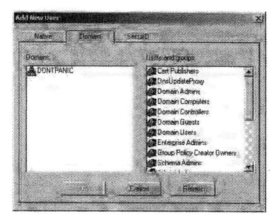

Figure 21.68: Adding a domain user

options are enabled and are at, or near, the top of the list. Click *Save* to confirm the change if required.

3. Go to the *Users* page and add a new user, as described previously. This time, however, you will have the option to select a domain user, as shown in Figure 21.68. Select your domain user and click *OK*. Click *Save* to confirm the addition of the domain user.

4. Open the *ACU* and edit the profile in use (the one you previously created). Switch to the *Network Security* tab and click the *Configure* button next to the Network Security Type drop-down (where *LEAP* is selected). Ensure that the *Use Windows User Name and Password* option is selected, as shown in Figure 21.69. In addition, ensure that a check is placed next to the *Include Windows Logon Domain with User Name* option.

5. Click *OK* three times to save and exit the profile configuration.

6. Log in to the network using LEAP and your domain user credentials.

21.5.6 LEAP Review

Now that you've had a chance to examine the workings of Cisco's LEAP, you should see quite a few benefits to be gained through its use. LEAP, implemented with Funk

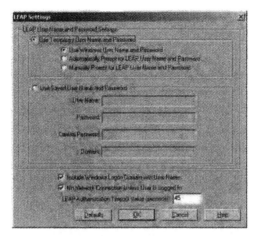

Figure 21.69: Configuring LEAP options for domain authentication

Software's Steel Belted Radius, is an ideal and very robust security solution for a wireless network of any size. By forcing users to authenticate to a back-end RADIUS server and creating per-user, per-session dynamic WEP keys, LEAP provides greatly enhanced authentication and security for your wireless network.

LEAP addresses all WEP's vulnerabilities and is being implemented into the 802.11b standard by the Wi-Fi Alliance (www.wi-fialliance.org), which has implemented LEAP into its standards under the name of Wi-Fi Protected Access (WPA). You can read more about WPA at the Wi-Fi Alliance Web site. In addition, Cisco has licensed the LEAP technology to several third-party vendors, so you can expect to see many more LEAP-compatible devices in the near future. For example, Apple's AirPort network adapter already supports LEAP with version 2 or better firmware.

21.6 Understanding and Configuring 802.1X RADIUS Authentication

To provide better security for wireless LANs, and in particular to improve the security of WEP, a number of existing technologies used on wired networks were adapted for this purpose, including:

- *Remote Authentication and Dial-In User Service (RADIUS).* Provides for centralized authentication and accounting.

- *802.1X.* Provides for a method of port-based authentication to LAN ports in a switched network environment.

These two services are used in combination with other security mechanisms, such as those provided by the Extensible Authentication Protocol (EAP), to further enhance the protection of wireless networks. Like MAC filtering, 802.1X is implemented at layer 2 of the Open System Interconnection (OSI) model: it will prevent communication on the network using higher layers of the OSI model if authentication fails at the MAC layer. However, unlike MAC filtering, 802.1X is very secure as it relies on mechanisms that are much harder to compromise than MAC address filters, which can be easily compromised through spoofed MAC addresses.

Although a number of vendors implement their own RADIUS servers, security mechanisms, and protocols for securing networks through 802.1X, such as Cisco's LEAP and Funk Software's EAP-TTLS, this section will focus on implementing 802.1X on a Microsoft network using Internet Authentication Services (IAS) and Microsoft's Certificate Services. Keep in mind, however, that wireless security standards are a moving target, and standards other than those discussed here, such as the PEAP, are being developed and might be available now or in the near future.

21.6.1 Microsoft RADIUS Servers

Microsoft's IAS provides a standards-based RADIUS server and can be installed as an optional component on Microsoft Windows 2000 and Net servers. Originally designed to provide a means to centralize the authentication, authorization, and accounting for dial-in users, RADIUS servers are now used to provide these services for other types of network access, including VPNs, port-based authentication on switches, and, importantly, wireless network access. IAS can be deployed within Active Directory to use the Active Directory database to centrally manage the login process for users connecting over a variety of network types. Moreover, multiple RADIUS servers can be installed and configured so that secondary RADIUS servers will automatically be used in case the primary RADIUS server fails, thus providing fault tolerance for the RADIUS infrastructure. Although RADIUS is not required to support the 802.1X standard, it is a preferred method for providing the authentication and authorization of users and devices attempting to connect to devices that use 802.1X for access control.

21.6.2 The 802.1X Standard

The 802.1X standard was developed to provide a means of restricting port-based Ethernet network access to valid users and devices. When a computer attempts to connect to a port on a network device, such as switch, it must be successfully authenticated before it can communicate on the network using the port. In other words, communication on the network is impossible without an initial successful authentication.

21.6.2.1 802.1X Authentication Ports

Two types of ports are defined for 802.1X authentication: *authenticator* or *supplicant.* The supplicant is the port requesting network access. The authenticator is the port that allows or denies access for network access. However, the authenticator does not perform the actual authentication of the supplicant requesting access. The authentication of the supplicant is performed by a separate authentication service, located on a separate server or built into the device itself, on behalf of the authenticator. If the authenticating server successfully authenticates the supplicant, it will communicate the fact to the authenticator, which will subsequently allow access.

An 802.1X-compliant device has two logical ports associated with the physical port: an *uncontrolled port* and a *controlled port.* Because the supplicant must initially communicate with the authenticator to make an authentication request, an 802.1X-compliant device will make use of a logical *uncontrolled port* over which this request can be made. Using the uncontrolled port, the authenticator will forward the authentication request to the authentication service. If the request is successful, the authenticator will allow communication on the LAN via the logical *controlled port.*

21.6.2.2 The Extensible Authentication Protocol

EAP is used to pass authentication requests between the supplicant and a RADIUS server via the authenticator. EAP provides a way to use different authentication types in addition to the standard authentication mechanisms provided by the Point-to-Point Protocol (PPP). Using EAR stronger authentication types can be implemented within PPP, such as those that use public keys in conjunction with smart cards. In Windows, there is support for two EAP types:

- *EAP MD-5 CHAP.* Allows for authentication based on a username/password combination. There are a number of disadvantages associated with using EAP MD-5 CHAP. First, even though it uses one-way hashes in combination with a challenge/response mechanism, critical information is still sent in the dear, making it vulnerable to compromise. Second, it does not provide mutual authentication between the client and the server; the server merely authenticates the client. Third, it does not provide a mechanism for establishing a secure channel between the client and the server.

- *EAP-TLS.* A security mechanism based on X.509 digital certificates that is more secure than EAP MD-5 CHAP. The certificates can be stored in the Registry or on devices such as smart cards. When EAP- TLS authentication is used, both the client and server validate one another by exchanging X.509 certificates as part of the authentication process. Additionally, EAP-TLS provides a secure mechanism for the exchange of keys to establish an encrypted channel. Although the use of EAP-TLS is more difficult to configure, in that it requires the implementation of a public key infrastructure (PKI)—not a trivial undertaking—EAP-TLS is recommended for wireless 802.1X authentication.

In a paper published in February, 2002 by William A. Arbaugh and Arunesh Mishra entitled "An Initial Security Analysis of the IEEE 802.1x Standard" the authors discuss how one-way authentication and other weaknesses made 802.1X vulnerable to man-in-the-middle and session-hijacking attacks. Therefore, while it might be possible to use EAP MD-5 CHAP for 802.1X wireless authentication on Windows XP (pre SP1), it is not recommended. EAP-TLS protects against the types of attacks described by this paper.

21.6.2.3 The 802.1X Authentication Process

For 802.1X authentication to work on a wireless network, the AP must be able to securely identify traffic from a particular wireless client. This identification is accomplished using authentication keys that are sent to the AP and the wireless client from the RADIUS server. When a wireless client (802.1X supplicant) comes within range of the AP (802.1X authenticator), the following simplified process occurs:

1. The AP point issues a challenge to the wireless client.

2. The wireless client responds with its identity.

3. The AP forwards the identity to the RADIUS server using the uncontrolled port.

4. The RADIUS server sends a request to the wireless station via the AP, specifying the authentication mechanism to be used (for example, EAPTLS).

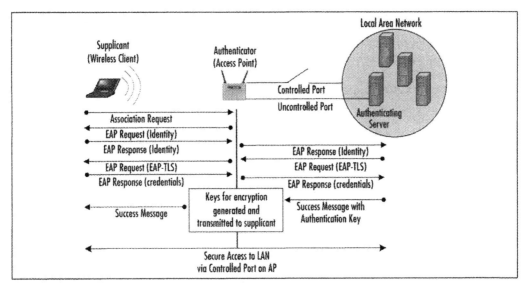

Figure 21.70: 802.1X authentication process using EAP-TLS

5. The wireless station responds to the KADIUS server with its credentials via the AP.

6. The RADIUS server sends an encrypted authentication key to the AP if the credentials are acceptable.

7. The AP generates a multicast/global authentication key encrypted with a per-station unicast session key, and transmits it to the wireless station.

Figure 21.70 shows a simplified version of the 802.1X authentication process using EAP-TLS.

When the authentication process successfully completes, the wireless station is allowed access to the controlled port of the AP, and communication on the network can occur. Note that much of the security negotiation in the preceding steps occurs on the 802.1X uncontrolled port, which is only used so that the AP can forward traffic associated with the security negotiation between the client and the RADIUS server. EAP-TLS is required for the process to take place. EAP-TLS, unlike EAP MD-5 CHAP, provides a mechanism to allow the secure transmission of the authentication keys from the RADIUS server to the client.

21.6.2.4 Advantages of EAP-TLS

There are a number of significant advantages to using EAP-TLS authentication in conjunction with 802.1X:

- The use of X.509 digital certificates for authentication and key exchange is very secure.

- EAP-TLS provides a means to generate and use dynamic one-time-per-user, session-based WEP keys on the wireless network.

- Neither the user nor the administrator know the WEP keys that are in use.

For these reasons, using EAP-TLS for 802.1X authentication removes much of the vulnerability associated with using WEP and provides a high degree of assurance.

In the following section, we will look at how to configure 802.1X using EAP-TLS authentication on a Microsoft-based wireless network. If you are using other operating systems and software, the same general principles will apply. However, you might have additional configuration steps to perform, such as the installation of 802.1X supplicant software on the client. Windows XP provides this software within the operating system.

21.6.3 Configuring 802.1X Using EAP-TLS on a Microsoft Network

Before you can configure 802.1X authentication on a wireless network, you must satisfy a number of prerequisites. At a minimum, you need the following:

- An AP that supports 802.1X authentication. You probably won't find these devices at your local computer hardware store. They are designed for enterprise-class wireless network infrastructures, and are typically higher priced. Note that some devices will allow the use of IP Sec between the AP and the wired network.

- Client software and hardware that supports 802.1X and EAP-TLS authentication and the use of dynamic WEP keys. Fortunately, just about any wireless adapter that allows the use of the Windows XP wireless interface will work. However, older wireless network adapters that use their own client software might not work.

- IAS installed on a Windows 2000 server to provide a primary RADIUS server and, optionally, installed on other servers to provide secondary RADIUS servers for fault tolerance.

- Active Directory

- A PKI using a Microsoft stand-alone or Enterprise Certificate server to support the use of X.509 digital certificates for EAP-TLS. More certificate servers can be deployed in the PKI for additional security. An Enterprise Certificate server can ease the burden of certificate deployment to clients and the RADIUS server through auto-enrollment of client computers that are members of the Windows 2000 domain.

- The most recent service packs and patches installed on the Windows 2000 servers and Windows XP wireless clients.

NOTES FROM THE UNDERGROUND

Beyond ISA Server

Certificates issued by a Microsoft certification authority (CA) will work for wireless authentication. However, certificates issued by other CAs probably will not work. Certificates that are used for wireless 802.1 X authentication *must* contain an optional field called Enhanced Key Usage (EKU). The field will contain one or more object identifiers (OIDs) that identify the purpose of the certificate. For example, the EKU of a typical client certificate used for multiple purposes might contain the following values;

- Encrypting File System (1.3,6.1.4.1.311.103.4)

- Secure Email (1.3.6.1.5.5.7.3.4)

- Client Authentication (1.3.6.1.5.5.7.3.2)

The EKU of the certificate installed on the IAS server and the wire-less client for computer authentication will contain a value for server authentication (1.3.6.1.5.5.7.3.1). Because the EKU is an optional field, it might be absent on certificates issued by non-Microsoft CAs, rendering them useless for 802.1 X authentication in a Microsoft infrastructure. Furthermore, the certificate must contain the fully qualified domain name (FQDN) of the computer on which it is installed in the Subject Alternate Name field, and, in the case of certificates used for user authentication, the user principal name (UPN). You can confirm whether these fields and values exist by viewing the properties of the certificate in

---

the **Certificates** snap-in of the MMC console. (Steps for loading this snap-in are detailed later in this chapter.) There are some other certificate requirements not mentioned here that must be also be satisfied. If you would like to use a third-party CA to issue client certificates for 802.1 X authentication, you should contact the vendor to see if it is supported for this purpose. If not, and you must use a third-party CA, you might need to look at solutions provided by other vendors of wireless hardware to use 802.1 X.

After configuring a PKI and installing IAS on your Windows 2000 network, there are three general steps to configure 802.1X authentication on your wireless network:

1. Install X.509 digital certificates on the wireless client and IAS servers.

2. Configure IAS logging and policies for 802.1X authentication.

3. Configure the wireless AP for 802.1X authentication.

4. Configure the properties of the client wireless network interface for dynamic WEP key exchange.

21.6.3.1 Configuring Certificate Services and Installing Certificates on the IAS Server and Wireless Client

After deploying Active Directory, the first step in implementing 802.1X is to deploy the PKI and install the appropriate X.509 certificates. You will have to install (at a minimum) a single certificate server, either a standalone or enterprise certificate server, to issue certificates. What distinguishes a standalone from an enterprise certificate server is whether it will depend on, and be integrated with, Active Directory. A standalone CA does not require Active Directory. This certificate server can be a *root* CA or a *subordinate* CA, which ultimately receives its authorization to issue certificates from a root CA higher in the hierarchy, either directly or indirectly through intermediate CAs, according to a *certification path*.

NOTE
The certification path can be viewed in the properties of installed certification.

The root CA can be a public or commercially available CA that issues an authorization to a subordinate CA, or one deployed on the Windows 2000 net-work. In enterprise networks that require a high degree of security, it is not recommended that you use the root CA to issue client certificates; for this purpose, you should use a subordinate CA authorized by the root CA. In very high-security environments, you should use intermediate CAs to authorize the CA that issues client certificates. Furthermore, you should secure the hardware and software of the root and intermediate CAs as much as possible, take them offline, and place them in a secure location. You would then bring the root and intermediate CAs online only when you need to perform tasks related to the management of your PKI.

In deploying your PKI, keep in mind that client workstations and the IAS servers need to be able to consult a *certificate revocation list* (CRL) to verify and validate certificates, especially certificates that have become compromised before their expiration date and have been added to a CRL. If a CRL is not available, authorization will fail. Consequently, a primary design consideration for your PKI is to ensure that the CRLs are highly available. Normally, the CRL is stored on the CA; however, additional distribution points for the CRL can be created to ensure a high degree of availability. The CA maintains a list of these locations and distributes the list in a field of the client certificate.

NOTE

It is beyond the scope of this book to discuss the implementation details of a PKI. For more information, please see the various documents available on the Microsoft Web site, in particular www.microsoft.com/windows2000/technologies/security/default. asp, www.microsoft.com/windows2000/techinfo/howitworks/security/pkiintro.asp, and www.microsoft.com/windows2000/techinfo/planning/security/pki.asp.

Whether you decide to implement a standalone or an enterprise CA to issue certificates, you will need to issue three certificates: for both the computer and the user account on the wireless client, as well as the RADIUS server. A certificate is required in all of these places because mutual authentication has to take place. The computer certificate provides initial access of the computer to the network, and the user certificate provides wireless access after the user logs in. While the RADIUS server will authenticate the client

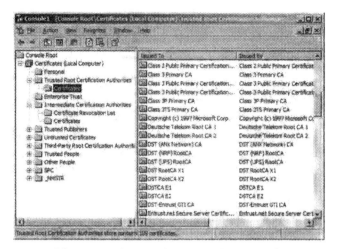

Figure 21.71: Certificate snap-in showing trusted root certification authorities

based on the wireless client's computer and user certificates, and the wireless client will authenticate the RADIUS server based on the server's certificate.

The certificates on the wireless client and the RADIUS server do not have to be issued by the same CA. However, both the client and the server have to trust each other's certificates. Within each certificate is information about the certificate path leading up to the root CA. If both the wireless client and the RADIUS server trust the root CA in each other's certificate, mutual authentication can successfully take place. If you are using a standalone CA that is not in the list of Trusted Root Certification Authorities, you will have to add it to the list. You can do this through a Group Policy Object, or you can do it manually. For information on how to add CAs to the Trusted Root Certification Authorities container, please see Windows 2000 and Windows XP help files. The container listing these trusted root certificates can be viewed in the Certificates snap-in of the MMC console, as shown in Figure 21.71.

Using an enterprise CA will simplify many of the tasks related to certificates that you have to perform. An enterprise CA is automatically listed in the Trusted Root Certification Authorities container. Furthermore, you can use auto-enrollment to issue computer certificates to the wireless client and the IAS server without any intervention on the part of the user. Using an enterprise CA and configuring auto-enrollment of computer certificates should be considered a best practice.

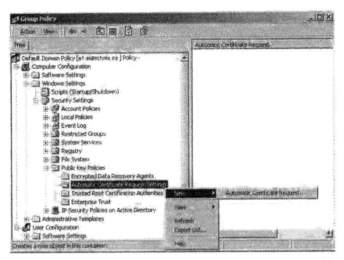

Figure 21.72: Configuring a domain group policy for auto-enrollment of computer certificates

If you put an enterprise CA into place, you will have to configure an Active Directory Group Policy to issue computer certificates automatically. You should use the *Default Domain Policy* for the domain in which your CA is located. To configure the *Group Policy* for auto-enrolment of computer certificates, do the following:

1. Access the *Properties* of the Group Policy object for the domain to which the enterprise CA belongs using *Active Directory Users and Computers*, and click *Edit*.

2. Navigate to *Computer Settings | Windows Settings | Security Settings | Public Key Policies | Automatic Certificate Request Settings*.

3. Right click the *Automatic Certificate Request Settings* with the, click *New*, and then click *Automatic Certificate Request*, as in Figure 21.72.

4. Click *Next* when the wizard appears. Click *Computer* in the *Certificate Templates*, as shown in Figure 21.73, and then click *Next*.

5. Click the enterprise CA, click *Next*, and then click *Finish*.

After you have configured a Group Policy for auto-enrollment of computer certificates, you can force a refresh of the group policy so that it will take effect immediately, rather

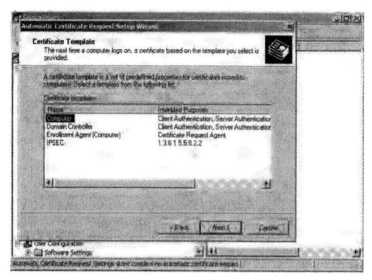

Figure 21.73: Choosing a computer certificate template for auto-enrollment

than waiting for the next polling interval for Group Policy Changes, which could take as long as 90 minutes. To force Group Policy to take effect immediately on a Windows XP computer, type the command *gpupdate/target: computer*.

NOTE

On a Windows 2000 client, group policy update is forced by using the *secedit/ refreshpolicy* command.

Once you have forced a refresh of group policy, you can confirm if the computer certificate is successfully installed. To confirm the installation of the computer certificate:

1. Type the command *mmc* and click *OK* from *Start | Run*.

2. Click *File* in the MMC console menu, and then click *Add/Remove Snap-in*.

3. Click *Add* in the *Add/Remove Snap-in* dialog box. Then, select *Certificates* from the list of snap-ins and check *Add*. You will be prompted to choose which certificate store the snap-in will be used to manage.

4. Select *computer account* when prompted about what certificate the snap-in will be used to manage, and then check **Next**. You will then be prompted to select the computer the snap-in will manage.

5. Select *Local computer (the computer this console is running on)* and check *Finish*. Then, check *Close* and check *OK* to close the remaining dialog boxes.

6. Navigate to the *Console Root | Certificates (Local Computer) | Personal | Certificates* container, as seen in a display similar to the one in Figure 21.73. The certificate should be installed there.

The next step is to install a user certificate on the client workstation and then map the certificate to a user account. There are a number of ways to install a user certificate: through Web enrollment: by requesting the certificate using the Certificates snap-in, by using a CAPICOM script (which can be executed as a login script to facilitate deployment), or by importing a certificate file.

The following steps demonstrate how to request the certificate using the Certificates snap-in:

1. Open an MMC console for *Certificates—Current User*. (To load this snap-in, follow the steps in the preceding procedure; however, at step 5, select *My user account*.)

2. Navigate to *Certificates | Personal* and check the container with the alternate mouse button. Highlight *All Tasks* and then check *Request New Certificate*, as shown in Figure 21.74. The *Certificate Request Wizard* appears.

3. Click *Next* on the *Certificate Request Wizard* welcome page.

4. Select *User* and click *Next* on the *Certificate Types*, as shown in Figure 21.75. You can also select the *Advanced* check box. Doing so will allow you to select from a number of different cryptographic service providers (CSPs), to choose a key length, to mark the private key as exportable (the option might not be available for selection), and to enable strong private key protection. The latter option will cause you to be prompted for a password every time the private key is accessed.

5. Type in a *Friendly Name* of your choosing and a *Description*, and then click *Next*.

6. Review your settings and click *Finish*.

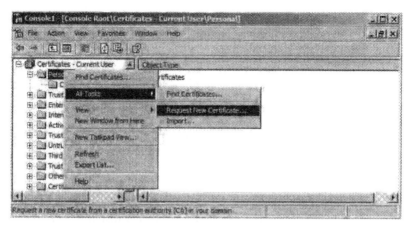

Figure 21.74: Requesting a user certificate

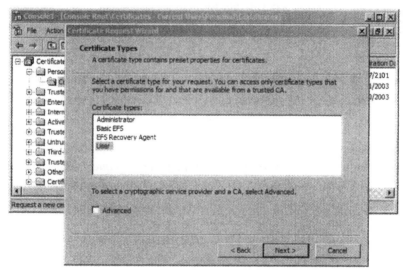

Figure 21.75: Choosing a certificate type

You now should have a user certificate stored on the computer used for wireless access. However, this user certificate will not be usable for 802.1X authentication unless it is mapped to a user account in Active Directory. By default, the certificate should be mapped to the user account. You can verify if it has been mapped by viewing the

Properties of the user account in Active *Directory Users* and Computers. The certificates that are mapped to the user account can be viewed in the *Published Certificates* tab of the Properties of the user account object.

After you configure certificate services and install computer and user certificates on the wireless client and a computer certificate on the RADIUS server, you must configure the RADIUS server for 802.1X authentication.

21.6.3.2 Configuring IAS Server for 802.1X Authentication

If you have configured RRAS for dial-in or VPN access, you will be comfortable with the IAS Server interface. It uses the same interfaces for configuring dial-in conditions and policies as does RRAS. You can use IAS to centralize dial-in access policies for your entire network, rather than have dial-in access policies defined on each RRAS server. A primary advantage of doing this is easier administration and centralized logging of dial-in access.

Installing an IAS server also provides a standards-based RADIUS server that is required for 802.1X authentication. As with configuring RRAS, you will need to add and configure a *Remote Access Policy* to grant access. A *Remote Access Policy* grants or denies access to remote users and devices based on matching conditions and a profile. For access to be granted, the conditions you define have to match. For example, the dial-in user might have to belong to the appropriate group, or connect during an allowable period. The profile in the *Remote Access Policy* defines such things as the authentication type and the encryption type used for the remote access. If the remote client is not capable of using the authentication methods and encryption strength defined in the profile, access is denied.

For 802.1X authentication, you will have to configure a *Remote Access Policy* that contains conditions specific to 802.1X wireless authentication and a *Profile* that requires the use of the *Extensible Authentication Protocol (EAP)* and strong encryption. After configuring the *Remote Access Policy*, you will have to configure the IAS server to act as a RADIUS server for the wireless AP, which is the RADIUS client.

Before installing and configuring the IAS server on your Windows 2000 or .NET/2003 network, you should consider whether you are installing it on a domain controller or member server (in the same or in a different domain). If you install it on a domain controller, the IAS server will be able to read the account properties in Active Directory. However, if you install IAS on a member server, you will have to perform an additional step to register the IAS server, which will give it access to Active Directory accounts.

There are a number of ways you can register the IAS server:

- The IAS snap-in

- The Active Directory Users and Computers admin tool

- The *netsh* command

NOTE

Perhaps the simplest way to register the IAS server is through the *netsh* command. To do this, log on to the IAS server, open a command prompt, and type the command *netsh ras add registeredserver*. If the IAS server is in a different domain, you will have to add arguments to this command. For more information on registering IAS servers, see Windows Help.

Once you have installed and, if necessary, registered the IAS server(s), you can configure the *Remote Access Policy*. Before configuring a *Remote Access Policy*, make sure that you apply the latest service pack and confirm that the IAS server has an X.509 computer certificate. In addition, you should create an Active Directory Global or Universal Group that contains your wireless users as members.

The *Remote Access Policy* will need to contain a condition for *NAS-Port-Type* that contains values for *Wireless-Other* and *Wireless-IEEE802.11* (these two values are used as logical OR for this condition) and a condition for *Windows-Groups=[the group created for wireless users]*. Both conditions have to match (logical AND) for access to be granted by the policy.

The *Profile* of the *Remote Access Policy* will need to be configured to use the *Extensible Authentication Protocol*, and the *Smart Card or Other Certificate* EAP type. Encryption in the *Profile* should be configured to force the strongest level of encryption, if supported by the AP. Depending on the AP you are using, you might have to configure vendor specific attributes (VSA) in the *Advanced* tab of the *Profile*. If you have to configure a VSA, you will need to contact the vendor of the AP to find out the value that should be used, if you can't find it in the documentation.

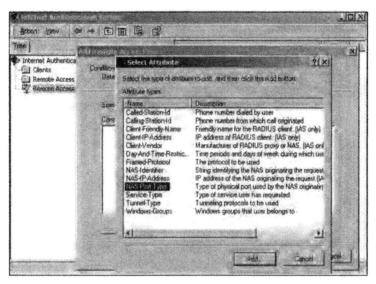

Figure 21.76: Adding a NAS-port-type condition to remote access policy

To configure the conditions for a *Remote Access Policy* on the IAS server:

1. Select Internet Authentication Services and open the IAS console from Start |Programs |Administrative Tools.

2. Right click Remote Access Policies, and from the subsequent context menu, click New Remote Access Policy.

3. Enter a friendly name for the policy and click Next.

4. Click Add in the Add Remote Access Policy Conditions dialog box. Then, select NAS-Port-Type in the Select Attribute dialog box and click Add, as shown in Figure 21.76.

5. Select *Wireless-IEEE 802.11* and *Wireless—Other* from the left-hand window in the *NAS-Port-Type* dialog, and click *Add >>* to move them to the Selected Types window, as shown in Figure 21.77. Click *OK*.

6. Add a condition for *Windows-Groups* that contains the group you created for wireless users after configuring the *NAS-Port-Type* conditions. Then, click *Next*.

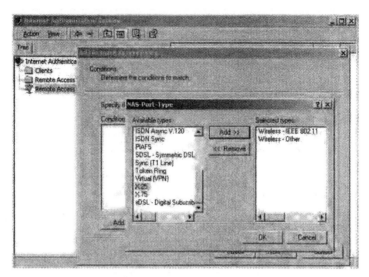

Figure 21.77: Adding wireless NAS-port-type conditions

7. Click the radio button to *Grant remote access permission if user matches conditions* in the subsequent *Permissions* page for the new policy. The next step is to configure the *Profile* to support EAP-TLS and force the strongest level of encryption (128 bit).

8. Click *Edit Profile* and click the *Authentication* tab.

9. Confirm that the checkbox for *Extensible Authentication Protocol* is selected and that *Smart Card or Other Certificate* is listed as the EAP type in the drop-down box. Clear all the other check boxes and click *Configure*.

10. Select the computer certificate you installed for use by the IAS server, and click *OK*. The resulting *Authentication* tab should look like the one in Figure 21.78.

11. Force the strongest level of encryption by clicking the *Encryption* tab and then clearing all the checkboxes except the one for *Strongest*.

12. Save the policy by clicking *OK* and then *Finish*. Make sure that the policy you created is higher in the list than the default Remote Access Policy. You can delete the default policy if you like.

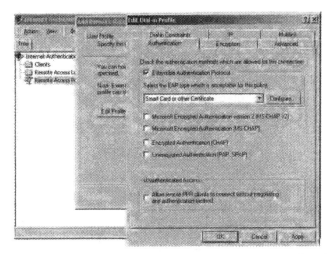

Figure 21.78: Configuring the dial-in profile for 802.1X authentication

Finally, you need to configure the IAS server for RADIUS authentication. To do this, you need to add a configuration for the RADIUS client—in this case, the AP—to the IAS server:

13. Right click the *Clients* folder in the IAS console, and click *New Client* from the context menu.

14. Supply a friendly name for the configuration and click *Next*. The screen shown in Figure 21.79 appears.

15. Configure the screen with the *Client address (IP or DNS)* of the wireless AE and click the checkbox indicating that the *Client must always send the signature attribute in the request.* For the *Shared secret*, add an alphanumeric password that is at least 22 characters long for higher security.

16. Check *Finish*.

You can change the port numbers for RADIUS accounting and authentication by obtaining the properties of the *Internet Authentication Service* container in the IAS console. You can also use these property pages to log successful and unsuccessful authentication attempts and to register the server in Active Directory.

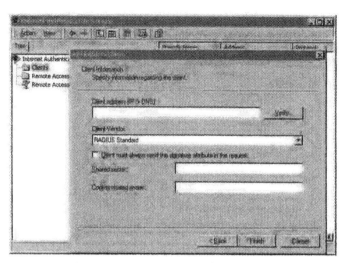

Figure 21.79: Adding a RADIUS client

After installing certificates on the wireless client and IAS server and configuring the IAS server for 802.1X authentication, you will need to configure the AP and the wireless client. The following text shows the typical steps to complete the configuration of your wireless network for 802.1X authentication.

21.6.3.3 Configuring an Access Point for 802.1X Authentication

Generally, only enterprise-class APs support 802.1X authentication; this is not a feature found in devices intended for the SOHO market. Enterprise-class APs are not likely to be found in your local computer store. If you want an AP that supports 802.1X, you should consult the wireless vendors' Web sites for information on the features supported by the APs they manufacture. Vendors that manufacture 802.1X-capable devices include: 3Com, Agere, Cisco, and others. The price for devices that support 802.1X authentication usually starts at $500 (USD) and can cost considerably more, depending on the vendor and the other features supported by the AP. If you already own an enterprise-class AP, such as an ORiNOCO Access Point 500 or Access Point 1000, 802.1X authentication might not be supported in the original firmware but can be added through a firmware update.

Regardless of the device you purchase, an 802.1X-capable AP will be configured similarly. The following text shows the typical configuration of 802.1X authentication on an ORiNOCO Access Point 500 with the most recent firmware update applied to it.

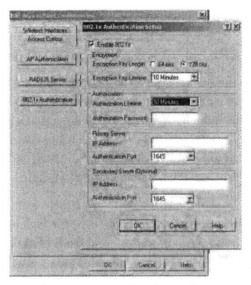

Figure 21.80: Configuring an ORiNOCO AP 500 for 802.1X Authentication

---

NOTE

For more information about the ORiNOCO device, see www.orinocowireless.com.

The configuration of the AP is straightforward and simple (see Figure 21.80). You will need to configure the following:

- *An encryption key length.* This can be either 64 or 128 bits (or higher if your hardware and software support longer lengths).

- *An encryption key lifetime.* When you implement 802.1X using EAP-TLS, WEP encryption keys are dynamically generated at intervals you specify. For higher-security environments, the encryption key life-time should be set to ten minutes or less.

- *An authorization lifetime.* This is the interval at which the client and server will re-authenticate with one another. This interval should be longer than the interval for the encryption key lifetime, but still relatively short in a high-security environment. A primary advantage here is that if a device is stolen, the certificates it uses can be immediately revoked. The next time it tries to authenticate, the CRL will be checked and authentication will fail.

- *An authorization password.* This is the shared-secret password you configured for RADIUS client authentication on the IAS server. This password is used to establish communication between the AP and the RADIUS server. Thus, it needs to be protected by being long and complex. This password should be at least 22 characters long and use mixed case, numbers, letters, and other characters. You might want to consider using a random string generation program to create this password for you.

- *An IP address o f a primary and, if configured for fault tolerance, a secondary RADIUS server.* If the AP is in a DMZ, and the RADIUS server is behind a firewall, this IP address can be the external IP address of the firewall.

- *A UDP port used for RADIUS authentication.* The default port for RADIUS is port 1645. However, you can change this port on the IAS server and the AP for an additional degree of security.

Depending on your AP, you might have to go through additional configuration steps. For example, you might have to enable the use of dynamic WEP keys. On the AP 500, this configuration is automatically applied to the AP when you finish configuring the 802.1X settings. Consult your AP's documentation for specific information on configuring it for 802.1X authentication.

21.6.3.4 Configuring the Wireless Interface on Windows XP for 802.1X Authentication

If you have been following the preceding steps in the same order for configuring 802.1X authentication, the final step is to configure the properties of the wireless interface in Windows XP. You will have to ensure that the properties for EAP-TLS authentication and dynamic WEP are configured. To do this, perform the following steps:

1. Obtain the *Properties* of the wireless interface and click the *Authentication* tab.

2. Ensure that the checkbox for *Enable access control for IEEE 802.1X* is checked and that *Smart Card or other Certificate* is selected as the EAP type, as shown in Figure 21.81.

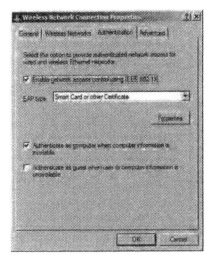

Figure 21.81: Authentication properties for wireless client

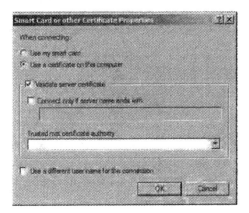

Figure 21.82: Configure Smart Card or other Certificate properties

3. Check *Properties* to view the *Smart Card or other Certificate Properties* window. Ensure that the checkbox for *Validate server certificate* is checked, as shown in Figure 21.82.

4. Select the root CA of the issuer of the server certificate in the *Trusted root certificate authority* drop-down box. If it is not already present, click *OK*. For

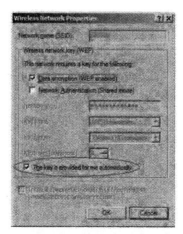

Figure 21.83: Configuring Windows XP wireless properties for 802.1X authentication

additional security, you could select the checkbox for *Connect only if server name ends with* and type in the root DNS name—for example, tacteam.net.

5. Obtain the properties of the wireless interface and click the *Wireless Network* tab.

6. Confirm that the checkbox for *Use Windows for my wireless network settings* is selected In the *Preferred networks* dialog box. Highlight the SSID of the 802.1X-enabled AP, and click *Properties*.

7. Click the checkbox for *The key is provided from me automatically*, as shown in Figure 21.83, and then click *OK*.

That's it. You're finished. The next time you attempt to authenticate and associate with the 802.1X-enabled AP, you might be presented with a prompt asking you to verify the identity of the IAS server certificate. By clicking *OK*, you will permit the authentication process to complete, thus allowing you secure access to the network.

Let's briefly review the steps to enable 802.1X authentication. We are assuming that you are using Active Directory, already have a PKI in place, can issue certificates from a Microsoft CA, and have installed and registered (if necessary) an IAS server. Your steps would be as follows:

1. Issue computer certificate to IAS server.

2. Issue computer certificate to wireless client.

3. Issue user certificate to wireless client user.

4. Create a Remote Access Policy on the IAS server for 802.1X authentication.

5. Configure RADIUS client settings on the IAS server.

6. Configure AP for 802.1X authentication.

7. Configure wireless client network interface for 802.1X authentication.

Although this might seem like a lot of work, the enhanced security provided by 802.1X might well justify the expense and effort of setting it up. Furthermore, much of the effort is up-front. Since you don't have to worry about frequently rotating static WEP keys, you will realize significant savings in effort and time later.

802.1X authentication in combination with EAP-TLS is not the final word in wireless security. It mitigates many of the vulnerabilities associated with wireless networks, but other types of attacks might still be possible.

21.7 Summary

Corporate or SOHO wireless networks require a level of security that goes beyond the basics. They have an obligation to protect their business proprietary and customer data. There are many different technologies that can be utilized to accomplish this. WiFi Protected Access (WPA) addresses many of the flaws inherent in WEP. WPA can utilize the Advanced Encryption Standard (AES) to encrypt wireless network transmissions.

Corporate wireless networks should never be deployed without a virtual private network. There are countless commercial VPN products available. Reef Edge Dolphin is a freeware wireless gateway that can be deployed with VPN capabilities. For SOHO users that don't have the time or the technical staff to deploy and configure a product like Dolphin, Linksys has developed the WRV54G VPN-Broadband Router. The WRV54G provides many enhanced security features. Designed specifically with the small business in mind, the WRV54G provides complete VPN support using IPSec tunnels.

802.1X was originally developed to provide a method for port-based authentication on wired networks. However, it was found to have significant application in wireless networks. With 802.1X authentication, a supplicant (a wireless work-station) needs to be authenticated by an authenticator (usually a RADIUS server) before access is granted to the network. The authentication process takes place over a logical uncontrolled port that is used only for the authentication process. If the authentication process is successful, access is granted to the network on the logical controlled port.

802.1X relies on the Extensible Authentication Protocol (EAP) to perform the authentication. The preferred EAP type for 802.1X is EAP-TLS. EAP-TLS provides the ability to use dynamic per-user, session-based WEP keys, thereby eliminating some of the more significant vulnerabilities associated with WEP. However, to use EAP-TLS, you must deploy a public key infrastructure (PKI) to issue digital X.509 certificates to the wireless clients and the RADIUS server.

21.8 Solutions Fast Track

21.8.1 Implementing WiFi Protected Access (WPA)

- WPA was developed to replace WEP because of the known insecurities associated with WEP's implementation of the RC4 encryption standard.

- Many of the newer access points support WPA; some require firmware upgrades in order to enable WPA functionality.

- Windows XP is WPA-ready with a patch; however, you must ensure that you have WPA drivers for your wireless card.

21.8.2 Implementing a Wireless Gateway with Reef Edge Dolphin

- Wireless gateways are implemented to control access to the network by authenticating users against an internal or external database.

- Wireless gateways can also perform other tasks, including enforcing security by group, implementing quality of service bandwidth controls, and many other advanced security functions such as VPN tunnels and mobile IP roaming between APs.

- Dolphin is a freeware wireless gateway that provides authentication of users against a local database and optional support for IPSec VPN tunnels for data protection. In a small, noncommercial environment, Dolphin can be quickly and economically put into use to increase network security by controlling wireless network access.

21.8.3 Implementing a VPN on a Linksys WRV54G VPN Broadband Router

- The Linksys WRV54G is a broadband router with a built-in access point and VPN.

- The WRV54G was specifically designed for Small Office/Home Office (SOHO) users that require more than the basic security protection for their wired and wireless networks.

- To utilize the VPN features on the WRV54G an IPSec tunnel must be established between the WRV54G and any clients that access it.

21.8.4 Implementing RADIUS with Cisco LEAP

- LEAP addresses all the problems inherent in the use of WEP in a wireless network. The largest vulnerabilities come from static WEP keys and the predictability of IVs.

- LEAP creates a per-user, per-session dynamic WEP key that is tied to the network logon, thereby addressing the limitations of static WEP keys. Since authentication is performed against a back-end RADIUS database, administrative overhead is minimal after initial installation and configuration.

- Policies can be set to force users to re-authenticate more often to the RADIUS server and thus receive fresh session keys. This can further reduce the window for network attacks because the WEP keys are rotated even more frequently.

- The IV is changed on a per-packet basis, so hackers cannot find a predetermined, predictable sequence to exploit. The capability to change the IV with every packet, combined with the dynamic keying and reauthentication, greatly increases security and makes it that much more difficult for an attacker to gain access to your wireless network.

21.8.5 Understanding and Configuring 802.1X RADIUS Authentication

- RADIUS provides for centralized authentication and accounting.

- 802.1X provides for a method of port-based authentication to LAN ports in a switched network environment.

- For 802.1X authentication to work on a wireless network, the AP must be able to securely identify traffic from a particular wireless client. This identification is accomplished using authentication keys that are sent to the AP and the wireless client from the RADIUS server.

Part 4
Other Wireless Technology

CHAPTER 22

Home Network Security

Tony Bradley

22.1 Introduction

I have a wireless network in my home. I am no "Mr. Fix-it" when it comes to home projects, so when I had to figure out how to run network cable from the router in the kitchen to my kids' rooms, going through walls and floors and around cinder blocks, I sprang for the wireless equipment instead. At first, it was primarily so that I could use my laptop from any room in the house, but as time went by we eventually switched almost every computer in the house to a wireless connection.

Wireless networks provide a great deal of convenience and flexibility, and are relatively easy to set up. Actually, they may be too easy to set up. Some pieces of wireless equipment are almost plug-and-play devices, which might explain why so many people don't read the manual or do their homework to figure out how to secure the network airwaves.

I took my laptop out with me today to work on this book. I had no particular need for a network connection since I was just using my word processor on my computer, but as I drove through the subdivision I watched as my wireless network adapter detected network after network completely insecure and announcing their presence to the world. Unfortunately, this seems to be the norm rather than the exception.

This chapter will take a look at wireless network security from two perspectives, and the steps you must take to use it securely in both. We will start with a brief overview of the wireless protocols and technology, and then look at what's required to secure your own home wireless network to keep unauthorized users out of your network. Lastly, we will examine the security precautions you should take to securely use a public wireless network.

22.2 The Basics of Wireless Networks

Think about how a wireless network affects the security of your network and your computers. When you have a wired network, you have only one way in more or less. If you put a firewall on the network cable between your computers and the public Internet, your computers are shielded from most unauthorized access. The firewall acts as a traffic cop, limiting and restricting access into your network through that single access point. Now you throw a wireless device on your network. It doesn't matter if it's one computer with a wireless network adapter or a wireless router or access point, the results are the same: you are now broadcasting data through the air. Your "access point" is now all around you. Rather than a single point of access that can be easily protected, your access point is now three dimensional, all around you, at various ranges, from the next room to the house next door to the roadside in front of your home.

Are You Owned?

Wardriving

The practice of cruising around in search of available wireless networks is known as "wardriving." The term derives from a similar activity to search for available modem connections by "wardialing," or automatically dialing phone numbers to identify which ones result in a dial-up modem connection.

Armed with a wireless device and antenna, wardrivers patrol city streets and neighborhoods and catalog the wireless networks they discover. Some sophisticated wardrivers also tie their wireless network discovery to a GPS to identify the exact coordinates of each wireless network.

For years, a group dedicated to demonstrating how insecure most wireless networks are and increasing awareness of wireless network security issues has organized something called the WorldWide WarDrive (WWWD). After four years, they have decided that the WWWD has done all it can to raise awareness and have moved on to other projects, but their efforts helped to spotlight the issues with insecure wireless networks.

For more information about wardriving and wireless network security in general, you can check out the book *WarDriving and Wireless Penetration Testing*.

Wireless equipment often boasts of ranges over 1,000 feet. The reality is that unless there are no obstructions, the temperature is above 75 and less than 78, the moon is in retrograde and it's the third Tuesday of the month, the range will be more like 100 feet. But if your wireless data can make it the 75 feet from your wireless router in the basement to where you are checking your e-mail while watching a baseball game as you sit on the couch in your living room, it can also make it the 60 feet over to your neighbor's house or the 45 feet out to the curb in front of your home. Although standard off-the-shelf equipment doesn't generally have tremendous range, the wardrivers, a term used to describe actively scouting areas specifically looking for insecure wireless networks to connect to, have homegrown super antennas made with Pringles cans and common household items from their garage that can help them detect your wireless network from a much greater range.

It is important that you take the time to understand the security features of your wireless equipment and make sure you take the appropriate steps to secure your network so that unauthorized users can't just jump onto your connection. Not only are your own computers exposed to hacking if an attacker can join your network, but they may initiate attacks or other malicious activity from your Internet connection which might have the local police or the FBI knocking on your door to ask some questions.

A wireless network uses radio or microwave frequencies to transmit data through the air. Without the need for cables, it is very convenient and offers the flexibility for you to put a computer in any room you choose without having to wire network connections. It also offers you the ability to roam through your home freely without losing your network connection.

In order to connect to the Internet, you will still need a standard connection with an ISP. Whether you use dial-up or a broadband connection like DSL or a cable modem, the data has to get to you some way before you can beam it into the air. Typically, you would connect your DSL or cable modem to a wireless router and from there the data is sent out into the airwaves. If you already have a wired router on your network and want to add wireless networking, you can attach a wireless access point to your router. Any computers that you wish to connect to the wireless network will need to have a wireless network adapter that uses a wireless protocol compatible with your router or access point.

A variety of wireless network protocols are currently in use. The most common equipment for home users tends to be either 802.11b or 802.11g with 802.11a equipment coming in a distant third. The most common protocol, particularly for home users, has been 802.11b; however, 802.11g is becoming the default standard because of its increased speed and compatibility with existing 802.11b networks. The following is a brief overview of the different protocols:

22.2.1 802.11b

Wireless network equipment built on the 802.11b protocol was the first to really take off commercially. 802.11b offers transmission speeds up to 11 mbps, which compares favorably with standard Ethernet networks—plus, the equipment is relatively inexpensive. One problem for this protocol is that it uses the unregulated 2.4 GHz frequency range, which is also used by many other common household items such as cordless phones and baby monitors. Interference from other home electronics devices may degrade or prevent a wireless connection.

22.2.2 802.11a

The 802.11a protocol uses a regulated 5 GHz frequency range, which is one contributing factor for why 802.11a wireless equipment is significantly more expensive than its counterparts. 802.11a offers the advantage of transmission speeds of up to 54 mbps; however, the increased speed comes with a much shorter range and more difficulty traversing obstructions, such as walls, due to the higher frequency range.

22.2.3 802.11g

The 802.11g protocol has emerged as the new standard at this time. It combines the best aspects of both 802.11b and 802.11a. It has the increased transmission speed of 54 mbps like 802.11a, but uses the unregulated 2.4 GHz frequency range, which gives it more range and a greater ability to go through walls and floors, and also helps keep the cost of the equipment down. 802.11g is also backwards-compatible with 802.11b, so computers with 802.11b wireless network adapters are still able to connect with 802.11g routers or access points.

22.2.4 Next-Generation Protocols

Wireless networking is relatively new and constantly evolving. A number of new protocols are currently being developed by the wireless industry, such as WiMax,

802.16e, 802.11n, and Ultrawideband. These protocols promise everything from exponentially increasing home wireless network speeds to allowing you to use a wireless connection to your ISP and even maintain a wireless network connection while in a moving vehicle.

Some of these concepts may not appear in the immediate future, but others are already in use in one form or another. Most wireless network equipment vendors have already begun producing Pre-N or Draft-N devices. These devices are based off of the 802.11n protocol, but have been produced before the 802.11n protocol has actually been finalized. They promise speeds 12 times faster than 802.11g, and a range up to four times that of 802.11g.

The major mobile phone carriers, such as Verizon, Cingular, and TMobile, all offer some sort of broadband wireless access which can be used virtually anywhere their cellular phone network can reach. Using a service like this can give you wireless access almost anywhere, any time, without restriction to any specific site.

22.3 Basic Wireless Network Security Measures

Regardless of what protocol your wireless equipment uses, some basic steps should be taken to make sure other users are not able to connect to your wireless network and access your systems or hijack your Internet connection for their own use.

22.3.1 Secure Your Home Wireless Network

To begin with, change the username and password required to access the administrative and configuration screens for your wireless router. Most home wireless routers come with a Web-based administrative interface. The default IP address the device uses on the internal network is almost always 192.168.0.1. Finding out what the default username and password are for a given manufacturer is not difficult. The equipment usually comes configured with something like "admin" for the username, and "password" for the password. Even without any prior knowledge about the device or the manufacturer defaults, an attacker could just blindly guess the username and password in fewer than ten tries. With a default IP address and default administrative username and password, your wireless router can be hacked into even by novices. Figure 22.1 shows the administration screen from a Linksys wireless router. This screen allows you to change the password for accessing the router management console.

Figure 22.1: The Administration screen from a Linksys wireless router

Make sure you change the username to something that only you would think of. Just like renaming the Administrator account on your computer, you want to choose a username that won't be just as easy to guess as "admin" or whatever the default username was. You also want to choose a strong password that won't be easily guessed or cracked. Lastly, you should change the internal IP subnet if possible. The 192.168.x.x address range is for internal use only. A large percentage of those who use this address range use 192.168.0.x as their subnet, which makes it easy to guess. You can use any number from 0 to 254 for the third octet, so choose something like 192.168.71.x so potential attackers will have to work a little harder.

Remember, the goal is to make it difficult for attackers or malware to penetrate your system. Nothing you do will make your network 100-percent impenetrable to a dedicated and knowledgeable attacker. But, by putting various layers of defense in place such as complex passwords, personal firewalls, antivirus software, and other security measures, you can make it sufficiently hard enough that no casual attacker will want to bother.

22.3.2 Change the SSID

Another big step in securing your home wireless network is not to announce that you have one. Public or corporate wireless networks may need to broadcast their existence so that new wireless devices can detect and connect to them. However, for your home, you are trying to prevent rogue wireless devices from detecting and connecting to your network.

The wireless router or access point has a Service Set Identifier (SSID). Basically, the SSID is the name of the wireless network. By default, wireless routers and access points will broadcast a beacon signal about every 1/10 of a second, which contains the SSID among other things. It is this beacon which wireless devices detect and which provides them with the information they need to connect to the network.

Your wireless network will most likely only have a handful of devices. Rather than relying on this beacon signal, you can simply manually enter the SSID and other pertinent information into each client to allow them to connect to your wireless network. Check the product manual that came with your wireless equipment to determine how to disable the broadcasting of the SSID.

Your device will come with a default SSID which is often simply the name of the manufacturer, such as Linksys or Netgear. Even with the SSID broadcasting turned off, it is important that you not use the default SSID. There are only a handful of manufacturers of home wireless equipment, so it wouldn't take long to guess at the possible SSIDs if you leave it set for the default. Therefore, you need to change this, and preferably not to something equally easy to guess, like your last name.

22.3.3 Configure Your Home Wireless Network

Next, you should configure your wireless network and any wireless network devices for infrastructure mode only. Two types of wireless networks are available for set up: infrastructure and ad hoc. In an infrastructure mode network, a router or access point is required, and all of the devices communicate with the network and with each other through that central point.

An ad hoc network, on the other hand, allows each device to connect to each other in an "ad hoc" fashion (hence the name). Since you are going through all of this effort to make your router or access point more secure, you also need to make sure that the wireless devices on your network are not configured for ad hoc mode and might be providing another means for rogue wireless devices to gain unauthorized access to your network.

By accessing the Properties for your wireless connection, you can click the **Advanced** button at the bottom of the Wireless Networks tab to configure whether your wireless

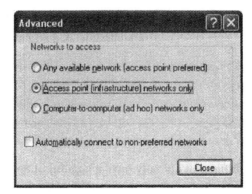

Figure 22.2: Configuring connections for your wireless adapter

adapter will connect to infrastructure, ad hoc, or both wireless network types
(see Figure 22.2).

22.3.4 Restrict Access to Your Home Wireless Network

To restrict access to your wireless network even further, you can filter access based
on the MAC (Media Access Code) addresses of your wireless devices. Each network
adapter has a unique MAC address that identifies it. As stated earlier in this chapter,
your network will most likely consist of only a handful of devices, so it wouldn't
require too much effort to enter the MAC address of each device into your wireless
router or access point and configure it to reject connections from any other MAC
addresses.

Even after you do all of these things, you're not completely secure. You're obscure,
but not secure. Using tools freely available on the Internet, a war-driver could still
intercept your wireless data packets as they fly through the air. They would be doing
so blindly because your wireless access point is no longer broadcasting its presence,
but it can still be done. Intercepting the traffic in this way can provide an attacker with
both the SSID and a valid MAC address from your network so that they could gain
access.

By adding the MAC addresses of the devices that you know you want to connect to
your wireless network, you can block access by other unknown devices and protect your
wireless network (see Figure 22.3).

Figure 22.3: Adding MAC addresses to your wireless router

22.3.5 Use Encryption in Your Home Wireless Network

To further protect your wireless communications, you should enable some form of encryption. Wireless manufacturers, in their haste to start selling equipment, rushed to create WEP (Wired Equivalent Privacy) encryption to provide some level of security while waiting for the official 802.1x security protocol to be standardized. It was quickly discovered that the underlying technology of WEP has a number of flaws which make it relatively easy to crack.

The wireless industry has since migrated to the newer WPA (Wi-Fi Protected Access) encryption, which offers a number of significant improvements over WEP yet remains backwards-compatible with WEP devices. In order to use WPA though, all devices on the network must be WPA-capable. If one device uses WEP, the network will not be able to use some of the improved security features of WPA and your network may still be vulnerable to being exploited by the weaknesses found in WEP.

WPA2 has recently emerged to replace even WPA. Devices that are WPA2-compliant meet stricter security requirements. Windows XP with Service Pack 2 (SP2) fully supports the features and functions of WPA2, allowing a higher level of wireless network security as long as all of your wireless network clients are capable of the same security level.

While a knowledgeable and dedicated attacker with the right tools can still crack the encryption and access your wireless data, this should not discourage you from enabling it. It would be unusual for someone to dedicate that much time and effort to get into your wireless network when they can probably find five more unprotected wireless networks on the next block. It isn't practical to think you will be 100-percent secure, but turning on some form of encryption combined with the other precautions listed previously will deter the casual hacker and curious passerby.

The more complex encryption schemes require more processing power to encode and decode, so you may consider sticking with the 40-bit (64-bit on some devices) WEP encryption rather than the stronger 128-bit, or even the WPA encryption, if you notice any performance issues. It is the difference between locking your house with a normal lock or using a deadbolt. Since an attacker can get past both with about the same effort, you may as well use the one that is easier for you but that still prevents most users from accessing your wireless network.

22.3.6 Review Your Logs

Most wireless routers keep logs of the devices that attach to them. Even if you have taken all of the preceding steps to secure your wireless network, it is a good idea to periodically review the logs from your wireless router and check for any rogue devices that may have gained access.

The other major points to consider regarding a secure home wireless network are the same as they are for a wired network or computer security in general. You should make sure you are using strong passwords that can't be easily guessed or cracked on all of your devices, and protect your computers with personal firewall software.

One final word of advice when it comes to securing your wireless network: a device that is not connected to the Internet can't be attacked or compromised *from* the Internet. You may want to consider turning off your wireless router or access point overnight or when you know that it won't be used for extended periods. If there are too many users trying to access the Internet and use their computers at varying hours, it may be impractical to turn off the wireless router, but you can still turn off any computers when not in use so that they are not exposed to any threats whatsoever.

22.3.7 Use Public Wireless Networks Safely

Public wireless networks, often referred to as hotspots, are springing up all over. National chains such as Starbucks Coffee, Borders Books, and McDonalds have started adding wireless network access to their establishments through services provided by companies like TMobile or Boingo. Major hotel chains have gone from no access to dial-up access to broadband access, and now many are offering wireless network access. Many airports and college campuses have wireless networks as well. It seems like every week someplace new pops up where you can surf the Web while you're out and about.

It is perilous enough jumping onto the Internet using your own network in the comfort of your home, but sharing an unknown network and not knowing if the network or the other computers are secure adds some new concerns. Some of the things you must do to use a public wireless network securely are just simple rules of computer security no matter what network you're connecting to, while others are unique to accessing a public wireless network.

22.3.8 Install Up-to-Date Antivirus Software

For starters, you should make sure you have antivirus software installed and that it is up-to-date. You don't know what, if any, protection the network perimeter offers against malware or exploits, or whether or not the other computers on the network with you are trying to propagate some malware. You also need to make sure that your operating system and applications are patched against known vulnerabilities to help protect you from attack.

22.3.9 Install a Personal Firewall

Your computer should have personal firewall software installed. Again, you have no way of knowing offhand if the network you are joining is protected by any sort of firewall or perimeter security at all. Even if it is, you need the personal firewall to protect you not only from external attacks, but also from attacks that may come from the other computers sharing the network with you.

As a standard rule of computer security, you should make sure that your critical, confidential, and sensitive files are password protected. In the event that any attacker or casual hacker happens to infiltrate your computer system, it is even more important that you protect these files when joining a public wireless network. Make sure you restrict access to only the User Accounts that you want to access those files and use a strong password that won't be easily guessed or cracked.

Tools & Traps...

AirSnarf

AirSnarf, a Linux-based program created to demonstrate inherent weaknesses in public wireless hotspots, can be used to trick users into giving up their usernames and passwords.

The AirSnarf program can interrupt wireless communications, forcing the computer to disconnect from the wireless network. Immediately following the service interruption, AirSnarf will broadcast a replica of the hotspot login page to lure the disconnected user to enter their username and password to reconnect.

The person sitting at the table next to you or sipping an iced latte in the parking lot could be running the program and it would be very difficult for you to realize what was going on. You should monitor your hotspot bill closely for excess usage or charges, and change your password frequently.

More importantly, it is vital that you disable file and folder sharing. This is even more critical if you happen to be using Windows XP Home edition because of the way Windows XP Home manages file and folder sharing and uses the Guest account with a blank password for default access to shared files and folders. Some attackers or malware may still find their way into your system, but that is no reason to leave the door unlocked and a big neon sign welcoming visitors.

22.4 Additional Hotspot Security Measures

All of the things I have mentioned so far are basic security measures that apply whether you are at home, at work, or connecting to a public wireless network while browsing books at Borders. Now let's take a look at some extra things you need to do or consider when connecting to a hotspot.

22.4.1 *Verify Your Hotspot Connection*

To begin with, you need to make sure you *are* connecting to a hotspot and not a malicious rogue access point. When you are connecting to a public wireless network, it will broadcast the SSID, or network name, along with other information your wireless adapter needs to know in order to connect. It is very easy though for an attacker to set up a rogue access point and use the same or similar SSID as the hotspot. They can then create a replica of the hotspot login Web site to lure users into giving up their usernames and passwords or possibly even get credit card numbers and other such information from users who think they are registering for access on the real site.

You should make sure that the location you are at even has a hotspot to begin with. Don't think that just because you happen to be at a coffee shop and a wireless network is available that it must be a free wireless hotspot.

If you are at a confirmed hotspot location and more than one SSID appears for your wireless adapter to connect to, you need to make sure you connect to the right one. Some attackers will set up rogue access points with similar SSIDs to lure unsuspecting users into connecting and entering their login or credit card information.

22.4.2 *Watch Your Back*

Once you take care of ensuring that you are connecting with a legitimate wireless network, you need to take stock of who may be sitting around you. Before you start entering your username and password to connect to the wireless network or any other usernames and passwords for things like your e-mail, your online bank account, and so on, you want to make sure that no overly curious neighbors will be able to see what you are typing.

After you have determined that nobody can see over your shoulder to monitor your typing and you have established that you are in fact connecting to a legitimate public wireless network, you can begin to use the Internet and surf the Web. You should always be aware though of the fact that your data can very easily be intercepted. Not only can other computers sharing the network with you use packet sniffer programs such as Ethereal to capture and analyze your data, but because your data is flying through the air in all directions even a computer in a nearby parking lot may be able to catch your data using programs like NetStumbler or Kismet.

22.4.3 Use Encryption and Password Protection

To prevent sensitive data or files from being intercepted, you should encrypt or protect them in some way. Compression programs, such as WinZip, offer the ability to password-protect the compressed file, providing you with at least some level of protection. You could also use a program such as PGP to encrypt files for even more security.

Password-protecting or encrypting individual files that you may want to send across the network or attach to an e-mail will protect those specific files, but they won't stop someone from using a packet sniffer to read everything else going back and forth on the airwaves from your computer. Even things such as passwords that obviously should be encrypted or protected in some way often are not. Someone who intercepts your data may be able to clearly read your password and other personal or sensitive information.

22.4.4 Don't Linger

One suggestion is to limit your activity while connected to a public wireless network. You should access only Web sites that have digital certificates and establish secure, encrypted connections using SSL (typically evidenced by the locked padlock icon and the URL beginning with "https:").

22.4.5 Use a VPN

For even greater security, you should use a VPN (virtual private network). By establishing a VPN connection with the computer or network on the other end, you create a secure tunnel between the two endpoints. All of the data within the tunnel is encrypted, and only the two ends of the VPN can read the information. If someone intercepts the packets midstream, all they will get is encrypted gibberish.

For SSL-based VPNs, just about any Web browser will do. However, a large percentage of the VPN technology in use relies on IPSec, which requires some form of client software on your computer to establish a connection. It is not important that the VPN software on your computer and that on the other end be the same or even from the same vendor, but it is a requirement that they use the same authentication protocol. Corporations that offer VPN access for their employees typically supply the client software, but you can also get VPN client software from Microsoft or from Boingo.

22.4.6 Use Web-Based E-mail

One final tip for using a public wireless network is to use Web-based e-mail. If you are connecting to a corporate network over an encrypted VPN connection and accessing a corporate mail server like Microsoft Exchange or Lotus Notes, you will be fine. But if you are using a POP3 e-mail account from your ISP or some other email provider, the data is transmitted in clear text for anyone to intercept and read. Web-based e-mail generally uses an encrypted SSL connection to protect your data in transit, and major Web-based mail providers such as Hotmail and Yahoo also scan e-mail file attachments for malware.

22.5 Summary

Wireless networks represent one of the greatest advances in networking in recent years, particularly for home users who want to share their Internet connection without having to run network cabling through the floors and walls. Unfortunately, if not properly secured, wireless networks also represent one of the biggest security risks in recent years.

In this chapter, you learned about the basic concepts of wireless networking and the key features of the main wireless protocols currently being used. We also covered some fundamental steps you need to do to protect your wireless network, such as changing default passwords and SSIDs, disabling the broadcasting of your SSID, or even filtering access to your wireless network by MAC address.

This chapter also discussed the strengths and weaknesses of the wireless encryption schemes such as WEP and WPA, and why you should ensure that your wireless data is encrypted in some way. You also learned that a layered defense, including components such as a personal firewall and updated antivirus software, is a key component of overall security, particularly when using public wireless hotspots.

The chapter ended by discussing some other security concerns that are unique to public wireless hotspots, such as ensuring that the wireless network you are connecting to is a legitimate one and not a rogue hotspot set up to steal your information. In addition, you learned that using a VPN for communications and utilizing Web-based e-mail can help improve your security and protect your information while using public wireless networks.

22.6 Additional Resources

The following resources provide more information on wireless network security:

- Bowman, Barb. *How to Secure Your Wireless Home Network with Windows XP*. Microsoft.com (www.microsoft.com/windowsxp/using/networking/learnmore/ bowman_05february10.mspx).

- Bradley,Tony, and Becky Waring. *Complete Guide to Wi-Fi Security*. Jiwire.com, September 20, 2005 (www.jiwire.com/wi-fi-security-travelerhotspot-1.htm).

- Elliott, Christopher. *Wi-Fi Unplugged: A Buyer's Guide for Small Businesses*. Microsoft.com (www.microsoft.com/smallbusiness/resources/technology/ broadband_mobility/wifi_unplugged_a_buyers_guide_for_small_businesses. mspx).

CHAPTER 23

Wireless Embedded System Security

Timothy Stapko

23.1 Wireless Technologies

Wireless applications essentially combine all of the reasons we need security for embedded systems with limited resources. By definition, many wireless devices will require limited resources, because they will be designed to run on batteries or in environments where available resources must be given to the communications hardware. In the world of inexpensive embedded systems (we exclude inexpensive consumer products from this category; we are referring more to industrial-type controllers), wireless technologies are only just starting to make inroads into the industry. Sure, cellular phones and PDAs have had wireless technologies built-in for some time, but those implementations are specialized and not widely available. For this reason, and the fact that many embedded wireless communications implementations are still jealously guarded by the companies that produce them, there are not a large number of available real-world embedded wireless devices to discuss. However, as wireless technologies decrease in cost and their implementations increase in number, we will begin to see more and more wireless devices available to smaller companies and hobbyists. Wireless is a radical change in the way devices communicate, so this infusion of technology will present some interesting challenges to embedded developers. In this chapter, we will look at a few of the most popular and up-and-coming wireless technologies and the security implications of changing our preferred communications medium from wires to the air.

As we discussed earlier, wired network hardware has dominated the Internet landscape for years. Wireless technologies predate the internet (think radio), but until recently, the technology had lagged behind wired technologies in providing the connectivity required by modern networked and Internet applications. Wi-Fi,[1] which follows the IEEE 802.11 standard protocols, has been around for a while, but the technology was typically reserved for consumer applications and PCs—the infrastructure and physical characteristics of the hardware just did not work with many embedded applications. We are just now seeing the 802.11 protocols make their way into the lowest reaches of the embedded realm, but Wi-Fi is by no means the only wireless technology available. In this chapter, we are going to discuss Wi-Fi, cellular technologies, ZigBee,[2] Bluetooth,[3] and other protocols that can be used for embedded applications. Before we get into the details, though, let's take a brief look at some of the other technologies for wireless communications.

While the Wi-Fi protocols dominate the computer industry, cellular communications is probably the most recognizable form of wireless technology. Cellular communications technologies have been around for a long time, and in some sense, cellular was one of the original "embedded" wireless protocols, since cell phones pretty much embody the principles of an embedded paradigm. Cellular communication is relatively old in the world of technology, but it was optimized for a single application—voice-based telecommunications. Recently, technologies such as GPRS have expanded cellular communications to more general data-driven applications, but it has been a little slow in coming, and any applications usually need to have the backing of one of the large cellular telecommunications companies. That being said, the cellular networking infrastructure is stable and ubiquitous and can be put to great use in applications where none of the other wireless protocols would be effective.

Cellular technology allows for devices to be connected to the global Internet from just about anywhere, but sometimes the scale (and cost) of cellular isn't needed or even justifiable for an embedded application. In recent years, two technologies have risen to the forefront of wireless communications specifically for the embedded world. The first of these to make a name for itself was Bluetooth, which is the technology behind hands-free mobile phone headsets, PDA keyboards, and a host of other consumer devices. However, perhaps even more exciting than Bluetooth for the embedded systems industry is the rise of ZigBee. ZigBee is an 802 protocol, like Ethernet and Wi-Fi (802.15.4 to be exact). It was specifically designed for low-power embedded applications. Bluetooth has some of the same properties, but has found a different niche as a sort of wireless USB protocol, where ZigBee is more

---

[1]The term "Wi-Fi" is a trademark of the Wi-Fi Alliance, www.wifialliance.com.

[2]"ZigBee" is a trademark of the ZigBee Alliance, www.zigbee.org.

[3]"Bluetooth" is a trademark of the Bluetooth Special Interest Group, www.bluetooth.com.

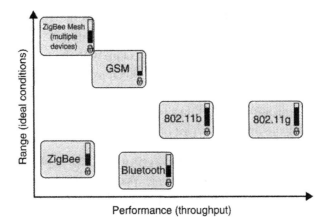

Figure 23.1: Comparison of Wi-Fi, ZigBee, Bluetooth, and Cellular/GSM

reminiscent of a wireless version of good old RS232 serial port. ZigBee is a relatively new protocol (the standard is still in the process of being finalized), but it is already being put to use in some exciting applications. In this chapter, we will look at both Bluetooth and ZigBee, and discuss the security features built right into both of these useful technologies.

23.1.1 Cellular Technologies

Cellular wireless technologies were created for mobile telephone communications but, like their wired counterparts, have diversified and evolved into general-purpose communications technologies. A few technologies are of interest when discussing the connection between cellular networks and digital communications, including GSM[4] (Global System for Mobile Communications, the base technology for a majority of cellular communications) and GPRS[5] (General Packet Radio Service), which adds data transfer capabilities to GSM and allows for services like text messaging and data communications. There are numerous other cellular wireless technologies, but we will keep our discussion to GPRS/GSM because of its widespread use.

---

[4]Originally, GSM referred to "Groupe Speciale Mobile", a European project for developing mobile technologies. Now it is managed and trademarked by The GSM Association (GSMA), www.gsmworld.com.

[5]GPRS and GSM have been expanded into new technologies that provide higher data rates and other services, first EDGE and later 3GSM. We focus on GPRS, since the new technologies just expand its basic functionality.

One of the largest barriers to using cellular technologies for inexpensive wireless communications (in our case, for embedded control applications) is that cellular networks are difficult to get on to, usually requiring a partnership with the organization that owns the network. For a lot of applications, the cost of this may not be practical. However, there do exist companies that do that part for you, and you can buy GPRS/GSM modems that will allow your application to be connected to a cellular network (the modem vendor will usually have a partnership with at least one or two carriers).

The closed nature of cellular networks makes security a difficult problem. The GSM and GPRS technologies have security built into their specifications, but the methods used are not the best. Poor encryption algorithms and questionable security design considerations mean that cellular communications may not be as secure as they could be. If you are going to use GPRS/GSM as a communications medium, it is recommended that you use a higher-level security protocol (SSL is a good choice) on top of the communications channel.

Cellular networking allows for a couple of features that are interesting for embedded applications. The networks are available nearly everywhere, so a cellular-enabled device would have a network connection nearly anywhere, and cellular networks are very good at providing roaming connections, so devices can move around. However, for a large number of embedded control applications, cellular technology is probably overkill. If the embedded device is in a warehouse somewhere and does not move around too much, but needs wireless connectivity, since wires are difficult to run, cellular is probably too slow (dial-up modem speeds are normal) or expensive. For this reason, we leave our discussion of cellular technologies and look at some more practical wireless technologies for limited-resource applications (and as a bonus, they all happen to be generally easier to secure than GPRS/GSM).

23.1.2 802.11 (Wi-Fi)

The wireless communications technologies usually referred to by the (trademarked) term "Wi-Fi" are those technologies based off of the IEEE 802.11 standards. Intended as a general wireless communications protocol (think of Ethernet without wires), 802.11 implementations are by far the most common form of wireless communication between PCs. Wi-Fi is characterized by having a medium range of communications capability (as compared to cellular) with a very large (relative) data rate.

802.11 wireless is a heavy-duty wireless protocol, supporting speeds that rival wired Ethernet (802.11b is capable of 11Mbits/second, and 802.11g is capable of 54Mbps). Its designers recognized the need for security early on and included a security protocol in the original specification: Wired Equivalent Privacy, or WEP. However, WEP was inherently flawed, due to the use of stream ciphers without accounting for some of the important issues inherent in using stream ciphers. One of the major issues was the use of short initialization vectors and infrequent changes of the master RC4 keys. Today, there are numerous implementations that show WEP can be broken in (literally) seconds on modern hardware. Despite this obvious flaw (which is basically a total lack of security), WEP will still stop eavesdroppers that don't know the technical details (i.e., stupid criminals), but it will not stop anyone else, so it should really never be used. One problem, however, was that thousands of (or more) applications were developed for Wi-Fi systems that used and relied on WEP, often in hardware or difficult-to-update firmware. As a result, fixing WEP was not an easy proposition, so a compromise was developed, Wi-Fi Protected Access (WPA).

WPA improved upon WEP by addressing the most grievous flaws exhibited by the original protocol, but retained a level of backward-compatibility that allowed WPA to easily be implemented for most systems that previously relied on WEP. One of the major improvements was to up the effective key size of 40 bits provided by WEP to a full 128 bits. Another major improvement was the use of automated dynamic key management to assure that the stream cipher keys (for RC4 in WEP) were changed on a regular basis to avoid the problem of key reuse (earlier we noted that if the same stream cipher key is used for two different messages, the plaintext was easily recoverable from the ciphertext without actually knowing the key). Authentication was also a problem with WEP, which used the WEP key itself, so WPA uses a separate authentication protocol (there are actually several protocols that can be used, as we will discuss).

WPA, though far superior to WEP (and still widely used), carries some of the concerns about WEP, since it provides backward-compatibility with WEP hardware. For this reason, follow on improvement to WPA was introduced by the Wi-Fi Alliance, creatively named WPA2,[6] which is based on the IEEE 802.11i standard. Both WPA and WPA2 improve security by moving to AES (and also supporting various key sizes larger than 128 bits as required by the US government), and differentiating between personal and

---

[6]WPA and WPA2 are trademarks of the Wi-Fi Alliance.

enterprise networks, which have different requirements. WPA-Personal is designed for home networks, where authentication is not as important as for a business. WPA-Enterprise is basically the same for protecting data (uses AES), but it provides much stronger authentication using a centralized, managed authentication server. The additional management requirements for the centralized server make it too cumbersome for home use, hence the split.

23.1.3 WPA Key Management

WPA and WPA2 (both Personal and Enterprise) utilize a key management mechanism called the Temporal Key Integrity Protocol, or TKIP. TKIP provides the dynamic key management that addressed the key reuse problems in WEP. TKIP is primarily used for WPA-Personal now, since it is based on the RC4 cipher, rather than the (assumed) more secure AES. In order to make deployment easier, WPA-Personal supports what is called a Pre-Shared Key, or PSK. The terms WPA-TKIP or WPA-PSK are often used to refer to WPA-Personal or WPA-Personal. For WPA2-Enterprise, the preferred method is called the CBC-MAC[7] Protocol (CCMP), based upon AES. In order to be compliant, WPA implementations must support TKIP, and WPA2 implementations must support both TKIP (for Personal) and CCMP (for Enterprise).

23.1.4 WPA Authentication

Authentication in WPA (and WPA2) is utilized to prevent unauthorized connections to a network and to help mitigate threats from rogue access points (so-called "evil twins" that trick the client into believing it has connected to the correct network). For WPA-Personal, authentication is not strictly required because of the work required to manage an authentication server. For WPA-Personal, the PSK is usually considered enough for authenticating home wireless networks, but the stronger methods could be used if desired.

For Enterprise authentication, both WPA and WPA2 utilize the same basic framework: 802.1X/EAP. The 802.1X protocol (note that the "X" is really an X, not a placeholder for a number) is part of the IEEE standards for managing both wired and wireless networks, and defines a secure transport mechanism for EAP messages. EAP, the Extensible Authentication Protocol, provides the basis for authentication in both WPA and WPA2.

---

[7]Counter-Mode Cipher Block Chaining (CBC) with Message Authentication Code (MAC). CBC is a method of using the AES block cipher as a stream cipher.

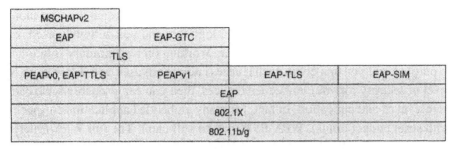

Figure 23.2: Wi-Fi authentication mechanisms

EAP was originally implemented for WPA2, and at the time the only EAP-variant required for Wi-Fi Alliance compliance (needed to use the trademarked Wi-Fi name and logo on a product) was called EAP-TLS, TLS being the Transport Layer Security protocol. However, the number of EAP variants has grown considerably, as various vendors have created their own mechanisms, all slightly different. Now, there are a total of 5 variants required to be compliant: EAP-TLS, EAP-TTLS/MSCHAPv2[8] (TTLS is simply "Tunnelled TLS"), PEAPv0/MSCHAPv2 (Protected EAP, which establishes a TLS connection over which EAP methods are used), PEAPv1/EAP-GTC (an EAP variant developed Cisco), and EAP-SIM which is essentially authentication using SIM cards for the telecom industry. In any case, Wi-Fi authentication is a dynamic and complex field and keeping up with it can be quite a challenge (by the time you are reading this it is likely that there have been a number of new protocols added and compliance requirements have changed). If you are interested in learning more about the wide array of authentication mechanisms for Wi-Fi, there are numerous online resources and there are even a few books on the subject. A good place to start is the Wi-Fi Alliance website itself: www.wi-fi.org. Figure 23.2 shows the relationships between the different authentication mechanisms used by Wi-Fi.

23.1.5 Drowning in Acronyms

The alphabet soup of Wi-Fi authentication protocols begs one important question: What do we need to support? Well, like we have talked about before, we need to adapt the protocol to our application. If you are implementing the latest and greatest consumer gadget and you must be compliant with every wireless access point under the sun, you

---

[8]MSCHAP is a variant of the PPP CHAP protocol developed by Microsoft. MSCHAPv2 is described in RFC 2759.

will likely need to implement most or all of the protocols and mechanisms (or more) described above (you probably want the Wi-Fi logo too, and you definitely need more information than this book provides). If you are working on a proprietary solution for a specific purpose and you can control what access points are used and just want to have some level of security for your embedded devices, then you can probably scale back to a lower level of authentication. In fact, for many embedded applications (especially those with strict budget limits), WPA-PSK may be sufficient. The full WPA-Enterprise authentication suite was designed for large organizations with numerous high-power devices such as laptops and expensive PDA's that need to be continuously updated. The level of security provided by an authentication server is probably overkill for an application that monitors the output of an oil well (for example).

Another important point that has not been addressed is the fact that Wi-Fi is a high-throughput, and therefore high-power wireless protocol. It is very likely that if you are developing an application using a low-power inexpensive microcontroller (which is why you picked up this book, right?), the amount of power required for the 802.11 radio probably exceeds your requirements. In other words, if you are looking to add wireless connectivity to an inexpensive embedded device, you probably do not want Wi-Fi.

23.1.6 Do You Really Need 802.11?

The extensive requirements of Wi-Fi, both in software support and in power consumption, make Wi-Fi a less attractive option for limited-resource systems. We mention it because it is so prevalent and it forms a majority of digital wireless connectivity today. It is possible to implement 802.11 wireless for inexpensive systems, but the functionality will likely need to be reduced to meet the system specs. Fortunately, more than a few people recognized the need for wireless protocols that provide connectivity without the resource requirements of full Wi-Fi. Two protocols have risen in recent years that promise the level of connectivity needed by low-power and inexpensive devices without having to support numerous security protocols and without having the power consumption associated with the higher bandwidth 802.11-based protocols. The first of these is Bluetooth, a standard that has come to be a household word due to its widespread use in mobile telephone headsets and various other consumer devices. The second protocol is a relative newcomer, ZigBee. Where Bluetooth provides a medium level of throughput and is suited for consumer applications (think of it as a wireless USB port as we mentioned at the beginning of the chapter), ZigBee is tailored specifically for embedded industrial applications which often have radically different requirements than consumer applications. We will finish out the chapter by looking at both Bluetooth and ZigBee and how their inherent security properties can be put to use in embedded applications.

23.2 Bluetooth

Named after a relatively obscure Scandinavian king, Bluetooth was one of the first wireless protocols to address the power consumption issues that are inherent in battery-powered consumer devices. By reducing the bandwidth and range requirements, the Bluetooth protocol lends itself to battery-powered applications that require a moderate level of throughput, such as wireless headsets for mobile phones and input devices (such as keyboards and mice) for PDAs. Originally developed by a consortium of technology corporations, Bluetooth has been widely adapted by vendors of consumer gadgets. Driven by widespread use, the Bluetooth physical layer specification was adapted by the IEEE to develop the 802.15.1 standard.

The security of Bluetooth, as with all the wireless protocols we are discussing in this chapter, is designed right into the standard itself. The standard was developed and is controlled by the Bluetooth Special Interest Group (www.bluetooth.com), and the security is based on a 3-mode model, with an unsecured mode, a "service level" secured mode, and a link-level secured mode (the entire connection is secured). According to the Bluetooth SIG, all known attacks against the Bluetooth protocol are actually against specific implementations and the protocol itself is secure.

The security of Bluetooth uses the concept of two separate keys, an authentication key and an encryption key. The authentication key is the master key, and encryption keys are regenerated with each new session. A random number, generated for each transaction, adds additional security. The basic cipher for data protection and the authentication mechanism are described in detail in the Bluetooth specification, should you choose to implement it yourself. However, there are a large number of vendors that supply complete Bluetooth solutions on a single chip, some of which may implement the security in hardware (or the entire Bluetooth stack, as National Semiconductor does with their Simply Blue modules). Due to the availability of such hardware (the Simply Blue modules sell for less than $30 each at the time of this writing), there is very little need to understand the Bluetooth stack in any detail, unless you want to try to implement your own Bluetooth solution.

It should suffice for our discussion to say that there are no known serious attacks on the protocol itself, but various implementations may be vulnerable to a few attacks, referred to as "bluejacking," "bluebugging," and "bluesnarfing."[9] All of these attacks relate to the ability of an attacker to connect to a Bluetooth device (in most instances a mobile phone with Bluetooth) without the knowledge of the device user. The bluejacking attack simply involves the sending of an unwanted message to the device user, which could be used to trick the user into providing sensitive information to the attacker (phishing). The other attacks involve the ability of an attacker to access the contents of a device, either being able to execute commands (bluebugging) or to download data from the device (bluesnarfing). In any case, these attacks require the attacker to be in close proximity (within a few meters) unless they have the equipment to boost the Bluetooth protocol's range. Apparently these issues have been addressed in newer implementations of the protocol but as always, it is a good idea to keep up with current developments, as a devastating attack can always be right around the corner.

Bluetooth provides a decent midrange protocol for embedded systems that need a moderate level of throughput, but it is a complex protocol (the specification is well over 1200 pages long), and the cost of a dedicated controller unit may be prohibitive depending on the application. In the next section, we will look at a relative newcomer to the wireless arena, the ZigBee protocol. Designed around the IEEE 802.15.4 low-power radio standard, ZigBee aims to be the go-to standard for industrial wireless communication where throughput is less of an issue, and flexibility, power consumption, and cost are primary concerns.

23.3 ZigBee

At the low end of the power-requirement spectrum for wireless devices, ZigBee[10] is the equivalent of a dripping faucet when compared to the garden hose of Bluetooth or the fire hose of 802.11 (in the case of 802.11g, a water cannon used for putting out aircraft fires), as seen in Figure 23.3. ZigBee is a relatively new standard developed and maintained by

---

[9]Terms and definitions for Bluetooth attacks adapted from the Bluetooth SIG overview of Bluetooth security (http://www.bluetooth.com/Bluetooth/Learn/Security/)

[10]Parts of the description of ZigBee in the section are adapted from the ZigBee Alliance ZigBee FAQ at www.zigbee.org.

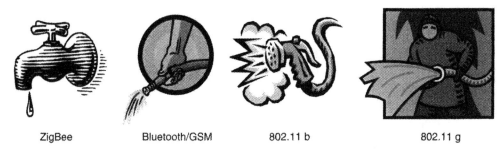

| ZigBee | Bluetooth/GSM | 802.11 b | 802.11 g |

Figure 23.3: Throughput comparison

the ZigBee Alliance. Like the Wi-Fi Alliance and the Bluetooth SIG, the ZigBee Alliance is a consortium of corporations that all utilize the protocol. ZigBee, as of this writing, has not been widely deployed due to it being so new. However, all indications point to ZigBee making a large splash in the industrial controls arena, since it is specifically tailored to such applications. As we mentioned before, ZigBee is characterized by low power consumption (able to run on batteries for extended periods of time due to the low duty cycle of its radio), low system resource requirements, and low throughput. The bandwidth of ZigBee is comparable to that of a dial-up modem (up to 250KB/s), but for the applications it was designed for, this will not be an issue. At first glance, ZigBee may seem very similar to Bluetooth, and in some respects it is, but the target applications for each protocol have led to some significant design differences. Bluetooth, being designed for consumer applications, focuses on higher bandwidth and convenience. ZigBee, on the other hand, is primarily concerned with flexibility of the network (ZigBee supports several network topologies that increase reliability of the entire network—we will look at these in a minute) and conservation of resources, especially power consumption. Geared toward industrial automation (as opposed to consumer connectivity like Bluetooth), ZigBee is definitely an industrial standard.

Like the other wireless protocols we have been discussing this chapter, ZigBee has security built into its specification. Some of the more interesting features of the security inherent in ZigBee have to do with self-healing mesh networks. ZigBee allows for thousands of nodes to be included in a single network, usually referred to as a Personal Area Network (PAN), which was coined to describe the networks specified in the IEEE 802.15 standards (ZigBee radios conform to 802.15.4). ZigBee actually supports several different network topologies, including mesh and clusters, as seen in Figure 23.4. The interesting thing about some of the topologies supported by ZigBee is that they are

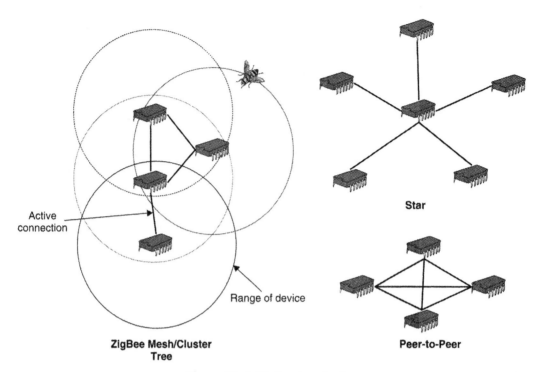

Active connection

Range of device

ZigBee Mesh/Cluster Tree

Star

Peer-to-Peer

Figure 23.4: ZigBee topologies

self-healing, in that the network is resilient and can deal with nodes coming in and out of the network, as would be expected in a noisy (RF noise) industrial environment. This self-healing property makes redundancy very easy to implement, and this can directly translate into a more secure application. Picture an attacker taking out nodes in a sensor network to prevent certain information to be collected. In a Wi-Fi network, once a node is dropped, it must reestablish the connection, including any authentication required. The attacker could take out a few key nodes and the entire network goes down. With a ZigBee network, there can be many more nodes (ZigBee is cheaper to implement) and if any nodes are dropped due to tampering, the network continues to function. The basic functionality of ZigBee is setup to use the radio as little as possible to conserve power, so the end result is that the nodes are usually "down" anyway. Without the need to keep close synchronization (as is the case with Wi-Fi, for example), ZigBee networks can deal with attacks inherently. Next we are going to look at some of the special needs of particular topologies and the security protocol built into the standard.

The network topologies supported by ZigBee vary, but depending on the type of network being deployed there are varying security considerations. In some ZigBee topologies, all nodes are considered equal and the network is basically an ad hoc peer-to-peer configuration. In some of the other topologies (star and tree forms), there must be a *coordinator* that facilitates the network. In such a topology, the end-nodes can be of reduced functionality to save cost, but there is an inherent problem: If the coordinator is disabled, reduced-functionality end nodes cannot reestablish the network. One way to get around this is to provide redundant full-function nodes that can all serve as a coordinator (this is a good idea for reliability, not just security). ZigBee nodes can also function as routers, directing communications between nodes that may not be able to communicate directly (as would be the case if the distance between those nodes was too great, but the router node bisected the path between them). If a ZigBee node is acting as a router, it is important that any information being passed along cannot be compromised if the router node is compromised.

The ZigBee protocol provides low-level security for communications between individual nodes using AES and a message authentication code scheme, but this only protects the data between nodes, not on the nodes themselves. Normally this would be OK if the nodes in question were functioning as end-nodes, but if they are functioning as coordinators or routers, it may be desirable to use a higher-level security scheme to protect that data. This presents another interesting challenge due to the limited bandwidth of ZigBee and the extremely limited resources that are likely to be found on ZigBee devices. A full-blown security protocol like SSL will probably not work because of the extra overhead (especially when you consider running RSA on a ZigBee node). In fact, a number of protocols, like IPSEC, do not even make sense, since ZigBee networks are not based upon TCP/IP. What would make more sense would actually be to use a simple AES-based scheme similar to what is done later in the PIC case study. Authentication could be provided through a password or key that is passed in encrypted form to the end device.

ZigBee is an exciting new technology that promises to bring a level of connectivity never before seen. Depending on how it is deployed, it may also represent a security challenge unlike any we have ever seen. What will be important to remember as ZigBee devices are deployed in factories and homes around the world is that these devices will probably not support a level of security equal to that of full-blown networking. Due to the extremely limited resources, some devices may have no security at all, so it will be vital to keep a constant vigil on what information is being sent over these networks. For these reasons, ZigBee will likely have a place in developing sensor networks and monitoring technologies, but it remains to be seen whether it will be used for anything more.

23.4 Wireless Technologies and the Future

The future of embedded technology lies side by side with the future of wireless. As wireless technologies are improved and are reduced in cost to the point they are as cheap to add to a system as a PIC, we will start to see an explosion of new applications as literally everything is connected to the global Internet. Embedded wireless communications will eventually make science fiction into reality as "smart" paint for bridges and buildings give real-time data on structural conditions and swarms of wirelessly connected micro-sensors show us the internal structure of tornados and hurricanes. Along with this diffusion of technology into every niche of our existences, security takes on a whole new meaning.

It is likely that engineers and corporations will push wireless technology to its boundaries, and security will be of the utmost importance. The lesson to be learned here is that you should consider security when choosing a technology for your particular application. When it comes to security, it never hurts to err on the side of caution and use far more computing power than you need to be sure that you can support the security you need both at deployment and years down the road.

CHAPTER 24

RFID Security

Frank Thornton

24.1 Introduction

In a broad context, radio transmissions containing some type of identifying information are considered Radio Frequency Identification (RFID). This can be a cab driver using his unit number over the air, or the call sign of a radio station. This chapter discusses the tools, applications, and security of RFID.

RFID is about devices and technology that use radio signals to exchange identifying data. In the usual context, this implies a small *tag* or *label* that identifies a specific object. The action receives a radio signal, interprets it, and then returns a number or other identifying information. (e.g., "What are you?" answered with "I am Inventory Item Number 12345").

Alternatively, it can be as complex as a series of cryptographically encoded challenges and responses, which are then interpreted through a database, sent to a global satellite communications system, and ultimately influence a backend payment system.

Some of the current uses of RFID technology include:

- Point of Sale (POS)
- Automated Vehicle Identification (AVI) systems
- Restrict access to buildings or rooms within buildings
- Livestock identification
- Asset tracking
- Pet ownership identification
- Warehouse management and logistics

- Product tracking in a supply chain

- Product security

- Raw material tracking/parts movement within factories

- Library books check-in/check-out

- Railroad car tracking

- Luggage tracking at airports

24.2 RFID Security in General

The multitude of questions regarding RFID applications are influenced by the policy decisions of implementing certain applications, and by the philosophical and religious outlook of the parties involved. Generally, those matters are not discussed, except where a security decision directly influences a privacy policy.

We often embrace new technology without understanding the security issues. We tend to cast a cynical eye at marketers' hyperbole concerning performance. Even so, sometimes we fail to be cynical regarding security claims (or lack thereof) surrounding new technology.

Security is often considered secondary to other issues of certain technologies. RFID is being used in multiple areas where little or no consideration was given to security issues.

Although RFID is a young technology, the security of some RFID systems has already been compromised. In January 2005, the encryption of ExxonMobil's SpeedPass and the RFID POS system was broken by a team of students (as an academic exercise at Johns Hopkins University), because common rules concerning strong encryption were not followed.

In February 2006, Adi Shamir, professor of Computer Science at the Weizmann Institute, reported that he could monitor power levels in RFID tags using a directional antenna and an oscilloscope. He said that patterns in the power levels can be used to determine when password bits are correctly and incorrectly received by an RFID device. Using that information, an attacker can compromise the Secure Hashing Algorithm 1 (SHA-1), which is used to cryptographically secure some RFID tags.

According to Shamir, a common cell phone can conduct an attack on RFID devices in a given area. (Shamir coauthored the Rivest, Shamir, & Adleman (RSA) public-key encryption in 1977.) Recently, a group at Amsterdam's Free University in the Netherlands created RFID viruses and worms as a "proof of concept." This group fit a malicious program (malware) onto the memory area of a programmable RFID chip (i.e., a tag). When the chip was queried by the reader, the malware passed from the chip to the backend database, from where the malware could be passed to other tags or used to carry out malevolent actions. The exploits employed, including Structured Query Language (SQL) and buffer overflow attacks, are generally used against servers.

Because RFID is based on radio waves, there is always the potential for unintended listeners. Even with the lowest powered radios, the distance that a signal travels can be many times more than considered the maximum (e.g., at the DefCon 13 security convention in Las Vegas, Nevada, in July 2005, some consultants received a response from an RFID device from 69 feet away, which is a considerable distance for a device designed to talk to its reader at less than 10 feet).

Additionally, radio waves can move in unexpected ways; they can be reflected off of some objects and absorbed by others. This unpredictability can cause information from an RFID tag to be read longer than intended, or it can prevent the information from being received.

The ability to receive RFID data further away than expected opens RFID to sniffing and spoofing attacks.

Being able to trigger a response from a tag beyond the expected distance makes RFID systems susceptible to denial-of-service (DOS) attacks, where radio signals are jammed with excessive amounts of data that overload the RFID reader.

Radio jamming, where the frequency is congested by a noisy signal, is still a destructive force to be considered when using modern RFID systems.

Much of the increased visibility of RFID within the last few years has been influenced by two things:

- In June 2003, Wal-Mart announced that it would begin using RFID in its supply chain by January 2005. A group of approximately 100 Wal-Mart vendors were selected to use RFID at the company's distribution centers. Those companies will use RFID-enabled cases and pallets, which will be scanned at the point of reception and departure from a given distribution center.

Figure 24.1: Various RFID tags

- The decision by the United States Department of Defense (DoD) to use RFID to improve data quality and management of inventories. In October 2003, the U.S. Acting Under-Secretary of Defense, Michael W. Wynne, issued a memo requiring military suppliers to use RFID tags on shipments to the military by January 2005. The goal is to have a real-time view of all materials.

The DoD has been using RFID to track freight containers since 1995. With a reported inventory of over $80 billion spread over much of the world, the ability to have a real-time view of the location of materials is a requirement.

The widespread use of RFID by both Wal-Mart and the DoD will make other people, companies, and groups aware of the benefits of using RFID. Also, their combined demand ensures that there will be an increase in RFID research and development, and a lowering of the overall prices of RFID equipment. Figure 24.1 shows various types of RFID tags.

As costs are driven down, other large retailers (e.g., Best Buy and Target) are starting to use RFID at the pallet level, or have RFID systems in the planning stage. The costs are

Figure 24.2: RFID reader including the antenna and electronics package

low enough so that smaller RFID units are attainable to hobbyists. Figure 24.2 is a photo of an RFID reader.

NOTES FROM THE UNDERGROUND...

Identification Friend or Foe (IFF)

The concept of automatic identification using a radio transponder originated in World War II as a way to distinguish friendly aircraft from the enemy; hence, the name Identification Friend or Foe (IFF). The "friendly" planes responded with the correct identification, while those that did not respond were considered "foes."

In principle, IFF operates much the same as RFID. A coded interrogation signal is sent out on a particular RF, which the transponder receives and decodes. The transponder then replies with encrypted identification information. Each transponder has a unique identifier; however, some secondary information can be manually set by the pilot.

IFF has expanded since WWII, and now includes several different identification modes for both civilian and military aircraft. These expanded modes add various additional pieces of information, such as the aircraft's altitude. Even though its modern role now includes civilian aircraft, the system is still commonly known as IFF.

24.3 RFID Radio Basics

The following section is a primer on radio waves. If you do not know much about radio, you are encouraged to read it. If you are a radio aficionado, it will seem simplistic; feel free to skip over it.

Radio is a small piece of the "electromagnetic spectrum" that covers all forms of radiation. Other parts of the electromagnetic spectrum that you may be familiar with are cosmic-ray photons, gamma rays, x-rays, and visible light. The Radio Frequency (RF) area is broken down into a number of "bands" (i.e., grouped frequencies) (e.g., the Very High Frequency (VHF) band covers from 30 megahertz (MHz) to 300 MHz. In the United States, using these bands is governed by the Federal Communications Commission (FCC), including who may use a given band, the power level they may transmit at, and how they modulate the signals. Most other countries have a similar regulatory body. Many European Union countries are regulated by the European Telecommunications Standards Institute (ETSI).

Tools and Traps...

It Hertz So Good

RFs are measured in hertz (Hz). Most of the measurements of radio waves for RFID occur in thousands of cycles per second (kilohertz [kHz]); millions of cycles per second (MHz); or billions of cycles per second (gigahertz [GHz]).

The term hertz is in honor of German physicist Heinrich Rudolf Hertz (1857–1894), who was a pioneer in electromagnetism. Hertz proved that electricity is transmitted in electromagnetic waves, and his discoveries helped lead to the development of radio.

Figure 24.3: Two different RFID tags and reader with integral antenna

For RFID, most systems utilize one of three general bands: low frequency (LF) at 125 kHz to 134 kHz, high frequency (HF) at 13.56 MHz, and ultra HF at 860 to 930 MHz. There may be some variation of frequency use, depending on the regulations in a particular locale. Manufacturers of RFID equipment usually choose a given band based on the physics of the band (e.g., how well the signal propagates in a specific environment).The properties of the band also influence the physical size of the antennas and what power transmission levels can be used. Conversely, physical limitations may influence which frequencies and RF bands are used for a given application. Figure 24.3 shows two different RFID tags and a reader.

24.4 Why Use RFID?

In the past few years, RFID has been largely seen as the next technology for pricing at the POS in retail stores. However, it has not replaced bar codes, mainly because the cost of individual tags is expensive. However, with the increased flexibility of being able to perform complete inventory tracking from manufacturer to warehouse to retailer, and with the economic influence of large retail chains, the cost of individual tags will soon become affordable.

Tools and Traps...

RFID Microchips for Pets

The act of placing a passive RFID tag under a pets skin, called "chipping" or "microchipping," has become more prevalent in recent years. A chip the size of a grain of rice is implanted via injection into the skin between the shoulders of the cat or dog. The chip is designed to supplement information used on traditional dog tags.

If a pet is lost and subsequently picked up by the animal control officer, it can be scanned at the animal shelter. If a chip is detected in the animal, shelter personnel obtain the owner information via a database provided by the microchip manufacturer. The owner is then notified that their pet has been impounded.

While excellent in theory, in practice it is not without its pitfalls. Since there are no industry standards for pet tags and readers, different manufacturers are using the same frequencies and encoding techniques. As a result, a scanner that reads chips from a given manufacturer cannot read a different brand of chip. Because of a lack of standardization, a pet was euthanized because the shelter could not read the tags. The detection failed because the shelter used a different brand of scanner than that used by the implanted chip.

Due to concerns about this type of event occurring again, "universal" readers that can read several different brands of chips are being developed and implemented. (For more information go to *www.npr.org/templates/story/story. php?storyId=4783788.*)

24.5 RFID Architecture

The RFID system architecture consists of a reader and a tag (also known as a *label* or *chip*).The reader queries the tag, obtains information, and then takes action based on that information. That action may display a number on a hand held device, or it may pass information on to a POS system, an inventory database, or relay it to a backend payment system thousands of miles away.

Let's look at some of the basic components of a typical RFID system.

24.5.1 Tag/Label

RFID units are in a class of radio devices known as *transponders*. A transponder is a combination transmitter and receiver, which is designed to receive a specific radio signal and automatically transmit a reply. In its simplest implementation, the transponder listens for a radio beacon, and sends a beacon of its own as a reply. More complicated systems may transmit a single letter or digit back to the source, or send multiple strings of letters and numbers. Finally, advanced systems may do a calculation or verification process and include encrypted radio transmissions to prevent eavesdroppers from obtaining the information being transmitted.

Transponders used in RFID are commonly called *tags*, *chips*, or *labels*, which are fairly interchangeable, although "chip" implies a smaller unit, and "tag" is used for larger devices. The designator label is mainly used for the labels that contain an RFID device. (The term "tag" is used for the purposes of this book.)

As a general rule, an RFID tag contains the following items:

- Encoding/decoding circuitry
- Memory
- Antenna
- Power supply
- Communications control

Tags fall into two categories: *active* and *passive* (see Figure 24.4).

24.5.1.1 Passive vs. Active Tags

Passive RFID tags do not contain a battery or other power source; therefore, they must wait for a signal from a reader. The tag contains a resonant circuit capable of absorbing power from the reader's antenna. Obtaining power from the reader device is done using an electromagnetic property known as the *Near Field*. As the name implies, the device must be relatively near the reader in order to work. The Near Field briefly supplies enough power to the tag so that it can send a response.

In order for passive tags to work, the antenna and the tag must be in close proximity to the reader, because the tags do not have an internal power source, and derive their power to transmit from coupling to the Near Field of the antenna. The Near Field takes

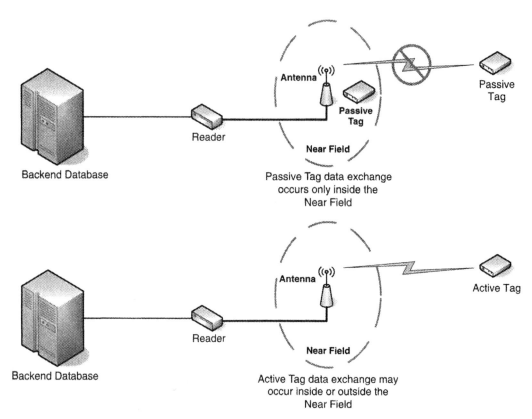

Figure 24.4: Passive and active tag processes

advantage of electromagnetic properties and generates a small, short-lived electrical pulse with the passive tag that can power a tag long enough for it to respond.

Tools and Traps...

Near Field

The Near Field is a phenomenon that occurs in a radio transmission, where the magnetic portion of the electromagnetic field is strong enough to induce an electrical field in a coil. As the name implies, the Near Field occurs in an area near to the antenna. Just how big the Near Field is, depends on the wavelength of the radio signal being used.

$$r = \lambda/2\pi$$

where λ is the wavelength.

For example, a common RFID frequency is 13.56 MHz and the wavelength of 13.56 MHz is approximately 22 meters. Therefore:

$$22/2\pi = 22/6.28 = 3.5 \text{ meters.}$$

The Near Field for an RFID device operating at 13.56 MHz is 3.5 meters or 11.5 feet. Passive tags requiring the Near Field have to be within that area in order to operate correctly.

The alternative to a *passive* tag is an *active* tag. Active tags have their own power source, usually an internal battery. Since they contain a battery to power the radio circuitry, they can actively transmit and receive on their own, without having to be powered by the Near Field of the reader's antenna. Because they do not have to rely on being powered by the reader, they are not limited to operating within the Near Field. They can be interrogated and respond at further distances away from the reader, which means that active tags (at a minimum) are able to transmit and receive over longer distances.

Semi-passive tags have a battery to power the memory circuitry, but rely on the Near Field to power the radio circuits during the receiving and sending of data.

24.5.2 Reader

The second component in a basic RFID system is the *interrogator* or *reader*. The term "reader" is a misnomer; technically, reader units are *transceivers* (i.e., a combination *transmitter* and *receiver*). But, because their usual role is to query a tag and receive data from it, they are seen as "reading the tag"; hence, the term "reader." Readers can have an integrated antenna, or the antenna can be separate. The antenna can be an integral part of the reader, or it can be a separate device. Handheld units are a combination reader/ antenna, while larger systems usually separate the antennas from the reader.

Other parts that a reader typically contains are a system interface such as an RS-232 serial port or Ethernet jack; cryptographic encoding and decoding circuitry; a power supply or battery; and communications control circuits.

The reader retrieves the information from the RFID tag. The reader may be self-contained and record the information internally; however, it may also be part of a localized system such as a POS cash register, a large Local Area Network (LAN), or a Wide Area Network (WAN). Readers that send data to a LAN or other system do so using a data interface such as Ethernet or serial RS-232.

Readers, and in particular their antenna arrays, can be different sizes, from postage stamp-sized to large devices with panels that are several feet wide and high.

24.5.3 Middleware

Middleware software manages the readers and the data coming from the tags, and passes it to the backend database system. Middleware sits in the middle of the data flow between the readers and the backend, and manages the flow of information between the readers and the backend. In addition to extracting data from the RFID tags and managing data flow to the backend, middleware performs functions such as basic filtering and reader integration and control.

As RFID matures, middleware will add features such as improved and expanded management capabilities for both readers and devices, and extended data management options.

The backend can be a standard commercial database such as SQL, My SQL, Oracle, Postgres, or similar product. Depending on the application, the backend database can run on a single PC in an office, to multiple mainframes networked together via global communications systems.

24.6 Data Communications

In the next few sections we'll look in detail at the data the tags are carrying, and how some of the more popular protocols work when they communicate the data to the reader. We'll also talk about the physical format of the cards, and how physical form can be adapted to the particular job.

24.6.1 Tag Data

Depending on the type of tag, the amount of data it can carry is anything from a few bytes up to several megabytes. The amount of data carried by a tag depends on the application and the individual tag.

The data carried in a tag can be in most formats, as long as both the tag and the reader agree on it. Many formats are proprietary, but standards are emerging. In the next section, we look at the Electronic Product Code™ (EPC™). The EPC™ is considered the RFID replacement for the Universal Product Code (UPC) barcode and, as such, will have a huge impact on retail sales in the future.

The UPC bar code has been the accepted means of conveying pricing at the POS in retail stores since the 1970s (see Figure 24.5).This particular UPC is from Syngress Publishing's *WarDriving: Drive, Detect, Defend.* Each UPC bar code contains basic information about the bar coding system, the manufacturer, the item, and a check digit. Because 5 digits are used for both the manufacturer and the item, the total number of manufacturers is limited to 100,000, each limited to 100,000 items. While this allows for 10,000,000,000 products, it is more restrictive than is obvious. As manufacturers add new items and close out old product lines, UPC numbers are quickly being used up. The UPC does not allow serial numbers to be encoded into the bar code.

24.6.1.1 Electronic Product Code

The new Electronic Product Code uses the EPCglobal organization's General Identifier (GID-96) format. GID-96 has 96 bits (12 bytes) of data. Under the GID-96 standard, every EPC™ consists of three separate fields: the 28-bit General Manager Number that identifies the company or organization; the 24-bit Object Class that breaks down products into groups; and the 36-bit serial number that is unique to the individual object. A fourth field consisting of an 8-bit header is used to guarantee the uniqueness of the EPC™

Figure 24.5: Typical UPC bar code

Table 24.1: EPC™ Fields

	Header	General Manager Number (Company)	Object Class (Groups)	Serial Numbers
Number of Bits:	8	28	24	36
Total numbers:		268,435,455	16,777,215	68,719,476,735

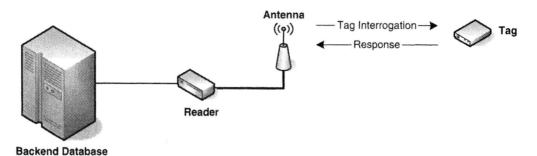

Figure 24.6: Reader and tag interaction

code (see Table 24.1). EPCglobal is a not-for-profit worldwide organization that assigns EPC™ to subscribers.

Each company or manufacturer is assigned a General Manager Number from EPCglobal. Each manufacturer assigns an Object Class number to each product line. Each individual item is identified by a Serial Number. Manufacturers can assign the product number and the serial number in any way they deem desirable. Potentially, this allows the manufacturer the ability to uniquely identify every single item.

This allows for a total of 30,939,155,745,879,204,468,201,375 unique items under the EPC™ system.

The EPC™ standards for data tags can be downloaded from:

www.epcglobalinc.org/standards_technology/EPC_Tag%20Data%20Specification%201. 1Rev%201.27.pdf.

24.6.2 Protocols

RFID systems work when a reader antenna transmits radio signals. Those signals are picked up by the tag, which answers with a responding radio signal (see Figure 24.6).

Table 24.2: RFID Tag Protocols

Protocol	Capabilities
EPC™ Generation 1	"Read Only," preprogrammed
Class 0	
EPC™ Generation 1	"Write" Once, "Read" Many
Class 1	
EPC™ Generation 2.0	"Write" Once, "Read" Many;
Class 1	A more globally accepted version of the Generation 1, Class 1 protocol.
ISO 18000 Standard	"Read-Only" tag identifier; may also contain rewritable memory available for user data. ISO 18000 has different subsections depending on the frequency used and the intended application.
ISO 15963	Unique Tag ID
ISO 15961	Data protocols: data encoding rules and logical memory functions
ISO 15962	Data protocols: application interface

That signal is then read by the reader's receiver. Depending on the tag's computational power (if any), the tag may perform some encryption or decryption functions.

Some tags are "read-only," while other tags have data "written" to them and "read" from them. Using a process similar to the "read" cycle, the reader can "write" data to the tag if it a data "write" operation is needed.

Some tag protocols are proprietary, but EPCglobal and the International Organization for Standardization (ISO) have defined several protocols (see Table 24.2).

ISO also has standards for supply chain applications, tag and reader performance and conformance, and product packaging tagging standards.

24.7 Physical Form Factor (Tag Container)

A tag can take almost any form desired to perform required functions. The design may be influenced by the type of antenna, which in turn may be dependant on the frequency used for the system. The tags may be standalone devices, or integrated into another object such as a car ignition key. Systems parameters, such as whether active or passive tags are required and whether a battery is on a tag, can also influence the design.

Figure 24.1 shows that tags can be put into packages of almost every conceivable shape. The rule is: The larger the tag, the further distance it may be "read."

The following sections discuss some typical tags.

24.7.1 Cards

RFID tags in a "credit card" physical format are usually used for purposes such as building access. This type typically involves security. Personnel that are allowed to enter, or restricted from entering, certain areas of the building are a given encoded cards. Readers are typically mounted next to a door where access is controlled. The reader relays the cardholder information to a database and the database determines whether the cardholder has line access to that particular area. If access is allowed, an electronic door lock is disengaged, allowing access to the building or to a particular room.

Some of the first commercial RFID applications were card-controlled entry systems using "proximity cards." Proximity cards do not carry as much information as newer RFID units and are about double or triple the thickness of a credit card. Newer RFID cards are the same thickness as a credit card.

The white rectangles seen in Figures 24.1 and 24.3 are RFID cards, each containing an electronic microchip with a serial number encoded.

Credit cards are seen as potential RFID tags. In late 2005, television viewers saw new credit card commercials showing the PayPass system and their "Tap 'N' Go" Tag line. The credit card becomes a tag, because it has an integral RFID chip. Instead of swiping the card through a traditional magnetic card reader, the user holds the credit card containing the RFID chip near the reader at the POS. The transaction is completed in a matter of seconds. According to the *RFID Gazette*, the tag conforms to the International Organization for Standardization (ISO)/IEC 14443 standard, uses Triple Data Encryption Standard (DES) and SHA-1 cryptography, and operates at 13.56 MHz.

The RFID technology is being pushed to the extent that the latest "dummy" cards used for American Express advertising show a fake RFID chip and antenna. The newest design calls for the card plastic to be clear. Figure 24.7 depicts a replica card recently received in a credit card application. The fake RFID chip and antenna are pointed out with arrows.

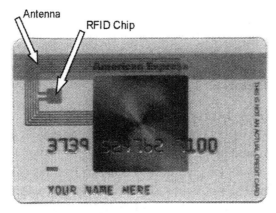

Figure 24.7: Fake credit card showing the RFID chip and antenna

Figure 24.8: A passive tag's internal components

24.7.2 Key Fobs

Key fobs are also popular for POS systems. The RFID tip is encapsulated in a small cylinder or other container designed to use on a key ring. This allows the tag to be conveniently located (e.g., the passive key fobs used as part of the ExxonMobil SpeedPass system are approximately 1-1/2″ long and 3/8″ in diameter). The internal electronics are even smaller; the glass-encased RFID chip and antenna assembly is approximately 7/8″ long by 5/32″ in diameter. Figure 24.8 shows an example of a passive tag's internal components.

The ExxonMobil SpeedPass is a passive tag, designed to be held in the user's hand, and waved within close proximity (>1″) in front of the gas pump's integral reader. ExxonMobil also makes active SpeedPass tags designed to be vehicle mounted.

24.7.3 Other Form Factors

In contrast to key fob tags, other tags may be designed very small to mount onto retail packages, or very large to mount onto vehicles (e.g., the tags used by the E-ZPass system, a toll .collection system used in the Northeast US, is a plastic box approximately 3-/2" wide × 3″ high × 5/8″ thick (see Figure 24.9). The E-ZPass tag is active, and designed to be carried on the windshield of a subscriber's vehicle. The reader antennas are either mounted on a tollbooth 6 to 10 feet from the vehicle, or on a gantry approximately 20 feet above the roadway (see Figure 24.10).

NOTES FROM THE UNDERGROUND...

How the ExxonMobil SpeedPass and E-ZPass Systems Work

The ExxonMobil SpeedPass employs RFID to speed customers through fuel purchases. Here's how it works:

1. An RFID tag mounted on the vehicle or attached to the consumer's key chain is activated by the reader. The reader is connected to the pump. The reader handshakes with the tag and reads the encrypted serial number.

2. Cables connect the reader and pump to a satellite transceiver in the gas station.

3. The transceiver sends the serial number from the RFID tag up to a Very Small Aperture Terminal Satellite (VSAT). The VSAT, in turn, relays the serial number to the earth station.

4. The serial number is sent to the ExxonMobil data center from the earth station. The data center verifies the serial number, and checks for authorization on the credit card that is linked to the account.

5. The authorization is sent back to the pump following the above route in reverse.

6. The pump turns once it receives the authorization, and allows the customer to gas up their vehicle.

ExxonMobil has extended the reader inside service stations and convenience stores. By placing a reader near the cash register, a customer can charge purchases made at an ExxonMobil store on the same charge system as their gasoline purchases.

The E-ZPass toll system works in a similar manner as the SpeedPass:

1. As the car enters the toll plaza, the car-mounted tag is activated by the reader antenna for that lane. Tags can be mounted on the windshield or the license plate.

2. An encoded number is sent from the tag back to the reader.

3. The reader transfers that information to the E-ZPass database.

4. The amount of the toll is deducted from the prepaid account, which is usually a fixed amount. However, on some highways such as the NY Thruway, the toll is based on the distance traveled, in which case the database tracks the entry and exit points, and the toll is computed based on those locations.

5. The database time- and date stamps the transaction, assigns a transaction number, and records the location of the tollbooth.

6. A green light, open gate, or text message (sometimes all three) tells the driver that they can pass through the toll booth. Other lights or messages may indicate errors or account problems.

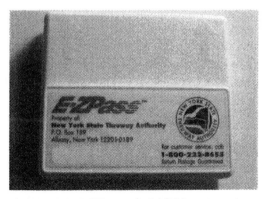

Figure 24.9: E-ZPass windshield-mounted tag

Figure 24.10: E-ZPass high-speed toll plaza–antenna array

24.8 Threat and Target Identification

So far, we have learned how Radio Frequency Identification (RFID) works and how it is applied in both theory and real-world operations. This chapter discusses how security is implemented in RFID, and the possible attacks that can occur on RFID systems and applications.

Before we can *analyze* possible attacks, we have to *identify* potential targets. A target can be an entire system (if the intent is to completely disrupt a business), or it can be any section of the overall system (from a retail inventory database to an actual retail item).

Those involved in information technology security tend to concentrate solely on "protecting the data." When evaluating and implementing security around RFID, it is important to remember that some physical assets are more important than the actual data. The data may never be affected, even though the organization could still suffer tremendous loss.

Consider the following example in the retail sector. If an individual RFID tag was manipulated so that the price at the Point of Sale (POS) was reduced from $200.00 to $19.95, the store would suffer a 90 percent loss of the retail price, but with no damage to the inventory database system. The database was not directly attacked and the data in the database was not modified or deleted, and yet, a fraud was perpetrated because part of the RFID system had been manipulated.

In many places, physical access is controlled by RFID cards called "proximity cards." If a card is duplicated, the underlying database is not affected, yet, whoever passes the counterfeit card receives the same access and privileges as the original cardholder.

24.8.1 Attack Objectives

To determine the type of an attack, you must understand the possible objectives of that attack, which will then help determine the possible nature of the attack.

Someone attacking an RFID system may use it to help steal a single object, while another attack might be used to prevent all sales at a single store or at a chain of stores. An attacker might want misinformation to be placed in a competitor's backend database so that it is rendered useless. Other people may want to outmaneuver physical access control, while having no interest in the data. Therefore, it is necessary for anyone looking at the security of an RFID system to identify how their assets are being protected and how they might be targets.

Just as there are several basic components to RFID systems, there are also several methods (or vectors) used for attacking RFID systems. Each vector corresponds to a portion of the system. The vectors are "on-the-air" attacks, manipulating data on the tag, manipulating middleware data, and attacking the data at the backend. The following sections briefly discuss each of these attacks.

24.8.1.1 Radio Frequency Manipulation

One of the simplest ways to attack an RFID system is to prevent the tag on an object from being detected and read by a reader. Since many metals can block radio frequency (RF) signals, all that is needed to defeat a given RFID system is to wrap the item in aluminum foil or place it in a metallic-coated Mylar bag. This technique works so well that New York now issues a metallic-coated Mylar bag with each E-ZPass.

From the standpoint of over-the-air attacks, the tags and readers are seen as one entity. Even though they perform opposite functions, they are essentially different faces of the same RF portion of the system.

An attack-over-the air-interface on tags and readers typically falls into one of four types of attacks: spoofing, insert, replay, and Denial of Service (DOS) attacks.

Spoofing

Spoofing attacks supply false information that looks valid and that the system accepts. Typically, spoofing attacks involve a fake domain name, Internet Protocol (IP) address, or Media Access Code (MAC). An example of spoofing in an RFID system is broadcasting an incorrect Electronic Product Code™ (EPC™) number over the air when a valid number was expected.

Insert

Insert attacks insert system commands where data is normally expected. These attacks work because it is assumed that the data is always entered in a particular area, and little to no validation takes place.

Insert attacks are common on Web sites, where malicious code is injected into a Web-based application. A typical use for this type of attack is to inject a Structured Query Language (SQL) command into a database. This same principle can be applied in an RFID situation, by having a tag carry a system command rather than valid data in its data storage area (e.g., the EPC number).

Replay

In a *replay* attack, a valid RFID signal is intercepted and its data is recorded; this data is later transmitted to a reader where it is "played back." Because the data appears valid, the system accepts it.

DOS

DOS attacks, also known as *flood* attacks, take place when a signal is flooded with more data than it can handle. They are well known because several large DOS attacks have impacted major corporations such as Microsoft and Yahoo. A variation on this is *RF jamming*, which is well known in the radio world, and occurs when the RF is filled with a noisy signal. In either case, the result is the same: the system is denied the ability to correctly deal with the incoming data. Either variation can be used to defeat RFID systems.

24.8.1.2 Manipulating Tag Data

We have learned how blocking the RF might work for someone attempting to steal a single item. However, for someone looking to steal multiple items, a more efficient way is to change the data on the tags attached to the items. Depending on the nature of the tag, the price, stock number, and any other data can be changed. By changing a price, a thief can obtain a dramatic discount, while still appearing to buy the item. Other changes to a tag's data can allow users' to buy age-restricted items such as X- or R-rated movies.

When items with modified tags are bought using a self-checkout cash register, no one can detect the changes. Only a physical inventory would reveal that shortages in a given item were not matching the sales logged by the system.

In 2004, Lukas Grunwald demonstrated a program he had written called RF Dump. RF Dump is written in Sun's Java language, and runs on either Debian Linux or Windows XP operating systems for PCs. The program scans for RFID tags via an ACG brand reader attached to the serial port of a computer. When the reader recognizes a card, the program presents the card data in a spreadsheet-like format on the screen. The user can then enter or change data and reflect those changes on the tag (see Figure 24.11). RF Dump also makes sure that the data written is the correct length for the tag's fields, by either padding zeros or truncating extra digits as needed.

Alternately, a personal digital assistant (PDA) program called RF Dump-PDA is available for use on PDAs such as the Hewlett-Packard iPAQ Pocket PC. RF Dump-PDA is written in Perl, and will run on Pocket PCs running the Linux operating system. Using a PDA and RF Dump-PDA, a thief can walk through a store and change the data on items with the ease of using a handheld Pocket PC.

Grunwald demonstrated the attack using the same EPC-based RFID system that the Future Store in Rheinberg, Germany, uses (see www.futurestore.org).The Future Store is designed to be a working supermarket and a live technology-demonstration store, and is owned and run by Metro AG, Germany's largest retailer and the fifth largest retail chain in the world.

24.8.1.3 Middleware

Middleware attacks can happen at any point between the reader and the backend. Let's look at a theoretical attack on the middleware of the Exxon Mobil SpeedPass system.

- The customer's SpeedPass RFID tag is activated by the reader over the air. The reader is connected to the pump or a cash register. The reader handshakes with the tag and reads the encrypted serial number.

- The reader and pump are connected to the gas station's data network, which in turn is connected to a very small aperture terminal (VSAT) satellite transceiver in the gas station.

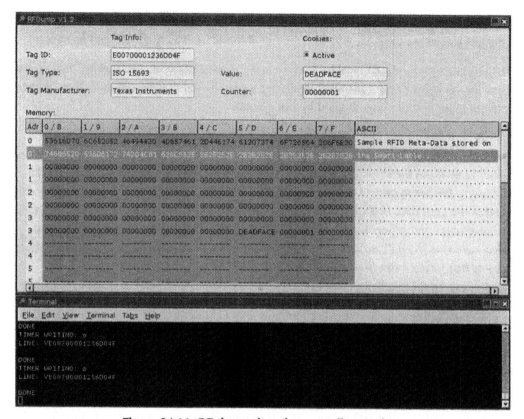

Figure 24.11: RF dump changing a retail tag's data

- The VSAT transceiver sends the serial number to an orbiting satellite, which in turn, relays the serial number to a satellite earth station.

- From the satellite earth station, the serial number is sent to ExxonMobil's data center. The data center verifies the serial number and checks for authorization on the credit card that is linked to the account.

- The authorization is sent back to the pump following the above route, but in reverse.

- The cash register or pump receives authorization and allows customers to make their purchases.

At any point in the above scenario, the system may be vulnerable to an outside attack. While requiring sophisticated transmitters systems, attacks against satellite systems have happened from as far back as the 1980s.

However, the weakest point in the above scenario is probably the local area network (LAN).This device could be sniffing valid data to use in a replay attack, or it could be injecting data into the LAN, causing a DOS attack against the payment system. This device could also be allowed unauthorized transmissions.

Another possibility might be a technically sophisticated person taking a job in order to gain access to the middleware. Some "social engineering" attacks take place when someone takes a low paying job that permits access to a target system.

Further along the data path, the connection between the satellite's earth station and the data center where the SpeedPass numbers are stored, is another spot where middleware can be influenced. The connections between the data center and the credit card centers are also points where middleware data may be vulnerable.

24.8.1.4 Backend

Because the backend database is often the furthest point away from the RFID tag, both in a data sense and in physical distance, it may seem far removed as a target for attacking an RFID system. However, it bears pointing out that they will continue to be targets of attacks because they are, as Willy Sutton said, "where the money is."

Databases may have some intrinsic value if they contain such things as customers' credit card numbers. A database may hold valuable information such as sales reports or trade secrets, which is invaluable to a business competitor.

Businesses that have suffered damage to their databases are at risk for losing the confidence of consumers and ultimately their market share, unless they can contain the damage or quickly correct it. The business sections of newspapers and magazines have reported many stories regarding companies suffering major setbacks because consumer confidence dropped due to an IT-related failure.

Manipulated databases can also have real-world consequences beyond the loss of consumers' buying power. It is conceivable that changing data in a hospital's inventory system could literally kill people or changing patient data on the patient records database could be deadly. A change of one letter involving a patient's blood type could put that person at risk if they received a transfusion. Hospitals have double and triple checks in place to combat these types of problems; however, checks will not stop bad things from happening due to manipulated data; they can only mitigate the risk.

24.8.2 Blended Attacks

Attacks can be used in combinations. The various attacks seen in opposition to RFID systems have also been made against individual subsystems. However, the increased cleverness of those who attack RFID systems will probably lead to *blended* attacks. An attacker might attack the RF interface of a retailer with a custom virus tag, which might then tunnel through the middleware, ultimately triggering the backend to dump credit card numbers to an unknown Internet site via an anonymous server.

24.9 Management of RFID Security

While sitting at your desk one morning, your boss walks in and announces that the company is switching to a new Radio Frequency Identification (RFID) setup for tracking products, which will add new equipment to the network and make it more secure. Your boss expects you to evaluate the new RFID equipment and devise an appropriate security plan.

The first thing you need to do is determine your security needs. You may be a position to influence the evaluations and purchasing of RFID applications and equipment; however, more than likely, you will be given a fixed set of parameters for applications and equipment.

In either case, the first thing you need to do is assess the vulnerabilities of the proposed RFID system. After you have assessed the RFID system it in detail, you can devise plans on how to manage system security.

24.9.1 Risk and Vulnerability Assessment

The assessment of risks and vulnerabilities go hand in hand. You have to make sure the obvious things are covered.

To begin evaluating your system, you need to ask questions regarding the assessment and tolerance of the risks: what types of information are you talking about at any given point in the system and what form is it in? How much of that information can potentially be lost? Will it be lost through the radio portion of the system, someplace in the middleware, or at the backend? Once these risks are evaluated, you can begin to plan how to secure it.

A good way to evaluate the risk is to ask the newspaper reporter's five classic investigative questions: "who?," "what?," "when?," "where?," and "how?"

- **Who** is going to conduct the attack or benefit from it? Will it be a competitor or an unknown group of criminals?

- **What** do they hope to gain from the attack? Are they trying to steal a competitor's trade secret? If it is a criminal enterprise, are they seeking customers' credit card numbers?

- **When** will the attack happen? When a business is open 24 hours a day, 7 days a week, it is easy to forget that attacks can occur when you are not there. If a business is not open 24 hours per day, some of the infrastructure (e.g., readers) may still be on during off-business hours and vulnerable to attack.

- **Where** will it take place? Will the attack occur at your company's headquarters or at an outlying satellite operation? Is the communications link provided by a third party vulnerable?

- **How** will they attack? If they attack the readers via an RF vulnerability, you need to limit how far the RF waves travel from the reader. If the attacker is going after a known vulnerability in the encryption used in the tag reader communications, you have to change the encryption type, and, therefore, also change all of the tags.

Asking these questions can help you focus and determine the risks of protecting your system and data.

The US military uses the phrase "hardening the target," which means designing a potential target such as a command bunker or missile silo to take hits from the enemy. The concept of hardening a target against an attack in the Information Technology (IT) sector is also valid, and further translates into the RFID area.

Basically, hardening the target means considering the types of specific attacks that can be brought against specific targets. When securing RFID systems, specific targets have specific attacks thrown at them.

Consider the following scenario. A warehouse has a palette tracking system where an RFID reader is mounted on a gantry over a conveyor belt. As pallets pass down the conveyor belt, they pass through the gantry, the reader's antennas activate the tags on each pallet, the tags are read, and the reader passes the information to the backend database.

In this situation, if you are concerned about potential attackers gleaning information from the radio waves emitted by the RFID reader station and the tags, you should harden it by limiting the RF waves from traveling beyond the immediate area of the reader. The easiest way is to lower the transmit power of the reader to the absolute minimum for triggering the tags. If that solution does not work or is not available, other options may include changing the position or orientation of the reader's antennas on the gantry, or constructing a Faraday cage around the reader. (A Faraday cage is an enclosure designed to prevent RF signals from entering or exiting an area, usually made from brass screen or some other fine metallic mesh.)

Consider whether other issues with the tags might cause problems. Is there is a repetition level for information hard coded into the tags? If you are using the codes for proximity entry control combined with a traditional key (e.g., in the Texas Instruments DST used with Ford car keys), a repeat of the serial numbers every 10,000 keys may be an acceptable risk. However, if it is being used as a pallet counting system, where 2000 pallets are processed daily, the same numbers will be repeated weekly, which may pose the risk of placing a rogue tag into a counting system. In this case, repeating a serial number every 10,000 times is probably not acceptable for that business model.

If you are concerned about attacks among the middleware and information being intercepted by an attacker, make sure that the reader's electronics or communications lines are not open to those who should not have access to them. In this case, hardening the target may be as simple as placing equipment (e.g., Ethernet switches) in locked communications closets, or performing a source code software review to ensure that an overloading buffer does not crash the reader.

Finally, hardening the target for the backend means preventing an attack on the database. In this regard, the security of a new RFID system should not cause anything new to a security professional, with the possible exception of a new attack vector in the form of a new communications channel.

A new channel may provide a challenge for securing previously unused Transmission Control Protocol (TCP) ports in the backend, by reexamining the database for the possibility of Structured Query Language (SQL) injection attacks. However, nothing at the backend is new to seasoned security professionals; therefore, standard risk evaluation practices for backend systems should prevail.

NOTES FROM THE UNDERGROUND...

Defaults Settings: Change Them!

Default passwords and other default security settings should be changed as soon as possible. This bears repeating, because many people do not make the effort to change their defaults.

You may think that your Acme Super RFID Reader 3000 is protected simply because no one else owns one; however, default settings are usually well known by the time new equipment is placed on the market. Most manufacturers place manuals on their Web sites in the form of either Web pages or Adobe Portable Document Format (PDF) files. Other Web sites contain pages full of default settings, ranging from unofficial tech support sites to sites frequented by criminals intent on cracking other people's security.

To learn how much of this information is available, type the name and model of a given device into your favorite search engine, followed by the words "default" and "passwords."

When evaluating the risks and vulnerabilities, the bottom line is this: Once you have determined the point of an attack and how it happened, you can decide what options are available for mitigating the attack. When these options are identified, you can begin formulating the management and policies that will hopefully minimize your exposure to an attack.

24.9.2 Risk Management

Once the risks and vulnerabilities are identified, begin managing the risks. Start by validating all of your equipment, beginning with the RFID systems and working down to the backend. At each stage, you should observe how a particular item works (both individually and in combination with other items), and how it fits into your proposed security model.

Let's look back at the warehouse example. A 900 MHz RFID tag is needed for tracking, because its RF properties work with the materials and products that are tracked to the warehouse. You need to decide if those same RF properties will cause a disruption in the security model. Will the 900 MHz signal travel further than expected compared to other frequencies? Can the signals be sniffed from the street in front of the warehouse? Managing this potential problem can be as simple as changing to a frequency with a shorter range, or as complicated as looking at other equipment with different capabilities.

Middleware management ensures that ensuing data is valid as it moves through the system. Receiving a text string instead of a numeric stock number may indicate that an attacker is attempting to inject a rogue tag command into the system. Checksums are also a common way to verify data, and may be required as part of the ongoing need to ensure that the data traveling through middleware applications is valid.

Managing middleware security usually involves using encryption to secure data, in which case, you need to consider the lifespan of the information in light of how long it would take an attacker to break the encryption. If your information becomes outdated within a week (e.g., shipment delivery information), it will probably take an attacker six months to break the encryption scheme. However, do not forget that increases in computing power and new encryption cracking techniques continually evolve. A strong encryption technique today may be a weak encryption tomorrow.

Managing a system also involves establishing policies for the users of that system. You can have the most secure encryption used today, but if passwords are posted on monitors, security becomes impossible. Make sure that the policies are realistic, and that they do not defeat security instead of enhancing it.

NOTES FROM THE UNDERGROUND...

Bad Policies May Unintentionally Influence Security

Do not assume that RFID security is just about databases, middleware, and radio transmissions. Policy decisions also have an impact on the security of an RFID system. Bad policies can increase risks (e.g., not patching a server against a known vulnerability).

In other areas, bad policies can directly affect security without being obvious. One state agency uses proximity cards as physical access control to enter its

---

building and to enter different rooms within the building. Like most of these types of systems, the card number is associated with the database containing the cardholder's name and the areas they are allowed to access. When the cardholder passes the card over the reader antenna associated with each door, the system looks in the database and makes a decision based on the privileges associated with that card.

Proximity cards are issued when an employee begins a new job, and are collected when the employee leaves the company. At this particular agency, the personnel department is responsible for issuing and collecting cards. Therefore, they implemented a policy that imposes a fine on employees that lose their card.

In one case, an employee lost a card, but did not report it to his superiors because he did not want to pay a fine. As a relatively low-level employee, reporting the loss and paying the fine would create a financial hardship.

The proximity card is the least costly part of the RFID-controlled entry system. However, because of a policy designed to discourage losing the cards, the entire building security could easily be compromised if someone found that particular card. The goal of securing physical access to the building was forgotten when the cost of the card replacement began to drive the policy. The people who wrote the policy assumed that if an employee lost a card, they would pay the fine.

At another agency, the people using the system issue the cards and control physical access to the building, taking great effort to password-protect the workstations that access the database. However, sometimes they forget to physically protect the control system. The RS-232 serial ports that directly control the system and the cables to each controlled door are accessible by anyone who wanders into the room. The room itself is accessible via an unlocked door to a room where visitors are allowed to roam unescorted.

This particular agency lacks policies regarding installing security equipment, the areas to secure, and the inability to fully understand the system, which all add up to a potential failure.

Review your policies and keep focused on the goal. Remember to asked questions like, "Are we trying to secure a building, or are we concerned about buying new cards?" "Are we leaving parts of a system vulnerable just because they are out of sight?" "Will people follow or evade this policy?"

24.9.3 Threat Management

When conducting threat management for RFID systems, monitor everything, which will help with any difficulties.

If you are performing information security, you may be overwhelmed by the large amount of data and communications that must be monitored. As a matter of routine, you should confirm the integrity of your systems via login access and Dynamic Host Configuration Protocol (DHCP) logs, and perform physical checks to make sure that new devices are not being added to the network without your knowledge.

Adding RFID systems to the list of systems to be monitored will increase the difficulty. In addition to physically checking the Ethernet connections, you will also have to perform RF sweeps for devices attempting to spoof tags, and keep an eye out for people with RF equipment who may attempt to sniff data from the airways.

You will need new equipment and training for the radio side of the system, since radio systems are usually outside the experience of most network professionals. You will also have new middleware connections that will add new channels, thus, introducing possible new threats and adding new vectors for the more routine threats such as computer viruses and spyware.

NOTES FROM THE UNDERGROUND...

Monitoring Isn't Just for Logs

Monitoring and tracking changes in files rather than logs is just as important. For example, suppose you have a program with the following RFID proximity cards and associated names:

```
Card1 DATA "8758176245"
Card2 DATA "4586538624"
Card3 DATA "7524985246"

Name1 DATA "George W. Bush", CR, 0
Name2 DATA "Dick Cheney", CR, 0
Name3 DATA "Condoleeza Rice", CR, 0
...
LOOKUP tagNum, [Name1, Name2, Name3]
```

If we make three small additions, it becomes easy to add a previously unauthorized user.

```
Card1 DATA "8758176245"
Card2 DATA "4586538624"
Card3 DATA "7524985246"
Card4 DATA "6571204348" '■

Name1 DATA "George W. Bush", CR, 0
Name2 DATA "Dick Cheney", CR, 0
Name3 DATA "Condoleeza Rice", CR, 0
Name4 DATA "Maxwell Smart", CR, 0 '■
...
LOOKUP tagNum, [Name1, Name2, Name3, Name4] ■
```

With the addition of 63 bytes of data, the security of this RFID card access system has been compromised. However, an increase of 63 bytes of data might not be noticed in a large database of cards comprising thousands of users.

Remember to periodically review the contents of databases with those people who know what the contents should be. Do not assume that all of data is valid.

---

*Code derived from the RFID.BS2 program written by Jon Williams, Parallax, Inc. www.parallax.com

When you are done securing your new RFID system and you think you have all the threats under control, go back to the beginning and start looking for new vulnerabilities, new risks, and new attacks. As previously mentioned, things such as increases in computing power and new encryption cracking techniques are constantly evolving, and may break a security model in short order. Keeping up with new security problems and the latest attack methods is an ongoing process—one that demands constant vigilance.

24.10 Summary

In this chapter, we discussed how RFID systems work; the various types of RFID tags, data formats, and tag protocols; and some typical applications. We also discussed some of the potential attacks that RFID systems are susceptible to. We learned that some of the attacks that are well known to IT professionals can also be applied to RFID.

With new technologies, we are often seduced by the grand vision of what "it" promises. Currently, RFID is one of the newest technologies offering this a grand vision. While RFID holds great promise in many applications, the last several years have proven that many aspects of RFID systems are insecure and new vulnerabilities are found daily.

The driving idea behind this chapter is applying information security (InfoSec) principles to RFID applications. What we [the authors] have attempted to do is show you some common pitfalls and their solutions, and get you started thinking about the security implications of installing and running an RFID system in your organization.

24.11 Links to Sites

- RFID Gazette—*www.rfidgazette.org*

- EPCglobal—*www.epcglobalus.org*

- ISO—*www.iso.org*

- RFID Buzz—*www.rfidbuzz.com*

- RFID Viruses—*www.rfidvirus.org*

Wireless Policy Essentials

John Rittinghouse
James F. Ransome

A.1 Wireless position statement

Over the last two years, articles have appeared in the press discussing security problems discovered in the WEP encryption scheme used on many 802.11b wireless networks. Although we are using a form of WEP on our wireless network, the security solution we are implementing uses Cisco technology that mitigates the flaws described in the press to a fairly significant extent.

Normal WEP encryption uses a single encryption key for all wireless transmissions. Current attacks on wireless security involve brute force hacking to obtain that key. Our system provides users with individual encryption keys that change each time they log into the wireless network. This means there is no one single key to hack, and because the keys are not static, the system is much harder to attack.

It is important to remember that WEP is not intended to be the only security used in a wireless network. WEP stands for Wired Equivalent Privacy and was just meant to try to make a wireless connection as hard to "sniff" as that of a wired network. In reality, the Cisco solution that we have deployed at ABC Inc. provides significantly more data privacy than a normal wired network connection.

As with the traditional wire-based network, additional security such as the use of encrypted Web pages using SSL and secure remote logins and file transfers using SSH should still be used for high-valued data transactions. The wireless encryption system only protects your data while it travels over the airwaves. As soon as your data hits the local wireless access point in your building, it flows over the building's standard wired network and is no longer protected by the wireless encryption system.

Two new wireless security solutions will be available over the next year and a half. The new solution, called WiFi Protected Access (WPA), is a subset of the still unfinished IEEE 802.11i security specification and will be usable by both home and enterprise wireless networks. Task Group I is working on 802.11i, and it is still on a path to be complete about this time next year with a fully ratified standard.

WPA will work with the majority of 802.11-based products out today once they've gone through a firmware/software upgrade. WPA is forward compatible with 802.11i. By the time 11i is ratified around September of next year, WPA version 2.0 is expected with full 802.11i support. Eventually, the Alliance expects to require WiFi products to shop with WPA turned on as a default. The way WPA will work in the enterprise is similar to the setup of any 802.1X authentication system. The clients and access points must have WPA enabled for encryption to and from an 802.1X with Extensible Authentication Protocol (EAP) authentication server of some sort, such as a RADIUS server, with centralized access management. WiFi Protected Access had several design goals:

- Be a strong security solution

- Interoperable

- Security replacement for WEP

- Be software ungradable to existing WiFi certified products

- Be applicable for both home and enterprise users and be available immediately

WiFi Protected Access was constructed to provide an improved data encryption, which was weak in WEP, and to provide user authentication, which was largely missing in WEP. To improve data encryption, WiFi Protected Access utilizes its Temporal Key Integrity Protocol (TKIP). TKIP provides important data encryption enhancements including a per-packet key mixing function, a message integrity check (MIC) named Michael, an extended initialization vector (IV) with sequencing rules, and a rekeying mechanism. Through these enhancements, TKIP addresses all WEP's known vulnerabilities. Enterprise-level User Authentication via 802.1x and EAP WEP has almost no user authentication mechanism. To strengthen user authentication, WiFi Protected Access implements 802.1x and the Extensible Authentication Protocol (EAP). Together, these implementations provide a framework for strong user authentication. This framework utilizes a central authentication server, such as RADIUS, to authenticate each user on the network before they join it, and also employs "mutual authentication" so that the wireless user doesn't accidentally join a rogue network that might steal its network credentials.

WiFi Protected Access will be forward-compatible with the IEEE 802.11i security specification currently under development by the IEEE. WiFi Protected Access is a subset of the current 802.11i draft, taking certain pieces of the 802.11i draft that are ready to bring to market today, such as its implementation of 802.1x and TKIP. These features can also be enabled on most existing WiFi certified products as a software upgrade. The main pieces of the 802.11i draft that are not included in WiFi Protected Access are secure IBSS, secure fast handoff, secure deauthentication and disassociation, as well as enhanced encryption protocols such as AES-CCMP. These features are either not yet ready for market or will require hardware upgrades to implement. The IEEE 802.11i specification is expected to be published at the end of 2003.

WiFi Protected Access effectively addresses the WLAN security requirements for the enterprise and provides a strong encryption and authentication solution before the ratification of the IEEE 802.11i standard. In an enterprise with IT resources, WiFi Protected Access should be used in conjunction with an authentication server such as RADIUS to provide centralized access control and management. With this implementation in place, the need for add-on solutions such as VPNs may be eliminated, at least for the express purpose of securing the wireless link in a network.

A.1.1 Typical Wireless Security Architectural Concerns

Normally, wireless networks are outside of the institutional firewall(s). In addition, they use static WEP keys on the WLAN to keep administrative costs low and provide a Network Intrusion Detection (NID) facility to monitor possible attacks emanating from the WLAN to the Internet and other networks. As part of the architecture, it is normally recommended that neither the IP address range nor the domain name of the wireless network be associated with any of the existing internal networks. This will allow for better segregation of wireless traffic and will assist in identifying and filtering traffic to and from this network.

WLANs are normally treated as though they are an untrusted network, like the Internet. Assuming that RF propagation is limited by a thorough site survey and the use of proper antenna and transmitter power settings, the WLAN does not represent any more significant a threat to internal networks than the Internet itself. Because roaming between

APs is still in the proprietary domain, it is highly recommended that all APs be purchased from the same vendor. This will ensure that an end station equipped with any 802.11-compatible NIC will be able to roam between APs. In addition, any new vendor-specific security improvements that are introduced may require homogenous APs.

Concerns over the usage of WEP and its ability to provide adequate security for a network have required additional measures to improve your security. It is useful to think of securing the wireless LAN as you would protect the internal LAN from the public Internet. Using this framework, you could install two firewalls: one at the gateway into your corporate LAN and another between the LAN and the wireless network. The wireless firewall can be configured to pass only VPN traffic. This allows a remote user to connect to the corporate LAN using the VPN. Likewise, a wireless user can authenticate to the wireless infrastructure while still having wireless data encrypted through the VPN tunnel.

By segregating the wireless infrastructure from your wired network, and enabling VPN traffic to pass between them, you create a buffer zone that increases network security. In addition, IPSec, the main IP Layerencryption protocol used in VPN technology, prevents productive traffic sniffing, which will thwart attacks that rely on using WEP for encryption, such as AirSnort. Another advantage of using the VPN approach is if you've already deployed a VPN, your remote users are already familiar with the limitations imposed by it. Getting wireless users to be comfortable with similar limitations should be relatively easy.

A.2 ABC Inc. InfoSec Risk Assessment Policy

Policy No. 1

Effective date Month/Day/Year

Implement by Month/Day/Year

1.0 Purpose

To empower InfoSec to perform periodic information security risk assessments (RAs) for the purpose of determining areas of vulnerability, and to initiate appropriate remediation.

2.0 Scope

Risk assessments can be conducted on any entity within ABC Inc. or any outside entity that has signed a Third Party Agreement <Insert Link> and the Acceptable Use Policy

<Insert Link> with ABC Inc. RAs can be conducted on any information system, to include applications, servers, and networks, and any process or procedure by which these systems are administered and/or maintained.

3.0 Policy

The execution, development, and implementation of remediation programs are the joint responsibility of InfoSec and the department responsible for the systems area being assessed. Employees are expected to cooperate fully with any RA being conducted on systems for which they are held accountable. Employees are further expected to work with the InfoSec Risk Assessment Team in the development of a remediation plan.

4.0 Enforcement

Any employee found to have violated this policy may be subject to disciplinary action, up to and including termination of employment.

5.0 Definitions

Entity: Any business unit, department, group, or third party, internal or external to ABC Inc., responsible for maintaining ABC Inc. assets.

Risk: Those factors that could affect confidentiality, availability, and integrity of ABC Inc.'s key information assets and systems. InfoSec is responsible for ensuring the integrity, confidentiality, and availability of critical information and computing assets, while minimizing the impact of security procedures and policies upon business productivity.

6.0 Exceptions

Exceptions to information system security policies exist in rare instances where a risk assessment examining the implications of being out of compliance has been performed, where a Policy Exception Form <Insert Link> has been prepared by the data owner or management, and where this form has been approved by both the CSO or Director of InfoSec and the Chief Information Officer (CIO).

7.0 Revision History

Date ___/___/_____

Version:_____

Author:_____

Summary:_____

A.3 ABC Inc. InfoSec Audit Policy

Policy No. 2

Effective date Month/Day/Year

Implement by Month/Day/Year

1.0 Purpose

To provide the authority for members of ABC Inc.'s InfoSec team to conduct a security audit on any system at ABC Inc. Audits may be conducted to:

- Ensure integrity, confidentiality, and availability of information and resources

- Investigate possible security incidents

- Ensure conformance to ABC Inc. security policies

- Monitor user or system activity where appropriate

- Measure and report on risk

2.0 Scope

This policy covers the following:

- All computer and communication devices that are part of, or associated with, the ABC Inc. Network

- All information stored on ABC Inc. media (digital and hard copy information)

3.0 Policy

When requested, and for the purpose of performing an audit, any access needed will be provided to members of ABC Inc.'s InfoSec team. This access may include:

- User level and/or system level access to any computing or communications device

- Access to information (electronic, hardcopy, etc.) that may be produced, transmitted, or stored on ABC Inc. equipment or premises

- Access to work areas (labs, offices, cubicles, storage areas, etc.)

- Access to interactively monitor and log traffic on ABC Inc. networks

4.0 Enforcement

Any employee found to have violated this policy may be subject to disciplinary action, up to and including termination of employment.

5.0 Definitions

6.0 Exceptions

Exceptions to information system security policies exist in rare instances where a risk assessment examining the implications of being out of compliance has been performed, where a Policy Exception Form <Insert Link> has been prepared by the data owner or management, and where this form has been approved by both the CSO or Director of InfoSec and the Chief Information Officer (CIO).

7.0 Revision History

Date ___/___/_____

Version:_____

Author:_____

Summary:_____

A.4 ABC Inc. InfoSec Acceptable Use Policy

Policy No. 3

Effective date Month/Day/Year

Implement by Month/Day/Year

1.0 Overview

Information Systems Security's intentions for publishing an Acceptable Use Policy are not to impose restrictions that are contrary to ABC Inc.'s established culture of openness, trust, and integrity. Information System Security is committed to protecting ABC Inc.'s employees, partners, and the company from illegal or damaging actions by individuals, either knowingly or unknowingly.

Internet/Intranet/Extranet-related systems, including but not limited to computer equipment, software, operating systems, storage media, network accounts providing electronic mail, WWW browsing, and FTP, are the property of ABC Inc. These systems are to be used for business purposes in serving the interests of the company, and of our clients and customers in the course of normal operations. Please review Human Resources policies at <Insert Link> for further details.

Effective security is a team effort involving the participation and support of every ABC Inc. employee and affiliate who deals with information and/or information systems. It is the responsibility of every computer user to know these guidelines, and to conduct their activities accordingly.

2.0 Purpose

The purpose of this policy is to outline the acceptable use of computer equipment at ABC Inc. These rules are in place to protect the employee and ABC Inc. Inappropriate use exposes ABC Inc. to risks including virus attacks, compromise of network systems and services, and legal issues.

3.0 Scope

This policy applies to employees, contractors, consultants, temporaries, and other workers at ABC Inc., including all personnel affiliated with third parties. This policy applies to all equipment that is owned or leased by ABC Inc.

4.0 Policy

4.1 General Use and Ownership

While ABC Inc.'s network administration desires to provide a reasonable level of privacy, users should be aware that the data they create on the corporate systems remains the property of ABC Inc. Because of the need to protect ABC Inc.'s network, management cannot guarantee the confidentiality of information stored on any network device belonging to ABC Inc. Employees are responsible for exercising good judgment regarding the reasonableness of personal use. Individual departments are responsible for creating guidelines concerning personal use of Internet/Intranet/Extranet systems. In the absence of such policies, employees should be guided by departmental policies on personal use, and if there is any uncertainty, employees should consult their supervisor or manager. Information System Security recommends that any information that users consider sensitive or vulnerable be encrypted. For security and network maintenance purposes, authorized individuals within ABC Inc. may monitor equipment, systems, and network traffic at any time, per Information System Security's Audit Policy. ABC Inc. reserves the right to audit networks and systems on a periodic basis to ensure compliance with this policy.

4.2 Security and Proprietary Information

The user interface for information contained on Internet/Intranet/Extranet- related systems should be classified as either confidential or not confidential, as defined by corporate confidentiality guidelines <Insert Link >, details of which can be found in Human Resources policies. Examples of confidential information include but are not limited to: company private, corporate strategies, competitor sensitive, trade secrets, specifications, customer lists, employee personal data, employee job data, and research data. Employees should take all necessary steps to prevent unauthorized access to this information. Keep passwords secure and do not share accounts. Authorized users are responsible for the security of their passwords and accounts.

All PCs, laptops, and workstations should be secured with a password-protected screensaver with the automatic activation feature set at 10 minutes or less, or by locking access to the computer (control-alt-delete for Window platforms users) when the host will be unattended. Because information contained on portable computers is especially vulnerable, special care should be exercised. Protect laptops in accordance with the "Laptop Security Guidelines" policy.

Postings by employees from an ABC Inc. email address to newsgroups are prohibited unless the posting is in the course of business duties. All hosts used by the employee that are connected to the ABC Inc. Internet/Intranet/ Extranet, whether owned by the employee or ABC Inc., shall be continually executing approved virus-scanning software with a current virus database, unless overridden by departmental or group policy.

Employees must use extreme caution when opening e-mail attachments received from unknown senders, which may contain viruses, e-mail bombs, or Trojan horse code.

4.3. Unacceptable Use

1. The following activities are, in general, prohibited. Employees may be exempted from these restrictions during the course of their legitimate job responsibilities (e.g., systems administration staff may have a need to disable the network access of a host if that host is disrupting production services).

2. Under no circumstances is an employee of ABC Inc. authorized to engage in any activity that is illegal under local, state, federal, or international law while utilizing ABC Inc.-owned resources.

3. The lists below are by no means exhaustive, but attempt to provide a framework for activities which fall into the category of unacceptable use.

System and Network Activities

The following activities are strictly prohibited, with no exceptions:

- Violations of the rights of any person or company protected by copyright, trade secret, patent, or other intellectual property, or similar laws or regulations, including, but not limited to, the installation or distribution of "pirated" or other software products that are not appropriately licensed for use by ABC Inc.

- Unauthorized copying of copyrighted material including, but not limited to, digitization and distribution of photographs from magazines, books or other copyrighted sources, copyrighted music, and the installation of any copyrighted software for which ABC Inc. or the end user does not have an active license is strictly prohibited.

- Exporting software, technical information, encryption software, or technology, in violation of international or regional export control laws, is illegal. The appropriate management should be consulted prior to export of any material that is in question.

- Introduction of malicious programs into the network or server (e.g., viruses, worms, Trojan horses, e-mail bombs, etc.).

- Revealing your account password to others or allowing use of your account by others. This includes family and other household members when work is being done at home.

- Using ABC Inc. computing assets to actively engage in procuring or transmitting material that is in violation of sexual harassment or hostile workplace laws in the user's local jurisdiction.

- Making fraudulent offers of products, items, or services originating from any ABC Inc. account.

- Making statements about warranty, expressly or implied, unless it is a part of normal job duties.

- Effecting security breaches or disruptions of network communication. Security breaches include, but are not limited to, accessing data of which the employee is not an intended recipient or logging into a server or account that the employee is not expressly authorized to access, unless these duties are within the scope of regular duties. For purposes of this section, "disruption" includes, but is not limited to, network sniffing, ping floods, packet spoofing, denial of service, and forged routing information for malicious purposes.

- Port scanning or security scanning is expressly prohibited unless prior notification to Information System Security is made.

- Executing any form of network monitoring, which will intercept data not intended for the employee's host, unless this activity is a part of the employee's normal job/duty.

- Circumventing user authentication or security of any host, network, or account.

- Interfering with or denying service to any user other than the employee's host (for example, denial of service attack).

- Using any program/script/command, or sending messages of any kind, with the intent to interfere with, or disable, a user's terminal session, via any means, locally or via the Internet/Intranet/Extranet.

- Providing information about, or lists of, ABC Inc. employees to parties outside ABC Inc.

Email and Communications Activities

- Sending unsolicited email messages, including the sending of "junk mail" or other advertising material to individuals who did not specifically request such material (email spam).

- Any form of harassment via email, telephone, or paging, whether through language, frequency, or size of messages.

- Unauthorized use, or forging, of email header information.

- Solicitation of email for any other email address, other than that of the poster's account, with the intent to harass or to collect replies.

- Creating or forwarding "chain letters," "Ponzi," or other "pyramid" schemes of any type.

- Use of unsolicited email originating from within ABC Inc.'s networks of other Internet/Intranet/Extranet service providers on behalf of, or to advertise, any service hosted by ABC Inc. or connected via ABC Inc.'s network.

- Posting the same or similar non-business-related messages to large numbers of Usenet newsgroups (newsgroup spam).

- Posting of ABC Inc. confidential information by employees is prohibited unless the posting is in the course of business duties.

- Transmission of ABC Inc.'s confidential information to unauthorized recipients (internal or external) by employees is prohibited unless the posting is in the course of business duties.

5.0 Enforcement

Any employee found to have violated this policy may be subject to disciplinary action, up to and including termination of employment.

6.0 Definitions

Spam: Unauthorized and/or unsolicited electronic mass mailings.

Junk Mail: Unsolicited email. It is also another term for Spam.

7.0 Exceptions

Exceptions to information system security policies exist in rare instances where a risk assessment examining the implications of being out of compliance has been performed, where a Policy Exception Form <Insert Link> has been prepared by the data owner or management, and where this form has been approved by both the CSO or Director of InfoSec and the Chief Information Officer (CIO).

8.0 Revision History

Date ___/____/_____

Version:_____

Author:_____

Summary:_____

A.5 ABC Inc. InfoSec Network Policy

Policy No. 4

Effective date Month/Day/Year

Implement by Month/Day/Year

1.0 Purpose

This policy establishes information security requirements for ABC Inc.'s network to ensure that ABC Inc.'s confidential information and technologies are not compromised, and that production services and other ABC Inc. interests are protected.

2.0 Scope

This policy applies to all internal networks, ABC Inc. employees, and third parties who access ABC Inc. networks. All existing and future equipment, which fall under the scope of this policy, must be configured according to the referenced documents.

3.0 Policy

3.1 Ownership Responsibilities

1. Network Operations is responsible for the security of the networks and the network's impact on the corporation. Network managers are responsible for adherence to

this policy and associated processes. Where policies and procedures are undefined, network managers must do their best to safeguard ABC Inc. from security vulnerabilities.

2. Network Operations is responsible for the network's compliance with all ABC Inc. security policies. The following are particularly important:

   - ABC Inc. Anti-Virus Policy
   - ABC Inc. Dial-in Policy
   - ABC Inc. Extranet Policy
   - ABC Inc. Password Policy
   - ABC Inc. Password Protection Policy
   - ABC Inc. Remote Access Policy
   - ABC Inc. Router Security Policy
   - ABC Inc. Server Security Policy
   - ABC Inc. VPN Security Policy
   - ABC Inc. Wireless Communications Policy
   - ABC Inc. Physical Security Policy

3. Network Operations is responsible for controlling network access. Access to any given network will only be granted by Network Operations to those individuals with an immediate business need within the network, either short-term or as defined by their ongoing job function. This includes continually monitoring the access list to ensure that those who no longer require access to the network have their access terminated.

4. Network Operations and/or InfoSec reserve the right to interrupt network connections that impact the corporate production network negatively or pose a security risk.

5. Network Operations must manage all network IP addresses, which are routed within ABC Inc. networks.

6. Any network that wants to add an external connection must provide a diagram and documentation to InfoSec with business justification, the equipment, and the IP address space information. InfoSec will review for security concerns and must

approve before such connections are implemented. Access to the ABC Inc. network from a third party must use ABC Inc.'s managed firewall.

7. All user passwords must comply with Password Policy. In addition, individual user accounts on any network device must be deleted when no longer authorized within three days. Group account passwords on network computers (Unix, windows, etc.) must be changed quarterly (once every 3 months). Groups accounts must be approved by the IT Network Operations group. For any network device that contains ABC Inc. proprietary information, group account passwords must be changed within three days following a change in group membership.

8. InfoSec will address non-compliance waiver requests on a case-by-case basis and approve waivers if justified through the completion of the Policy Exception Form.

3.2 General Configuration Requirements

1. All external network IP traffic must go through a Network Operations maintained firewall.

2. Original firewall configurations and any changes must be reviewed and approved through the proper IT Operations Change Control process. InfoSec may require security improvements as needed.

3. Networks are prohibited from engaging in port scanning, network auto-discovery, traffic spamming/flooding, and other similar activities that negatively impact the corporate network and/or non-ABC Inc. networks.

4. InfoSec reserves the right to audit all network-related data and administration processes at any time, including but not limited to, inbound and outbound packets, firewalls, and network peripherals.

5. Network gateway devices are required to comply with all ABC Inc. product security advisories and must authenticate against the Corporate Authentication servers.

6. The password for all network gateway devices must be different from all other equipment passwords in the network. The password must be in accordance with Password Policy. The password will only be provided to those who are authorized to administer the network. There will be no group accounts, and anyone accessing these devices will have an individual account.

7. In networks where non-ABC Inc. personnel have physical access (e.g., training networks), direct connectivity to the corporate production network is not allowed. Additionally, no ABC Inc. confidential information can reside on any computer equipment in these networks. Connectivity for authorized personnel from these networks can be allowed to the corporate production network only if authenticated against the Corporate Authentication servers, temporary access lists (lock and key), SSH, client VPNs, or similar technology approved by InfoSec.

8. All network external connection requests must be reviewed and approved by InfoSec. Analog or ISDN lines must be configured to only accept trusted call numbers. Strong passwords must be used for authentication.

9. All networks with external connections must not be connected to ABC Inc. corporate production network or any other internal network directly or via a wireless connection, or via any other form of computing equipment. A waiver from InfoSec is required where air-gapping is not possible (e.g., Partner Connections to third party networks).

4.0 Enforcement

Any employee found to have violated this policy may be subject to disciplinary action, up to and including termination of employment.

5.0 Definitions

DMZ (De-Militarized Zone): This describes networks that exist outside of primary corporate firewalls, but are still under ABC Inc. administrative control.

External Connections: Connections that include (but are not limited to) third-party connections, such as a DMZ, data network-to-network, analog and ISDN data lines, or any other Telco data lines.

Extranet: Connections between third parties that require access to connections non-public ABC Inc. resources, as defined in InfoSec's Extranet policy (link).

Firewall: A device that controls access between networks. It can be a PIX, a router with access control lists, or similar security devices approved by InfoSec.

Internal: A network that is within ABC Inc.'s corporate firewall and connected to ABC Inc.'s corporate production network.

Network: A network is any non-production environment, intended specifically for developing, demonstrating, training, and/or testing of a product.

Network Manager: The individual who is responsible for all network activities and personnel.

Network Owned Gateway Device: A network owned gateway device is the network device that connects the network to the rest of ABC Inc.'s network. All traffic between the network and the corporate production network must pass through the network owned gateway device unless approved by InfoSec.

Network Support Organization: Any InfoSec approved ABC Inc. support organization that manages the networking of non-network networks.

Telco: A Telco is the equivalent to a service provider. Telcos offer network connectivity, e.g., T1, T3, OC3, OC12 or DSL. Telcos are sometimes referred to as "baby bells," although Sprint and AT&T are also considered Telcos. Telco interfaces include BRI, or Basic Rate Interface, a structure commonly used for ISDN service, and PRI, Primary Rate Interface, a structure for voice/dial-up service.

Traffic: Mass volume of unauthorized and/or unsolicited network Spamming/Flooding traffic.

6.0 Exceptions

Exceptions to information system security policies exist in rare instances where a risk assessment examining the implications of being out of compliance has been performed, where a Systems Security Policy Exception Form has been prepared by the data owner or management, and where this form has been approved by both the CSO or Director of InfoSec and the Chief Information Officer (CIO).

7.0 Revision History

Date ___/____/_____

Version:_____

Author:_____

Summary:_____

A.6 ABC Inc. InfoSec De-Militarized Zone (DMZ) Policy

Policy No. 5

Effective date Month/Day/Year

Implement by Month/Day/Year

1.0 Purpose

This policy establishes information security requirements for all networks and equipment deployed and located in the ABC Inc. "De-Militarized Zone" (DMZ) as well as screened subnets. Adherence to these requirements will minimize the potential risk to ABC Inc. from the damage to public image caused by unauthorized use of ABC Inc. resources, and the loss of sensitive/company confidential data and intellectual property.

2.0 Scope

ABC Inc. networks and devices (including but not limited to routers, switches, hosts, etc.) that are Internet facing and located outside ABC Inc. corporate Internet firewalls are considered part of the DMZ and are subject to this policy. This includes DMZ in primary Internet Service Provider (ISP) locations and remote locations. All existing and future equipment, which falls under the scope of this policy, must be configured according to the referenced documents. This policy does not apply to information systems and components which reside inside ABC Inc.'s corporate Internet firewalls. Standards for these are defined in the Internal Network Security Policy <Link>.

3.0 Policy

3.1. Ownership and Responsibilities

1. All new DMZs must present a business justification with sign-off at the business unit Vice President level. InfoSec must keep the business justifications on file.

2. DMZ system owning organizations are responsible for assigning managers, point of contact (POC), and back up POC, for each system. The DMZ owners must maintain up to date POC information with InfoSec and the corporate enterprise management system, if one exists. DMZ system managers or their backup must be available around-the-clock for emergencies.

3. Changes to the connectivity and/or purpose of existing DMZ system/application and establishment of new DMZ system/applications must be requested through an ABC Inc. Network Support Organization and approved by InfoSec.

4. All ISP connections must be maintained by an ABC Inc. Network Support Organization.

5. A Network Support Organization must maintain a firewall device between the DMZ and the Internet.

6. The Network Support Organization and InfoSec reserve the right to interrupt connections if a security concern exists.

7. The Network Support Organization will provide and maintain network devices deployed in the DMZ up to the Network Support Organization point of demarcation.

8. The Network Support Organization must record all DMZ address spaces and current contact information must be stored in a secure location.

9. The Network Support Organization is ultimately responsible for their DMZ complying with this policy.

10. Immediate access to equipment and system logs must be granted to members of InfoSec and the Network Support Organization upon request, in accordance with the Audit Policy.

11. Individual accounts must be deleted within three days when access is no longer authorized. Group account passwords must comply with the Password Policy and must be changed within three days from a change in the group membership.

12. InfoSec will address non-compliance waiver requests on a case-by-case basis through the submission of a Policy Exception Form.

3.2. General Configuration Requirements

1. Internal production resources must not depend upon resources on the DMZ networks.

2. DMZs must be connected through a firewall to access ABC Inc.'s corporate internal networks. Any form of cross-connection which bypasses the firewall device is strictly prohibited.

3. DMZs should be in a physically separate room from any internal networks. If this is not possible, the equipment must be in a locked rack or cage with limited access. In addition, the DMZ Manager must maintain a list of who has access to the equipment.

4. DMZ Managers are responsible for complying with the following related policies:
   a. Password Policy
   b. Wireless Communications Policy
   c. Anti-Virus Policy

5. The Network Support Organization maintained firewall devices must be configured in accordance with least-access principles and the DMZ business needs. All firewall filters will be maintained by InfoSec.

6. Original firewall configurations and any changes must be reviewed and approved through proper IT Operations change control processes (including both general configurations and rule sets). InfoSec may require additional security measures as needed.

7. Traffic from DMZ to the ABC Inc. internal network, including VPN access, falls under the Remote Access Policy

8. All routers and switches not used for testing and/or training must conform to the DMZ Router and Switch standardization documents.

9. Operating systems of all hosts internal to the DMZ running Internet Services must be configured to the secure host installation and configuration standards. [Add URL link to internal configuration standards].

10. Current applicable security patches/hot-fixes for any applications that are Internet services must be applied. Administrative owner groups must have processes in place to stay current on appropriate patches/hot-fixes at the first available opportunity.

11. All applicable security patches/hot-fixes recommended by the vendor must be installed. Administrative owner groups must have processes in place to stay current on appropriate patches/hot-fixes at the first available opportunity.

12. Services and applications not serving business requirements must be disabled.

13. ABC Inc. confidential information is prohibited on equipment in DMZs where non-ABC Inc. personnel have physical access.

14. Remote administration must be performed over secure channels (e.g., encrypted network connections using SSH or IPSEC) or console access independent from the DMZ networks.

4.0 Enforcement

Any employee found to have violated this policy may be subject to disciplinary action up to and including termination of employment.

5.0 Definitions

Access Control List (ACL): Lists kept by routers to control access to or from the router for a number of services (for example, to prevent packets with a certain IP address from leaving a particular interface on the router).

DMZ (de-militarized zone): Networking that exists outside of ABC Inc. primary corporate firewalls, but is still under ABC Inc. administrative control.

Network Support Organization: Any InfoSec-approved support organization that manages the networking of non-lab networks.

Least Access Principle: Access to services, hosts, and networks is restricted unless otherwise permitted.

Internet Services: Services running on devices that are reachable from other devices across a network. Major Internet services include DNS, FTP, HTTP, etc.

Point of Demarcation: The point at which the networking responsibility transfers from a Network Support Organization to the DMZ. Usually a router or firewall.

Screened Subnet: Screened subnets, or perimeter networks, are networks separated from the internal network by a screening router.

6.0 Exceptions

Exceptions to information system security policies exist in rare instances where a risk assessment examining the implications of being out of compliance has been performed, where a Systems Security Policy Exception Form has been prepared by the data owner or management, and where this form has been approved by both the CSO or Director of InfoSec and the Chief Information Officer (CIO).

7.0 Revision History

Date ___/____/_____

Version:_____

Author:_____

Summary:_____

A.7 ABC Inc. InfoSec Router Policy

Policy No. 6

Effective date Month/Day/Year

Implement by Month/Day/Year

1.0 Purpose

This document describes a required minimal security configuration for all routers and switches connecting to a production network or used in a production capacity at or on behalf of ABC Inc.

2.0 Scope

All routers and switches connected to ABC Inc. production networks are affected. Routers and switches within the internal networks are not affected. Routers and switches within DMZ areas fall under the DMZ Policy.

3.0 Policy

Every router must meet the following configuration standards:

1. No local user accounts are configured on the router. Routers must use TACACS+ for all user authentication.

2. The enable password on the router must be kept in a secure encrypted form. The router must have the enable password set to the current production router password from the router's support organization.

3. Disallow the following:
   a. IP directed broadcasts
   b. Incoming packets at the router sourced with invalid addresses such as RFC1918 address
   c. TCP small services
   d. UDP small services
   e. All source routing
   f. All web services running on router

4. Use corporate standardized SNMP community strings.

5. Access rules are to be added as business needs arise.

6. The router must document router configurations and point of contact(s).

7. Each router must have the following statement posted in clear view:

   "UNAUTHORIZED ACCESS TO THIS NETWORK DEVICE IS PROHIBITED. You must have explicit permission to access or configure this device. All activities performed on this device may be logged, and violations of this policy may result in disciplinary action, and may be reported to law enforcement. There is no right to privacy on this device."

4.0 Enforcement

Any employee found to have violated this policy may be subject to disciplinary action, up to and including termination of employment.

5.0 Definitions

Production Network: The "production network" is the network used in the daily business of ABC Inc. Any network connected to the corporate backbone, either directly or indirectly, which lacks an intervening firewall device. Any network whose impairment would result in direct loss of functionality to ABC Inc. employees or impact their ability to do work.

Lab Network: A "lab network" is defined as any network used for the purposes of testing, demonstrations, training, etc. Any network that is stand-alone or firewalled off from the production network(s) and whose impairment will not cause direct loss to ABC Inc. nor affect the production network.

6.0 Exceptions

Exceptions to information system security policies exist in rare instances where a risk assessment examining the implications of being out of compliance has been performed, where a Systems Security Policy Exception Form has been prepared by the data owner or management, and where this form has been approved by both the CSO or Director of InfoSec and the Chief Information Officer (CIO).

7.0 Revision History

Date ___/____/_____

Version:_____

Author:_____

Summary:_____

A.8 ABC Inc. InfoSec Extranet Policy

Policy No. 7

Effective date Month/Day/Year

Implement by Month/Day/Year

1.0 Purpose

This document describes the policy under which third party organizations connect to ABC Inc. networks for the purpose of transacting business related to ABC Inc.

2.0 Scope

Connections between third parties that require access to non-public ABC Inc. resources fall under this policy, regardless of whether a Telco circuit (such as frame relay or ISDN) or VPN technology is used for the connection. Connectivity to third parties such as the Internet Service Providers (ISPs) that provide Internet access for ABC Inc. or to the Public Switched Telephone Network does NOT fall under this policy.

3.0 Policy

3.1 Prerequisites

The following conditions (prerequisites) must be satisfied before extranet usage is granted:

- Security Review

- Third Party Connection Agreement

- Business Case

- Point of Contact

3.1.1 Security Review

All new extranet connectivity will go through a security review with the InfoSec department. The reviews are to ensure that all access matches the business requirements in the best possible way, and that the principle of least access is followed.

3.1.2 Third Party Connection Agreement

All new connection requests between third parties and ABC Inc. require that the third party and ABC Inc. representatives agree to and sign the Third Party Agreement. This agreement must be signed by the Vice President of the Sponsoring Organization as well as a representative from the third party who is legally empowered to sign on behalf of the third party. The signed document is to be kept on file with the relevant extranet group. Documents pertaining to connections into ABC Inc. DMZs are to be kept on file with the IT Operations Department.

3.1.3 Business Case

All production extranet connections must be accompanied by a valid business justification, in writing, that is approved by a project sponser. DMZ connections must be approved by the InfoSec Department. Typically this function is handled as part of the Third Party Agreement.

3.1.4 Point Of Contact

The Sponsoring Organization must designate a person to be the Point of Contact (POC) for the extranet connection. The POC acts on behalf of the Sponsoring Organization, and is responsible for those portions of this policy and the Third Party Agreement that pertain to it. In the event that the POC changes, the relevant extranet organization must be informed promptly. A POC must also be identified for the external party to the extranet connection.

3.2 Establishing Connectivity

Sponsoring Organizations within ABC Inc. that wish to establish connectivity to a third party are to file a new site request <Check for correct terminology> with the IT Operation group. The extranet group will engage InfoSec to address security issues inherent in the project. If the proposed connection is to terminate within a DMZ at ABC Inc., the Sponsoring Organization must engage the InfoSec department. The Sponsoring Organization must provide full and complete information as to the nature of the proposed access to the extranet group and InfoSec department, as requested. All connectivity established must be based on the least-access principle, in accordance with the approved business requirements and the security review. In no case will ABC Inc. rely upon the third party to protect ABC Inc.'s network or resources.

3.3 Modifying or Changing Connectivity and Access

All changes in access must be accompanied by a valid business justification, and are subject to security review. Changes are to be implemented via corporate change management process. The Sponsoring Organization is responsible for notifying the IT Operations group and/or InfoSec when there is a material change in their originally provided information so that security and connectivity evolve accordingly.

3.4 Terminating Access

When access is no longer required, the Sponsoring Organization within ABC Inc. must notify the extranet team responsible for that connectivity, which will then terminate the access. This may mean a modification of existing permissions up to terminating the circuit, as appropriate. The IT Operations group must conduct an audit of their respective connections on an annual basis to ensure that all existing connections are still needed, and that the access provided meets the needs of the connection. Connections that are found to be depreciated, and/or are no longer being used to conduct ABC Inc. business, will be terminated immediately. Should a security incident or a finding that a circuit has been depreciated and is no longer being used to conduct ABC Inc. business necessitate a modification of existing permissions, or termination of connectivity, InfoSec and/or the IT Operations group will notify the POC or the Sponsoring Organization of the change prior to taking any action.

4.0 Enforcement

Any employee found to have violated this policy may be subject to disciplinary action, up to and including termination of employment.

5.0 Definitions

Circuit: For the purposes of this policy, circuit refers to the method of network access, whether it's through traditional ISDN, Frame Relay etc., or via VPN/Encryption technologies.

Sponsoring Organization: The ABC Inc. organization who requested that the third party have access into ABC Inc.

Third Party: A business that is not a formal or subsidiary part of ABC Inc.

6.0 Exceptions

Exceptions to information system security policies exist in rare instances where a risk assessment examining the implications of being out of compliance has been performed, where a Systems Security Policy Exception Form has been prepared by the data owner or management, and where this form has been approved by both the CSO or Director of InfoSec and the Chief Information Officer (CIO).

7.0 Revision History

Date ___/___/_____

Version:_____

Author:_____

Summary:_____

A.9 ABC Inc. InfoSec Remote Access Policy

Policy No. 8

Effective date Month/Day/Year

Implement by Month/Day/Year

1.0 Purpose

The purpose of this policy is to define standards for connecting to ABC Inc.'s network from any host. These standards are designed to minimize the potential exposure to ABC Inc. from damages which may result from unauthorized use of ABC Inc. resources. Damages include the loss of sensitive or company confidential data, intellectual property, damage to public image, damage to critical ABC Inc. internal systems, etc.

2.0 Scope

This policy applies to all ABC Inc. employees, contractors, vendors, and agents with an ABC Inc.-owned or personally-owned computer or workstation used to connect to the ABC Inc. network. This policy applies to remote access connections used to do work on behalf of ABC Inc., including reading or sending email and viewing intranet web resources. Remote access implementations that are covered by this policy include, but are not limited to, dial-in modems, frame relay, ISDN, DSL, VPN, SSH, and cable modems, etc.

3.0 Policy

3.1 General

1. It is the responsibility of ABC Inc. employees, contractors, vendors, and agents with remote access privileges to ABC Inc.'s corporate network to ensure that their remote access connection is given the same consideration as the user's on-site connection to ABC Inc.

2. General access to the Internet for recreational use by immediate household members through the ABC Inc. Network on personal computers is permitted for employees that have flat-rate services. The ABC Inc. employee is responsible to ensure the family member does not violate any ABC Inc. policies, does not perform illegal activities,

and does not use the access for outside business interests. The ABC Inc. employee bears responsibility for the consequences should the access be misused.

3. Please review the following policies for details of protecting information when accessing the corporate network via remote access methods, and acceptable use of ABC Inc.'s network:

   a. Acceptable Use Policy
   b. Virtual Private Network (VPN) Policy
   c. Wireless Communications Policy

4. For additional information regarding ABC Inc.'s remote access connection options, including how to order or disconnect service, cost comparisons, troubleshooting, etc., go to the Remote Access Services website.

3.2 Requirements

1. Secure remote access must be strictly controlled. Control will be enforced via one-time password authentication or public/private keys with strong passphrases. For information on creating a strong passphrase, see the Password Policy.

2. At no time should any ABC Inc. employee provide their login or email password to anyone, not even family members.

3. ABC Inc. employees and contractors with remote access privileges must ensure that their ABC Inc.-owned or personal computer or workstation, which is remotely connected to ABC Inc.'s corporate network, is not connected to any other network at the same time, with the exception of personal networks that are under the complete control of the user.

4. ABC Inc. employees and contractors with remote access privileges to ABC Inc.'s corporate network must not use non-ABC Inc. email accounts (i.e., Hotmail, Yahoo, AOL), or other external resources to conduct ABC Inc. business, thereby ensuring that official business is never confused with personal business.

5. Routers for dedicated ISDN lines configured for access to the ABC Inc. network must meet minimum authentication requirements of CHAP.

6. Reconfiguration of a home user's equipment for the purpose of split-tunneling or dual homing is not permitted at any time.

7. Frame Relay must meet minimum authentication requirements of Data Link Connection Identifier (DLCI) standards.

8. Non-standard hardware configurations must be approved by Remote Access Services, and InfoSec must approve security configurations for access to hardware.

9. All hosts that are connected to ABC Inc. internal networks via remote access technologies must use the most up-to-date antivirus software (place URL to corporate software site here), this includes personal computers. Third party connections must comply with requirements as stated in the Third Party Agreement.

10. Personal equipment that is used to connect to ABC Inc.'s networks must meet the requirements of ABC Inc.-owned equipment for remote access.

11. Organizations or individuals who wish to implement nonstandard Remote Access solutions to the ABC Inc. production network must obtain prior approval from Remote Access Services and InfoSec.

4.0 Enforcement

Any employee found to have violated this policy may be subject to disciplinary action, up to and including termination of employment.

5.0 Definitions

Cable Modem: Cable companies such as AT&T Broadband provide Internet access over Cable TV coaxial cable. A cable modem accepts this coaxial cable and can receive data from the Internet at over 1.5 Mbps. Cable is currently available only in certain communities.

CHAP: Challenge Handshake Authentication Protocol (CHAP) is an authentication method that uses a one-way hashing function. The Data Link Connection Identifier (DLCI) is a unique number assigned to a Permanent Virtual Circuit (PVC) endpoint in a frame relay network. DLCI identifies a particular PVC endpoint within a user's access channel in a frame relay network, and has local significance only to that channel.

Dial-in Modem: A peripheral device that connects computers to each other for sending communications via the telephone lines. The modem modulates the digital data of computers into analog signals to send over the telephone lines, then demodulates back into digital signals to be read by the computer on the other end; thus the name "modem" for modulator/ demodulator.

Dual Homing: Having concurrent connectivity to more than one network from a computer or network device. Examples include: Being logged into the corporate network via a local Ethernet connection, and dialing into AOL or other Internet service provider (ISP). Being on an ABC Inc.-provided Remote Access home network, and connecting to another network, such as a spouse's remote access. Configuring an ISDN router to dial into ABC Inc. and an ISP, depending on packet destination.

DSL: Digital Subscriber Line (DSL) is a form of high-speed Internet access competing with cable modems. DSL works over standard phone lines and supports data speeds of over 2 Mbps downstream (to the user) and slower speeds upstream (to the Internet).

Frame Relay: A method of communication that incrementally can go from the speed of an ISDN to the speed of a T1 line. Frame Relay has a flat-rate billing charge instead of a per time usage. Frame Relay connects via the telephone company's network.

ISDN: There are two flavors of Integrated Services Digital Network or ISDN: BRI and PRI. BRI is used for home office/remote access. BRI has two "Bearer" channels at 64kbit (aggregate 128kb) and 1 D channel for signaling info. PRI is short for Primary-Rate Interface, a type of ISDN service designed for larger organizations. PRI includes 23 B-channels (30 in Europe) and one D-Channel.

One-time password (OTP): A security system that requires a new password every time a user authenticates themselves. OTP generates these passwords using either the MD4 or MD5 hashing algorithms.

Remote Access: Any access to ABC Inc.'s corporate network through a non-ABC Inc.-controlled network, device, or medium.

Split-tunneling: Simultaneous direct access to a non-ABC Inc. network (such as the Internet, or a home network) from a remote device (PC, PDA, WAP phone, etc.) while connected into ABC Inc.'s corporate network via a VPN tunnel. Virtual Private Network (VPN) is a method for accessing a remote network via "tunneling" through the Internet.

6.0 Exceptions

Exceptions to information system security policies exist in rare instances where a risk assessment examining the implications of being out of compliance has been performed, where a Systems Security Policy Exception Form has been prepared by the data owner or management, and where this form has been approved by both the CSO or Director of InfoSec and the Chief Information Officer (CIO).

7.0 Revision History

Date ___ / ___ / _____

Version: _____

Author: _____

Summary: _____

A.10 ABC Inc. InfoSec Dial-In Access Policy

Policy No. 9

Effective date Month/Day/Year

Implement by Month/Day/Year

1.0 Purpose

The purpose of this policy is to protect ABC Inc.'s electronic information from being inadvertently compromised by authorized personnel using a dial-in connection.

2.0 Scope

The scope of this policy is to define appropriate dial-in access and its use by authorized personnel.

3.0 Policy

1. ABC Inc. employees and authorized third parties (customers, vendors, etc.) can use dial-in connections to gain access to the corporate network through vendor solutions

approved and provided by IT Operations. Dial-in access should be strictly controlled, using one-time password authentication. Dial-in access should be requesting using the corporate account request process.

2. It is the responsibility of employees with dial-in access privileges to ensure a dial-in connection to ABC Inc. is not used by non-employees to gain access to company information system resources. An employee who is granted dial-in access privileges must remain constantly aware that dial-in connections between their location and ABC Inc. are literal extensions of ABC Inc.'s corporate network, and that they provide a potential path to the company's most sensitive information. The employee and/or authorized third party individual must take every reasonable measure to protect ABC Inc.'s assets.

3. Only IT Operations approved dial-in numbers will be used.

4. Analog and non-GSM digital cellular phones cannot be used to connect to ABC Inc.'s corporate network, as their signals can be readily scanned and/or hijacked by unauthorized individuals. Only GSM standard digital cellular phones are considered secure enough for connection to ABC Inc.'s network. For additional information on wireless access to the ABC Inc. network, consult the InfoSec Wireless Communications Policy.

5. For a third party using dial-in or remote access:
    *   All connections or accounts must have an expiry date with a duration of 12 months or end of contract, whichever comes first.
    *   A new network access request must to be submitted to extend the access time period beyond the expiration date.
    *   There will be no auto-renewal upon expiration. Connection will be automatically disabled upon expiration date.

Note: Dial-in accounts are considered 'as needed' accounts. Account activity is monitored, and if a dial-in account is not used for a period of six months, the account will expire and no longer function. If dial-in access is subsequently required, the individual must request a new account as described above.

4.0 Enforcement

Any employee found to have violated this policy may be subject to disciplinary action, up to and including termination of employment.

5.0 Definitions

6.0 Exceptions

Exceptions to information system security policies exist in rare instances where a risk assessment examining the implications of being out of compliance has been performed, where a Systems Security Policy Exception Form has been prepared by the data owner or management, and where this form has been approved by both the CSO or Director of InfoSec and the Chief Information Officer (CIO).

7.0 Revision History

Date ___/____/_____

Version:_____

Author:_____

Summary:_____

A.11 ABC Inc. InfoSec VPN Communication Policy

Policy No. 10

Effective date Month/Day/Year

Implement by Month/Day/Year

1.0 Purpose

The purpose of this policy is to provide guidelines for Remote Access IPSec or L2TP Virtual Private Network (VPN) connections to the ABC Inc. corporate network.

2.0 Scope

This policy applies to all ABC Inc. employees, contractors, consultants, temporaries, and other workers including all personnel affiliated with third parties utilizing VPNs to access the ABC Inc. network. This policy applies to implementations of VPN that are directed through an IPSec Concentrator. Site-to-site VPN connection policies are covered in the Extranet Policy.

3.0 Policy

Approved ABC Inc. employees and authorized third parties (customers, vendors, etc.) may utilize the benefits of VPNs, which are a "user managed" service. This means that the user is responsible for selecting an Internet Service Provider (ISP), coordinating installation, installing any required software, and paying associated fees. Further details may be found in the Remote Access Policy. Additionally:

1. It is the responsibility of employees with VPN privileges to ensure that unauthorized users are not allowed access to ABC Inc. internal networks.

2. VPN use is to be controlled using either a one-time password authentication such as a token device or a public/private key system with a strong passphrase.

3. When actively connected to the corporate network, VPNs will force all traffic to and from the PC over the VPN tunnel: all other traffic will be dropped.

4. Dual (split) tunneling is NOT permitted; only one network connection is allowed.

5. VPN gateways will be set up and managed by ABC Inc. network operational groups.

6. All computers connected to ABC Inc. internal networks via VPN or any other technology must use the most up-to-date anti-virus software that is the corporate standard (provide URL to this software); this includes personal computers.

7. VPN users will be automatically disconnected from ABC Inc.'s network after 30 minutes of inactivity. The user must then logon again to reconnect to the network. Pings or other artificial network processes are not to be used to keep the connection open.

8. The VPN concentrator is limited to an absolute connection time of 24 hours.

9. Users of computers that are not ABC Inc.-owned equipment must configure the equipment to comply with ABC Inc.'s VPN and Network policies.

10. Only InfoSec-approved VPN clients may be used.

11. By using VPN technology with personal equipment, users must understand that their machines are a de facto extension of ABC Inc.'s network, and as such are subject to the same rules and regulations that apply to ABC Inc.-owned equipment, i.e., their machines must be configured to comply with InfoSec's Security Policies.

4.0 Enforcement

Any employee found to have violated this policy may be subject to disciplinary action, up to and including termination of employment.

5.0 Definitions

6.0 Exceptions

Exceptions to information system security policies exist in rare instances where a risk assessment examining the implications of being out of compliance has been performed, where a Systems Security Policy Exception Form has been prepared by the data owner or management, and where this form has been approved by both the CSO or Director of InfoSec and the Chief Information Officer (CIO).

7.0 Revision History

Date ___/____/_____

Version:_____

Author:_____

Summary:_____

A.12 ABC Inc. InfoSec Wireless Communication Policy

Policy No. 11

Effective date Month/Day/Year

Implement by Month/Day/Year

1.0 Purpose

This policy prohibits access to ABC Inc. networks via unsecured wireless communication mechanisms. Only wireless systems that meet the criteria of this policy or have been granted an exclusive waiver by InfoSec are approved for connectivity to ABC Inc.'s networks.

2.0 Scope

This policy covers all wireless data communication devices (e.g., personal computers, cellular phones, PDAs, Blackberries, etc.) connected to any of ABC Inc.'s internal networks. This includes any form of wireless communication device capable of transmitting packet data. Wireless devices and/or networks without any connectivity to ABC Inc.'s networks do not fall under the purview of this policy.

3.0 Policy

3.1 Register Access Points and Cards

All wireless Access Points/Base Stations connected to the corporate network must be registered and approved by InfoSec. These Access Points/Base Stations are subject to periodic penetration tests and audits. All wireless Network Interface Cards (i.e., PC cards) used in corporate laptop or desktop computers must be registered with InfoSec

3.2 Approved Technology

All wireless LAN access must use corporate-approved vendor products and security configurations.

3.3 VPN Encryption and Authentication

All computers with wireless LAN devices must utilize a corporate-approved Virtual Private Network (VPN) configured to drop all unauthenticated and unencrypted traffic. To comply with this policy, wireless implementations must maintain point to point hardware encryption of at least 56 bits. All implementations must support a hardware address that can be registered and tracked, i.e., a MAC address. All implementations must support and employ strong user authentication which checks against an external database such as TACACS+, RADIUS, or something similar.

3.4 Setting the SSID

The SSID shall be configured so that it does not contain any identifying information about the organization, such as the company name, division title, employee name, or product identifier.

4.0 Enforcement

Any employee found to have violated this policy may be subject to disciplinary action, up to and including termination of employment.

5.0 Definitions

User Authentication: A method by which the user of a wireless system can be verified as a legitimate user independent of the computer or operating system being used.

6.0 Exceptions

Exceptions to information system security policies exist in rare instances where a risk assessment examining the implications of being out of compliance has been performed, where a Systems Security Policy Exception Form has been prepared by the data owner or management, and where this form has been approved by both the CSO or Director of InfoSec and the Chief Information Officer (CIO).

7.0 Revision History

Date ___/____/_____

Version:_____

Author:_____

Summary:_____

A.13 ABC Inc. InfoSec Server Policy

Policy No. 12

Effective date Month/Day/Year

Implement by Month/Day/Year

1.0 Purpose

The purpose of this policy is to establish standards for the base configuration of internal server equipment that is owned and/or operated by ABC Inc. Effective implementation of this policy will minimize unauthorized access to ABC Inc. proprietary information and technology.

2.0 Scope

This policy applies to server equipment owned and/or operated by ABC Inc., and to servers registered under any ABC Inc.-owned internal network domain. This policy is specifically for equipment on the internal ABC Inc. network. For secure configuration of equipment external to ABC Inc. on the DMZ, refer to the DMZ Policy.

3.0 Policy

3.1 Ownership and Responsibilities

All internal servers deployed at ABC Inc. must be owned by an operational group that is responsible for system administration. Approved server configuration guides must be established and maintained by each operational group, based on business needs and approved by InfoSec. Operational groups should monitor configuration compliance and implement an exception policy tailored to their environment. Each operational group must establish a process for changing the configuration guides, which includes review and approval by InfoSec.

Servers must be registered within the corporate IT Network Operations. At a minimum, the following information is required to positively identify the point of contact:

- Server contact(s) and location, and a backup contact

- Hardware and Operating System/Version

- Main functions and applications, if applicable

- Information in the corporate enterprise management system must be kept up-to-date.

- Configuration changes for production servers must follow the appropriate change management procedures.

3.2 General Configuration Guidelines

- Operating System configuration should be in accordance with approved InfoSec guidelines.

- Services and applications that will not be used must be disabled where practical.

- Access to services should be logged and/or protected through access-control methods such as TCP Wrappers, if possible.

- The most recent security patches must be installed on the system as soon as practical, the only exception being when immediate application would interfere with business requirements.

- Trust relationships between systems are a security risk, and their use should be avoided. Do not use a trust relationship when some other method of communication will do.

- Always use standard security principles of least required access to perform a function.

- Do not use root when a non-privileged account will do.

- User accounts that have system-level privileges granted through group memberships or programs such as "sudo" must have a unique password from all other accounts held by that user.

- If a methodology for secure channel connection is available (i.e., technically feasible), privileged access must be performed over secure channels (e.g., encrypted network connections using SSH or IPSec).

- Servers should be physically located in an access-controlled environment.

- Servers are specifically prohibited from operating from uncontrolled cubicle areas.

- Current applicable security patches/hot-fixes for any applications that are Internet services must be applied. Administrative owner groups must have processes in place to stay current on appropriate patches/ hot-fixes at the first available opportunity.

3.3 Monitoring

All security-related events on critical or sensitive systems must be logged and audit trails saved as follows:

- All security related logs will be kept online for a minimum of 1 week.

- Daily incremental tape backups will be retained for at least 1 month.

- Weekly full tape backups of logs will be retained for at least 1 month.

- Monthly full backups will be retained for a minimum of 2 years.

Security-related events will be reported to InfoSec, who will review logs and report incidents to IT management. Corrective measures will be prescribed as needed. Security-related events include, but are not limited to:

- Port-scan attacks

- Evidence of unauthorized access to privileged accounts

- Anomalous occurrences that are not related to specific applications on the host

3.4 Compliance

Audits will be performed on a regular basis by authorized organizations within ABC Inc. Audits will be managed by the internal audit group or InfoSec, in accordance with the Audit Policy. InfoSec will filter findings not related to a specific operational group and then present the findings to the appropriate support staff for remediation or justification. Every effort will be made to prevent audits from causing operational failures or disruptions.

4.0 Enforcement

Any employee found to have violated this policy may be subject to disciplinary action, up to and including termination of employment.

5.0 Definitions

DMZ: De-militarized Zone. A network segment external to the corporate production network.

Server: For purposes of this policy, a Server is defined as an internal ABC Inc. Server. Desktop machines and Lab equipment are not relevant to the scope of this policy.

6.0 Exceptions

Exceptions to information system security policies exist in rare instances where a risk assessment examining the implications of being out of compliance has been performed, where a Systems Security Policy Exception Form has been prepared by the data owner or management, and where this form has been approved by both the CSO or Director of InfoSec and the Chief Information Officer (CIO).

7.0 Revision History

Date ___/___/_____

Version:_____

Author:_____

Summary:_____

A.14 ABC Inc. InfoSec Password Policy

Policy No. 13

Effective date Month/Day/Year

Implement by Month/Day/Year

1.0 Overview

Passwords are an important aspect of computer security. They are the front line of protection for user accounts. A poorly chosen password may result in the compromise of ABC Inc.'s entire corporate network. As such, all ABC Inc. employees (including contractors and vendors with access to ABC Inc. systems) are responsible for taking the appropriate steps, as outlined below, to select and secure their passwords.

2.0 Purpose

The purpose of this policy is to establish a standard for creation of strong passwords, the protection of those passwords, and the frequency of change.

3.0 Scope

The scope of this policy includes all personnel who have or are responsible for an account (or any form of access that supports or requires a password) on any system that resides at any ABC Inc. facility, has access to the ABC Inc. network, or stores any non-public ABC Inc. information.

4.0 Policy

4.1 General

- All system-level passwords (e.g., root, enable, NT admin, application administration accounts, etc.) must be changed on at least a quarterly basis.

- All production system-level passwords must comply with IT Operations and/or Application Support procedures.

- All user-level passwords (e.g., email, web, desktop computer, etc.) must be changed at least every six months. The recommended change interval is every four months.

- User accounts that have system-level privileges granted through group memberships or programs such as "sudo" must have a unique password from all other accounts held by that user.

- Passwords should not be inserted into email messages or other forms of electronic communication in cleartext. When passwords must be sent through email, they must be temporary in nature (e.g., forced password change within a 24-hour period).

- Default passwords must not be used.

- All user-level and system-level passwords must conform to the guidelines described below.

4.2 Guidelines

A. General Password Construction Guidelines

Passwords are used for various purposes at ABC Inc. Some of the more common uses include: user-level accounts, web accounts, email accounts, screensaver protection, voicemail password, and local router logins. Since very few systems have support for onetime tokens (i.e., dynamic passwords which are only used once), everyone should be aware of how to select strong passwords.

Poor, weak passwords have the following characteristics:

- The password contains less than eight characters.

- The password is a word found in a dictionary (English or foreign).

- The password is a common usage word such as:

  - Names of family, pets, friends, co-workers, fantasy characters, etc.
  - Computer terms and names, commands, sites, companies, hardware, software.
  - The words "ABC Inc.", "sanjose", "sanfran" or any derivation.
  - Birthdays and other personal information such as addresses and phone numbers.
  - Word or number patterns like aaabbb, qwerty, zyxwvuts, 123321, etc.
  - Any of the above spelled backwards.
  - Any of the above preceded or followed by a digit (e.g., secret1, 1secret).

Strong passwords have the following characteristics:

- Contain both upper and lower case characters (e.g., a-z, A-Z).

- Have digits and punctuation characters as well as letters (0-9, !@#$%^&*()_ + /~ − =\`{}[]:";' < >?,./).

- Are at least eight alphanumeric characters long.

- Are not a word in any language, slang, dialect, jargon, etc.

- Are not based on personal information, names of family, etc.

Passwords should never be written down or stored on-line. Try to create passwords that can be easily remembered. One way to do this is create a password based on a song title, affirmation, or other phrase. For example, the phrase might be: "This May Be One Way To Remember" and the password could be: "TmB1w2R!" or "Tmb1W>r~" or some other variation. NOTE: Do not use these examples as passwords!

B.   Password Protection Standards

- Do not use the same password for ABC Inc. accounts as for other non-ABC Inc. access (e.g., personal ISP account, option trading, benefits, etc.). Where possible, don't use the same password for various ABC Inc. access needs. For example, select one password for the Engineering systems and a separate password for IT systems. Also, select a separate password to be used for an NT account and a UNIX account.

- Do not share ABC Inc. passwords with anyone, including administrative assistants or secretaries. All passwords are to be treated as sensitive, confidential ABC Inc. information. Here is a list of "don'ts":
  - Don't reveal a password over the phone to ANYONE.
  - Don't reveal a password in an email message.
  - Don't reveal a password to the boss.
  - Don't talk about a password in front of others.
  - Don't hint at the format of a password (e.g., "my family name").
  - Don't reveal a password on questionnaires or security forms.
  - Don't share a password with family members.
  - Don't reveal a password to co-workers while on vacation.
  - If someone demands a password, refer them to this document or have them call someone in the InfoSec Department.
  - Do not use the "Remember Password" feature of applications (e.g., Eudora, OutLook, Netscape Messenger). Again, do not write passwords down and store them anywhere in your office. Do not store passwords in a file on ANY computer system (including Palm Pilots or similar devices) without encryption.
  - Change passwords at least once every six months (except system-level passwords, which must be changed quarterly). The recommended change interval is every four months.
  - If an account or password is suspected to have been compromised, report the incident to InfoSec and change all passwords.
  - Password cracking or guessing may be performed on a periodic or random basis by InfoSec or its delegates. If a password is guessed or cracked during one of these scans, the user will be required to change it.

C. Application Development Standards

Application developers must ensure their programs contain the following security precautions. Applications:

- should support authentication of individual users, not groups.

- should not store passwords in cleartext or in any easily reversible form.

- should provide for some sort of role management, such that one user can take over the functions of another without having to know the other's password.

- should support TACACS+, RADIUS and/or X.509 with LDAP security retrieval, wherever possible.

- must support dynamic passwords

Note: See the Application Password Policy

D. Use of Passwords and Passphrases for Remote Access Users

Access to the ABC Inc. Networks via remote access is to be controlled using either a one-time password authentication or a public/private key system with a strong passphrase.

E. Passphrases

Passphrases are generally used for public/private key authentication. A public/private key system defines a mathematical relationship between the public key that is known by all, and the private key, that is known only to the user. Without the passphrase to "unlock" the private key, the user cannot gain access.

Passphrases are not the same as passwords. A passphrase is a longer version of a password and is, therefore, more secure. A passphrase is typically composed of multiple words. Because of this, a passphrase is more secure against "dictionary attacks." A good passphrase is relatively long and contains a combination of upper and lowercase letters and numeric and punctuation characters. An example of a good passphrase:

"The*?#>*@TrafficOnThe101Was*&#!#ThisMorning".

All of the rules above that apply to passwords apply to passphrases.

5.0 Enforcement

Any employee found to have violated this policy may be subject to disciplinary action, up to and including termination of employment.

6.0 Definitions

Application Administration Account: Any account that is for the administration of an application (e.g., Oracle database administrator, ISSU administrator).

7.0 Exceptions

Exceptions to information system security policies exist in rare instances where a risk assessment examining the implications of being out of compliance has been performed, where a Systems Security Policy Exception Form has been prepared by the data owner or management, and where this form has been approved by both the CSO or Director of InfoSec and the Chief Information Officer (CIO).

8.0 Revision History

Date ___/___/_____

Version:_____

Author:_____

Summary:_____

A.15 ABC Inc. InfoSec Application Password Policy

Policy No. 14

Effective date Month/Day/Year

Implement by Month/Day/Year

1.0 Purpose

This policy states the requirements for securely storing and retrieving application usernames and passwords (i.e., application credentials) for use by a program that will access an application running on one of ABC Inc.'s networks.

Computer programs running on ABC Inc.'s networks often require the use of one of the many internal application servers. In order to access one of these applications, a program must authenticate to the application by presenting acceptable credentials. The application privileges that the credentials are meant to restrict can be compromised when the credentials are improperly stored.

2.0 Scope

This policy applies to all software that will access an ABC Inc., multi-user production application.

3.0 Policy

3.1 General

In order to maintain the security of ABC Inc.'s internal applications, access by software programs must be granted only after authentication with credentials. The credentials used for this authentication must not reside in the main, executing body of the program's source code in cleartext. Application credentials must not be stored in a location that can be accessed through a web server.

3.2 Specific Requirements

3.2.1. Storage of Database User Names and Passwords

- Application user names and passwords may be stored in a file separate from the executing body of the program's code. This file must not be world readable.

- Application credentials may reside on the application server. In this case, a hash number identifying the credentials may be stored in the executing body of the program's code.

- Application credentials may be stored as part of an authentication server (i.e., an entitlement directory), such as an LDAP server used for user authentication. Application authentication may occur on behalf of a program as part of the user authentication process at the authentication server. In this case, there is no need for programmatic use of application credentials.

- Application credentials may not reside in the documents tree of a web server.

- Pass through authentication (i.e., Oracle OPS$ authentication) must not allow access to the application based solely upon a remote user's authentication on the remote host.

- Passwords or passphrases used to access an application must adhere to the Password Policy.

3.2.2. Retrieval of Application User Names and Passwords

If stored in a file that is not source code, then application user names and passwords must be read from the file immediately prior to use. Immediately following application authentication, the memory containing the user name and password must be released or cleared.

The scope into which you may store application credentials must be physically separated from the other areas of your code, e.g., the credentials must be in a separate source file. The file that contains the credentials must contain no other code but the credentials (i.e., the user name and password) and any functions, routines, or methods that will be used to access the credentials.

For languages that execute from source code, the credentials' source file must not reside in the same browseable or executable file directory tree in which the executing body of code resides.

3.2.3. Access to Application User Names and Passwords

Every program or every collection of programs implementing a single business function must have unique application credentials.

Application passwords used by programs are system-level passwords as defined by the Password Policy. Developer groups must have a process in place to ensure that application passwords are controlled and changed in accordance with the InfoSec Password Policy. This process must include a method for restricting knowledge of application passwords to a need-to-know basis.

4.0 Enforcement

Any employee found to have violated this policy may be subject to disciplinary action, up to and including termination of employment.

5.0 Definitions

Application: Any program, or database which provides access to information (e.g. SAP, Siebel, Oracle, SQL Server...).

Computer language: A language used to generate programs. Credentials: Something you know (e.g., a password or pass phrase) and/or something that identifies you (e.g., a user name, a fingerprint, voiceprint, retina print) are presented for authentication.

Entitlement: The level of privilege that has been authenticated and authorized. The privileges level at which to access resources.

Executing body: The series of computer instructions that the computer executes to run a program.

Hash: An algorithmically generated number that identifies a datum or its location.

LDAP: Lightweight Directory Access Protocol, a set of protocols for accessing information directories.

Module: A collection of computer language instructions grouped together either logically or physically. A module may also be called a package or a class, depending upon which computer language is used.

Name space: A logical area of code in which the declared symbolic names are known and outside of which these names are not visible.

Production: Software that is being used for a purpose other than when software is being implemented or tested.

6.0 Exceptions

Exceptions to information system security policies exist in rare instances where a risk assessment examining the implications of being out of compliance has been performed, where a Policy Exception Form has been prepared by the data owner or management, and where this form has been approved by both the CSO or Director of InfoSec and the Chief Information Officer (CIO).

7.0 Revision History

Date ___/___/_____

Version:_____

Author:_____

Summary:_____

A.16 ABC Inc. InfoSec Anti-Virus Policy

Policy No. 15

Effective date Month/Day/Year

Implement by Month/Day/Year

1.0 Purpose

To establish requirements which must be met by all computers connected to ABC Inc. lab networks to ensure effective virus detection and prevention.

2.0 Scope

This policy applies to all ABC Inc. computers that are Windows-based or utilize Windows-based file directory sharing. This includes, but is not limited to, desktop computers, laptop computers, file/ftp/tftp/proxy servers, and any Windows-based lab equipment such as traffic generators.

3.0 Policy

All ABC Inc. PC-based lab computers must have ABC Inc.'s standard, supported anti-virus software installed and scheduled to run at regular intervals. In addition, the anti-virus software and the virus pattern files must be kept up-to-date. Virus-infected computers must be removed from the network until they are verified as virus-free. Information System Admins and Managers are responsible for creating procedures that ensure anti-virus software is run at regular intervals, and computers are verified as virus-free. Any activities with the intention to create and/or distribute malicious programs into ABC Inc.'s networks (e.g., viruses, worms, Trojan horses, e-mail bombs, etc.) are prohibited, in accordance with the Acceptable Use Policy. Refer to ABC Inc.'s IT Anti-Virus Web page to help prevent virus problems. Noted exceptions: Machines with operating systems other than those based on Microsoft products are exempted at the current time.

4.0 Enforcement

Any employee found to have violated this policy may be subject to disciplinary action, up to and including termination of employment.

5.0 Definitions

6.0 Exceptions

Exceptions to information system security policies exist in rare instances where a risk assessment examining the implications of being out of compliance has been performed, where a Policy Exception Form has been prepared by the data owner or management, and where this form has been approved by both the CSO or Director of InfoSec and the Chief Information Officer (CIO).

7.0 Revision History

Date ____/____/_____

Version:_____

Author:_____

Summary:_____

A.17 ABC Inc. InfoSec Policy Exception Form

Requestor's Name:

Requestor's Phone Number:

Date:

Policy for which Exception is being requested:

A brief description of the justification for the request, including the organizations that would benefit from the exception:

Compensating procedures to be implemented to mitigate risk:

A technical description of the situation that is to exist after grant of the exception:

A risk analysis, including the organizations that might be put at risk by the exception:

The organizations responsible for implementing the exception

Requestor_____

Signature_____

Date_____

CIO_____

Signature_____

Date_____

CIO_____

Signature_____

Date_____

Glossary

Tony Bradley

ActiveX: ActiveX is a Microsoft creation designed to work in a manner similar to Sun Microsystems' Java. The main goal is to create platform-independent programs that can be used continually on different operating systems. ActiveX is a loose standards definition, not a specific language. An ActiveX component or control can be run on any ActiveX-compatible platform.

ActiveX defines the methods with which these COM objects and ActiveX controls interact with the system; however, it is not tied to a specific language. ActiveX controls and components can be created in various programming languages such as Visual C++, Visual Basic, or VBScript.

Active Scripting: Active scripting is the term used to define the various script programs that can run within and work with Hypertext Markup Language (HTML) in order to interact with users and create a dynamic Web page. By itself, HTML is static and only presents text and graphics. Using active scripting languages such as JavaScript or VBScript, developers can update the date and time displayed on the page, have information pop up in a separate window, or create scrolling text to go across the screen.

Adware: While not necessarily malware, adware is considered to go beyond the reasonable advertising one might expect from freeware or shareware. Typically, a separate program that is installed at the same time as a shareware or similar program, adware will usually continue to generate advertising even when the user is not running the originally desired program.*

Antivirus Software: Antivirus software is an application that protects your system from viruses, worms, and other malicious code. Most antivirus programs monitor traffic while you surf the Web, scan incoming e-mail and file attachments, and periodically check all local files for the existence of any known malicious code.

Application Gateway: An application gateway is a type of firewall. All internal computers establish a connection with the proxy server. The proxy server performs all communications with the Internet. External computers see only the Internet Protocol (IP) address of the proxy server and never communicate directly with the internal clients. The application gateway examines the packets more thoroughly than a circuit-level gateway when making forwarding decisions. It is considered more secure; however, it uses more memory and processor resources.

Attack: The act of trying to bypass security controls on a system. An attack may be active, resulting in the alteration of data; or passive, resulting in the release of data. Note: The fact that an attack is made does not necessarily mean that it will succeed. The degree of success depends on the vulnerability of the system and the effectiveness of the existing countermeasures. Attack is often used as a synonym for a specific exploit.*

Authentication: One of the keys in determining if a message or file you are receiving is safe is to first authenticate that the person who sent it is who they say they are. Authentication is the process of determining the true identity of someone. Basic authentication is using a password to verify that you are who you say you are. There are also more complicated and precise methods such as biometrics (e.g., fingerprints, retina scans).

Backbone: The backbone of the Internet is the collection of major communications pipelines that transfer the data from one end of the world to the other. Large Internet service providers (ISPs) such as AT&T and WorldCom make up the backbone. They connect through major switching centers called Metropolitan Area Exchange (MAE) and exchange data from each others' customers through peering agreements.

Backdoor: A backdoor is a secret or undocumented means of gaining access to a computer system. Many programs have backdoors placed by the programmer to allow them to gain access in order to troubleshoot or change a program. Other backdoors are placed by hackers once they gain access to a system, to allow for easier access into the system in the future or in case their original entrance is discovered.

Biometrics: Biometrics is a form of authentication that uses unique physical traits of the user. Unlike a password, a hacker cannot "guess" your fingerprint or retinal scan pattern. Biometrics is a relatively new term used to refer to fingerprinting, retinal scans, voice wave patterns, and various other unique biological traits used to authenticate users.

Broadband: Technically, broadband is used to define any transmission that can carry more than one channel on a single medium (e.g., the coaxial cable for cable TV carries many channels and can simultaneously provide Internet access). Broadband is also often used to describe high-speed Internet connections such as cable modems and digital subscriber lines (DSLs).

Bug: In computer technology, a bug is a coding error in a computer program. After a product is released or during public beta testing, bugs are still apt to be discovered. When this occurs, users have to either find a way to avoid using the "buggy" code or get a patch from the originators of the code.

Circuit-level Gateway: A circuit-level gateway is a type of firewall. All internal computers establish a "circuit" with the proxy server. The proxy server performs all communications with the Internet. External computers see only the IP address of the proxy server and never communicate directly with the internal clients.

Compromise: When used to discuss Internet security, compromise does not mean that two parties come to a mutually beneficial agreement. Rather, it means that the security of your computer or network is weakened. A typical security compromise can be a third party learning the administrator password of your computer.

Cross Site Scripting: Cross site scripting (XSS) refers to the ability to use some of the functionality of active scripting against the user by inserting malicious code into the HTML that will run code on the users' computers, redirect them to a site other than what they intended, or steal passwords, personal information, and so on.

XSS is a programming problem, not a vulnerability of any particular Web browser software or Web hosting server. It is up to the Web site developer to ensure that user input is validated and checked for malicious code before executing it.

Cyberterrorism: This term is more a buzzword than anything and is used to describe officially sanctioned hacking as a political or military tool. Some hackers have used stolen information (or the threat of stealing information) as a tool to attempt to extort money from companies.

DHCP: Dynamic Host Configuration Protocol (DHCP) is used to automate the assignment of IP addresses to hosts on a network. Each machine on a network must have a unique address. DHCP automatically enters the IP address, tracks which ones are in use, and remembers to put addresses back into the pool when devices are removed. Each device that is configured to use DHCP contacts the DHCP server to request an IP address. The DHCP server then assigns an IP address from the range it has been configured to use. The IP address is leased for a certain amount of time. When the device is removed from the network or when the lease expires, the IP address is placed back into the pool to be used by another device.

Demilitarized Zone: The demilitarized zone (DMZ) is a neutral zone or buffer that separates the internal and external networks and usually exists between two firewalls. External users can access servers in the DMZ, but not the computers on the internal network. The servers in the DMZ act as an intermediary for both incoming and outgoing traffic.

DNS: The Domain Name System (DNS) was created to provide a way to translate domain names to their corresponding IP addresses. It is easier for users to remember a domain name (e.g., yahoo.com) than to try and remember an actual IP address (e.g., 65.37.128.56) of each site they want to visit. The DNS server maintains a list of domain names and IP addresses so that when a request comes in it can be pointed to the correct corresponding IP address.

Keeping a single database of all domain names and IP addresses in the world would be exceptionally difficult, if not impossible. For this reason, the burden has been spread around the world. Companies, Web hosts, ISPs, and other entities that choose to do so can maintain their own DNS servers. Spreading the workload like this speeds up the process and provides better security instead of relying on a single source.

Denial of Service: A Denial-of-Service (DoS) attack floods a network with an overwhelming amount of traffic, thereby slowing its response time for legitimate traffic or grinding it to a halt completely. The more common attacks use the built-in features of the Transmission Control Protocol (TCP)/IP to create exponential amounts of network traffic.

E-mail Spoofing: E-mail spoofing is the act of forging the header information on an e-mail so that it appears to have originated from somewhere other than its true source. The protocol used for e-mail, Simple Mail Transfer Protocol (SMTP), does not have any authentication to verify the source. By changing the header information, the e-mail can appear to come from someone else.

E-mail spoofing is used by virus authors. By propagating a virus with a spoofed e-mail source, it is more difficult for users who receive the virus to track its source. E-mail spoofing is also used by distributors of spam to hide their identity.

Encryption: Encryption is when text, data, or other communications are encoded so that unauthorized users cannot see or hear it. An encrypted file appears as gibberish unless you have the password or key necessary to decrypt the information.

Firewall: Basically, a firewall is a protective barrier between your computer (or internal network) and the outside world. Traffic into and out of the firewall is blocked or restricted as you choose. By blocking all unnecessary traffic and restricting other traffic to those protocols or individuals that need it, you can greatly improve the security of your internal network.

Forensic: Forensic is a legal term. At its root it means something that is discussed in a court of law or that is related to the application of knowledge to a legal problem.

In computer terms, forensic is used to describe the art of extracting and gathering data from a computer to determine how an intrusion occurred, when it occurred, and who the intruder was. Organizations that employ good security practices and maintain logs of network and file access are able to accomplish this much easier. But, with the right knowledge and the right tools, forensic evidence can be extracted even from burned, waterlogged, or physically damaged computer systems.

Hacker: Commonly used to refer to any individual who uses their knowledge of networks and computer systems to gain unauthorized access to computer systems. While often used interchangeably, the term hacker typically applies to those who break in out of curiosity or for the challenge itself, rather than those who actually intend to steal or damage data. Hacker purists claim that true hacking is benign and that the term is misused.

Heuristic: Heuristics uses past experience to make educated guesses about the present. Using rules and decisions based on analysis of past network or e-mail traffic, heuristic scanning in antivirus software can self-learn and use artificial intelligence to attempt to block viruses or worms that are not yet known and for which the antivirus software does not yet have a filter to detect or block.

Hoax: A hoax is an attempt to trick a user into believing something that is not true. It is mainly associated with e-mails that are too good to be true or that ask you to do things like "forward this to everyone you know."

Host: As far as the Internet is concerned, a host is essentially any computer connected to the Internet. Each computer or device has a unique IP address which helps other devices on the Internet find and communicate with that host.

HTML: HTML is the basic language used to create graphic Web pages. HTML defines the syntax and tags used to create documents on the World Wide Web (WWW). In its basic form, HTML documents are static, meaning they only display text and graphics. In order to have scrolling text, animations, buttons that change when the mouse pointer is over them, and so on, a developer needs to use active scripting like JavaScript or VBScript or use third-party plug-ins like Macromedia Flash.

There are variations and additions to HTML as well. Dynamic Hypertext Markup Language (DHTML) is used to refer to pages that include things like JavaScript or CGI scripts in order to dynamically present information unique to each user or each time the user visits the site. Extensible Markup Language (XML) is gaining in popularity because of its ability to interact with data and provide a means for sharing and interpreting data between different platforms and applications.

ICMP: Internet Control Message Protocol (ICMP) is part of the IP portion of TCP/IP. Common network testing commands such as PING and Trace Route (TRACERT) rely on the ICMP.

Identity Theft: Use of personal information to impersonate someone, usually for the purpose of fraud.*

IDS: An Intrusion Detection System (IDS) is a device or application that is used to inspect all network traffic and to alert the user or administrator when there has been unauthorized access or an attempt to access a network. The two primary methods of monitoring are signature based and anomaly based. Depending on the device or application used, the IDS can alert either the user or the administrator or set up to block specific traffic or automatically respond in some way.

Signature-based detection relies on the comparison of traffic to a database containing signatures of known attack methods. Anomaly-based detection compares current network traffic to a known good baseline to look for anything out of the ordinary. The IDS can be placed strategically on the network as a Network-based Intrusion Detection System (NIDS), which will inspect all network traffic, or it can be installed on each individual system as a Host-based Intrusion Detection System (HIDS), which inspects traffic to and from that specific device only.

Instant Messaging: Instant messaging (IM) offers users the ability to communicate in real time. Starting with Internet Relay Chat (IRC), users became hooked on the ability to "chat" in real time rather than sending emails back and forth or posting to a forum or message board.

Online service providers such as America Online (AOL) and CompuServe created proprietary messaging systems that allow users to see when their friends are online and available to chat (as long as they use the same instant messaging software). ICQ introduced an IM system that was not tied to a particular ISP and that kicked off the mainstream popularity of instant messaging.

Internet: The Internet was originally called Arpanet, and was created by the United States government in conjunction with various colleges and universities for the purpose of sharing research data. As it stands now, there are millions of computers connected to the Internet all over the world. There is no central server or owner of the Internet; every computer on the Internet is connected with every other computer.

Intranet: An Intranet is an Internet with restricted access. Corporate Intranets generally use the exact same communication lines as the rest of the Internet, but have security in place to restrict access to the employees, customers, or suppliers that the corporation wants to have access.

IP: The IP is used to deliver data packets to their proper destination. Each packet contains both the originating and the destination IP address. Each router or gateway that receives the packet will look at the destination address and determine how to forward it. The packet will be passed from device to device until it reaches its destination.

IP Address: An IP Address is used to uniquely identify devices on the Internet. The current standard (IPv4) is a 32-bit number made up of four 8-bit blocks. In standard decimal numbers, each block can be any number from 0 to 255. A standard IP address would look something like "192.168.45.28."

Part of the address is the network address which narrows the search to a specific block, similar to the way your postal mail is first sent to the proper zip code. The other part of the address is the local address that specifies the actual device within that network, similar to the way your specific street address identifies you within your zip code. A subnet mask is used to determine how many bits make up the network portion and how many bits make up the local portion.

The next generation of IP (IPv6 or [IP Next Generation] IPng) has been created and is currently being implemented in some areas.

IP Spoofing: IP spoofing is the act of replacing the IP address information in a packet with fake information. Each packet contains the originating and destination IP address. By replacing the true originating IP address with a fake address, a hacker can mask the true source of an attack or force the destination IP address to reply to a different machine and possibly cause a DoS.

IPv4: The current version of IP used on the Internet is version 4 (IPv4). IPv4 is used to direct packets of information to their correct address. Due to a shortage of available addresses and to address the needs of the future, an updated IP is being developed (IPv6).

IPv6: To address issues with the current IP in use (IPv4) and to add features to improve the protocol for the future, the Internet Engineering Task Force (IETF) has introduced IP version 6 (IPv6) also known as IPng.

IPv6 uses 128-bit addresses rather than the current 32-bit addresses, allowing for an exponential increase in the number of available IP addresses. IPv6 also adds new security and performance features to the protocol. IPv6 is backwards compatible with IPv4 so that different networks or hardware manufacturers can choose to upgrade at different times without disrupting the current flow of data on the Internet.

ISP: An ISP is a company that has the servers, routers, communication lines, and other equipment necessary to establish a presence on the Internet. They in turn sell access to their equipment in the form of Internet services such as dial-up, cable modem, Digital Subscriber Line (DSL), or other types of connections. The larger ISPs form the backbone of the Internet.

JavaScript: JavaScript is an active scripting language that was created by Netscape and based on Sun Microsystems' platform-independent programming language, Java. Originally named LiveScript, Netscape changed the name to JavaScript to ride on the coattails of Java's popularity. JavaScript is used within HTML to execute small programs, in order to generate a dynamic Web page. Using JavaScript, a developer can make text or graphics change when the mouse points at them, update the current date and time on the Web page, or add personal information such as how long it has been since that user last visited the site. Microsoft Internet Explorer supports a subset of JavaScript dubbed JScript.

Malware: Malicious Code (malware) is a catch-all term used to refer to various types of software that can cause problems or damage your computer. The common types of malware are viruses, worms, Trojan horses, macro viruses, and backdoors.

NAT: Network Address Translation (NAT) is used to mask the true identity of internal computers. Typically, the NAT server or device has a public IP address that can be seen by external hosts. Computers on the local network use a completely different set of IP addresses. When traffic goes out, the internal IP address is removed and replaced with the public IP address of the NAT device. When replies come back to the NAT device, it determines which internal computer the response belongs to and routes it to its proper destination.

An added benefit is the ability to have more than one computer communicate on the Internet with only one publicly available IP address. Many home routers use NAT to allow multiple computers to share one IP address.

Network: Technically, it only takes two computers (or hosts) to form a network. A network is any two or more computers connected together to share data or resources. Common network resources include printers that are shared by many users rather than each user having their own printer. The Internet is one large network of shared data and resources.

Network Security: This term is used to describe all aspects of securing your computer or computers from unauthorized access. This includes blocking outsiders from getting into the network, as well as password protecting your computers and ensuring that only authorized users can view sensitive data.

P2P: Peer-to-peer Networking (P2P) applies to individual PCs acting as servers to other individual PCs. Made popular by the music file swapping service, Napster, P2P allows users to share files with each other through a network of computers using that same P2P client software. Each computer on the network has the ability to act as a server by hosting files for others to download, and as a client by searching other computers on the network for files they want.

Packet: A packet, otherwise known as a datagram, is a fragment of data. Data transmissions are broken up into packets. Each packet contains a portion of the data being sent as well as header information, which includes the destination address.

Packet Sniffing: Packet sniffing is the act of capturing packets of data flowing across a computer network. The software or device used to do this is called a packet sniffer. Packet sniffing is to computer networks what wire tapping is to a telephone network.

Packet sniffing is used to monitor network performance or to troubleshoot problems with network communications. However, it is also widely used by hackers and crackers to illegally gather information about networks they intend to break into. Using a packet sniffer, you can capture data such as passwords, IP addresses, protocols being used on the network, and other information that will help an attacker infiltrate the network.

Patch: A patch is like a Band-Aid®. When a company finds bugs and defects in their software, they fix them in the next version of the application. However, some bugs make the current product inoperable or less functional, or may even open security vulnerabilities. For these bugs, users cannot wait until the next release to get a fix; therefore, the company must create a small interim patch that users can apply to fix the problem.

Phishing: Posting of a fraudulent message to a large number of people via spam or other general posting asking them to submit personal or security information, which is then used for further fraud or identity theft. The term is possibly an extension of trolling, which is the posting of an outrageous message or point of view in a newsgroup or mailing list in the hope that someone will "bite" and respond to it.*

Port: A port has a dual definition in computers. There are various ports on the computer itself (e.g., ports to plug in your mouse, keyboards, Universal Serial Bus [USB] devices, printers, monitors, and so forth). However, the ports that are most relevant to information security are virtual ports found in TCP/IP. Ports are like channels on your computer. Normal Web or Hypertext Transfer Protocol (HTTP) traffic flows on port 80. Post Office Protocol version 3 (POP3) e-mail flows on port 110. By blocking or opening these ports into and out of your network, you can control the kinds of data that flows through your network.

Port Scan: A port scan is a method used by hackers to determine what ports are open or in use on a system or network. By using various tools, a hacker can send data to TCP or User Datagram Protocol (UDP) ports one at a time. Based on the response received, the port scan utility can determine if that port is in use. Using this information, the hacker can then focus his or her attack on the ports that are open and try to exploit any weaknesses to gain access.

Protocol: A protocol is a set of rules or agreed-upon guidelines for communication. When communicating, it is important to agree on how to do so. If one party speaks

French and one German, the communications will most likely fail. If both parties agree on a single language, communications will work.

On the Internet, the set of communications protocols used is called TCP/IP. TCP/IP is actually a collection of various protocols that have their own special functions. These protocols have been established by international standards bodies and are used in almost all platforms and around the globe to ensure that all devices on the Internet can communicate successfully.

Proxy Server: A proxy server acts as a middleman between your internal and external networks. It serves the dual roles of speeding up access to the Internet and providing a layer of protection for the internal network. Clients send Internet requests to the proxy server, which in turn initiates communications with actual destination server.

By caching pages that have been previously requested, the proxy server speeds up performance by responding to future requests for the same page, using the cached information rather than going to the Web site again.

When using a proxy server, external systems only see the IP address of the proxy server so the true identity of the internal computers is hidden. The proxy server can also be configured with basic rules of what ports or IP addresses are or are not allowed to pass through, which makes it a type of basic firewall.

Rootkit: A rootkit is a set of tools and utilities that a hacker can use to maintain access once they have hacked a system. The rootkit tools allow them to seek out usernames and passwords, launch attacks against remote systems, and conceal their actions by hiding their files and processes and erasing their activity from system logs and a plethora of other malicious stealth tools.

Script Kiddie: Script kiddie is a derogatory term used by hackers or crackers to describe novice hackers. The term is derived from the fact that these novice hackers tend to rely on existing scripts, tools, and exploits to create their attacks. They may not have any specific knowledge of computer systems or why or how their hack attempts work, and they may unleash harmful or destructive attacks without even realizing it. Script kiddies tend to scan and attack large blocks of the Internet rather than targeting a specific computer, and generally don't have any goal in mind aside from experimenting with tools to see how much chaos they can create.

SMTP: Simple Mail Transfer Protocol (SMTP) is used to send e-mail. The SMTP protocol provides a common language for different servers to send and receive e-mail messages. The default TCP/IP port for the SMTP protocol is port 25.

SNMP: Simple Network Management Protocol (SNMP) is a protocol used for monitoring network devices. Devices like printers and routers use SNMP to communicate their status. Administrators use SNMP to manage the function of various network devices.

Stateful Inspection: Stateful inspection is a more in-depth form of packet filter firewall. While a packet filter firewall only checks the packet header to determine the source and destination address and the source and destination ports to verify against its rules, stateful inspection checks the packet all the way to the Application layer. Stateful inspection monitors incoming and outgoing packets to determine source, destination, and context. By ensuring that only requested information is allowed back in, stateful inspection helps protect against hacker techniques such as IP spoofing and port scanning

TCP: The TCP is a primary part of the TCP/IP set of protocols, which forms the basis of communications on the Internet. TCP is responsible for breaking large data into smaller chunks of data called packets. TCP assigns each packet a sequence number and then passes them on to be transmitted to their destination. Because of how the Internet is set up, every packet may not take the same path to get to its destination. TCP has the responsibility at the destination end of reassembling the packets in the correct sequence and performing error-checking to ensure that the complete data message arrived intact.

TCP/IP: TCP/IP is a suite of protocols that make up the basic framework for communication on the Internet.

TCP helps control how the larger data is broken down into smaller pieces or packets for transmission. TCP handles reassembling the packets at the destination end and performing error-checking to ensure all of the packets arrived properly and were reassembled in the correct sequence.

IP is used to route the packets to the appropriate destination. The IP manages the addressing of the packets and tells each router or gateway on the path how and where to forward the packet to direct it to its proper destination.

Other protocols associated with the TCP/IP suite are UDP and ICMP.

Trojan: A Trojan horse is a malicious program disguised as a normal application. Trojan horse programs do not replicate themselves like a virus, but they can be propagated as attachments to a virus.

UDP: UDP is a part of the TCP/IP suite of protocols used for communications on the Internet. It is similar to TCP except that it offers very little error checking and does not establish a connection with a specific destination. It is most widely used to broadcast a message over a network port to all machines that are listening.

VBScript: VBScript is an active scripting language created by Microsoft to compete with Netscape's JavaScript. VBScript is based on Microsoft's popular programming language, Visual Basic. VBScript is an active scripting language used within HTML to execute small programs to generate a dynamic Web page. Using VBScript, a developer can cause text or graphics to change when the mouse points at them, update the current date and time on the Web page, or add personal information like how long it has been since that user last visited the site.

Virus: A virus is malicious code that replicates itself. New viruses are discovered daily. Some exist simply to replicate themselves. Others can do serious damage such as erasing files or rendering a computer inoperable.

Vulnerability: In network security, a vulnerability refers to any flaw or weakness in the network defense that could be exploited to gain unauthorized access to, damage, or otherwise affect the network

Worm: A worm is similar to a virus. Worms replicate themselves like viruses, but do not alter files. The main difference is that worms reside in memory and usually remain unnoticed until the rate of replication reduces system resources to the point that it becomes noticeable.

---

*These definitions were derived from Robert Slade's *Dictionary of Information Security* (Syngress. ISBN: 1-59749-115-2).With over 1,000 information security terms and definitions, Slade's book is a great resource to turn to when you come across technical words and acronyms you are not familiar with.

Index

Printed and bound by CPI Group (UK) Ltd, Croydon, CR0 4YY

03/10/2024

01040342-0006